Lecture Notes in Computer Science 5203

Commenced Publication in 1973
Founding and Former Series Editors:
Gerhard Goos, Juris Hartmanis, and Jan van Leeuwen

Solomon W. Golomb   Matthew G. Parker
Alexander Pott   Arne Winterhof (Eds.)

Sequences and Their Applications – SETA 2008

5th International Conference
Lexington, KY, USA, September 14-18, 2008
Proceedings

 Springer

Volume Editors

Solomon W. Golomb
University of Southern California
USC Viterbi School of Engineering
Los Angeles, CA 90089-2565, USA
E-mail: sgolomb@usc.edu

Matthew G. Parker
University of Bergen
The Selmer Center, Department of Informatics
5020 Bergen, Norway
E-mail: matthew.parker@ii.uib.no

Alexander Pott
Otto-von-Guericke-University
Institute for Algebra and Geometry
39016 Magdeburg, Germany
E-mail: alexander.pott@ovgu.de

Arne Winterhof
Johann Radon Institute for Computational and Applied Mathematics (RICAM)
Austrian Academy of Sciences
4040 Linz, Austria
E-mail: arne.winterhof@oeaw.ac.at

Library of Congress Control Number: 2008933733

CR Subject Classification (1998): G.2, I.1, F.2, E.4, E.3

LNCS Sublibrary: SL 1 – Theoretical Computer Science and General Issues

ISSN   0302-9743
ISBN-10   3-540-85911-X Springer Berlin Heidelberg New York
ISBN-13   978-3-540-85911-6 Springer Berlin Heidelberg New York

Springer is a part of Springer Science+Business Media

springer.com

© Springer-Verlag Berlin Heidelberg 2008

Typesetting: Camera-ready by author, data conversion by Scientific Publishing Services, Chennai, India
Printed on acid-free paper   SPIN: 12511037   06/3180   5 4 3 2 1 0

Preface

This volume contains the refereed proceedings papers of the *Fifth International Conference on Sequences and their Applications (SETA 2008)*, held in Lexington, Kentucky (USA), September 14–18, 2008. The conference SETA is well established in the mathematics and computer science community. Topics of "SETA" include

- Randomness of sequences
- Correlation (periodic and aperiodic types) and combinatorial aspects of sequences (difference sets)
- Sequences with applications in coding theory and cryptography
- Sequences over finite fields/rings/function fields
- Linear and nonlinear feedback shift register sequences
- Sequences for radar distance ranging, synchronization, identification, and hardware testing
- Sequences for wireless communication
- Pseudorandom sequence generators
- Correlation and transformation of Boolean functions
- Multidimensional sequences and their correlation properties
- Linear and nonlinear complexity of sequences.

This year's proceedings contain 32 contributed papers. All papers have been thoroughly refereed by at least two referees. Most of the refereeing was done by members of the program committee. We thank all of them for their help. We are also grateful to Serdar Boztaş, Nina Brandstätter, Nicolas Courtois, Fangwei Fu, Honggang Hu, Alexander Kholosha, Emmanuel Prouff, Martin Rötteler, Ayineedi Venkateswarlu and Bo-Yin Yang for their assistance in the reviewing process.

In addition to the contributed papers, we had four invited lectures given by Claude Carlet (University of Paris 8, France), Pierre L'Ecuyer (Université de Montréal, Canada), Guang Gong (University of Waterloo, Canada) and Robert McEliece (California Institute of Technology, USA). You find papers based on these invited talks in this volume.

We would like to thank Ramakanth Kavuluru, Diane Mier, Paul Linton and Mark Goresky for their help in organizing this conference. We also thank Faruk Göloğlu and Tan Yin who helped to prepare this proceedings volume.

July 2008

Solomon Golomb
Matthew G. Parker
Alexander Pott
Arne Winterhof

Organization

General Chair

Andrew Klapper   University of Kentucky, USA

Program Co-chairs

Solomon Golomb   University of Southern California, USA
Alexander Pott   Otto-von-Guericke-University Magdeburg, Germany

Conference Proceedings Co-editors

Solomon Golomb   University of Southern California, USA
Matthew G. Parker   University of Bergen, Norway
Alexander Pott   Otto-von-Guericke-University Magdeburg, Germany
Arne Winterhof   Austrian Academy of Sciences, Austria

Program Committee

K.T. Arasu   Wright State University, USA
Claude Carlet   INRIA and University of Paris 8, France
Pascale Charpin   INRIA, France
Habong Chung   Hongik University, Korea
Cunsheng Ding   The Hong Kong University of Science and Technology, China

Tuvi Etzion   Technion Israel Institute of Technology, Israel
Pingzhi Fan   Southwest Jiaotong University, China
Harold Fredricksen   Naval Postgraduate School, USA
Guang Gong   University of Waterloo, Canada
Mark Goresky   Institute for Advanced Study, USA
Tor Helleseth   University of Bergen, Norway
Kathryn Horadam   RMIT, Australia
Tom Høholdt   Technical University of Denmark, Denmark
Thomas Johansson   Lunds Universitet, Sweden
P. Vijay Kumar   Indian Institute of Science, India
Gohar Kyureghyan   Otto-von-Guericke-University Magdeburg, Germany

Wilfried Meidl   Sabanci University, Turkey
Oscar Moreno   University of Puerto Rico, Puerto Rico

Jong-Seon No	Seoul National University, Korea
Udaya Parampalli	University of Melbourne, Australia
Matthew G. Parker	University of Bergen, Norway
Bernhard Schmidt	Nanyang Technological University, Singapore
Hong-Yeop Song	Yonsei University, Korea
Arne Winterhof	Austrian Academy of Sciences, Austria
Kyeongcheol Yang	Pohang University of Science and Technology, Korea
Amr Youssef	Concordia University, Canada

Sponsoring Institutions

The University of Kentucky College of Engineering

Table of Contents

Combinatorial and Algebraic Foundations

Security Aspects of Sequences

Algorithms

Correlation of Sequences over Rings

Nonlinear Functions over Finite Fields

Comparison of Point Sets and Sequences for Quasi-Monte Carlo and for Random Number Generation*
(Invited Paper)

Pierre L'Ecuyer

DIRO, CIRRELT, and GERAD,
Université de Montréal, Canada
http://www.iro.umontreal.ca/~lecuyer

Abstract. Algorithmic random number generators require recurring sequences with very long periods and good multivariate uniformity properties. Point sets and sequences for quasi-Monte Carlo numerical integration need similar multivariate uniformity properties as well. It then comes as no surprise that both types of applications share common (or similar) construction methods. However, there are some differences in both the measures of uniformity and the construction methods used in practice. We briefly survey these methods and explain some of the reasons for the differences.

1   Introduction

1.1   Random Number Generators

(Pseudo)Random number generators (RNGs) are typically defined by a (deterministic) recurring sequence in a finite state space $\mathcal{S}$, usually a finite field or ring, and an output function mapping each state to an output value in $\mathcal{U}$, which is often either a real number in the interval $(0, 1)$ or an integer in some finite range [1,2,3,4]. We shall assume here that each output is a real number in $(0, 1)$ and that the purpose of the RNG is to mimic a sequence of independent $U(0, 1)$ random variables (i.e., uniformly distributed over $(0, 1)$). Of course, with such an algorithmic construction, this can only be a fake. The quality of the fake should be measured in a way that depends on the application.

One could argue that physical noise would provide a safer source of (true) randomness. With careful design, it does, and it is appropriate for certain applications such as generating random keys in cryptology and for gaming machines, for example, where unpredictability is a major requirement. In this paper, we focus on RNGs for simulation applications, where fast algorithmic RNGs are much more convenient than physical sources, because they are faster, they permit one to replicate the same exact sequence several times (this is often needed in

---

* This work was supported NSERC-Canada Grant Number ODGP0110050 and by a Canada Research Chair to the author.

S.W. Golomb et al. (Eds.): SETA 2008, LNCS 5203, pp. 1–17, 2008.

simulation, for example for comparing similar systems and in optimization and variance reduction algorithms [5,6,7]), and they do not require special hardware.

Let $s_0, s_1, s_2, \ldots$ denote the successive states of our RNG, and $u_0, u_1, u_2, \ldots$ be the corresponding sequence of output values. After selecting s_0 (which might be random), the successive states follow the deterministic recurrence $s_i = f(s_{i-1})$ for a given *transition function* $f : \mathcal{S} \to \mathcal{S}$, and the *output* at step i is $u_i = g(s_i)$ for some *output function* $g : \mathcal{S} \to (0,1)$. This output sequence is necessarily (eventually) periodic, with period ρ that cannot exceed the number of distinct states, $|\mathcal{S}|$. When the state occupies b bits of memory, we have $\rho \leq 2^b$ and we usually require that ρ be close to 2^b, to avoid wasting memory. Values of b for certain practical RNGs can be as much as 20,000 or even more [8,9,10], but for simulation, a few hundred is probably large enough.

Besides a long period, other standard requirements for RNGs include high running speed (in 2008, fast RNGs can produce over 100 million $U(0,1)$ random numbers per second on a standard computer), reproducibility and portability (the ability to reproduce the same sequence several times on the same computer, and also on any standard computing platform), and the possibility to quickly jump ahead by an arbitrarily large number of steps in the sequence. The latter is frequently used to split the sequence into several shorter (but still very long) disjoint subsequences, in order to provide an arbitrary number of virtual generators (often called RNG streams) [5,6].

However, these basic properties are not sufficient To see why, just note that an RNG defined by $u_i = s_i = (i+1/2)/2^{10000} \bmod 1$ easily satisfies all the above properties, and the values cover the interval $(0,1)$ very evenly in the long run, but this RNG certainly fails to provide a good imitation of independent $U(0,1)$ random variables. The problem is the lack of (apparent) *independence* between successive values.

A key observation is that we have both uniformity and independence if and only if for any integer $s > 0$, $(u_0, \ldots, u_{s-1})$ is a random vector with the uniform distribution over the s-dimensional unit hypercube $(0,1)^s$. We want the RNG to provide a good approximation of this property. If the seed s_0 is selected at random in $\mathcal{S}$, then the vector $(u_0, \ldots, u_{s-1})$ has the uniform distribution over the finite set $\Psi_s = \{(u_0, \ldots, u_{s-1}) : s_0 \in \mathcal{S}\}$, which can be viewed as a discrete approximation of the uniform distribution over $(0,1)^s$. For this approximation to be good, Ψ_s must provide a dense and uniform coverage of the unit hypercube $(0,1)^s$, at least for moderate values of s. This is possible only if $\mathcal{S}$ has large cardinality. In fact, this is a more important reason for having a long period than the risk of exhausting the RNG cycle. More generally, we want high uniformity of the sets of the form $\Psi(u) = \{(u_{i_1}, \ldots, u_{i_d}) : s_0 \in \mathcal{S}\}$, where $u = \{i_1, \ldots, i_d\}$ is an arbitrary set of integers such that $0 \leq i_1 < \cdots < i_d$.

But how should we measure the uniformity of Ψ_s (and of the sets $\Psi(u)$)? There are many ways. But a crucial requirement is that the selected measure of uniformity must be computable efficiently without generating the random numbers (or enumerating the points of Ψ_s) explicitly, because there are just too many of them. For nonlinear RNGs, easily computable measures of uniformity are difficult to find. This is the main reason why most good RNGs used in practice are based on linear recurrences [3,9]. Specific measures for linear RNGs are discussed later in this paper. The choice generally depends on the type of RNG, for computability reasons.

To construct a new RNG, one would usually specify a parameterized class of (large-period) constructions (based on the availability of a fast implementation) and a figure of merit, and then perform a computer search for the construction with the largest figure of merit that can be found within that class. Then, the selected RNG is implemented and tested empirically. There is a large variety of *empirical statistical tests* for testing the statistical behavior of RNGs [1,9].

1.2   Low-Discrepancy Sets and Sequences

In *quasi-Monte Carlo* (QMC) numerical integration, we want a set of n points, $P_n = \{\mathbf{u}_0, \ldots, \mathbf{u}_{n-1}\}$, that cover the s-dimensional unit hypercube $[0,1)^s$ very evenly. These points are used to approximate the integral of some function $f : [0,1)^s \to \mathbb{R}$, say

$$\mu = \int_{[0,1)^s} f(\mathbf{u})d\mathbf{u},$$

by the average

$$\bar{\mu}_n = \frac{1}{n} \sum_{i=0}^{n-1} f(\mathbf{u}_i). \tag{1}$$

This μ can be interpreted as the mathematical expectation $\mu = \mathbb{E}[f(\mathbf{U})]$, where $\mathbf{U}$ is a random vector uniformly distributed over $[0,1)^s$. Here, the cardinality n of the point set is much smaller than for RNGs, because we need to evaluate f at each of these points. It rarely exceeds a million.

Again, a key question is: How do we measure the uniformity of the point set P_n? Standard practice is to use a figure of merit that measures the *discrepancy* between the empirical distribution of P_n and the uniform distribution over $[0,1)^s$ [11,12,13,14,4]. Many of these discrepancies are actually defined as the worst-case integration error, with P_n, over all functions f whose *variation* $V(f) = \|f - \mu\|$ does not exceed 1, in a given Banach (or Hilbert) space $\mathcal{H}$ of functions [11,15]. In this setting, the worst-case error can be bounded by the product of the discrepancy $D(P_n)$, that depends only on P_n, and the function's variation, that depends only on f:

$$|\bar{\mu}_n - \mu| \leq D(P_n)V(f) \tag{2}$$

for all $f \in \mathcal{H}$. This is a generalized form of the Koksma-Hlawka inequality [4]. Generally speaking, for a given point set, the more restricted (or smoother) the class of functions f for which $V(f) \leq 1$, the smaller will be the discrepancy, and the faster $\min_{P_n} D(P_n)$ will converge to 0 as a function of n. That is, the worst-case error bound for the best possible point set will converge to 0 more quickly. Specific examples of discrepancies are mentioned in Section 2.

Here, because n is not so large, a computing time of $O(n)$ or even $O(n^2)$ for the discrepancy is acceptable (in contrast to RNGs). The choice of discrepancy can be different for this reason.

In practice, the worst-case error bound (2) is usually much too difficult to compute or even to approximate, and may be very loose for our specific function f. For these reasons, one would rather turn the deterministic approximation $\bar{\mu}_n$ into a randomized unbiased estimator, and replace the error bound by a probabilistic error estimate obtained by estimating the variance of the estimator. This is achieved by randomizing the point set P_n so that [16,17,13,18]:

(a) it retains its high uniformity when taken as a set and
(b) each individual point is a random vector with the uniform distribution over $(0,1)^s$.

A *randomized QMC* (RQMC) estimator of μ, denoted $\hat{\mu}_{n,\mathrm{rqmc}}$, is defined as the average (1) in which the $\mathbf{u}_i$ are replaced by the n randomized points. By performing m independent replicates of this randomization, and computing the sample mean and sample variance of the m independent realizations of $\hat{\mu}_{n,\mathrm{rqmc}}$, one obtains an unbiased estimator of μ and an unbiased estimator of the variance of this estimator. This permit one to obtain an asymptotically valid confidence interval on μ [17,13].

A simple randomization that satisfies these conditions is a *random shift modulo 1* [19,17,20]: Generate a single point $\mathbf{U}$ uniformly over $(0,1)^s$ and add it to each point of P_n, modulo 1, coordinate-wise. Another one is a *random digital shift in base b* [21,13,22]: generate $\mathbf{U}$ uniformly over $(0,1)^s$, expand each of its coordinates in base b, and add the digits, modulo b, to the corresponding digits of each point of P_n. For $b = 2$, this amounts to applying a coordinate-wise exclusive-or (xor) of $\mathbf{U}$ to all the points.

The variance of the RQMC estimator is the same as its mean square error, because it is unbiased, so by squaring on each side of (2) we obtain that

$$\mathrm{Var}[\hat{\mu}_{n,\mathrm{rqmc}}] \leq V^2(f) \cdot \mathbb{E}[D^2(P_n)],$$

so the variance converges at least as fast as the mean square discrepancy.

A *low-discrepancy sequence* is an infinite sequence $\{\mathbf{u}_0, \mathbf{u}_1, \mathbf{u}_2, \ldots\}$ such that the point set P_n formed by the first n points of the sequence has low discrepancy for all large enough n. Often, the discrepancy is lower along a subsequence of values of n; e.g., if n is a power of a given integer. Such sequences are convenient if we want to increase n (adding more points to the set) until some error bound or error estimate is deemed small enough, instead of fixing n a priori. A low discrepancy point set of sequence can also be *infinite-dimensional*; i.e., each point has an infinite sequence of coordinates. In this case, the sequence of coordinates are usually defined by a recurrence. This is convenient when $f(\mathbf{U})$ depends on a random and unbounded number of uniform random variables, which is frequent.

1.3   Importance of Low-Dimensional Projections

When the dimension s is large, filling up the s-dimensional unit cube evenly requires too many points. For $s = 100$, for example, we already need 2^{100} points just to have one point near each corner of the hypercube. For quasi-Monte Carlo, this is impractical. For RNGs, we can easily have more than 2^{100} points in Ψ_s, but the high-dimensional uniformity eventually breaks down as well for a larger s.

In QMC, we are saved by the fact that f is often well approximated by a sum of low-dimensional functions, that depend only on a small number of coordinates of $\mathbf{u}$; that is,

$$f(\mathbf{u}) \approx \sum_{\mathfrak{u} \subseteq \mathcal{J}} f_{\mathfrak{u}}(\mathbf{u}), \tag{3}$$

where each $f_{\mathfrak{u}} : (0,1)^s \to \mathbb{R}$ depends only on $\{u_j, j \in \mathfrak{u}\}$, and $\mathcal{J}$ is a family of small-cardinality subsets of $\{1, \ldots, s\}$. Then, to integrate f with small error, it suffices to integrate with small error the low-dimensional functions $f_{\mathfrak{u}}$ making up the approximation. For that, we only need high uniformity of the projections $P_n(\mathfrak{u})$ of P_n over the low-dimensional sets of coordinates $\mathfrak{u} \in \mathcal{J}$. This suggests a figure of merit defined as a weighted sum (or supremum) of discrepancy measures computed over the sets $P_n(\mathfrak{u})$ for $\mathfrak{u} \in \mathcal{J}$. The figures of merit used to select QMC point sets are typically of that form.

This heuristic interpretation can be made rigorous via a functional ANOVA decomposition of f [23,18,24]. When (3) holds for $\mathcal{J} = \{\mathfrak{u} : |\mathfrak{u}| \leq d\}$ for a small d, we say that f has low effective dimension in the superposition sense, while if it holds for $\mathcal{J} = \{\mathfrak{u} \subseteq \{1, \ldots, d\}\}$ for a small d, we say that f has low effective dimension in the *truncation sense* [25,18]. Low effective dimension can often be achieved by redefining f without changing its expectation, via a change of variables [26,25,27,28,14,29,30]. That is, we change the way the uniforms (the coordinates of $\mathbf{u}$) are transformed in the simulation. There are important applications in computational finance, for example, where after such a change of variable, the one- and two-dimensional functions $f_{\mathfrak{u}}$ account for more than 99% of the variability of f [14,30]. For these applications, it is important that the one- and two-dimensional projections $P_n(\mathfrak{u})$ have very good uniformity, and there is no need to care much about the high-dimensional projections.

It makes sense that the figures of merit for the point sets Ψ_s produced by RNGs also take the low-dimensional projections into account, as suggested in [3,31,17], for example. In fact, the standardized figures of merit based on the spectral test, as defined in [32,33,34] for example, already do this to a certain extent by giving more weight to the low-dimensional projections in the truncation sense (where $\mathfrak{u} = \{1, \ldots, d\}$ for small d).

2 Examples of Discrepancies for QMC

Discrepancies and the corresponding variations are often defined via reproducing kernel Hilbert spaces (RKHS). An RKHS is constructed by selecting a *kernel* $K : [0,1]^{2s} \to \mathbb{R}$, which is a symmetric and positive semi-definite function. The kernel determines in turn a set of basis functions and a scalar product, that define a Hilbert space $\mathcal{H}$ [35]. For $f \in \mathcal{H}$ and a point set P_n, (2) holds with

$V(f) = \|f - \mu\|_K$, where $\|\cdot\|_K$ is the norm that corresponds to the scalar product of $\mathcal{H}$, and $D(P_n)$ that satisfies

$$
\begin{aligned}
&D^2(P_n) \\
&= \frac{1}{n^2} \sum_{i=0}^{n-1} \sum_{j=0}^{n-1} K(\mathbf{u}_i, \mathbf{u}_j) - \frac{2}{n} \sum_{i=0}^{n-1} \int_{[0,1]^s} K(\mathbf{u}_i, \mathbf{v}) d\mathbf{v} + \int_{[0,1]^{2s}} K(\mathbf{u}, \mathbf{v}) d\mathbf{u} d\mathbf{v} \quad (4)
\end{aligned}
$$

(see [11,12]). When $K(\mathbf{u}, \mathbf{v})$ can be computed in $O(s)$ time for an arbitrary $(\mathbf{u}, \mathbf{v})$, then this $D(P_n)$ can be computed in $O(n^2 s)$ time, but this assumption does not always hold.

One important type of kernel has the form

$$
K(\mathbf{u}, \mathbf{v}) = \sum_{\mathbf{h} \in \mathbb{Z}^s} w(\mathbf{h}) e^{2\pi \iota \mathbf{h}^t (\mathbf{u} - \mathbf{v})} \quad (5)
$$

where $\iota = \sqrt{-1}$, the t means "transposed", and the $w(\mathbf{h})$ are non-negative *weights* such that $\sum_{\mathbf{h} \in \mathbb{Z}^s} w(\mathbf{h}) < \infty$. The corresponding square variation is

$$
V^2(f) = \sum_{0 \neq \mathbf{h} \in \mathbb{Z}^s} |\hat{f}(\mathbf{h})|^2 / w(\mathbf{h}),
$$

where the $\hat{f}(\mathbf{h})$ are the Fourier coefficients of f. The corresponding discrepancy is easily computable only for special shapes of the weights.

For example, if

$$
w(\mathbf{h}) = \prod_{j=1}^{s} \min(1, \gamma_j |h_j|^{-2\alpha}), \quad (6)
$$

for some integer $\alpha \geq 1$ and positive real numbers (weights) γ_j, then the kernel becomes

$$
K_\alpha(\mathbf{u}, \mathbf{v}) = \prod_{j=1}^{s} \left[1 - \gamma_j \frac{(-4\pi^2)^\alpha}{(2\alpha)!} B_{2\alpha}((u_j - v_j) \bmod 1) \right], \quad (7)
$$

where $B_{2\alpha}$ is the Bernoulli polynomial of degree 2α [20]. This kernel can be computed in $O(s)$ time, so the discrepancy can be computed in $O(n^2 s)$ time. The corresponding $V(f)$ in this case satisfies

$$
V^2(f) = \sum_{\phi \neq \mathbf{u} \subseteq S} \left(\prod_{j \in \mathbf{u}} \gamma_j^{-\alpha} \right) (4\pi^2)^{-\alpha|\mathbf{u}|} \int_{[0,1]^{|\mathbf{u}|}} \left| \int_{[0,1]^{s-|\mathbf{u}|}} \frac{\partial^{\alpha|\mathbf{u}|} f}{\partial \mathbf{u}_\mathbf{u}^\alpha} (\mathbf{u}) d\mathbf{u}_{\bar{\mathbf{u}}} \right|^2 d\mathbf{u}_\mathbf{u},
$$

$$(8)$$

where $\mathbf{u}_\mathbf{u}$ represents the coordinates of $\mathbf{u}$ whose indices are in the set $\mathbf{u}$ and $\mathbf{u}_{\bar{\mathbf{u}}}$ represents those whose indices are not in $\mathbf{u}$. This RKHS contains only periodic functions f of period 1, with $f(0) = f(1)$, and the variability measures the smoothness via the partial derivatives of f.

Slight variations of this discrepancy have interesting geometric interpretations. For example, if we replace $B_{2\alpha}((u_j - v_j) \bmod 1)$ in (7) by an appropriate (simple) polynomial in u_j and v_j, we obtain a weighted $\mathcal{L}_2$-*unanchored discrepancy* whose interpretation is the following [11,12]. For each subset $\mathbf{u}$ of coordinates and any $\mathbf{u}, \mathbf{v} \in [0,1]^{|\mathbf{u}|}$, let $D(P_n(\mathbf{u}), \mathbf{u}, \mathbf{v})$ be the absolute difference between the volume of the $|\mathbf{u}|$-dimensional box with opposite corners at $\mathbf{u}$ and $\mathbf{v}$, and the fraction of the points of P_n that fall in that box. The square discrepancy is then defined as

$$[D_2(P_n)]^2 = \sum_{\phi \neq \mathbf{u} \subseteq S} \left(\prod_{j \in \mathbf{u}} \gamma_j \right) \int_{[0,1]^{|\mathbf{u}|}} \int_{[0,\mathbf{v}]} D^2(P_n(\mathbf{u}), \mathbf{u}, \mathbf{v}) d\mathbf{u} d\mathbf{v}. \qquad (9)$$

Other similarly defined RKHSs contain non-periodic functions f. For example, by taking again the appropriate (simple) function in place of $B_{2\alpha}((u_j - v_j) \bmod 1)$ in (7), we obtain a weighted $\mathcal{L}_2$-*star discrepancy*, whose square can be written as

$$[D_2(P_n)]^2 = \sum_{\phi \neq \mathbf{u} \subseteq S} \left(\prod_{j \in \mathbf{u}} \gamma_j \right) \int_{[0,1]^{|\mathbf{u}|}} D^2(P_n(\mathbf{u}), \mathbf{0}, \mathbf{v}) d\mathbf{v} \qquad (10)$$

and the square variation is

$$V^2(f) = \sum_{\phi \neq \mathbf{u} \subseteq S} \left(\prod_{j \in \mathbf{u}} \gamma_j^{-1} \right) \int_{[0,1]^{|\mathbf{u}|}} \left| \frac{\partial^{|\mathbf{u}|}}{\partial \mathbf{u}_\mathbf{u}} f_\mathbf{u}(\mathbf{u}_\mathbf{u}) \right|^2 d\mathbf{u}_\mathbf{u}$$

(see [11,12]). All these discrepancies can be computed in $O(n^2 s)$ time for general point sets.

It is known that there exists point sets P_n for which the discrepancy based on (7) converges as as $O(n^{-\alpha+\delta})$ for any $\delta > 0$, and point sets for which the discrepancy (10) converges as $O(n^{-1+\delta})$.

3   RNGs Based on Linear Recurrences Modulo a Large Integer m

An important (and widely used) class of RNGs is based on the general linear recurrence

$$x_i = (a_1 x_{i-1} + \cdots + a_k x_{i-k}) \bmod m, \qquad (11)$$

where k and m are positive integers, $a_1, \ldots, a_k \in \{0, 1, \ldots, m-1\}$, and $a_k \neq 0$. The state at step i is $s_i = \mathbf{x}_i = (x_{i-k+1}, \ldots, x_i)$. Suppose the output is $u_i = x_i/m$ (in practice it is slightly modified to make sure that $0 < u_i < 1$). This RNG is called a *multiple recursive generator*. For $k = 1$, it is known as a *linear congruential generator* (LCG). For prime m and well-chosen coefficients, the period length can reach $m^k - 1$ [1], which can be made arbitrarily large by increasing k. Because of the linearity, jumping ahead from $\mathbf{x}_i$ to $\mathbf{x}_{i+\nu}$ is easy, regardless of ν: One can write $\mathbf{x}_{i+\nu} = \mathbf{A}_\nu \mathbf{x}_i \bmod m$, where $\mathbf{A}_\nu$ is a $k \times k$ matrix that can be precomputed once for all [3].

It is well-known that in this case, Ψ_s is the intersection of a lattice

$$L_s = \left\{ \mathbf{v} = \sum_{j=1}^{s} h_j \mathbf{v}_j \text{ such that each } h_j \in \mathbb{Z} \right\} \tag{12}$$

with the unit hypercube $[0,1)^s$ [1,36], where a lattice basis $\{\mathbf{v}_1, \ldots, \mathbf{v}_s\}$ is easy to obtain [36]. Theoretical measures of uniformity used in practice are defined in terms of the geometry of this lattice. The lattice structure implies in particular that Ψ_s is contained in a family of equidistant parallel hyperplanes. For the family for which the successive hyperplanes are farthest apart, the inverse of the distance between the successive hyperplanes is equal to the Euclidean length of the shortest nonzero vector in the dual lattice L_s^*, defined by $L_s^* = \{\mathbf{h} \in \mathbb{R}^s : \mathbf{h}^t \mathbf{v} \in \mathbb{Z} \text{ for all } \mathbf{v} \in L_s\}$. The $\mathcal{L}_1$ length of the shortest nonzero vector in L_s^* gives the number of hyperplanes required to cover the points. The computing time of these shortest vectors is exponential in s (in the worst case, with the best known algorithms), but depends very little on m and k. In practice, one can handle values of s up to 50 (or more, for easier lattices), even with $m^k > 2^{1000}$ [32].

Note that for every subset of coordinates $\mathbf{u} = \{i_1, \ldots, i_d\}$, the projection set $\Psi(\mathbf{u})$ is also the intersection of a lattice with the unit hypercube $[0,1)^d$, and one can compute the length $\ell(\mathbf{u})$ of a shortest nonzero vector in the corresponding dual lattice $L^*(\mathbf{u})$. Moreover, for any given number of points $n = m^k$ and a given dimension d, there is a theoretical upper bound $\ell_d^*(n)$ on this length $\ell(\mathbf{u})$. One can divide $\ell(\mathbf{u})$ by such an upper bound to obtain a standardized value between 0 and 1, and take the worst case over a selected class $\mathcal{J}$ of index sets $\mathbf{u}$. This gives a figure of merit of the form

$$\min_{\mathbf{u} \in \mathcal{J}} \ell(\mathbf{u})/\ell_{|\mathbf{u}|}^*(n). \tag{13}$$

This type of criterion has been proposed in [17]. Simplified versions, where $\mathcal{J}$ contains only the subsets of successive coordinates up to a given dimension, have been used for a long time to measure the quality of RNGs [34,1,32,3].

It is important to look at the projections $\Psi(\mathbf{u})$, because fast long-period (but otherwise poorly designed) RNGs often have some very bad low-dimensional projections. For example, *lagged-Fibonacci* generators follow the recurrence (11) with only two nonzero coefficients, say a_r and a_k, both equal to ± 1. It turns out that for these generators, with $\mathbf{u} = \{0, k-r, k\}$, all the points of $\Psi_s(\mathbf{u})$ lie in only two parallel planes in the three-dimensional cube. This type of structure can have a disastrous impact on certain simulations. The add-with-carry and subtract-with-borrow generators, proposed in [37] and still available in some popular software, have exactly the same problem. A well-chosen figure of merit of the form (13) should give high penalties to these types of bad projections.

A variant of the multiple recursive generator is the multiply-with-carry generator, based on a linear recurrence with carry:

$$x_i = (a_1 x_{i-1} + \cdots + a_k x_{i-k} + c_{i-1})d \bmod b,$$
$$c_i = \lfloor (a_0 x_i + a_1 x_{i-1} + \cdots + a_k x_{i-k} + c_{i-1})/b \rfloor,$$
$$u_i = x_i/b + x_{i+1}/b^2 + \cdots$$

where $b \geq 2$ is an integer, a_0 is relatively prime to b, and d is the multiplicative inverse of $-a_0$ modulo b. In practice, the expansion that defines u_i is truncated to a few terms. An important result, if we neglect this truncation, states that this RNG is exactly equivalent to an LCG with modulus $m = \sum_{\ell=0}^{k} a_\ell b^\ell$ and multiplier a equal to the inverse of b modulo m [38,39,40]. This means that this RNG can be seen as a clever way of implementing an LCG with very large modulus and large period (this RNG can be quite fast if b is a power of two, for example), and that its uniformity can be measured in terms of the lattice structure of this LCG [39].

4   RNGs Based on Linear Recurrences Modulo 2

Another important class of construction methods uses a linear recurrence as in (11), but with $m = 2$ [41,42]. That is, all operations are performed in the finite field $\mathbb{F}_2 = \{0, 1\}$. This general construction can be written in matrix form as [3,31]:

$$\mathbf{x}_i = \mathbf{A}\mathbf{x}_{i-1}, \quad \mathbf{y}_i = \mathbf{B}\mathbf{x}_i, \quad u_i = \sum_{\ell=1}^{w} y_{i,\ell-1} 2^{-\ell},$$

where $\mathbf{x}_i = (x_{i,0}, \ldots, x_{i,k-1})^t$ is the *state* at step i, $\mathbf{A}$ is a $k \times k$ binary matrix, $\mathbf{y}_i = (y_{i,0}, \ldots, y_{i,w-1})^t$, $\mathbf{B}$ is a $w \times k$ binary matrix, k and w are positive integers, and $u_i \in [0, 1)$ is the *output* at step i. (In practice, the output can be modified slightly to avoid returning 0.)

This framework covers several well-known generators such that the Tausworthe, polynomial LCG, generalized feedback shift register (GFSR), twisted GFSR, Mersenne twister, WELL, xorshift, linear cellular automaton, and combinations of these [43,3,31,44,42]. With a careful design, for which $\mathbf{A}$ has a primitive characteristic polynomial over $\mathbb{F}_2$, the period length can reach $2^k - 1$. The matrices $\mathbf{A}$ and $\mathbf{B}$ should be constructed so that the products (14) and (14) can be computed very quickly by a few simple binary operations on blocks of bits, such as or, exclusive-or, shift, and rotation. A compromise must be made between the number of such operations and a good uniformity of the point sets Ψ_s (with too few operations, there are in general limitations on the quality of the uniformity that can be reached). For these types of generators, the uniformity of the point sets Ψ_s is measured by their *equidistribution* properties, as explained in Section 6.

Combined generators of this type, defined by xoring the output vectors $\mathbf{y}_i$ of the components, are equivalent to yet another generator from the same class.

Such combinations provide efficient ways of implementing RNGs with larger state spaces [45,46,31].

5   Lattice Rules for QMC

When a lattice L_s as in (12) contains $\mathbb{Z}^s$, it is called an *integration lattice*, and the QMC approximation (1) with $P_n = L_s \cap [0, 1)^s$ is a *lattice rule* [4,20]. The most frequently used lattice rules are of rank 1: we have $\mathbf{v}_1 = \mathbf{a}_1/n$ where $\mathbf{a}_1 = (a_1, \ldots, a_s)$, and $\mathbf{v}_j = \mathbf{e}_j$ (the jth unit vector) for $j \geq 2$. A special case is when P_n is the point set Ψ_s that correspond to a LCG; we then have a Korobov lattice rule.

Integration lattices are usually randomized by a random shift modulo 1. The shift preserves the lattice structure. It turns out that the variance of the corresponding RQMC estimator can be written explicitly as

$$\text{Var}[\hat{\mu}_{n,\text{rqmc}}] = \sum_{0 \neq \mathbf{h} \in L_s^*} |\hat{f}(\mathbf{h})|^2, \tag{14}$$

where the $\hat{f}(\mathbf{h})$ are the coefficients in the Fourier expansion of the function f [17]. Assuming that we want to minimize the variance, (14) gives the ultimate discrepancy measure of a lattice point set to integrate a given function f. The square Fourier coefficients are generally unknown, but they can be replaced by weights $w(\mathbf{h})$ that try to approximate their expected behavior for a given class of functions of interest. It might be difficult to obtain such an approximation. Another problem is that computing (14) with the weights $w(\mathbf{h})$, and searching for lattices that minimize its value, may be difficult unless the weights have a special form. In [17], the authors argue that since the Fourier coefficients for the short vectors $\mathbf{h}$ represent the main trends of the function's behavior, they are likely to be those having the most impact on the variance, so we should try to keep them out of the dual lattice to eliminate them from the sum in (14). If we do this for a selected class of low-dimensional projections of L_s, and weight these projections, this leads to the same criterion as in (13). Specific Korobov lattice parameters found on the basis of this criterion are given in [17].

In these criteria, the Euclidean length of $\mathbf{h}$ could also be replaced by other notions of length; for example, the $\mathcal{L}_1$ length, as discussed earlier, of the product length $\prod_{j=1}^{s} \max(1, |h_j|)$, for which the length of the shortest vector in L_s^* is called the *Zaremba index* [20].

For lattice rules, discrepancies based on kernels of the general form (5) admit simplified formulas, thanks to the fact that $\sum_{i=0}^{n-1} e^{2\pi \iota \mathbf{h}^t \mathbf{u}_i} = n$ for $\mathbf{h} \in L_s^*$, and 0 otherwise [20]. In particular, the square discrepancy for the kernel (7) simplifies to

$$D^2(P_n) = -1 + \frac{1}{n} \sum_{i=0}^{n-1} \prod_{j=1}^{s} \left(1 + \gamma_j \frac{(-4\pi^2)^\alpha}{(2\alpha)!} B_{2\alpha}(u_j)/2\right), \tag{15}$$

which can be computed in $O(ns)$ time. This discrepancy is a weighted version of a criterion known as $P_{2\alpha}$ [11,20], widely used for lattice rules.

It is known that for any given dimension s and arbitrarily small $\delta > 0$, there exist lattice rules whose discrepancy (15) is $O(n^{-\alpha+\delta})$ [47,4,20]. Until very recently, it was unclear if those lattices were easy to find, but explicit construction methods are now available.

Indeed, for a large s and moderate n, trying all possibilities for the basis vectors is just impractical. Even if we restrict ourselves to a rank-1 lattice, there is already $(n-1)^s$ possibilities. So one must either search at random, or perform a more restricted search. One possibility is to limit the search to Korobov rules, so that there is only the parameter a_1 to select. Another possibility is to adopt a greedy component-by-component (CBC) construction of a rank-1 lattice, for a given n, by selecting the components a_j of the vector $\mathbf{a}_1 = (a_1, \ldots, a_s)$ iteratively as follows [48,49]. Start with $a_1 = 1$. At step j, $a_1, \ldots, a_{j-1}$ are fixed and we select a_j to optimize a given discrepancy measure for the j-dimensional rank-1 lattice with generating vector $\mathbf{v}_1 = \mathbf{a}_1/n$. At each step, there is at most $n-1$ choices to examine for a_j, so at most $O(ns)$ discrepancies need to be computed to construct a lattice of dimension s. For a discrepancy of the form (15), since each discrepancy is computable in $O(ns)$ time, we can conclude that computing $\mathbf{a}_1 = (a_1, \ldots, a_s)$ requires at most $O(n^2 s^2)$ time. But in fact, faster algorithms have been designed that can compute $\mathbf{a}_1$ in $O(n \log(n)s)$ time using $O(n)$ memory [50,48,51].

The remarkable feature of these CBC constructions is that for a large variety of discrepancies, defined mostly via RKHS, it has been proved that one obtains rank-1 lattices whose discrepancy converges at the same rate, as a function of n, as the best possible lattice constructions [50,52,47,53]. In other words, for these discrepancies, CBC provides a practical way of constructing lattices with optimal convergence rate of the discrepancy. These results provide supporting arguments for the use of these types of discrepancies as figures of merit.

6   Digital Nets for QMC

We start with an arbitrary integer $b \geq 2$, usually a prime. An s-dimensional *digital net in base* b, with $n = b^k$ points, is a point set $P_n = \{\mathbf{u}_0, \ldots, \mathbf{u}_{n-1}\}$ defined by selecting s generator matrices $\mathbf{C}_1, \ldots, \mathbf{C}_s$ with elements in $\mathbb{Z}_b = \{0, \ldots, b-1\}$, where each $\mathbf{C}_j$ is $\infty \times k$. The points $\mathbf{u}_i$ are defined as follows. For $i = 0, \ldots, b^k - 1$, we write

$$i = \sum_{\ell=0}^{k-1} a_{i,\ell} b^\ell,$$

$$\begin{pmatrix} u_{i,j,1} \\ u_{i,j,2} \\ \vdots \end{pmatrix} = \mathbf{C}_j \begin{pmatrix} a_{i,0} \\ a_{i,1} \\ \vdots \\ a_{i,k-1} \end{pmatrix} \mod b, \tag{16}$$

$$u_{i,j} = \sum_{\ell=1}^{\infty} u_{i,j,\ell} b^{-\ell}, \quad \text{and} \quad \mathbf{u}_i = (u_{i,1}, \ldots, u_{i,s}). \quad (17)$$

In practical implementations, only the first r rows of the $\mathbf{C}_j$'s are nonzero, for some positive integer r (for example, with $b^r \approx 2^{31}$ on 32-bit computers).

It is usually the case that the first k rows of each $\mathbf{C}_j$ form a nonsingular $k \times k$ matrix, so each one-dimensional projection $P_n(\{j\})$ truncated to the first k digits is equal to $\mathbb{Z}_n/n = \{0, 1/n, \ldots, (n-1)/n\}$. The role of each $\mathbf{C}_j$ is to define a permutation of $\mathbb{Z}_n/n$ so that these numbers are enumerated in a different order for the different coordinates. The choice of these permutations determines the uniformity of P_n and of its projections $P_n(\mathbf{u})$ (which are also digital nets).

In a more general definition of digital net [4], one can also apply bijections (or permutations) to the digits of $\mathbb{Z}_b$ before and after the multiplication by $\mathbf{C}_j$. These multiplications are performed in an arbitrary ring R of cardinality b, and the bijections may depend on ℓ and j. The bijections provide additional opportunity for improving the uniformity.

A *digital sequence in base b* is defined by selecting matrices $\mathbf{C}_j$ with an infinite number of columns. This gives an infinite sequence of points. For each k, the first k columns determine the first b^k points, which form a digital net. Widely-used instances of digital sequences are those of Sobol' [54] in base 2, of Faure [55] in prime base b, of Niederreiter [56], and of Niederreiter and Xing [57]. With an infinite sequence of matrices $\mathbf{C}_j$, we have an infinite-dimensional digital net. These infinite sequences of columns and matrices are often defined via recurrences (each column and matrix being a function of the previous ones).

Polynomial lattice rules use point sets defined by a lattice as in (12), but where the h_j are polynomials over $\mathbb{Z}_b$, and the coordinates of the $\mathbf{v}_j$ are polynomials over $\mathbb{Z}_b$ divided by a common polynomial of degree k. The output is produced simply by "evaluating" each $\mathbf{v}(z)$, which is a vector of formal series in z, at $z = b$. These polynomial lattice rules turn out to be special cases of digital nets in base b. Moreover, much of the theory developed for ordinary lattice rules has a counterpart for those other types of lattice rules [58,59].

Important parts of this theory also extends to digital nets in general [60,61]. In particular, there is a dual space $\mathcal{C}_s^*$ that plays the same role as the dual lattice L_s^* (in the case of polynomial lattice rules, $\mathcal{C}_s^*$ is also a lattice). For a digital net with a random digital shift in base b, in analogy with (14), the RQMC estimator has variance

$$\text{Var}[\hat{\mu}_{n,\text{rqmc}}] = \sum_{0 \neq \mathbf{h} \in \mathcal{C}_s^*} |\tilde{f}(\mathbf{h})|^2, \quad (18)$$

where the $\tilde{f}(\mathbf{h})$ are the coefficients of the Walsh expansion of f. For a given f, this expression provides an ultimate discrepancy measure for a digital net with a random digital shift. This discrepancy has the same limitations as (14) for ordinary lattices (the Walsh coefficients are usually unknown, etc), and the practical alternatives are analogous. They lead to figures of merit as in (13), but where the $\ell(\mathbf{u})$ are lengths of shortest vectors in the dual spaces $\mathcal{C}^*(\mathbf{u})$ associated with the projections $P_n(\mathbf{u})$ instead of in the dual lattices $L^*(\mathbf{u})$.

Do the discrepancies discussed in Section 2 have simplified expressions for digital nets, as was the case for lattice rules? Those based on the kernel (5) do not, but they do if we take a slightly different kernel, based on Walsh expansions in base b instead of Fourier expansions. This can be exploited to develop practical figures of merit analogous to those used for lattice rules.

The most widely used class of figures of merit, for both RNG and QMC point sets, are those based on the notion of *equidistribution*, defined as follows. For a vector of non-negative integers $(q_1, \ldots, q_s)$, we partition the jth axis into b^{q_j} equal parts for each j. This partitions the hypercube $[0, 1)^s$ into $b^{q_1 + \cdots + q_s}$ rectangular boxes of the same size and shape. A point set P_n of cardinality $n = b^k$ is $(q_1, \ldots, q_s)$-*equidistributed in base* b if each box of this partition contains exactly b^t points of P_n, where $t = k - q_1 - \cdots - q_s$. For a digital net in base b, this property holds if and only if the set of $k - q = q_1 + \cdots + q_t$ rows that comprise the first q_j rows of $\mathbf{C}_j$, for $j = 1, \ldots, s$, is linearly independent in the finite ring R [4].

The set P_n is a (t, k, s)-*net in base* b if it is $(q_1, \ldots, q_s)$-equidistributed whenever $q_1 + \cdots + q_s \leq k - t$ [4]. The smallest such t is the t-*value* of the net. A digital sequence $\{\mathbf{u}_0, \mathbf{u}_1, \ldots, \}$ in s dimensions is a (t, s)-*sequence in base* b if for all integers $k > 0$ and $\nu \geq 0$, the point set $Q(k, \nu) = \{\mathbf{u}_i : i = \nu b^k, \ldots, (\nu + 1)b^k - 1\}$ is a (t, k, s)-net in base b. The t-value is a widely used figure of merit for digital nets. Ideally, we would like it to be zero, but there are theoretical lower bounds on it. In particular, a $(0, k, s)$-net in base b can exist only if $b \geq s - 1$, and a $(0, t)$-sequence in base b can exist only if $b \geq t$. Lower bounds for general pairs (b, s), together with the best values achieved by known constructions, are tabulated in [62]. For example, for $b = 2$, $k = 16$, and $s = 20$, the t-value cannot be smaller than 9. A t-value of 9 in this case only guarantees equidistribution for a partition in $2^7 = 128$ boxes, which must contain $2^9 = 512$ points each. For a given partition, this is a weak requirement.

The problem is that a small t-value would require equidistribution for a very rich family of partitions into rectangular boxes, and this becomes impossible when t is too small. To get around this, we may restrict our consideration to a smaller family of partitions; for example, only cubic boxes. We then want P_n to be $(\ell, \ldots, \ell)$-equidistributed for the largest possible ℓ, which obviously cannot exceed $\lfloor k/s \rfloor$. We want to minimize the *resolution gap* $\delta = \lfloor k/s \rfloor - \ell$.

These definitions apply to the projections $P_n(\mathbf{u})$ as well. Let $t_\mathbf{u}$ and $\delta_\mathbf{u}$ denote the t-value and the resolution gap for $P_n(\mathbf{u})$, and $t^*_{|\mathbf{u}|}$ a theoretical lower bound on $t_\mathbf{u}$. Discrepancy measures for digital nets can be defined by

$$\max_{\mathbf{u} \in \mathcal{J}} \gamma_\mathbf{u} \delta_\mathbf{u}, \quad \text{or} \quad \sum_{\mathbf{u} \in \mathcal{J}} \gamma_\mathbf{u} \delta_\mathbf{u}, \quad \text{or}$$

$$\max_{\mathbf{u} \in \mathcal{J}} \gamma_\mathbf{u} \left[t_\mathbf{u} - t^*_{|\mathbf{u}|} \right], \quad \text{or} \quad \sum_{\mathbf{u} \in \mathcal{J}} \gamma_\mathbf{u} \left[t_\mathbf{u} - t^*_{|\mathbf{u}|} \right],$$

for some non-negative weights $\gamma_\mathbf{u}$ and a preselected class $\mathcal{J}$ of index sets $\mathbf{u}$ [58,13,59,63]. The choice of $\mathcal{J}$ is a matter of compromise. With a larger (richer) $\mathcal{J}$, the criterion is more expensive to compute, and its best possible value is generally larger, so it will have less discriminating power for the more important low-dimensional projections. In practice, the weights are often taken all equal to 1.

Specific instances of these criteria, and search results for good parameters for specific types of digital nets, are reported in [64,21,65,63]. A special case of the first of these four criteria has been widely used to assess the uniformity of $\mathbb{F}_2$-linear RNGs [45,66,31,67,10,42].

7   Conclusion

Low-discrepancy point sets and sequences used for QMC, and the point sets formed by vectors of successive output values produced by RNGs, have much in common. They are often defined via similar linear recurrences. We also want both of them to be highly uniform in the unit hypercube. However, the figures of merit commonly used to measure their uniformity are slightly different. One of the reasons for this difference is the difference of cardinality between those types of point sets: For RNGs, the cardinality is huge and we must restrict ourselves to criteria that can be computed without enumerating the points explicitly, for example. For QMC, on the other hand, certain discrepancies are motivated by the fact that they provide explicit error bounds or variance bounds on certain classes of functions.

References

1. Knuth, D.E.: The Art of Computer Programming, 3rd edn. Seminumerical Algorithms, vol. 2. Addison-Wesley, Reading (1998)
2. L'Ecuyer, P.: Uniform random number generation. Annals of Operations Research 53, 77–120 (1994)
3. L'Ecuyer, P.: Uniform random number generation. In: Henderson, S.G., Nelson, B.L. (eds.) Simulation. Handbooks in Operations Research and Management Science, ch.3, pp. 55–81. Elsevier, Amsterdam (2006)
4. Niederreiter, H.: Random Number Generation and Quasi-Monte Carlo Methods. SIAM CBMS-NSF Regional Conference Series in Applied Mathematics, vol. 63. SIAM, Philadelphia (1992)
5. Law, A.M., Kelton, W.D.: Simulation Modeling and Analysis, 3rd edn. McGraw-Hill, New York (2000)
6. L'Ecuyer, P., Buist, E.: Simulation in Java with SSJ. In: Kuhl, M.E., Steiger, N.M., Armstrong, F.B., Joines, J.A. (eds.) Proceedings of the 2005 Winter Simulation Conference, pp. 611–620. IEEE Press, Pistacaway (2005)
7. L'Ecuyer, P.: Pseudorandom number generators. In: Platen, E., Jaeckel, P. (eds.) imulation Methods in Financial Engineering. Encyclopedia of Quantitative Finance, Wiley, Chichester (forthcoming, 2008)
8. Deng, L.Y.: Efficient and portable multiple recursive generators of large order. ACM Transactions on Modeling and Computer Simulation 15(1), 1–13 (2005)
9. L'Ecuyer, P., Simard, R.: TestU01: A C library for empirical testing of random number generators. ACM Transactions on Mathematical Software, Article 22 33(4) (2007)

10. Panneton, F., L'Ecuyer, P., Matsumoto, M.: Improved long-period generators based on linear recurrences modulo 2. ACM Transactions on Mathematical Software 32(1), 1–16 (2006)
11. Hickernell, F.J.: A generalized discrepancy and quadrature error bound. Mathematics of Computation 67, 299–322 (1998)
12. Hickernell, F.J.: What affects the accuracy of quasi-Monte Carlo quadrature? In: Niederreiter, H., Spanier, J. (eds.) Monte Carlo and Quasi-Monte Carlo Methods 1998, pp. 16–55. Springer, Berlin (2000)
13. L'Ecuyer, P., Lemieux, C.: Recent advances in randomized quasi-Monte Carlo methods. In: Dror, M., L'Ecuyer, P., Szidarovszky, F. (eds.) Modeling Uncertainty: An Examination of Stochastic Theory, Methods, and Applications, pp. 419–474. Kluwer Academic, Boston (2002)
14. L'Ecuyer, P., Lécot, C., Tuffin, B.: A randomized quasi-Monte Carlo simulation method for Markov chains. Operations Research (to appear, 2008)
15. Hickernell, F.J., Sloan, I.H., Wasilkowski, G.W.: On strong tractability of weighted multivariate integration. Mathematics of Computation 73(248), 1903–1911 (2004)
16. Ben-Ameur, H., L'Ecuyer, P., Lemieux, C.: Combination of general antithetic transformations and control variables. Mathematics of Operations Research 29(4), 946–960 (2004)
17. L'Ecuyer, P., Lemieux, C.: Variance reduction via lattice rules. Management Science 46(9), 1214–1235 (2000)
18. Owen, A.B.: Latin supercube sampling for very high-dimensional simulations. ACM Transactions on Modeling and Computer Simulation 8(1), 71–102 (1998)
19. Cranley, R., Patterson, T.N.L.: Randomization of number theoretic methods for multiple integration. SIAM Journal on Numerical Analysis 13(6), 904–914 (1976)
20. Sloan, I.H., Joe, S.: Lattice Methods for Multiple Integration. Clarendon Press, Oxford (1994)
21. L'Ecuyer, P., Lemieux, C.: Quasi-Monte Carlo via linear shift-register sequences. In: Proceedings of the 1999 Winter Simulation Conference, pp. 632–639. IEEE Press, Los Alamitos (1999)
22. Matoušek, J.: Geometric Discrepancy: An Illustrated Guide. Springer, Berlin (1999)
23. Liu, R., Owen, A.B.: Estimating mean dimensionality (manuscript, 2003)
24. Wang, X., Sloan, I.H.: Why are high-dimensional finance problems often of low effective dimension? SIAM Journal on Scientific Computing 27(1), 159–183 (2005)
25. Caflisch, R.E., Morokoff, W., Owen, A.: Valuation of mortgage-backed securities using Brownian bridges to reduce effective dimension. The Journal of Computational Finance 1(1), 27–46 (1997)
26. Avramidis, T., L'Ecuyer, P.: Efficient Monte Carlo and quasi-Monte Carlo option pricing under the variance-gamma model. Management Science 52(12), 1930–1944 (2006)
27. Glasserman, P.: Monte Carlo Methods in Financial Engineering. Springer, New York (2004)
28. Imai, J., Tan, K.S.: A general dimension reduction technique for derivative pricing. Journal of Computational Finance 10(2), 129–155 (2006)
29. Morokoff, W.J.: Generating quasi-random paths for stochastic processes. SIAM Review 40(4), 765–788 (1998)
30. Wang, X., Sloan, I.H.: Brownian bridge and principal component analysis: Toward removing the curse of dimensionality. IMA Journal of Numerical Analysis 27, 631–654 (2007)

31. L'Ecuyer, P., Panneton, F.: $\mathbf{F}_2$-linear random number generators. In: Alexopoulos, C., Goldsman, D. (eds.) Advancing the Frontiers of Simulation: A Festschrift in Honor of George S. Fishman. Springer, New York (to appear, 2007)

32. L'Ecuyer, P.: Good parameters and implementations for combined multiple recursive random number generators. Operations Research 47(1), 159–164 (1999)

33. L'Ecuyer, P.: Tables of linear congruential generators of different sizes and good lattice structure. Mathematics of Computation 68(225), 249–260 (1999)

34. Fishman, G.S.: Monte Carlo: Concepts, Algorithms, and Applications. Springer Series in Operations Research. Springer, New York (1996)

35. Wahba, G.: Spline Models for Observational Data. SIAM CBMS-NSF Regional Conference Series in Applied Mathematics, vol. 59. SIAM, Philadelphia (1990)

36. L'Ecuyer, P., Couture, R.: An implementation of the lattice and spectral tests for multiple recursive linear random number generators. INFORMS Journal on Computing 9(2), 206–217 (1997)

37. Marsaglia, G., Zaman, A.: A new class of random number generators. The Annals of Applied Probability 1, 462–480 (1991)

38. Tezuka, S., L'Ecuyer, P., Couture, R.: On the add-with-carry and subtract-with-borrow random number generators. ACM Transactions of Modeling and Computer Simulation 3(4), 315–331 (1994)

39. Couture, R., L'Ecuyer, P.: Distribution properties of multiply-with-carry random number generators. Mathematics of Computation 66(218), 591–607 (1997)

40. Goresky, M., Klapper, A.: Efficient multiply-with-carry random number generators with maximal period. ACM Transactions on Modeling and Computer Simulation 13(4), 310–321 (2003)

41. Golomb, S.W.: Shift-Register Sequences. Holden-Day, San Francisco (1967)

42. Tezuka, S.: Uniform Random Numbers: Theory and Practice. Kluwer Academic Publishers, Norwell (1995)

43. L'Ecuyer, P.: Tables of maximally equidistributed combined LFSR generators. Mathematics of Computation 68(225), 261–269 (1999)

44. Matsumoto, M., Nishimura, T.: Mersenne twister: A 623-dimensionally equidistributed uniform pseudo-random number generator. ACM Transactions on Modeling and Computer Simulation 8(1), 3–30 (1998)

45. L'Ecuyer, P.: Maximally equidistributed combined Tausworthe generators. Mathematics of Computation 65(213), 203–213 (1996)

46. L'Ecuyer, P., Panneton, F.: A new class of linear feedback shift register generators. In: Joines, J.A., Barton, R.R., Kang, K., Fishwick, P.A. (eds.) Proceedings of the 2000 Winter Simulation Conference, pp. 690–696. IEEE Press, Pistacaway (2000)

47. Dick, J., Sloan, I.H., Wang, X., Wozniakowski, H.: Good lattice rules in weighted Korobov spaces with general weights. Numerische Mathematik 103, 63–97 (2006)

48. Nuyens, D., Cools, R.: Fast algorithms for component-by-component construction of rank-1 lattice rules in shift-invariant reproducing kernel Hilbert spaces. Mathematics and Computers in Simulation 75, 903–920 (2006)

49. Sloan, I.H., Kuo, F.Y., Joe, S.: On the step-by-step construction of quasi-Monte Carlo rules that achieve strong tractability error bounds in weighted Sobolev spaces. Mathematics of Computation 71, 1609–1640 (2002)

50. Cools, R., Kuo, F.Y., Nuyens, D.: Constructing embedded lattice rules for multivariate integration. SIAM Journal on Scientific Computing 28(16), 2162–2188 (2006)

51. Nuyens, D., Cools, R.: Fast component-by-component construction of rank-1 lattice rules with a non-prime number of points. Journal of Complexity 22, 4–28 (2006)

52. Dick, J., Sloan, I.H., Wang, X., Wozniakowski, H.: Liberating the weights. Journal of Complexity 20(5), 593–623 (2004)
53. Kuo, F.Y., Sloan, I.H.: Lifting the curse of dimensionality. Notices of the AMS 52(11), 1320–1328 (2005)
54. Sobol', I.M.: The distribution of points in a cube and the approximate evaluation of integrals. U.S.S.R. Comput. Math. and Math. Phys. 7, 86–112 (1967)
55. Faure, H.: Discrépance des suites associées à un système de numération en dimension s. Acta Arithmetica 61, 337–351 (1982)
56. Niederreiter, H.: Point sets and sequences with small discrepancy. Monatshefte für Mathematik 104, 273–337 (1987)
57. Niederreiter, H., Xing, C.: The algebraic-geometry approach to low-discrepancy sequences. In: Hellekalek, P., Larcher, G., Niederreiter, H., Zinterhof, P. (eds.) Monte Carlo and Quasi-Monte Carlo Methods 1996. Lecture Notes in Statistics, vol. 127, pp. 139–160. Springer, New York (1998)
58. L'Ecuyer, P.: Polynomial integration lattices. In: Niederreiter, H. (ed.) Monte Carlo and Quasi-Monte Carlo Methods 2002, pp. 73–98. Springer, Berlin (2004)
59. Lemieux, C., L'Ecuyer, P.: Randomized polynomial lattice rules for multivariate integration and simulation. SIAM Journal on Scientific Computing 24(5), 1768–1789 (2003)
60. L'Ecuyer, P., Touzin, R.: On the Deng-Lin random number generators and related methods. Statistics and Computing 14, 5–9 (2004)
61. Niederreiter, H., Pirsic, G.: Duality for digital nets and its applications. Acta Arithmetica 97, 173–182 (2001)
62. Schmid, W.C., Schürer, R.: MinT, the database for optimal (t, m, s)-net parameters (2005), http://mint.sbg.ac.at
63. Panneton, F., L'Ecuyer, P.: Infinite-dimensional highly-uniform point sets defined via linear recurrences in $\mathbb{F}_{2^w}$. In: Niederreiter, H., Talay, D. (eds.) Monte Carlo and Quasi-Monte Carlo Methods 2004, pp. 419–429. Springer, Berlin (2006)
64. Joe, S., Kuo, F.Y.: Constructing Sobol sequences with better two-dimensional projections. SIAM Journal on Scientific Computing (to appear, 2008)
65. Lemieux, C.: L'utilisation de règles de réseau en simulation comme technique de réduction de la variance. PhD thesis, Université de Montréal (May 2000)
66. L'Ecuyer, P., Panneton, F.: Fast random number generators based on linear recurrences modulo 2: Overview and comparison. In: Proceedings of the 2005 Winter Simulation Conference, pp. 110–119. IEEE Press, Los Alamitos (2005)
67. Panneton, F., L'Ecuyer, P.: On the xorshift random number generators. ACM Transactions on Modeling and Computer Simulation 15(4), 346–361 (2005)

On Independence and Sensitivity of Statistical Randomness Tests

Meltem Sönmez Turan[1], Ali Doğanaksoy[1], and Serdar Boztaş[2]

[1] Institute of Applied Mathematics
Middle East Technical University, Ankara, Turkey
{msonmez,aldoks}@metu.edu.tr
[2] School of Mathematical and Geospatial Sciences, RMIT University, Australia
serdar.boztas@ems.rmit.edu.au

Abstract. Statistical randomness testing has significant importance in analyzing the quality of random number generators. In this study, we focus on the independence of randomness tests and its effect on the coverage of test suites. We experimentally observe that frequency, overlapping template, longest run of ones, random walk height and maximum order complexity tests are correlated for short sequences. We also proposed the concept of sensitivity, where we analyze the effect of simple transformations on output p-values. We claim that whenever the effect is significant, the composition of the transformation and the test may be included to the suite as a new test.

Keywords: Randomness testing, Coverage, Independence.

1 Introduction

Random numbers are widely used in many applications such as statistical sampling, Monte Carlo simulation, numerical analysis, game theory, cryptography, etc. In cryptography, the need for random numbers arises in key generation, authentication protocols, digital signature schemes, zero-knowledge protocols, etc. Using weak random numbers may result in an adversary ability to break the whole cryptosystem. However, for many simulation applications strong random numbers are not required, in other words, different applications may require different levels of randomness.

Generating high quality random numbers is a very serious problem–and is very difficult if deterministic methods are used. The best way to generate unpredictable random numbers is to use physical processes such as radioactive decay, thermal noise or sound samples from a noisy environment. However, generating random numbers using these physical processes is extremely inefficient. Therefore, most systems use Pseudo Random Number Generators (PRNGs) based on deterministic algorithms. Desired properties of PRNGs are (i) good randomness properties of the output sequence, (ii) reproducibility, (iii) speed or efficiency and (iv) large period. The unpredictability of random sequences is established using a random *seed* that is obtained by a physical source like timings of keystrokes.

S.W. Golomb et al. (Eds.): SETA 2008, LNCS 5203, pp. 18–29, 2008.

Without the seed, the attacker must not be able to make any predictions about the output bits, even when all details of the generator is known.

A theoretical proof for the randomness of a generator is impossible to give, therefore statistical inference based on observed sample sequences produced by the generator seems to be the best option. Considering the properties of binary random sequences, various statistical tests can be designed to evaluate the assertion that the sequence is generated by a perfectly random source. Test suites [1,2,3,4,5] define a collection of tests to evaluate randomness of generators extensively.

One important attribute of test suites is the variety or coverage of tests that they include. Coverage can be defined as the sequences that fail any of the tests in the suite for a pre-specified type I error. A generator may behave randomly based on a number of tests, and fail when it is evaluated by another test. So, to have more confidence in the randomness of generators, coverage of test suite should be as high as possible.

Test suites produce many p-values while evaluating a random number generator. Interpretation of these p-values is sometimes complicated and requires careful attention. According to the NIST test suite, these p-values are evaluated based on (i) their uniformity and (ii) the proportion of p-values that are less than a predefined threshold value which is typically 0.01 [1]. As Soto stated in [6] to achieve reliable results, the statistical tests in a suite should be independent, in other words the results of the tests should be uncorrelated. In [7], the relation between *approximate entropy, overlapping serial and universal test* is analyzed and highly correlated results are obtained using defective sources.

There is a strong relation between the coverage of the suite and independence of tests. In this paper, we study the independence of some commonly used randomness tests and present some theoretical and experimental results. Moreover, we define the concept of sensitivity, the effect of some transformations to sequences on test results. We propose to add the composition of the transformation and the test to the suite, whenever the effect of the transformation is significant.

The organization of the paper is as follows. In the next section, basic background information about randomness tests and test suites are presented. In Sect. 3, theoretical and experimental results on the independence of randomness tests are presented for a subset of tests. In Sect. 4, the concept of sensitivity is defined and effects of some basic transformations are presented. In the last section, we give some concluding remarks and possible future work directions.

2   Randomness Tests

Let L be the set of n bit binary sequences, trivially $|L| = 2^n$. A statistical test T is a deterministic algorithm that takes a sample sequence and produces a decision regarding the randomness of a sequence, as $T : L \rightarrow \{\text{accept}, \text{reject}\}$. Therefore, T divides the set L of binary sequences $S_n = (s_1, s_2, \ldots, s_n)$ of length n into two sets;

$$L_A = \{S_n \in L : T(S_n) = \text{accept}\} \subseteq L$$
$$L_R = \{S_n \in L : T(S_n) = \text{reject}\} \subseteq L$$

The size of these two sets is determined by the type I error α, i.e. $|L_R|/|L| = \alpha$.

The most commonly used test is probably the *frequency test* [1] that evaluates the randomness of a given sequence based on the number of ones (that is, the weight w) of the sequence. A trivial extension of the frequency test can be given as the *equidistribution test* that focuses on the frequencies of k-bit tuples throughout the sequence. *Overlapping* and *Non-overlapping template matching tests* [1] only consider the number of occurrences of pre-specified templates in the sequence.

Another commonly used statistics in randomness testing is about the changes from 1 to 0 or visa versa, which is called a *run*. Runs test [1] focuses on the number of runs in the sequence which is expected to be around $\frac{(n+1)}{2}$ for a random sequence of length n. Alternatively, there are tests that concentrate on length of runs, particularly *length of the longest run of ones* [1] in the sequence.

Especially in cryptographic applications, the complexity of sequences -the ability to reproduce the sequence- is of interest. Sequences with low complexity measures can be easily generated with alternative machines. Some examples of randomness tests based on complexity measures can be given as *linear complexity* [1], *Lempel-Ziv* [8] and *maximum order complexity* [9].

Classification of Tests. While selecting a subset of randomness tests to be included in a test suite, it is necessary to understand what exactly the test measures. Tests that focus on similar properties should not be included in the test suite, but one of such a class should be chosen.

In [10], two important properties of random sequences are emphasized. The first is about the appearance of the sequence which is related to the *ability of the attacker to guess the next bit better than random guessing*. The second property is about the *ability to reproduce the sequence*, i.e., the complexity of the sequence. Considering these properties, we give a rough classification of randomness tests into two categories:

Tests based on k-tuple pattern frequencies aim to detect the biases in the appearance of the sequences. Weaknesses in the appearance of sequences can be considered as the deviations in the frequencies of k-tuples ($1 \leq k \leq log_2 n$) which are expected to occur nearly equally many times throughout the sequence. An example for this category with $k = 1$ is the most commonly used frequency test. A trivial extension of the frequency test is the equidistribution test with $k = 2, 3, \ldots$. Other examples can be given as runs, overlapping template, serial, coupon collectors, etc.

Tests based on ordering of k-tuples aim to detect weaknesses based on the reproducibility of sequences. Most of the test statistics in this category are related to the size of simpler systems that generate the sequence. If the size of these machines are less than expected, it is possible to use these machines to regenerate the sequence. Tests based on complexity measures like linear complexity, maximum order complexity can be given as examples.

These categories are closely related and obviously not disjoint. It is possible to find randomness tests that can be included in more than one category. As an example, consider the runs test which can be included to the category of *tests based on ordering of $k = 1$-tuples* and also to the category of *tests based on $k = 2$-tuple pattern frequency* since it actually counts the number of 01 and 10 patterns throughout the sequence.

Even for applications that do not require strong random numbers, generators should produce uniform outputs, in other words, should pass the tests in the first category. The tests in the second category usually take more time compared to the first category, but they are more important for applications that require strong randomness such as cryptographic applications where randomness is used in key, nonce and initial vector generation.

Test Suites. A randomness test suite is a collection of randomness tests that are selected to analyze the randomness properties of generators. Here we give some information about widely used test suites.

- *Knuth Test Suite* [4], developed in 1998, presents several empirical tests including frequency, serial, gap, poker, coupon collector's, permutation, run, maximum-of-t, collision, birthday spacings and serial correlation.
- *DIEHARD Test Suite* [2] consists of 18 different, independent statistical tests including; birthday spacings, overlapping 5-permutations, binary rank, bitstream test, monkey tests on 20-bit Words, monkey tests OPSO, OQSO, DNA, count the 1's in a stream of bytes, count the 1's in specific bytes, parking lot, minimum distance, 3D spheres, squeeze, overlapping sums, runs and craps.
- *Crypt-XS* [3] suite which was developed in the Information Security Research Centre at Queensland University of Technology consists of frequency, binary derivative, change point, runs, sequence complexity and linear complexity tests.
- *NIST Test Suite* [1] consists of 16 tests namely frequency, block frequency, cumulative sums, runs, long runs, rank, spectral, nonoverlapping template matchings, overlapping template matchings, Maurer's universal statistical, approximate entropy, random excursions, Lempel-Ziv complexity, linear complexity, and serial. During the evaluation of block ciphers presented for AES, Soto [6] proposed nine different ways to generate large number of data streams from a block cipher and tested these streams using the NIST test suite to evaluate the security of AES candidates.
- *TestU01 Suite* [5] is another test suite for empirical testing of random number generators. This suite consists of many randomness tests and it is also suitable to test sequences that take real values.

Multi Level Testing. In [11], to increase the power of tests, it is proposed to apply the test N times to disjoint parts of the sequence, yielding N different test

One Level: $\underbrace{(s_1, \ldots, s_n)}_{T} \rightarrow$ Evaluate T

Two Level: $\underbrace{(s_1, \ldots, s_m)}_{T_1}, \ldots \underbrace{(s_{n-m+1}, \ldots, s_n)}_{T_N} \rightarrow$ Evaluate $T_1, \ldots, T_N$

Fig. 1. Multi Level Testing

statistics, $t_1, t_2, \ldots, t_N$. Then, using standard goodness of fit tests, the empirical distribution of t_i values is compared to the the theoretical distribution of the test statistic under H_0. This is called level-2 testing (See Fig. 1). To increase the power, the level of tests may be increased further. The level-2 version of frequency test is the *frequency test within a block* [1] that focuses on the weight of disjoint parts of a given sequences.

3   Independence of Tests

There are extensive number of randomness tests in the literature and to design a test suite a careful selection should be done. Many generators may appear to be random according to a number of tests, but may be non-random when subjected to another test, therefore the variety or the coverage of the tests used in the test suite should be high enough. However, including dependent tests may result in wrong conclusions about the generators.

Two tests T_1 and T_2 are considered to be independent if the distribution of their test statistics t_1 and t_2 (and corresponding p-values) are independent, that is

$$Pr(t_1|t_2) = Pr(t_1), \tag{2}$$

and visa versa.

In this section, we analyze the relation between some of the commonly used tests and try to observe if there exists any statistically significant correlation. We consider ten level-1 tests, which are also suitable for testing to short sequences. Given a sequence $(s_1, s_2, \ldots, s_n)$ of length n, each test defines a test statistic as described below.

- *Frequency Test:* The test statistic is the weight of the sequence, that is $t = s_1 + \ldots + s_n$, taking values between 0 and n.
- *Overlapping Template Test:* Test statistic is the number of occurrences of a m bit pattern **a** throughout the sequence, that is

$$t = |\{i|(s_i, \ldots, s_{i+m-1}) = \mathbf{a}, 1 \leq i \leq n - m + 1\}|. \tag{3}$$

For our experiments, **a** is chosen to be 111.
- *Longest Run of Ones Test:* Test statistic is the length of the longest run of ones, that is

$$t = \max\{m|s_i = s_{i+1} \ldots = s_{i+m} = 1, 1 \leq i \leq n\}, \tag{4}$$

taking values between 0 and n.

- *Runs Test:* Test statistic is the number runs throughout the sequence, taking values between 1 and n.
- *Random Walk Height Test:* Test statistic is the height of random walk, that is

$$t = \max_{i=1,...,n} |\sum_{j=1}^{i} (2s_j - 1)| \qquad (5)$$

  taking values between 1 and n.
- *Random Walk Excursion Test:* Test statistic is the number of excursions in the random walk, that is

$$t = |\{i| \sum_{j=1}^{i} (2s_j - 1) = 0, 1 \leq i \leq n\}|, \qquad (6)$$

  taking values between 0 and $n/2$
- *Linear Complexity Test:* Test statistic is the linear complexity of the sequence, that is, the length of the shortest LFSR that generates the sequence, taking values between 0 and n. Linear complexity of a sequence can efficiently be calculated using the Berlekamp-Massey algorithm.
- *k-error Linear Complexity Test:* Test statistic is the k-error linear complexity of the sequence that is the length of the shortest LFSR that generate the sequence with at most k bit difference. In our experiments, we focused on the $k = 1$ case, in which the test statistic takes values between 0 and $n/2$.
- *Maximum Order Complexity Test:* Test statistic is the maximum order complexity of the sequence that is the length of the shortest feedback shift register that generates the sequence, taking values between 0 and $n - 1$.
- *Lempel-Ziv Test:* Test statistic is the Lempel-Ziv complexity of the sequence that takes its maximum value as $n/2$. For instance, Lempel-Ziv complexity of the sequence 010101001001011 is 7, since different patterns observed are 0|1|01|010|0100|10|11.

3.1 Theoretical Results

In this section, to analyze the relation of two tests T_1 and T_1, we present some theoretical bounds on the maximum and minimum values of t_1 as a function of t_2.

Frequency versus Runs Test. Given a sequence of length n and weight w, the maximum possible number of runs R is

$$\max\{R\} = \begin{cases} n & \text{if } w = \frac{n}{2} \\ \min\{2w + 1, 2(n - w) + 1\} & \text{if } w \neq \frac{n}{2} \end{cases}$$

whereas the minimum number of runs R is

$$\min\{R\} = \begin{cases} 2 \text{ if } 1 \leq w < n \\ 1 \ w = 0 \ or \ w = n \end{cases}.$$

For balanced sequences, the number of runs takes values between 1 and n, but as the weight of the sequence deviates from $n/2$, the maximum possible number of runs decreases. Sequences with weight less than $n/4$, it is not possible to achieve expected number of runs. Conversely, given the number of runs R, $w \geq \lfloor \frac{R}{2} \rfloor$.

Frequency versus Random Walk Height Test. Given a sequence of length n and weight w, the random walk height test statistic H attains the maximum value as $\max\{w, n - w\}$ and minimum value

$$\min\{H\} = \begin{cases} 1 & \text{if } w = \lfloor \frac{n}{2} \rfloor \\ |n - 2w| & \text{otherwise} \end{cases}.$$

Given H, the weight of the sequence is at least $\min\{H, n - H\}$. From this property, if a sequence fails random walk height test, it is very likely that it also fails the frequency test. The relation is more significant for short sequences.

Frequency versus Longest Run of Ones. Given a sequence of length n and weight w, the longest run of ones test statistic L takes its maximum value as w and minimum value as $\left\lceil \frac{w}{n-w-1} \right\rceil$.

Frequency versus Random Walk Excursion Test. Given the weight w of a n-bit sequence, the number of random walk excursion test statistic E takes its maximum value as $\min\{w, n - w\}$ and minimum value as

$$\min\{E\} = \begin{cases} 1 & \text{if } w = \frac{n}{2} \\ 0 & \text{otherwise} \end{cases}.$$

Runs versus Random Walk Excursion. Both tests are related to the speed of changes from 0 to 1 (or from 1 to 0). Sequences with large number of excursions are expected to have large number of runs, similarly sequences with small number of runs, are expected to have less number of excursions. Each excursion consists of at least two runs, therefore number of runs is at least twice the number of excursions.

Frequency versus Linear Complexity. There is no direct relation between the weight and the linear complexity of the sequence. Even with very low weight, it is possible to achieve high linear complexity values. As an example, consider the sequences with $w = 1$, location of the bit 1 determines the linear complexity. There is however a strong relation between the weight of the Discrete Fourier Transform of the sequence and its linear complexity, so-called Blahut's theorem [12].

3.2   Experiments on Short Sequences

As an alternative definition of independence, tests T_1 and T_2 can be considered independent, if their rejection regions are independent for all selection of α. In this part of the study, we analyze the relations of 10 tests given in Sect. 3, focusing the rejection regions with size approximately 0.01.

Considering all binary sequences of length $n = 20$ and 30, we formed the rejection regions $R_n^{T_i}$ of each test, that is, the set of all n bit sequences that fail T_i. The upper and lower acceptable limits for test statistics are calculated so that $\alpha \approx |R_n^1| \approx \ldots \approx |R_n^{10}| \approx 0.01$. If the test statistic is more extreme than the lower or upper limit given in Table 1, the sequence is assumed to be non-random.

Table 1. Lower limits (LLs) and upper limits (ULs) of the test statistics for 20 bit sequences and corresponding type I error, α

Test	$n = 20$			$n = 30$		
	LL	UL	α	LL	UL	α
Frequency	5	15	0.011818	9	21	0.016125
Overlapping Template	0	8	0.014478	0	10	0.017750
Longest Run of Ones	1	8	0.012691	1	9	0.010727
Runs	5	15	0.011818	9	21	0.016125
RW Height	1	11	0.014395	1	14	0.010446
RW Excursion	0	7	0.022461	0	10	0.011818
LC	8	12	0.031250	13	17	0.031250
1-error LC	6	10	0.012996	11	15	0.019407
Maximum Order Complexity	2	10	0.017889	4	12	0.016291
Lempel-Ziv	8	10	0.026367	11	13	0.031378

The $(i, j)^{th}$ entry of Tables 2 and 3 represents the proportion of sequences that fail T_i and T_j to the sequences that fail T_i, that is, $\frac{|R_n^i \cap R_n^j|}{|R_n^i|}$, for $n = 20, 30$. The expected value of this proportion is 0.01 and tables are expected to be symmetric for larger n values. In the tables, the percentages that significantly deviate (> 0.10) from expected values are highlighted.

According to Table 2 and 3, frequency, overlapping template (with input template 111), longest run of ones, random walk height tests and maximum order complexity tests are closely related. Also, there is a correlation between the

Table 2. Relation of tests for all sequences of length $n = 20$ for $\alpha = 0.01$

Test	Frequency	Overlapping Template	Longest Run of Ones	Runs	RW Height	RW Excursion	LC	1-error LC	MOC	Lempel-Ziv
Frequency	-	0.4334	0.2012	0.04341	1	0	0.1011	0.0657	0.1785	0.3689
Overlapping Template	0.3538	-	0.4391	0.0516	0.4171	0	0.0491	0.0234	0.1699	0.1632
Longest Run of Ones	0.1874	0.5009	-	0.0634	0.2370	0	0.0385	0.0225	0.1740	0.1040
Runs	0.04341	0.0632	0.0681	-	0.0618	0.1933	0.0933	0.0531	0.1635	0.1228
RW Height	0.8210	0.4195	0.2089	0.0507	-	0	0.1050	0.0542	0.1972	0.3108
RW Excursion	0	0	0	0.1017	0	-	0.0351	0.0160	0.0369	0.0348
LC	0.0383	0.0227	0.0156	0.0353	0.0484	0.0252	-	0.1339	0.1737	0.0424
1-error LC	0.0598	0.0260	0.0220	0.0483	0.0601	0.0277	0.3220	-	0.0961	0.0643
MOC	0.1179	0.1375	0.1235	0.1080	0.1586	0.0464	0.3035	0.0698	-	0.0951
Lempel-Ziv	0.1654	0.0896	0.05006	0.0550	0.1697	0.0296	0.05031	0.0317	0.0645	-

Table 3. Relation of tests for all sequences of length $n = 30$ for $\alpha = 0.01$

Test	Frequency	Overlapping Template	Longest Run of Ones	Runs	RW Height	RW Excursion	LC	k-error LC	MOC	Lempel-Ziv
Frequency	-	0.3853	0.1231	0.0409	0.5591	0	0.0339	0.0207	0.1035	0.3399
Overlapping Template	0.3500	-	0.3096	0.0733	0.2490	0	0.0309	0.0185	0.0875	0.1631
Longest Run of Ones	0.1851	0.5124	-	0.0869	0.1602	0	0.0303	0.0172	0.1026	0.0991
Runs	0.0409	0.0807	0.0578	-	0.0441	0.0876	0.0343	0.0212	0.1016	0.1308
RW Height	0.8630	0.4231	0.1645	0.0681	-	0	0.0355	0.0204	0.1498	0.3520
RW Excursion	0	0	0	0.1195	0	-	0.0320	0.0192	0.0328	0.0395
LC	0.0175	0.0175	0.0104	0.0177	0.0119	0.0121	-	0.1398	0.0250	0.0328
1-error LC	0.0172	0.0170	0.0095	0.0176	0.0110	0.0117	0.2251	-	0.0184	0.0330
MOC	0.1025	0.0953	0.0676	0.1005	0.0960	0.0238	0.0480	0.0219	-	0.0917
Lempel-Ziv	0.1747	0.0922	0.0339	0.0672	0.1172	0.0149	0.0327	0.0204	0.0476	-

results of linear complexity and 1-error linear complexity tests. Moreover, a significant relation is observed between Lempel-Ziv and frequency test. As also mentioned in the theoretical results part in Sect. 3.1, no correlation between the weight and the linear complexity is observed.

Another interesting result is that none of the sequences that fail the random walk excursion test, fail any of the (i) frequency; (ii) overlapping template; (iii) longest run of ones or (iv) random walk height tests for sequences of length $n = 20$ and 30. This means that including the random walk excursion test increases the coverage of test suites significantly. To measure the effect of each test to the coverage, we present the number of 20 bit sequences that only fail the given test in Table 4. The *tests based on ordering of k-tuples* seem to increase the coverage of the selection more compared to *tests based on k-tuple pattern frequency* and this is mainly due to the correlation of these tests presented in above tables. Also, it is observed that all sequences that fail frequency test also fail any of the other tests in our scheme. So, there is no contribution of frequency tests to the coverage of selected tests, for sequences of length 20.

We also calculated the coverage, that is $|\cup_{i=1}^{10} R_n^i|$, for $n = 20$ and 30 as 0.122948 and 0.134930, respectively. Whenever $R_n^{T_i}$ sets are disjoint, coverage takes its

Table 4. Number of sequences that only fail the given test (but pass all other tests)

Tests	Number of Sequences
Linear Complexity	22436
Lempel-Ziv Complexity	19680
Random Walk Excursion	19428
Maximum Order Complexity	8419
$k = 1$-error Linear Complexity	7895
Runs	6454
Longest Run of Ones	6196
Overlapping Template	4765
Random Walk Height	1163
Frequency	0

maximum value as $\sum \alpha_i$, which is 0.176163 for 20 and 0.181317 for 30 bit sequences. Due to the correlations in this suite, coverage reduces around 30%.

Testing short sequences, these correlations should be considered. As the length of the sequences increase, it is possible to observe weaker correlation between tests. For instance, in Table 2 for $n = 20$, the number of highlighted cells (proportion > 0.10) is 37, whereas this number decreases to 27 for $n = 30$ given in Table 3. It should be noted that in case of testing longer sequences by level-2 version of these tests, correlations still exist whenever the input block size is small.

4   Sensitivity of Tests

To have more confidence in a random number generator, it is advantageous to use as many randomness tests as possible. In this part of the study, we propose to apply simple transformations to input sequences that significantly change the output p-value of a randomness test as an alternative to developing more tests.

Definition 1. *Consider a randomness test T and a one-to-one transformation $\sigma : L \to L$ where L is the set of all n bit binary sequences. T is said to be invariant under σ if for any S in L, $T(S) = T(\sigma(S))$.*

Here, we define a new concept of *sensitivity* to measure the effect of a transformation to output p-values. If a test T is invariant under σ, sensitivity of T to σ is represented by 0. If the transformation has small effect on the test results, that is, there is a significant correlation between $T(S)$ and $T(\sigma(S))$, sensitivity is represented by 1. Whenever $T(S)$ and $T(\sigma(S))$ are statistically independent, sensitivity is represented by 2, in those cases $T(\sigma(.))$ can be added to the test suite as a new test.

The transformation σ can be chosen in various ways, in this section, we consider a few of them as examples.

- *Complementation* is applying the unary bitwise NOT operation, that is $\sigma_c(s_1, \ldots, s_n) = (s_1 \oplus 1, \ldots, s_n \oplus 1)$.
- *l-Rotation* is a circular shift operation commonly used in cryptography, that is $\sigma_{l-rot}(s_1, \ldots, s_n) = (s_{l+1}, \ldots, s_n, s_1, \ldots, s_l)$. Most of the level-2 tests are invariant to l-rotation, when l is equal to the block length.
- i^{th} *Bit flip* is simply flipping i^{th} bit of the sequence, that is $\sigma_{f_i}(s_1, \ldots, s_n) = (s_1, \ldots, s_i \oplus 1, \ldots, s_n)$.
- *Reversing* is simply considering the sequence backwards, that is $\sigma_{rvs}(s_1, \ldots, s_n) = (s_n, \ldots, s_1)$.
- l^{th} *Derivative* of a sequence is the summation of the sequence and its l-bit rotation, that is $\sigma_{d_l}(s_1, \ldots, s_n) = (s_1 \oplus s_{l+1}, \ldots, s_{n-l} \oplus s_n, \ldots, s_n \oplus s_l)$. This transformation is not one-to-one and applying such transformations results in changes in the α values. To make the transformation one-to-one, following simple modification can be done:

$$\sigma_{d_l}(s_1, \ldots, s_n) = (s_1 \oplus s_{l+1}, \ldots, s_{n-l} \oplus s_n, \ldots, s_{n-1} \oplus s_{l-1}, s_n).$$

Table 5. Sensitivity of randomness tests toward some transformations

Tests	σ_c	σ_{l-rot}	σ_{f_i}	σ_{rvs}	σ_d
Frequency	0	0	1	0	2
Overlapping template	2	1	1	0	2
Longest run of ones	2	1	1	0	2
Runs	0	1	1	0	2
Random walk height	0	1	1	2	2
Random walk excursion	0	1	1	2	2
Linear complexity	1	2	2	2	2
1-error linear complexity	1	2	0	2	2
Maximum order complexity	0	1	2	2	2
Lempel-Ziv	0	1	1	2	2

In Table 5, sensitivity of the tests selected in previous section are given according to $\sigma_c, \sigma_{l-rot}, \sigma_{f_i}, \sigma_{rvs}$ and σ_{d_l}. As observed from the Table 5, some of the transformations do not have any effect on the test results. Complementing the input sequences only affect the results of overlapping template and longest run of ones tests. However, for the overlapping template test, instead of complementing the whole sequence, it is also possible to complement the input template which would result in the same p-value. l-rotation affects the results of linear complexity and 1-error linear complexity tests, whereas flipping i^{th} bit affects the results of linear complexity and maximum order complexity tests. Reversing the sequences significantly changes the outputs of tests based on random walks and complexity measures. However, for balanced sequences reversing the output of random excursion test does not affect the output. Taking the lth derivative of sequences significantly affect all test results available in our set. So, taking the lth derivative seems to be a good choice of transformation to design new tests.

It is obvious that the independence of $T(\sigma(S))$ and $T(S)$ is not enough to justify adding $T(\sigma(\cdot))$ to the suite. It should also be independent of other tests in the suite. As an example, applying the frequency tests to the first derivative of the sequence is equivalent to applying the runs test.

5   Conclusion

Statistical testing plays an important role in analyzing the strength of random number generators and there are various test suites in the literature for this purpose. Generators which pass suites with high coverage engender more confidence. In this study, we emphasize the importance of independence of randomness tests in test suites and present some theoretical and experimental results. We experimentally observe that frequency, overlapping template (with input template 111), longest run of ones, random walk height tests and maximum order complexity tests produce correlated results for short sequences. These correlations should be considered while analyzing generators using short sequences. The strength of these correlations is likely to decrease as the input lengths increase, but in the case of testing longer sequences by level-2 version of these tests, correlations still exist whenever the input block size is small.

We also defined the concept of sensitivity, where we analyze the effect of simple transformations on test results. If a transformation significantly changes the output p-values, then the composition of the transformation and the test may be included in the suite to increase the coverage. Ideally, we would like to have each test applied to a transformed sequence $\sigma(S)$ to be independent of all different tests applied to the original sequence. Clearly, as the set of allowable transformations grows, this becomes harder to achieve. By choosing a good set of allowable transformations, one can use a given set of tests in a more powerful fashion. For example, one should not introduce unnatural transformations of the data, but stick to a set of transformations which are generated by a small set of basic transformations, such as the ones given here as examples. It is of interest to investigate this problem further in future work.

References

1. Rukhin, A., Soto, J., Nechvatal, J., Smid, M., Barker, E., Leigh, S., Levenson, M., Vangel, M., Banks, D., Heckert, A., Dray, J., Vo, S.: A Statistical Test Suite for Random and Pseudorandom Number Generators for Cryptographic Applications (2001), http://www.nist.gov
2. Marsaglia, G.: The Marsaglia Random Number CDROM including the DIEHARD Battery of Tests of Randomness (1996)
3. Caelli, W., Dawson, E., Nielsen, L., Gustafson, H.: CRYPT–X Statistical Package Manual, Measuring the Strength of Stream and Block ciphers (1992)
4. Knuth, D.E.: Seminumerical Algorithms. The Art of Computer Programming, vol. 2. Addison-Wesley, Reading (1981)
5. L'Ecuyer, P., Simard, R.: TestU01: A C library for Empirical Testing of Random Number Generators. ACM Transactions on Mathematical Software (to appear, 2006)
6. Soto, J.: Randomness Testing of the AES Candidate Algorithms (1999)
7. Hellekalek, P., Wegenkittl, S.: Empirical Evidence Concerning AES. ACM Trans. Model. Comput. Simul. 13(4), 322–333 (2003)
8. Doğanaksoy, A., Göloğlu, F.: On Lempel-Ziv Complexity of Sequences. In: Gong, G., Helleseth, T., Song, H.-Y., Yang, K. (eds.) SETA 2006. LNCS, vol. 4086, pp. 180–189. Springer, Heidelberg (2006)
9. Kasselman, P.: A Statistical Test for Stream Ciphers Based on the Maximum Order Complexity. In: South African Symposium On Communication and Signal Processing, pp. 213–218 (1998)
10. Robshaw, M.: Stream Ciphers. Technical Report TR - 701 (1994)
11. L'Ecuyer, P.: Testing Random Number Generators. In: Proceedings of the 1992 Winter Simulation Conference, pp. 305–313. IEEE Press, Los Alamitos (1992)
12. Massey, J.L., Serconek, S.: A Fourier Transform Approach to the Linear Complexity of Nonlinearly Filtered Sequences. In: Desmedt, Y.G. (ed.) CRYPTO 1994. LNCS, vol. 839, pp. 332–340. Springer, Heidelberg (1994)

New Distinguishers Based on Random Mappings against Stream Ciphers

Meltem Sönmez Turan[1], Çağdaş Çalık[1], Nurdan Buz Saran[2], and Ali Doğanaksoy[1]

[1] Institute of Applied Mathematics,
Middle East Technical University, Ankara, Turkey
[2] Computer Engineering Department, Çankaya University, Ankara, Turkey
{msonmez, ccalik, e135760, aldoks}@metu.edu.tr

Abstract. Statistical randomness testing plays an important role in security analysis of cryptosystems. In this study, we aim to propose a new framework of randomness testing based on random mappings. Considering the probability distributions of coverage and ρ-lengths, we present three new distinguishers; (i) coverage test, (ii) ρ-test and (iii) DP-coverage test and applied them on Phase III Candidates of eSTREAM project. We experimentally observed some statistical weaknesses of Pomaranch using the coverage test.

Keywords: Random Mappings, TMTO Attacks, Randomness tests, Stream Ciphers.

1 Introduction

Random mappings are functions from a finite set of n elements into itself. They are widely used in many combinatorial problems such as (i) proportion of empty urns after throwing n balls into n urns, (ii) probability that two persons have the same birthday among a group of n people, (iii) the required number of random selections among n coupons to obtain a full collection, etc. Using a subset of key and IV bits as input and a subset of keystream bits as output, it is possible to form random mappings by using stream ciphers.

The problem of inverting (finding the pre-image of a given point) random mappings has a great practical importance. For mappings generated by stream ciphers, this problem corresponds to obtaining secret state bits using the output keystream sequence, i.e. breaking the cipher. However, there is no known efficient algorithm to find the pre-image of a given point for a random function. Trivially, using exhaustive search, it is possible to check all possible inputs until desired output is obtained, leading an excessive time requirement. Alternatively, a lookup table that contains the pre-images of all points can be constructed, resulting in a large memory requirement. To balance solution time and required memory, Time Memory Tradeoff (TMTO) attacks are proposed as generic methods to invert random mappings [1].

S.W. Golomb et al. (Eds.): SETA 2008, LNCS 5203, pp. 30–41, 2008.
© Springer-Verlag Berlin Heidelberg 2008

It is possible to apply TMTO attacks to stream ciphers by various different approaches [2,3,4,5]. The success probabilities of the attacks are calculated under the assumption that the cipher behaves like a random mapping. To avoid the attacks, some conditions for stream ciphers are given as (i) IV size should be at least equal to key size, (ii) state size should be at least the size of key plus IV and (iii) key size should be at least 80 bits [5].

In this study, we aim to analyze the security of stream ciphers based on some properties of random mappings. By focusing on different TMTO attacks, we try to find suitable test statistics. First, we consider the coverage of mappings generated using a subset of IV bits, then we analyze the index of the first repetition when the random mapping is iteratively applied. Finally, we consider the distinguished point method against stream ciphers and analyze the coverage properties of random mappings that are followed by a special keystream pattern. Using these test statistics, we propose three new distinguishers namely (i) coverage test, (ii) ρ-test and (iii) distinguished point (DP) coverage test and apply these tests to the Phase III Candidates of eSTREAM project [6].

The organization of the paper is as follows. In the following section, some preliminary information about random mappings is presented and then background knowledge on TMTO attacks focusing on stream ciphers are presented. In Sect. 4, the new distinguishers are presented and in Sect. 5 experimental results on eSTREAM candidates are given. Finally, we conclude the paper in the last section.

2 Preliminaries

Let $f(x)$ be a random mapping $X \rightarrow X$ where X is a finite set of n elements. Random mappings independently assign one of the image points $y \in X$ for each input $x \in X$. The sample space consists of n^n random mappings. For each mapping f, we can associate the random variable $C = |\text{image of } f|$.

Proposition 1. *Let us have n independent and identically distributed random variables $X_1, X_2, \ldots X_n$, each uniformly selected from the set $\{1, 2, \ldots, n\}$. Let A_k be the number of selections that contain k different elements. The recursive formulation of A_k is given as*

$$A_k = \binom{n}{k} \left[k^n - \sum_{i=1}^{k-1} \binom{k}{i} \frac{A_{k-i}}{\binom{n}{k-i}} \right] \tag{1}$$

where $A_1 = \binom{n}{1}$.

Example 1. Let $n = 3$. The total number of selections is $3^3 = 27$. The selections with $k = 1$ distinct elements are $\{111, 222, 333\}$, with $k = 2$ distinct elements are $\{112, 121, 211, 113, 131, 311, 221, 212, 122, 223, 232, 322, 331, 313, 133, 332, 323, 233\}$ and with $k = 3$ distinct elements are $\{123, 132, 213, 231, 312, 321\}$. Then, $A_1 = 3$, $A_2 = 18$, $A_3 = 6$, with a total of 27.

Following the above proposition, the probability distribution of C is obtained as

$$Pr(C = k) = \frac{A_k}{n^n},\qquad(2)$$

for $k = 1, 2, \ldots, n$. The expected value of C is $n - n(1 - \frac{1}{n})^n$ which is approximately $n(1 - e^{-1})$ and for large n, hence $C \approx 0.63n$.

Iterative application of f to $x_0 \in X$ yields the sequence

$$\{x_0, x_1 = f(x_0), x_2 = f(x_1), \ldots, x_n = f(x_{n-1})\}.\qquad(3)$$

In Fig. 1, the typical behavior of an iteration operation is given. Since the set X is finite, after some iterations, we will encounter a point that has occurred before. Starting with a point x_0, let x_m be the point that the iteration enters a loop to form a cycle. The path between x_0 and x_m is called the *tail length*. The sum of the tail length and cycle length is defined as the *ρ-length*.

Fig. 1. Graphical representation of an iteration

Proposition 2. *The probability distribution of the ρ-length for a random mapping of n elements is given as*

$$Pr(\rho - length = k) = (\frac{k-1}{n}) \prod_{i=1}^{k-2} (\frac{n-i}{n}), \; for \; k \geq 2.\qquad(4)$$

3   Time Memory Tradeoff Attacks

TMTO attacks aim to speed up the exhaustive key search at the expense of memory usage and the success rate depends on the time and memory allocated for cryptanalysis. The attacks consist of *offline* (or pre-computation) and *online* phases. In the offline phase, a look up table based on the cipher is constructed. In the online phase, a given target is searched using the previously constructed table. In the attack,

- N is the size of the search space,
- P is the complexity reserved for pre-calculations,
- M is the amount of memory available,
- T is the online time complexity and
- D is the amount of data available.

Complexity of the TMTO attack is usually assumed to be the maximum of T and M and the attack is considered to be successful if the complexity is less than N. Generally, the pre-computation complexity P is not included to the attack complexity.

In 1980, Hellman presented a probabilistic tradeoff attack that recovers the secret key in $N^{2/3}$ operations using $N^{2/3}$ words of memory after N operations of precomputation, for an arbitrary block cipher with N keys [1]. The success rate of the attack depends on the assigned memory and time. Unlike exhaustive key search that uses an arbitrary known plaintext and ciphertext pair, tradeoff attacks against block ciphers require a chosen plaintext block. Given a plaintext and ciphertext pair (P_0, C_0), the aim is to find the secret key k that satisfies $C_0 = E_k(P_0)$.

In the offline phase, chains of length t are generated using the function f which is the composition of the *reduction function* R and the encryption function E. Reduction functions are necessary for cases where the input and output size of f are different. Successive keys are generated using the following equation

$$x_{j\ i+1} = f(x_{j\ i}),$$
$$= R(E_{x_{j\ i}}(P_0)),\ 0 \le i \le t,$$

where x_{j0} $(1 \le j \le m)$ is selected randomly. Starting from m random points x_{j0}, the target P_0 is encrypted using the successive keys and starting points (SPs) and end points (EPs) of m chains are stored in a table. The construction of the table is summarized in Fig. 2.

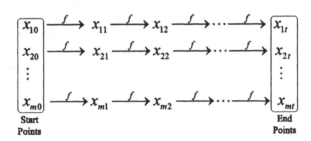

Fig. 2. Construction of the lookup table in the offline phase

In the online phase, the aim is to find the key that generates (P_0, C_0) pair, assuming it is one of the keys used in the offline phase. Since only SPs and EPs of the table are stored, a similar chain for C_0 is generated and after each encryption, the obtained value is compared to the EPs of the table. If a match is obtained, then the whole chain corresponding to the EP is regenerated and the key just before reduction of C_0 is obtained as the secret key. Therefore, the success probability of the attack is closely related to the percentage of the keys that are covered in the offline phase. It should be noted that sometimes the desired key is a part of a chain that is merged with another chain of the table which causes a false alarm.

The efficiency of TMTO attacks can be improved using different approaches. Using distinguished points approach, instead of generating fixed length chains, the chains are terminated whenever a distinguished x_{ij} (e.g. having last m bits as zero) is obtained. This improvement reduces the memory access in the online phase, since a value is compared to the end points only if it is a distinguished point. In [7], the idea of distinguished points is used to attack 40-bit DES and the key is recovered in 10 seconds with a success rate of 72%. Another improvement is to use Rainbow tables [8] where different reduction functions are utilized. In this approach, instead of having t different tables, only one table of size $mt \times t$ is generated.

TMTO Attacks Against Stream Ciphers. TMTO attacks against stream ciphers are firstly proposed by Babbage [2] and Golić [3] independently. In the pre-computation phase, the function $f : \mathbb{F}_2^{logN} \rightarrow \mathbb{F}_2^{logN}$ that inputs the $logN$ bit state and outputs $logN$ keystream bits is considered and the following table is generated:

$$(S_1, f(S_1))$$
$$(S_2, f(S_2))$$
$$\vdots$$
$$(S_M, f(S_M))$$

using M randomly selected initial states S_i. Then, the table is sorted with complexity $O(MlogM)$ based on the output keystream bits. Given the output keystream of length $D + logN - 1$, D overlapping $logN$ bit subsequences are generated and compared to the table generated in the offline phase. It should be noted that unlike block ciphers, the TMTO attacks against stream ciphers do not require chosen plaintext.

In [4], Biryukov and Shamir combined the tradeoff attacks presented by Hellman and Babbage-Golić. Using the same f function, chains of length t are generated by assigning the output keystream as the new internal state as presented in Fig. 2. Instead of using output size of n bits (internal state size), for stream ciphers using k bit key and v bit IVs the output size is taken as $k+v$ bits can be used. Using this approach, it is possible to recover the key with time and memory complexity $T = M = 2^{\frac{1}{2}(k+v)}$ if $D = 2^{\frac{1}{4}(k+v)}$ frames are available [5]. For IVs shorter than key ($v < k$), the stream ciphers are vulnerable to the TMTO attack. Therefore, to avoid the attack, IV size should be at least equal to the key size and IVs should not be used in a predictable way and the state size should be at least the size of key plus IV.

According to [4], it is also possible to use distinguished points idea against stream ciphers. Similar to Babbage-Golić scheme, the states and the corresponding keystream with distinguished property are stored. This is especially important in online phase where different keystream portions are compared to the table. Since there is no need to check keystream portions that do not satisfy the distinguishing property, this reduces the complexity of checking memory significantly.

In [5], the following remarks are given; (i) putting restriction on the number of resynchronizations using a fixed key does not increase the security of the cipher against TMTO, (ii) the complexity of the initialization phase has no effect on the efficiency of the attack, and (iii) the attack may be applied to any keystream positions as long as they are known.

4   Chosen IV Statistical Attacks

In the following subsections, we present three new distinguishers against synchronous stream ciphers, but they can also be used to test the security of other cryptosystems such as block ciphers and hash functions. For a synchronous stream cipher with k bit key $K = (k_1, k_2, \dots, k_k)$ and v bit $IV = (iv_1, iv_2, \dots, iv_v)$, let $Z = z_0, z_1, \dots$ denote the keystream sequence.

It is possible to test stream ciphers statistically in various ways. In a classical way, a long keystream is generated and standard randomness tests such as frequency, runs, etc. are applied [9]. Alternatively, it is also possible to use a chosen IV approach and analyze the cipher using tests such as d-monomial [10] and correlation tests [11]. In this study, we also considered the chosen IV approach where the attacker has access to a number of different keystream sequences generated using different (possibly chosen) IV values and same key.

In our tests, we used *Pearson's Chi Square* (χ^2) *test* that involves grouping data into classes and comparing observed outcomes to the expected figures under the null distribution. The test statistic is defined as

$$\chi^2 = \sum_{i=1}^{k}(o_i - e_i)^2/e_i \tag{6}$$

where o_i and e_i is the observed and expected frequency for group i, respectively. If the null hypothesis is true, then χ^2 is distributed according to $\chi^2(k-1-p)$ where k is the number of groups and p is the number of parameters estimated from the data and the hypothesis is rejected if the test statistics χ^2 is greater than tabulated $\chi^2(1-\alpha; k-1-p)$ value for some significance level α with $k-1-p$ degrees of freedom.

4.1   Coverage Test

The coverage test is a new probabilistic distinguisher against stream ciphers where a table similar to the approach of Babbage-Golić is used. In the original approach, output keystreams of length n (state size) are generated. Since for our statistical test, it is impractical to use keystream bits of size as large as n, we focus on a subset of IV bits (l out of v) and generate l bit keystreams.

First, we select l random (active) positions from IV and fix the rest (inactive) bits to a random value. Then, we synchronize the cipher for all possible 2^l IVs and generate l bit keystream $(z_1^{(i)}, z_2^{(i)}, \dots, z_l^{(i)})$ for each IV as given in Fig. 3. Then,

$$\overbrace{IV^{(1)} = (\underbrace{* * * \ldots *}_{v-l \text{ bits}} | \underbrace{000 \ldots 0}_{l \text{ bits}}) \to Z^{(1)} = (z_1^{(1)}, z_2^{(1)}, \ldots, z_l^{(1)})}$$

$$IV^{(2)} = (* * * \ldots * | 000 \ldots 1) \to Z^{(2)} = (z_1^{(2)}, z_2^{(2)}, \ldots, z_l^{(2)})$$

$$\vdots$$

$$IV^{(2^l)} = (* * * \ldots * | 111 \ldots 1) \to Z^{(2^l)} = (z_1^{(2^l)}, z_2^{(2^l)}, \ldots, z_l^{(2^l)})$$

Fig. 3. The table generated in the coverage test

we calculate the number of distinct $Z^{(i)}$'s and denote it as C_1 which is expected to be around 0.63×2^l. We repeat the experiment for a number of times with different assignments of the inactive IV bits and obtain a coverage variable for each trial, then evaluate the randomness of the cipher based on the distribution of C_i's. The pseudocode of the coverage test is given in Algorithm 4.1.

Using the recursive formula (2), the probability distributions of C_i for 12 and 14 bits are calculated and categorized into 5 groups with approximately equal probability. The limit of the groups and corresponding probabilities are given in Table 1.

Algorithm 4.1: COVERAGE TEST(R, l)

Randomly select l positions $p_1, p_2, \ldots, p_l$ from v bits of IV;
for $i \leftarrow 0$ **to** R
$\quad \Big\lceil$ Randomly select $IV = (iv_1, iv_2, \ldots, iv_v)$;
$\quad \Big|$ **for** $j \leftarrow 0$ **to** $2^l - 1$
$\quad \Big|$ $\quad \Big\lceil J = (j_1, j_2, \ldots, j_l)$ binary representation of j;
$\quad \Big|$ $\quad \Big| (iv_{p_1}, iv_{p_2}, \ldots, iv_{p_l}) = J$;
$\quad \Big|$ $\quad \Big\lfloor Z^{(j)} =$ First l keystream bits using K and IV;
$\quad \Big\lfloor Coverage_i =$ Number of distinct $Z^{(0)}, \ldots, Z^{(2^l - 1)}$;
Evaluate $(Coverage_1, \ldots, Coverage_R)$ using χ^2 test;
return $(p - value)$

If the coverage test returns low p-value (< 0.01), it means that the coverage of the corresponding mapping is statistically different than the expected values. Obtaining a low coverage value means that the first keystream bits that are generated using different IVs are similar, it is obviously a threat for frequently resynchronized ciphers. This is also an indication of low diffusion properties. Obtaining a high coverage value means that the mapping is close to a permutation. This may be interpreted as follows; whenever a subpart of the secret bits is recovered and the rest of the bits form a permutation, to identify unknown state bits, the required number of keystream bits is equal to the number of

Table 1. Interval and probability values of coverage test using 12 and 14 IV bits

12 IV Bits		14 IV Bits	
Category Limits	Probability	Category Limits	Probability
0-2571	0.199139	0-10322	0.201591
2572-2583	0.204674	10323-10345	0.195966
2584-2593	0.197856	10346-10366	0.207519
2594-2605	0.203225	10367-10389	0.195253
2606-4096	0.195106	10390-16384	0.199671

unknown state bits. For mappings close to permutation, cipher is more vulnerable to TMTO attacks.

4.2 ρ-Test

ρ-test is another probabilistic distinguisher against stream ciphers where the encryption function is iteratively applied and a sequence of l bits keystreams $Z^{(1)}, Z^{(2)}, \ldots$ are generated until one of the entries is repeated (See Fig. 4). The index of the last entry, ρ-length, is used to evaluate the randomness of the cipher.

$$Z^{(0)} \xrightarrow{IV^{(0)}=(***|Z^{(0)})} Z^{(1)} \xrightarrow{IV^{(1)}=(***|Z^{(1)})} \ldots \xrightarrow{IV^{(R)}=(***|Z^{(R)})} Z^{(R)}$$

Fig. 4. The rows generated in the ρ-test

First, we select l random positions from IV and fix the rest (inactive) bits to a random value, then initialize the cipher with this IV and secret key K and generate l bit keystream, $Z^{(1)}$. Then, $Z^{(1)}$ is assigned to the variable part of IV and iteratively l bit keystreams are generated until one of the $Z^{(i)}$ is repeated and the index of the last entry is stored as $Index_1$. This is repeated for different assignments of inactive IV bits, then $Index_i$ values are compared to their theoretical distribution using χ^2 goodness of fit tests. The pseudocode of ρ test is given in Algorithm 4.2.

Algorithm 4.2: ρ-TEST(R, l)

Randomly select l positions $p_1, p_2, \ldots, p_l$ from v bits of IV;
for $i \leftarrow 0$ **to** R
$\Big($ Randomly select $IV = (iv_1, iv_2, \ldots, iv_v)$;
$\Big|$ $index_i = 0$;
$\Big|$ **repeat**
$\Big\{$ $\quad Z_{index_i} = $ First l keystream bits using K and IV;
$\Big|$ $\quad (iv_{p_1}, iv_{p_2}, \ldots, iv_{p_l}) = (z_1, z_2, \ldots, z_l)$;
$\Big|$ $\quad Index_i + +$;
$\Big($ **until** *a Z value is repeated*
Evaluate $(Index_1, \ldots, Index_R)$ using χ^2 test;
return $(p - value)$

Table 2. Interval and probability values of ρ-test using 15 and 20 IV bits

15 IV Bits		20 IV Bits	
Category Limits	Probability	Category Limits	Probability
2- 122	0.201906	2-685	0.200258
123 - 184	0.200448	686-1036	0.200124
185 - 246	0.199904	1037-1386	0.199400
247 - 325	0.198270	1387-1838	0.200518
326 - 32768	0.199472	1839 -1048576	0.199700

Using the recursive formula (4), the probability distribution of R_i for 15 and 20 bits is calculated and categorized into 5 groups with approximately equal probability. The limit of the groups and corresponding probabilities are given in Table 2.

Rows of the Hellman tables are generated by applying encryption and reduction functions iteratively. This test generates rows similar to Hellman's table and calculates their ρ-length. Obtaining a low p-value from the ρ test means that length of iterations is statistically different than the expected values. Having short cycles results low coverage, which motives us to use smaller number of iterations t.

4.3   DP-Coverage Test

The last distinguisher is very similar to the coverage test described in Sect. 4.1. The only difference is instead of considering the coverage of first l keystream bits, we find the coverage of l bit keystream after the first k bit DP, as given in Fig. 5.

$$(\underbrace{0,0,\ldots,0}_{k \; bits}, \underbrace{*, *, \ldots, *}_{l \; bits})$$

Fig. 5. Distinguished Points

First, we select l random positions (active bits) from IV and fix the rest (inactive) bits to a random value. Then, we synchronize the cipher for all possible 2^l IVs and generate l bit distinguished keystreams. Then, we calculate the number of different l bit keystreams and denote it as C_1. We repeat the experiment for a number of times with different assignments of the inactive IV bits and obtain a coverage variable for each trial, then evaluate the randomness of the cipher based on the distribution of C_i's. The pseudocode of the DP coverage test is given in Algorithm 4.3. To evaluate the output coverage values, the theoretical distribution given in the coverage test is used.

Distinguished points in TMTO attacks are used to reduce the number of memory checks, since only distinguished keystream portions with a special property are checked. These distinguished portions are assumed to be uniformly distributed throughout the keystream, otherwise it is possible to distinguish the cipher using the Overlapping Template Matching Test from the randomness test suite of NIST [9].

Algorithm 4.3: DP-COVERAGE TEST(R, l, k)

Randomly select K;
Randomly select l positions $p_1, p_2, \ldots, p_l$ from v bits of IV;
for $i \leftarrow 0$ **to** R
$\left\{\begin{array}{l} \text{Randomly select } IV = (iv_1, iv_2, \ldots, iv_v); \\ \textbf{for } j \leftarrow 0 \textbf{ to } 2^l - 1 \\ \left\{\begin{array}{l} J = (j_1, j_2, \ldots, j_l) \text{ binary representation of } j; \\ (iv_{p_1}, iv_{p_2}, \ldots, iv_{p_l}) = J; \\ Z^{(j)} = l \text{ bit keystream after } k \text{ bit distinguisher using } K \text{ and } IV; \end{array}\right. \\ Coverage_i = \text{Number of distinct } Z^{(0)}, \ldots, Z^{(2^l - 1)}; \end{array}\right.$
Evaluate $(Coverage_1, \ldots, Coverage_R)$ using χ^2 test;
return $(p - value)$

An important criterion against stream ciphers is that the initial states (states that are generated after key/IV initialization phase) should be uniformly distributed throughout the keystream. If for any key, there exist IV_1 and IV_2 that identify close starting points, most important assumption of stream ciphers *keystream must be used only once* may be violated. A similar observation is pointed out by Biryukov et al. [12] and used to attack A5/1 by TMTO attacks. The ciphers having close starting points are expected to reach the same distinguished keystream portions resulting in low coverage.

5   Experimental Results

We applied the three distinguishers to the Phase III candidates of eSTREAM project with the following parameters: Coverage(100,12), Coverage(100,14), ρ(100,15), ρ(100,20), DP-Coverage(100,12,10), DP-Coverage(100,14,10). The last bits of the IVs are chosen to be active. Each test is repeated 100 times using random keys and the average values are tabulated in Table 3. Since the p-values are expected to distribute uniformly between 0 and 1, the average of 100 p-values are expected to be distributed normally with mean 0.5 and standard deviation 0.0289.

Most significant deviation from 0.5 is obtained from the cipher Pomaranch [13] using the coverage test with 14 variable IV bits. For a secure cipher, although the probability that the average is less than 0.320433 is negligible, we repeated the experiment 450 times and obtained the following histogram. As seen from the figure, the distribution of p-values significantly deviates from uniform distribution.

Table 3. The average p-values obtained from Coverage, ρ and DP-Coverage tests using 100 different keys (eSTREAM API-compliant source codes are used to test candidates)

Cipher	Coverage Test		ρ Test		DP Coverage Test	
	12	14	15	20	12	14
Crypt.MT.v3	0.421172	0.502259	0.438767	0.434522	0.491562	0.496520
Decim.v2	0.483673	0.498243	0.515213	0.561964	0.507519	0.434853
Dragon	0.467800	0.524956	0.531447	0.508126	0.475863	0.490167
Edon80	0.503102	0.504606	0.496585	0.506221	0.458080	0.511870
FFCSR-16	0.501281	0.536393	0.522143	0.505929	0.487775	0.501770
Grain128	0.507743	0.546400	0.521265	0.473777	0.514002	0.481063
HC-128	0.453212	0.502525	0.489393	0.475678	0.516889	0.513914
Lex	0.472844	0.497004	0.500221	0.475852	0.473196	0.497984
Mickey-128.v2	0.490894	0.499849	0.510405	0.479021	0.434051	0.544828
NLS.v2	0.508358	0.483571	0.474961	0.477997	0.516875	0.468336
Pomaranch.v1	0.433858	0.320433	0.506190	0.520325	0.483106	0.513709
Rabbit	0.512423	0.473658	0.522667	0.518543	0.470558	0.470504
Salsa20	0.485817	0.527911	0.533091	0.501501	0.430587	0.498490
Sosemanuk	0.439461	0.487562	0.497158	0.527524	0.555262	0.531222
Trivium	0.413683	0.500991	0.491455	0.495252	0.458730	0.508625

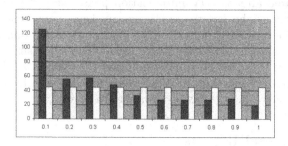

Fig. 6. The number of p-values in intervals of length 0.1 versus expected values for Pomaranch

6   Conclusion

Statistical randomness testing plays an important role in analyzing the security of cryptosystems. In this study, we propose a new framework of randomness testing based on some properties of random mappings, focusing on TMTO attacks against stream ciphers. Here, we present three chosen IV distinguishers namely; (i) coverage test, (ii) ρ-test and (iii) DP-coverage test. We experimentally observed some statistical deviations in the distribution of p-values in Pomaranch. We observe that Pomaranch fails the Coverage(100,14) test with probability approximately equal to 0.3125 using a type I error $\alpha = 0.01$. Applying the test a number of times(≈ 500) using the same secret key, the cipher can be distinguished from a random mapping.

References

1. Hellman, M.E.: A cryptanalytic time-memory trade off. IEEE Trans. Inform. Theory IT-26, 401–406 (1980)
2. Babbage, H.: Improved exhaustive search attacks on stream ciphers. In: European Convention on Security and Detection, IEE Conference publication (408), pp. 161–166 (1995)
3. Golić, J.D.: Cryptanalysis of alleged A5 stream cipher. In: Fumy, W. (ed.) EUROCRYPT 1997. LNCS, vol. 1233, pp. 239–255. Springer, Heidelberg (1997)
4. Biryukov, A., Shamir, A.: Cryptanalytic time/memory/data tradeoffs for stream ciphers. In: Okamoto, T. (ed.) ASIACRYPT 2000. LNCS, vol. 1976, pp. 1–13. Springer, Heidelberg (2000)
5. Hong, J., Sarkar, P.: Rediscovery of time memory tradeoffs. Cryptology ePrint Archive, Report 2005/090 (2005), http://eprint.iacr.org/
6. eStream: ECRYPT Stream Cipher Project. IST-2002-507932 (2004), http://www.ecrypt.eu.org/stream
7. Quisquater, J., Standaert, F., Rouvroy, G., David, J., Legat, J.: A cryptanalytic time-memory tradeoff: First FPGA implementation. In: Glesner, M., Zipf, P., Renovell, M. (eds.) FPL 2002. LNCS, vol. 2438, pp. 780–789. Springer, Heidelberg (2002)
8. Oechslin, P.: Making a faster cryptanalytic time-memory trade-off. In: Boneh, D. (ed.) CRYPTO 2003. LNCS, vol. 2729, pp. 617–630. Springer, Heidelberg (2003)
9. Rukhin, A., Soto, J., Nechvatal, J., Smid, M., Barker, E., Leigh, S., Levenson, M., Vangel, M., Banks, D., Heckert, A., Dray, J., Vo, S.: A Statistical Test Suite for Random and Pseudorandom Number Generators for Cryptographic Applications (2001), http://www.nist.gov
10. Englund, H., Johansson, T., Sonmez Turan, M.: A framework for chosen IV statistical analysis of stream ciphers. In: Srinathan, K., Rangan, C.P., Yung, M. (eds.) INDOCRYPT 2007. LNCS, vol. 4859, pp. 268–281. Springer, Heidelberg (2007)
11. Sonmez Turan, M., Doöanaksoy, A., Çalik, Ç.: Statistical analysis of synchronous stream ciphers. In: SASC 2006: Stream Ciphers Revisited (2006)
12. Biryukov, A., Shamir, A., Wagner, D.: Real time cryptanalysis of A5/1 on a PC. In: Schneier, B. (ed.) FSE 2000. LNCS, vol. 1978, pp. 1–18. Springer, Heidelberg (2001)
13. Jansen, C., Kolosha, A.: Cascade jump controlled sequence generator. eSTREAM, ECRYPT Stream Cipher Project, Report 2005/022 (2005)

A Probabilistic Approach on Estimating the Number of Modular Sonar Sequences

Ki-Hyeon Park and Hong-Yeop Song

Department of Electrical and Electronic Engineering
Yonsei University, Seoul, 121-749, Korea
{kh.park, hysong}@yonsei.ac.kr

Abstract. We report some results of an extensive computer search for $m \times n$ modular sonar sequences and estimate the number of inequivalent examples of size $m \times n$ using a probabilistic approach. Evidence indicates strongly that a full size example exists with extremely small probability for large m.

1 Introduction

A sonar sequence is an integer sequence that has some interesting properties for use in communication applications. Its mathematical concept was well described in [2] and the original motivation and application to some communication problems can be found in [1,6].

Recently, [8] has discussed a search for a 35×35 modular sonar sequence and, in general, $m \times m$ examples where $m = p(p+2)$ is a product of twin primes. It was for their application to the design of CDMA sequences, but they failed to find any single example beyond $m = 15$. This paper is an attempt to continue this effort, and shows some results of an extensive computer search for small values of m. Based on the search result, we now believe that no $m \times (m + 1)$ modular sonar sequence exists, except for those given by the algebraic constructions. To explain this, we use some probabilistic approaches for estimating the number of $m \times (m + 1)$ modular sonar sequences.

An $m \times n$ sonar sequence is defined as a function from the set of integers $\{1, 2, ..., n\} \triangleq A_n$ to the set of integers $\{1, 2, ..., m\} \triangleq A_m$ with the following distinct difference property (DDP) [3].

Definition 1. *(DDP) A function $f : A_n \to A_m$ has a distinct difference property if for all integers h, i, and j, with $1 \le h \le n - 1$ and $1 \le i, j \le n - h$,*

$$f(i + h) - f(i) = f(j + h) - f(j) \quad implies \quad i = j. \tag{1}$$

An $m \times n$ sonar sequence is a function $f : A_n \to A_m$ with DDP. The main problem in sonar sequences research is to determine the maximum value n for each given m such that an $m \times n$ sonar sequence exists. For values of m up to 100, the best known n is reported in [3]. To obtain these values, they have introduced "modular sonar sequences." A modular sonar sequence is a sonar

S.W. Golomb et al. (Eds.): SETA 2008, LNCS 5203, pp. 42–50, 2008.
© Springer-Verlag Berlin Heidelberg 2008

sequence $f : A_n \rightarrow A_m$ with the condition (1) replaced by the distinct modular difference property (DMDP):

$$f(i+h) - f(i) = f(j+h) - f(j) \pmod{m} \quad \text{implies} \quad i = j. \tag{2}$$

Note that, obviously, DMDP implies DDP, but not conversely. A trivial upper bound on the maximum length n for a given m for sonar sequences is $2m$, since the maximum number of differences with $h = 1$ in (1) is $2m - 1$. Similarly for modular sonar sequences, this upper bound is given as $m+1$ since the maximum number of differences with $h = 1$ in (2) is m.

If an f is a modular sonar sequence, the function g given by

$$g(i) = uf(i) + si + a, \quad i = 1, 2, ..., n \tag{3}$$

is also a modular sonar sequence for all integer s and a, and for all integer u relatively prime to m [3]. Two $m \times n$ modular sonar sequences with this relation are said to be equivalent.

There are essentially three algebraic methods constructing an $m \times (m + 1)$ modular sonar sequence for certain values of m. These are Quadratic Method [4] and Extended Exponential Welch Method [7] both for m being a prime and Shift Sequence Method [5] for m being one less than a prime power.

Given an $m \times n$ (modular) sonar sequence, we can always have $m \times (n - 1)$ (modular) sonar sequence by deleting the last term. It works since the condition (1) or (2) remains satisfied when the domain of f is restricted to $\{1, 2, ..., n-1\}$. We call it "Reduction." Conversely, if there is no $m \times n$ (modular) sonar sequence, then there is no $m \times (n + 1)$ (modular) sonar sequence.

2   Back-Track Search and Results

This section reports some results from an exhaustive back-track search for $m \times n$ modular sonar arrays for some small values of m.

The algorithm recursively builds up a set of GOOD symbols for the current position t based on a modular difference triangle (MDT) of depth $t - 2$ constructed from the sequence $f(1), f(2), ..., f(t - 1)$ of length $t - 1$ in order to

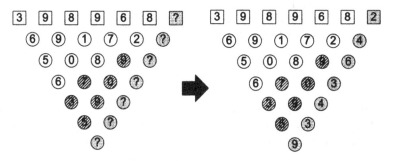

Fig. 1. Determining $f(7)$ by back-track algorithm for $m=10$

assign a symbol to $f(t)$. This is done by removing all the symbols from A_m, which will violate DMDP when it reaches the t-th position. When this set of GOOD symbols for the current position becomes empty, the algorithm will output the sequence constructed so far (if it has a new longer length) and then back-track. The algorithm will stop when the set of GOOD symbols for the first position becomes empty.

Figure 1 shows a situation for $m = 10$, in which the algorithm has filled up 6 terms and seeks to assign a symbol to $f(7)$. Symbols (or numbers) in squares at the top are the sequence $f(i)$ for $i = 1, 2, ..., 6$, and those in circles at level h deep are the differences mod 10 of terms in the distance h, i.e., $f(i + h) - f(i)$ (mod 10). They are said to form a modular difference triangle (MDT) with no symbol repeating in any row (except for the top row that corresponds to the sequence itself). The algorithm will remove a symbol from A_{10} (to build a set of GOOD symbols for $f(7)$) if it does not satisfy DMDP with respect to the given MDT of depth 5. Observe for example in this case that the symbol 4 will be removed since $f(2) - f(1) = 6 = 4 - f(6)$ (mod 10) from the rows of level 1 deep. Similarly, because of the differences 9, 1, 7, and 2 at level 1 deep, the symbols 7, 9, 5 and 10 will also be removed. Because of the differences $5, 0, 8, 9$ at level 2 deep and $f(5) = 6$, the symbols $1, 6, 4, 5$ will also be removed. We note that the symbol 5 had already been removed.

Fact 1. Observe that the difference 9 at level 2 does not produce any new constraint because if $f(7) - f(5) = 9 = f(6) - f(4)$ then $f(7) - f(6) = f(5) - f(4)$. Thus, the update process will become simpler when it considers only those differences in the un-shaded circles.

We have focused on all existing examples of maximum length except for those given by the algebraic constructions mentioned in the previous section and their reductions. The initial search was to answer the following two questions:

Q.1. Determine the maximum length n_{max} such that an $m \times n_{max}$ modular sonar array exists. What would be the maximum length n_e if we count only those that are NOT equivalent to any examples constructed by the three algebraic methods mentioned in the previous section and/or their reductions?

Q.2. How many inequivalent sequences of length n_e exist for a given m, excluding those which are equivalent to the one given by the three algebraic constructions and/or their reductions?

The result of the search is shown in Table 1, from which we were able to detect the following behaviors of values m and n_e.

Observation 1. $m - n_e$ is monotonically non-decreasing as m is increasing.

Observation 2. The number of inequivalent sequences of length n_e (not equivalent to ones from the algebraic constructions) is decreasing as m is increasing for the range where the value $m - n_e$ remains the same.

Next section will be devoted to describing the above behaviors and more.

Table 1. Result of an initial search

m	Description	n_{max}	n_e	$m - n_e$	Answer to **Q.2**
5	Q,EEW	6	0	5	0
6	SS	7			
7	Q,EEW	8	8	-1	1
8	SS	9			
9	U	10	10	-1	3
10	SS	11			
11	Q,EEW	12	11	0	30
12	SS	13			
13	Q,EEW	14	13	0	17
14	U	14	14	0	2
15	SS	16	15	0	1
16	SS	17	16	0	1
17	Q,EEW	18	16	1	33
18	SS	19			
19	Q,EEW	20	17	2	321
20	U	18	18	2	136
21	U	19	19	2	17
22	SS	23			
23	Q,EEW	24			
24	SS	25			
25	U	22	22	3	4
26	SS	27			
27	U	?			
28	SS	29			
29	Q,EEW	30			
30	SS	31			
31	Q,EEW,SS	32			
32	U	?			
33	U	?			
34	U	?			
35	U	?			

SS:Shift Sequence **Q**:Quadratic **EEW**:Extended Exponential Welch **U**:Unidentified

3   Probabilistic Approach

The idea is in the back-tracking algorithm. The algorithm must check each differential value whether it (potentially) violates DMDP or not. Checking can be performed "independently" with regard to all the possible differences that have already appeared in MDT. Even if the constraints are not independent, we can over-estimate the situation and we may assume they are so. Specifically, we are trying to estimate the number of $m \times t$ modular sonar sequences that can be constructed from a given $m \times (t-1)$ modular sonar sequence by adjoining one last symbol. It would be equal to the size of the set $S(m,t)$ of GOOD symbols for $f(t)$. Thus, the goal is to estimate or to give some bound on $|S(m,t)|$ when we are given an $m \times (t-1)$ modular sonar sequence.

We will make some reasonable assumptions under which we could recursively estimate the number $N(m,t)$ of $m \times t$ modular sonar sequences as the number $N(m,t-1)$ of $m \times (t-1)$ sequences times $|S(m,t)|$.

Obviously, there are $N(m,1) = m$ sequences of size $m \times 1$. For $2 \leq t \leq m+1$, we need to estimate the fraction $p(m,t) = |S(m,t)|/m$. We claim that

$$p(m,t) \approx \prod_{h=1}^{\lfloor \frac{t}{2} \rfloor} q(m,t,h), \quad \text{for } 2 \leq t \leq m+1, \tag{4}$$

where

$$q(m,t,h) \approx 1 - \frac{m(t-2h)}{(m-h+1)^2}, \quad \text{for } 1 \leq h \leq \lfloor \tfrac{t}{2} \rfloor. \tag{5}$$

From the value given in (4), we may obtain the following:

$$N(m,1) = m,$$
$$N(m,n) \approx N(m,n-1)mp(m,n)$$
$$= m^n \prod_{t=1}^{n} p(m,t), \quad 2 \leq n \leq m+1. \tag{6}$$

The key to the derivation of (6) is to identify the quantities $p(m,t)$ and $q(m,t,h)$ as certain probabilities of related models which simulate the back-track algorithm of the search. The probability model in reality must consist of a set of events, each of whose probabilities are heavily inter-dependent with one another. To make things simple, we use three assumptions discussed below so that the dependence disappears, and the result becomes a simple multiplication of individual probabilities. In doing so, we will adjust a bit further so that the approximation becomes reasonably meaningful. The value $p(m,t)$ will be identified with the probability that an arbitrarily selected symbol at position t satisfies DMDP with respect to all the previous entries of the MDT constructed so far. The first assumption is given as follows:

Assumption 1. For any m and t, the value $p(m,t)$ remains the same no matter which $m \times (t-1)$ modular sonar sequence might be given.

Taking Assumption 1 into account, we will have a sequence of length t with probability $p(m,t)$ given ANY sequence of length $t-1$. The second assumption enables us to factor $p(m,t)$ as a product of some individual probabilities:

Assumption 2. In figuring out the size $|S(m,t)|$, the number of constraints (DMDP) is independent with the depth parameter h (in the definition of DMDP) when a suitable range for h is taken into consideration.

Following Assumption 2, $p(m,t)$ is a product of probabilities of individual events related to the depth parameter h of a given MDT. Note that the RHS of (4) is the product of $q(m,t,h)$'s in the range of h from 1 to $\lfloor \frac{t}{2} \rfloor$. The value $q(m,t,h)$ will be identified with the probability that an arbitrarily selected symbol at position t satisfies DMDP of level h deep of the entries of the MDT constructed so far. This quantity is still too complicated to calculate exactly, and we need the following third assumption:

Assumption 3. The probability $q(m,t,h)$ can be approximated as the conditional probability that an arbitrarily selected symbol at position t satisfies DMDP of level h deep with regard to the entries of the MDT constructed so far, given the condition that it satisfies all the DMDP of level $k < h$ deep.

Under Assumption 3, the value $q(m,t,h)$ can be approximated as the conditional probability that there is no j such that $h < j < t$ and $f(t) - f(t-h) = f(j) - f(j-h)$ (mod m), given the condition that, for each and every k with $1 \le k < h$, there is no j such that $k < j < t$ and $f(t) - f(t-k) = f(j) - f(j-k)$ (mod m). To find this conditional probability and show that it is given as in RHS of (5), we claim that the complementary event has an approximated probability

$$1 - q(m,t,h) \approx \frac{(t-2h)}{(m-h+1)} \cdot \frac{m}{(m-h+1)}. \tag{7}$$

We may easily determine an upper and lower bound on $1 - q(m,t,h)$ which is the fraction of symbols that violates DMDP. Recall that the current position is t, and we are given a sequence of length $t-1$ and the corresponding MDT of depth $t-2$. There are $t-(h+1)$ symbols which violate DMDP for fixed h. It is just the number of entries of MDT at level h deep. Thus, at most $\frac{t-(h+1)}{m}$ of A_m will be BAD for $f(t)$ from row h of MDT, and hence, $\frac{t-(h+1)}{m}$ is an upper bound on $1 - q(m,t,h)$. When we use Fact 1 and Assumption 3, we see that there are at least $\frac{t-(h+1)-(h-1)}{m-(h-1)} = \frac{t-2h}{m-h+1}$ of A_m, which will be BAD for $f(t)$, since $h-1$ symbols (shaded area of the row h in Fig. 1, for example, and using Fact 1) have already been taken care of with regard to the DMDP of level $k < h$ deep. Thus, $\frac{t-2h}{m-h+1}$ is a lower bound on $1 - q(m,t,h)$. Therefore, we have

$$\frac{t-2h}{m-h+1} \le 1 - q(m,t,h) \le \frac{t-(h+1)}{m}.$$

By carefully examining the situation further, we have chosen a factor as shown in (7) and obtained the result given in (5).

In order to check the validity of the estimated number $N(m,n)$ of $m \times n$ modular sonar sequences, we have done a second round search for the values $N(m,n)$ for m up to 14 and n up to $m+1$. These are shown in Table 2 with the calculated number from (5). Further, this relation is plotted in Fig. 2.

Table 2. Comparison of the true and estimated values of $N(m,n)$

m	n	Search	Estimate	m	n	Search	Estimate	m	n	Search	Estimate
7	4	5	5.00	10	9	707	895	13	5	100	100
	5	16	16.1		10	63	103		6	729	738
	6	27	29.5		11	2	0.857		7	3712	3842
	7	16	17.7	11	4	9	9.00		8	12433	13492
	8	2	1.73		5	64	64.1		9	22983	26184
8	4	10	10.5		6	343	350		10	20198	25321
	5	43	43.9		7	1152	1215		11	5922	8835
	6	108	118		8	2209	2479		12	481	852
	7	128	140		9	1857	2190		13	22	11.5
	8	50	54.3		10	533	670		14	1	0.0073
	9	2	2.26		11	35	37.1	14	4	26	26.0
9	4	9	9.33		12	1	0.142		5	262	262
	5	48	48.1	12	4	27	27.5		6	2160	2188
	6	167	173		5	222	223		7	12896	13362
	7	292	326		6	1399	1430		8	53373	57579
	8	249	271		7	5848	6187		9	130547	147892
	9	37	54.2		8	15324	17022		10	168576	209626
	10	3	1.12		9	20155	23392		11	87718	127056
10	4	18	18.0		10	10199	13865		12	14775	27624
	5	110	110		11	1351	2297		13	615	1362
	6	480	499		12	25	67.7		14	2	9.14
	7	1216	1325		13	4	0.0996		15	0	0.0022
	8	1619	1845	13	4	11	11.0				

Fig. 2 shows that the estimated value of $N(m,n)$ fits well with the true value. So, we can say estimation function shows the value's tendency similarly. That is, the tendency of Fig. 3 can be a partial explanation of the decaying tendency of modular sonar sequences.

Remark 1. Estimated values in Table 2 and two figures represent fractions of the value given in (6) divided by $m^2\phi(m)$, since there are at most $m^2\phi(m)$ equivalent but possibly distinct modular sonar sequences on A_m with very high probability. Exception occurs when $g = f$ in Eq.(3) although $u \neq 1$. This event occurs rarely even at small m. We ignored the exceptions.

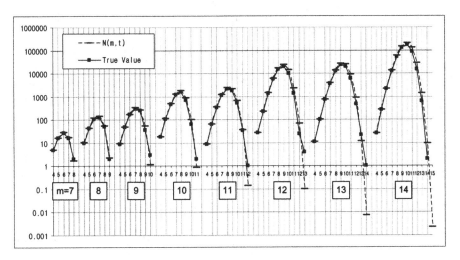

Fig. 2. Comparison of the true and estimated values of $N(m, t)$. From left to right, m runs from 7 to 14, and t runs from 4 to $m + 1$ in each group.

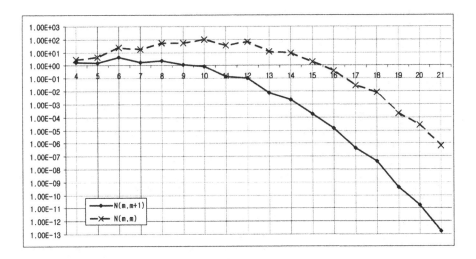

Fig. 3. Estimated values of $N(m, m + 1)$ and $N(m, m)$ for some small m

Note that $\frac{N(25,22)}{25^2 \phi(25)} \doteqdot 20.1$ and $\frac{N(25,23)}{25^2 \phi(25)} \doteqdot 0.003$. We already know that there is no 25×23 modular sonar sequence from Table 1.

4   Conclusion

We have checked the existence of $m \times n$ modular sonar sequences by computer search for some small values of m, and estimated the number of inequivalent examples for various values of m by carefully examining the back-track algorithm for the search.

From this estimate, we could have concluded that no full-size modular sonar sequence exists for m beyond a certain value. This is, however, not true, since there are some algebraic constructions which give full size examples (of length $m+1$ on m symbols) for infinite values of m. We could safely guess that any full-size example for large values of m must be either from an algebraic construction, or else the probability that it exists is extremely small.

We still leave the following problems unsolved:

Unsolved Problem 1. Find an example of 35×35 modular sonar sequences (mod 35) or prove that none exists.

Unsolved Problem 2. Generalize the above to the case of $m = p(p+2)$ being a product of twin primes.

Unsolved Problem 3. Find infinitely many values of m for which an $m \times (m+1)$ modular sonar sequences do not exist.

Unsolved Problem 4. Except for m being a prime or one less than a prime power, would the fact that the value in (6) is close to zero imply non-existence?

Unsolved Problem 5. How accurate is the estimate in (6)?

Unsolved Problem 6. Could a similar approach be used to estimate the number of Costas arrays, see [7] ?

References

1. Costas, J.P.: Medium constraints on sonar design and performance. FASCON CONV. Rec. 68A–68L (1975)
2. Erdös, P., Graham, R.L., Ruzsa, I.Z., Taylor, H.: Bounds for arrays of dots with distinct slopes or lengths. Combinatorica 12, 1–6 (1992)
3. Moreno, O., Games, R.A., Taylor, H.: Sonar sequences from costas arrays and the best known sonar sequences with up to 100 symbols. IEEE Trans. Inform. Theory 39(6) (November 1987)
4. Gagliardi, R., Robbins, J., Taylor, H.: Acquisition sequences in PPM communications. IEEE Trans. Inform. Theory IT-33, 738–744 (1987)
5. Games, R.A.: An algebraic construction of sonar sequences using M-sequences. SIAM J. Algebraic Discrete Methods 8, 753–761 (1987)
6. Golomb, S.W., Gong, G.: Signal Design for Good Correlation. Cambridge University Press, Cambridge (2005)
7. Golomb, S.W., Taylor, H.: Two-dimensional synchronization patterns for minimum ambiguity. IEEE Trans. Inform. Theory IT-28, 263–272 (1982)
8. Yoon., S.-J., Song, H.-Y.: Existence of Modular Sonar Sequences of Twin-Prime Product Length. In: Golomb, S.W., Gong, G., Helleseth, T., Song, H.-Y. (eds.) SSC 2007. LNCS, vol. 4893, pp. 184–191. Springer, Heidelberg (2007)
9. Silverman, J., Vickers, V.E., Mooney, J.M.: On the number of Costas arrays as a function of array size. Proceedings of the IEEE 76(7) (July 1988)

A Study on the Pseudorandom Properties of Sequences Generated Via the Additive Order

Honggang Hu and Guang Gong

Department of Electrical and Computer Engineering, University of Waterloo,
Waterloo, Ontario N2L 3G1, Canada
h7hu@uwaterloo.ca, ggong@calliope.uwaterloo.ca

Abstract. In this paper, we study the randomness properties of sequences generated by a function via the additive order. We derive some conditions under which such sequences have the maximum period. The autocorrelation is also studied. For the case of large p, nontrivial upper bounds are given; for the binary case, experimental results show that the autocorrelation of two types of sequences is small compared with the period of sequences.

1 Introduction

For any $n \geq 1$, let $\mathbb{F}_{p^n}$ be the finite field of order p^n, where p is a prime number, and let $\{\alpha_1, ..., \alpha_n\}$ be an ordered basis of $\mathbb{F}_{p^n}$ over $\mathbb{F}_p$. For any $0 \leq i < p^n$, we define ξ_i by

$$\xi_i = i_1 \alpha_1 + ... + i_n \alpha_n,$$

if

$$i = i_1 + i_2 p + ... + i_n p^{n-1}, \ 0 \leq i_k < p, \ k = 1, 2, ..., n.$$

Furthermore, we obtain the sequence $\{\xi_i\}_{i=0}^{\infty}$ by extending $\{\xi_i\}_{i=0}^{p^n-1}$ with period p^n, i.e., $\xi_{i+p^n} = \xi_i$ for any $i \geq 0$. For any function $A(x)$ from $\mathbb{F}_{p^n}$ to $\mathbb{F}_p$, we define the sequence $S = \{s_i\}_{i=0}^{\infty}$ by

$$s_i = A(\xi_i), \ i = 0, 1, 2, \dots . \tag{1}$$

The sequence S in (1) is called a sequence generated by $A(x)$ via the additive order.

The additive order is related to the counter-mode encryption (CTR mode encryption) of block ciphers [5], and it is also related to some Golay complementary sequences [2,10]. These two applications are the motivations of our study for this topic.

To study the pseudorandom properties of sequences generated via the additive order is also interesting itself [3]. The additive order is different from the conventional order in sequence design, and the randomness properties of the sequences from this order are hardly to determine. Suppose that $A(x)$ is a function which generates a sequence with good pseudorandom properties using the conventional

S.W. Golomb et al. (Eds.): SETA 2008, LNCS 5203, pp. 51–59, 2008.
© Springer-Verlag Berlin Heidelberg 2008

order. Then a sequence generated by $A(x)$ via the additive order may not have good pseudorandom properties.

In this paper, we study the least period and the autocorrelation of the sequence S defined by (1). Some conditions are given under which they have the maximum period. For the autocorrelation, we present some nontrivial upper bounds when p is large. When $p = 2$ and $A(x) = Tr(x^{2^n-2})$ or $A(x) = Tr(x^{2^n-4})$, where $Tr(\cdot)$ is the trace function from $\mathbb{F}_{2^n}$ to $\mathbb{F}_2$, some numerical results are given. The data show that S has good autocorrelation in these two cases. One conjecture on the upper bound of the autocorrelation in these cases is also given.

This paper is organized as follows. In Section 2, we present some background which will be used in this paper. In Section 3, we study the least period of the sequence S defined by (1). In Section 4, we present some results on the autocorrelation of the sequence S defined by (1). Finally, Section 5 concludes this paper.

2   Preliminaries

Definition 1. *For any polynomial* $f(x) = \sum_{i=0}^M f_i x^i \in \mathbb{F}_{p^n}[x]$, *if* $f_i = 0$ *for any* $p \mid i$, *then* $f(x)$ *is called nondegenerate.*

For any polynomial $f(x) \in \mathbb{F}_{p^n}[x]$, we can find a nondegenerate polynomial $g(x) \in \mathbb{F}_{p^n}[x]$ such that

$$Tr(f(\xi)) = Tr(g(\xi)), \ \xi \in \mathbb{F}_{p^n}.$$

Such $g(x)$ is called the nondegenerate polynomial associated with $f(x)$.

Let $\Gamma(n)$ denote the set of cosetleaders of $p^n - 1$ with respect to p. For any $k \in \Gamma(n)$, we denote the coset containing k by C_k. The the function $A(x)$ in (1) can be written as (see [3])

$$A(x) = \sum_{k \in \Gamma(n)} Tr_1^{n_k}(A_k x^k) + A_{p^n-1} x^{p^n-1},$$

where $Tr_1^{n_k}(\cdot)$ is the trace function from $\mathbb{F}_{p^{n_k}}$ to $\mathbb{F}_p$, n_k is the cardinality of C_k, $A_k \in \mathbb{F}_{p^{n_k}}$, and $A_{p^n-1} \in \mathbb{F}_p$.

For any $a \in \mathbb{F}_{p^n}$, the map $\psi_a(x) = e^{2\pi i Tr(ax)/p}$ is an additive character of $\mathbb{F}_{p^n}$, and all the additive characters of $\mathbb{F}_{p^n}$ can be written in this form [4]. If $a \neq 0$, then ψ_a is called a nontrivial character.

Lemma 1 ([4]). *Let* ψ *be a nontrivial additive character of* $\mathbb{F}_{p^n}$. *Then for any* $a \in \mathbb{F}_{p^n}$, *we have*

$$\sum_{\mu \in \mathbb{F}_{p^n}} \psi(a\mu) = \begin{cases} p^n, & \text{if } a = 0; \\ 0, & \text{if } a \neq 0. \end{cases}$$

The following bound for exponential sums generalizes that for the well known Weil exponential sums.

Lemma 2 ([8]). *Let ψ be a nontrivial additive character of $\mathbb{F}_{p^n}$, and let the expression $f(x)/g(x)$ be a rational function over $\mathbb{F}_{p^n}$. Let v be the number of distinct roots of the polynomial $g(x)$ in the algebraic closure $\overline{\mathbb{F}}_{p^n}$ of $\mathbb{F}_{p^n}$. Suppose that $f(x)/g(x)$ is not of the form $h^p(x) - h(x)$, where $h(x)$ is a rational function over $\overline{\mathbb{F}}_{p^n}$. Then*

$$
\left| \sum_{\xi \in \mathbb{F}_{p^n}, g(\xi) \neq 0} \psi\left(\frac{f(\xi)}{g(\xi)}\right) \right| \leq (\max(\deg(f), \deg(g)) + v^* - 2)p^{n/2} + \delta,
$$

where $v^* = v$ and $\delta = 1$ if $\deg(f) \leq \deg(g)$, and $v^* = v + 1$ and $\delta = 0$ otherwise.

Niederreiter and Winterhof proved the following result which will be used in the next section.

Lemma 3 ([9]). *Let $f(x)/g(x)$ be a rational function over $\mathbb{F}_{p^n}$ such that $g(x)$ is not divisible by the pth power of a nonconstant polynomial over $\mathbb{F}_{p^n}$, $f(x) \neq 0$, and $\deg(f) - \deg(g) \not\equiv 0 \mod p$ or $\deg(f) < \deg(g)$. Then $f(x)/g(x)$ is not of the form $h^p(x) - h(x)$, where $h(x)$ is a rational function over $\overline{\mathbb{F}}_{p^n}$.*

3   Period

In this section we will derive some conditions under which the least period of S defined by (1) is p^n.

For any polynomial $f(x)$ over $\mathbb{F}_{p^n}$, in order to investigate the period of $\{s_i\}_{i=0}^{\infty}$ given by (1) with $A(x) = Tr(f(x))$, we only need to consider the case $f(x) - f(x + \alpha_n)$ where α_n is the element in the basis $\{\alpha_1, ..., \alpha_n\}$.

Theorem 1. *For any polynomial $f(x)$ over $\mathbb{F}_{p^n}$, let $g(x)$ be the nondegenerate polynomial associated with $f(x) - f(x + \alpha_n)$. If $g(x) \neq c^p - c$ for any $c \in \mathbb{F}_{p^n}$, then the least period of the sequence $S = \{s_i\}_{i=0}^{\infty}$ defined by (1) with $A(x) = Tr(f(x))$ is p^n.*

Proof. Let T denote the least period of S. Suppose that $T \neq p^n$. Then we have $T \mid p^{n-1}$ because p^n is a period of S. Hence $s_i = s_{i+p^{n-1}}$ for any $0 \leq i < p^n$. By the definition of S, we have

$$
Tr(f(\xi_i)) = Tr(f(\xi_{i+p^{n-1}})), \text{ for any } 0 \leq i < p^n.
$$

Hence

$$
Tr(f(\xi_i) - f(\xi_i + \alpha_n)) = 0, \text{ for any } 0 \leq i < p^n.
$$

Then there exists $c \in \mathbb{F}_{p^n}$ such that $g(x) = c^p - c$. It is a contradiction. Thus the least period of S is p^n. $\qquad\square$

Lemma 4. *Let $f(x) = x^{p^n - 2}$. If $p^n > 4$, then for any $\alpha \in \mathbb{F}_{p^n}^*$, the nondegenerate polynomial $g(x)$ associated with $f(x) - f(x + \alpha)$ is not of the form $c^p - c$ for any $c \in \mathbb{F}_{p^n}$.*

Proof. For any $\alpha \in \mathbb{F}_{p^n}^*$, suppose that the nondegenerate polynomial $g(x)$ associated with $f(x) - f(x + \alpha)$ is of the form $c^p - c$ for some $c \in \mathbb{F}_{p^n}$. Then there exists $h(x) \in \mathbb{F}_{p^n}[x]$ such that $f(x) - f(x + \alpha) = h^p(x) - h(x)$. It follows that

$$\frac{\alpha}{x(x + \alpha)} = h^p(x) - h(x).$$

By Lemma 3, it is a contradiction. □

Corollary 1. *Let* $f(x) = x^{p^n - 2}$. *If* $p^n > 4$, *then the least period of the sequence* $S = \{s_i\}_{i=0}^{\infty}$ *defined by (1) with* $A(x) = Tr(f(x))$ *is* p^n.

Proof. By Theorem 1 and Lemma 4, the result follows. □

4   Autocorrelation

In this section, we investigate the autocorrelation of the sequence S defined in (1). For any $0 \leq \tau < p^n$, the autocorrelation of S at shift τ is defined by

$$C_S(\tau) = \sum_{i=0}^{p^n - 1} e^{2\pi i(s_{i+\tau} - s_i)/p} = \sum_{i=0}^{p^n - 1} e^{2\pi i(A(\xi_{i+\tau}) - A(\xi_i))/p}.$$

In order to study $C_S(\tau)$, we need to know the relationship between ξ_i and $\xi_{i+\tau}$. For any $0 \leq i, \tau < p^n$, let

$$i = i_1 + i_2 p + \ldots + i_n p^{n-1}, \quad 0 \leq i_k < p, \quad k = 1, 2, \ldots, n,$$

and

$$\tau = \tau_1 + \tau_2 p + \ldots + \tau_n p^{n-1}, \quad 0 \leq \tau_k < p, \quad k = 1, 2, \ldots, n.$$

Put $\omega_1 = 0$, and for $1 \leq k \leq n - 1$ define recursively

$$\omega_{k+1} = \begin{cases} 1, & \text{if } i_k + \tau_k + \omega_k \geq p; \\ 0, & \text{otherwise.} \end{cases}$$

Then we have

$$\xi_{i+\tau} = \xi_i + \xi_\tau + \omega,$$

where $\omega = \sum_{k=2}^{n} \omega_k \alpha_k$. There are at most 2^{n-1} choices for ω. For fixed τ, we define the sets

$$P_\omega = \{\xi_i \in \mathbb{F}_{p^n} \mid \xi_{i+\tau} = \xi_i + \xi_\tau + \omega\}$$

for all possible ω. Then the sets P_ω is a partition of $\mathbb{F}_{p^n}$. For fixed $\omega = \sum_{k=2}^{n} \omega_k \alpha_k$, the set P_ω can be written in the form

$$P_\omega = \{ \sum_{k=1, \omega_{k+1}=1}^{n-1} (p - (\tau_k + \omega_k))\alpha_k + \sum_{k=1}^{n} u_k \alpha_k \mid 0 \leq u_k \leq a_k - 1, k = 1, 2, \ldots, n\},$$

where

$$
a_k = \begin{cases} p - (\tau_k + \omega_k), & \omega_{k+1} = 0, 1 \le k < n, \\ \tau_k + \omega_k, & \omega_{k+1} = 1, 1 \le k < n, \\ p, & k = n. \end{cases}
$$

Henceforth we denote the additive canonical character ψ_1 by ψ for simplicity.

Lemma 5 ([7]). *With the notation as above, we have*

$$
\sum_{\eta \in \mathbb{F}_{p^n}} \left| \sum_{\xi \in P_\omega} \psi(\eta\xi) \right| \le q(1 + \ln p)^{n-1}.
$$

Theorem 2. *For any polynomial $f(x)$ over $\mathbb{F}_{p^n}$, and any $\xi \in \mathbb{F}_{p^n}^*$, let $g_\xi(x)$ be the nondegenerate polynomial associated with $f(x) - f(x + \xi)$. If $\deg(g_\xi) > 1$ for any $\xi \in \mathbb{F}_{p^n}^*$, then for any $0 < \tau < p^n$, we have the following bound for the autocorrelation of the sequence $S = \{s_i\}_{i=0}^\infty$ defined by (1) with $A(x) = Tr(f(x))$*

$$
|C_S(\tau)| \le 2(2^{n-1} - 1) \max_{\xi \ne 0}(\deg(g_\xi) - 1)p^{n/2}(1 + \ln p)^{n-1} + \max_{\xi \ne 0}(\deg(g_\xi) - 1)p^{n/2},
$$

where $\ln(\cdot)$ is the natural logarithm.

Proof. For any $0 < \tau < p^n$, we have

$$
C_S(\tau) = \sum_{i=0}^{p^n-1} e^{2\pi i(s_{i+\tau} - s_i)/p} = \sum_{i=0}^{p^n-1} \psi(f(\xi_{i+\tau}) - f(\xi_i))
$$

$$
= \sum_{\xi \in \mathbb{F}_{p^n}} \psi(f(\xi + \xi_\tau) - f(\xi))
$$

$$
+ \sum_{\omega \ne 0} \sum_{\xi \in P_\omega} (\psi(f(\xi + \xi_\tau + \omega)) - \psi(f(\xi + \xi_\tau)))\psi(-f(\xi)).
$$

By Lemma 2,

$$
|C_S(\tau)| \le \max_{\xi \ne 0}(\deg(g_\xi) - 1)p^{n/2} + \sum_{\omega \ne 0} \left| \sum_{\xi \in P_\omega} \psi(f(\xi + \xi_\tau + \omega) - f(\xi)) \right|
$$

$$
+ \sum_{\omega \ne 0} \left| \sum_{\xi \in P_\omega} \psi(f(\xi + \xi_\tau) - f(\xi)) \right|. \tag{2}
$$

Now we derive an upper bound for

$$
\left| \sum_{\xi \in P_\omega} \psi(f(\xi + \xi_\tau + \omega) - f(\xi)) \right|.
$$

By Lemma 1,

$$\psi(f(\xi + \xi_\tau + \omega) - f(\xi)) = \frac{1}{p^n} \sum_{\beta \in \mathbb{F}_{p^n}} \psi(f(\beta + \xi_\tau + \omega) - f(\beta)) \sum_{\eta \in \mathbb{F}_{p^n}} \psi(\eta(\xi - \beta)).$$

We get

$$\left| \sum_{\xi \in P_\omega} \psi(f(\xi + \xi_\tau + \omega) - f(\xi)) \right|$$

$$= \frac{1}{p^n} \left| \sum_{\xi \in P_\omega} \sum_{\beta \in \mathbb{F}_{p^n}} \psi(f(\beta + \xi_\tau + \omega) - f(\beta)) \sum_{\eta \in \mathbb{F}_{p^n}} \psi(\eta(\xi - \beta)) \right|$$

$$\leq \frac{1}{p^n} \sum_{\eta \in \mathbb{F}_{p^n}} \left| \sum_{\beta \in \mathbb{F}_{p^n}} \psi(f(\beta + \xi_\tau + \omega) - f(\beta) - \eta\beta) \right| \left| \sum_{\xi \in P_\omega} \psi(\eta\xi) \right|.$$

By Lemmas 2 and 5,

$$\left| \sum_{\xi \in P_\omega} \psi(f(\xi + \xi_\tau + \omega) - f(\xi)) \right| \leq \max_{\xi \neq 0}(\deg(g_\xi) - 1) p^{n/2} \sum_{\eta \in \mathbb{F}_{p^n}} \left| \sum_{\xi \in P_\omega} \psi(\eta\xi) \right| / p^n$$

$$\leq \max_{\xi \neq 0}(\deg(g_\xi) - 1) p^{n/2} (1 + \ln p)^{n-1}.$$

Similarly, we have

$$\left| \sum_{\xi \in P_\omega} \psi(f(\xi + \xi_\tau) - f(\xi)) \right| \leq \max_{\xi \neq 0}(\deg(g_\xi) - 1) p^{n/2} (1 + \ln p)^{n-1}.$$

Finally, by (2),

$$|C_S(\tau)| \leq 2(2^{n-1} - 1) \max_{\xi \neq 0}(\deg(g_\xi) - 1) p^{n/2} (1 + \ln p)^{n-1} + \max_{\xi \neq 0}(\deg(g_\xi) - 1) p^{n/2}.$$

$\square$

Remark 1. The bound in Theorem 2 is applicable only when p is large. The basic idea in the proof above is from [7]. Note that the approach for establishing Theorem 3 in [1] is also from [7]. Unfortunately, this technique cannot be extended to the case of $p = 2$, because one needs to consider the number of nonzero ω which is 2^{n-1}.

Theorem 3. *For any $0 < \tau < p^n$, we have the following bound for the autocorrelation of the sequence $S = \{s_i\}_{i=0}^\infty$ defined by (1) with $A(x) = Tr(x^{p^n - 2})$*

$$|C_S(\tau)| < 4(2^{n-1} - 1)(2 \cdot p^{n/2} + 1)(1 + \ln p)^{n-1} + 2 \cdot p^{n/2} + 3.$$

Table 1. The generating polynomials

n	The generating polynomial for $\mathbb{F}_{2^n}$
5	$x^5 + x^3 + 1$
6	$x^6 + x^5 + 1$
7	$x^7 + x + 1$
8	$x^8 + x^4 + x^3 + x + 1$
9	$x^9 + x^4 + 1$
10	$x^{10} + x^3 + 1$
11	$x^{11} + x^2 + 1$
12	$x^{12} + x^3 + 1$
13	$x^{13} + x^4 + x^3 + x + 1$
14	$x^{14} + x^5 + x^3 + x + 1$
15	$x^{15} + x + 1$
16	$x^{16} + x^5 + x^3 + x^2 + 1$
17	$x^{17} + x^3 + 1$
18	$x^{18} + x^5 + x^2 + x + 1$

Table 2. The case of $\{\alpha_1, ..., \alpha_n\} = \{1, x, x^2, ..., x^{n-1}\}$

n	period	$\max_{\tau \neq 0} \|C_S(\tau)\|$ of $A(x) = Tr(x^{2^n-2})$	$\max_{\tau \neq 0} \|C_S(\tau)\|$ of $A(x) = Tr(x^{2^n-4})$
5	32	16	8
6	64	24	20
7	128	28	36
8	256	40	40
9	512	64	72
10	1024	104	88
11	2048	152	144
12	4096	208	264
13	8192	336	348
14	16384	452	468
15	32768	700	688
16	65536	1000	1088
17	131072	1592	1684
18	262144	2184	2224

Proof. Let $f(x) = x^{p^n-2}$. Then by Lemma 2, we have

$$\left| \sum_{\xi \in \mathbb{F}_{p^n}} \psi(f(\xi) - f(\xi + \xi_\tau)) \right| \leq 2 + \left| \sum_{\xi \neq 0, -\xi_\tau} \psi\left(\frac{\xi_\tau}{\xi(\xi + \xi_\tau)} \right) \right| \leq 2 \cdot p^{n/2} + 3,$$

and for any $\eta \neq 0$, we have

$$\left| \sum_{\xi \in \mathbb{F}_{2^n}} \psi(f(\xi) - f(\xi + \xi_\tau) - \eta\xi) \right| \leq 2 + \left| \sum_{\xi \neq 0, -\xi_\tau} \psi\left(\frac{\xi_\tau}{\xi(\xi + \xi_\tau)} - \eta\xi \right) \right| \leq 4 \cdot p^{n/2} + 2.$$

Table 3. The case of $\{\alpha_1, ..., \alpha_n\} = \{x^{n-1}, x^{n-2}, ..., x, 1\}$

n	period	$\max_{\tau \neq 0} \lvert C_S(\tau) \rvert$ of $A(x) = Tr(x^{2^n-2})$	$\max_{\tau \neq 0} \lvert C_S(\tau) \rvert$ of $A(x) = Tr(x^{2^n-4})$
5	32	12	12
6	64	12	24
7	128	24	28
8	256	48	36
9	512	64	56
10	1024	104	108
11	2048	140	172
12	4096	204	304
13	8192	304	352
14	16384	488	456
15	32768	712	736
16	65536	1100	1472
17	131072	1548	1548
18	262144	2272	2656

The rest of the proof is the same as that of Theorem 2, so we omit it here.   □

Remark 2. In [9], Niederreiter and Winterhof studied the distribution of $(\xi_i, \xi_{i \oplus \tau})$ implicitly for $\tau > 0$, where $i \oplus j = h$ if and only if $\xi_i + \xi_j = \xi_h$. In general, $\xi_{i+\tau} \neq \xi_{i \oplus \tau}$. Hence, we can not derive an upper bound on $C_S(\tau)$ using the discrepancy bound in [9] together with the result in [6].

For many different bases, we computed $\max_{\tau \neq 0} \lvert C_S(\tau) \rvert$ from $\mathbb{F}_{2^5}$ to $\mathbb{F}_{2^{18}}$ in cases of $A(x) = Tr(x^{2^n-2})$ and $A(x) = Tr(x^{2^n-4})$. We found that $\max_{\tau \neq 0} \lvert C_S(\tau) \rvert$ is small compared with the period of sequences; namely,

$$2.12 \times 2^{n/2} \leq \max_{\tau \neq 0} \lvert C_S(\tau) \rvert \leq 5.75 \times 2^{n/2}$$

in the case of $5 \leq n \leq 18$ and $A(x) = Tr(x^{2^n-2})$ or $Tr(x^{2^n-4})$. Based on these data, we present the following conjecture.

Conjecture 1. For any sequence S defined by (1) with $p = 2$ and $A(x) = Tr(x^{2^n-2})$ or $A(x) = Tr(x^{2^n-4})$, there exists one constant a such that

$$\max_{\tau \neq 0} \lvert C_S(\tau) \rvert \leq a \cdot 2^{n/2}$$

holds for all n.

The experimental results are shown in Tables 2 and 3. Table 2 is for the case of $\{\alpha_1, ..., \alpha_n\} = \{1, x, x^2, ..., x^{n-1}\}$, and Table 3 is for the case of $\{\alpha_1, ..., \alpha_n\} = \{x^{n-1}, x^{n-2}, ..., x, 1\}$.

5 Conclusion

The randomness properties of sequences generated by a function via the additive order are studied in this paper. Some simple conditions are derived under which such sequences have the maximum period. The autocorrelation of such sequences is also studied, and one conjecture is presented. A nice approach for proving this conjecture is desirable.

Acknowledgments

The work is supported by NSERC SPG Grant. The authors wish to thank the anonymous reviewers for their helpful and valuable comments and suggestions. Especially, the authors wish to thank Arne Winterhof for his comments which made the proof of Theorem 1 extremely shorten and established the bound given by Theorem 2.

References

1. Chen, Z.: Finite binary sequences constructed by explicit inversive methods. Finite Fields Appl. (to appear)
2. Davis, J.A., Jedwab, J.: Peak-to-mean power control in OFDM, Golay complementary sequences and Reed-Muller codes. IEEE Trans. Inform. Theory 45(7), 2397–2417 (1999)
3. Golomb, S.W., Gong, G.: Signal Design for Good Correlation-For Wireless Communication, Cryptography and Radar. Cambridge Univ. Press, Cambridge (2005) (Section 6.6)
4. Lidl, R., Niederreiter, H.: Finite Fields. Addison-Wesley, Reading (1983) (now distributed by Cambridge Univ. Press)
5. Lipmaa, H., Rogaway, P., Wagner, D.: Comments to NIST concerning AES modes of operations: CTR-mode encryption,
   http://www.cs.ucdavis.edu/rogaway/papers/ctr.pdf
6. Mauduit, C., Niederreiter, H., Sárközy, A.: On pseudorandom $(0,1)$ and binary sequences. Publ. Math. Debrecen. 71(3-4), 305–324 (2007)
7. Meidl, W., Winterhof, A.: On the autocorrelation of cyclotomic generator. In: Mullen, G.L., Poli, A., Stichtenoth, H. (eds.) Fq7 2003. LNCS, vol. 2948, pp. 1–11. Springer, Heidelberg (2003)
8. Moreno, C.J., Moreno, O.: Exponential sums and Goppa codes: I. Proc. Amer. Math. Soc. 111, 523–531 (1991)
9. Niederreiter, H., Winterhof, A.: Incomplete exponential sums over finite fields and their applications to new inversive pseudorandom number generators. Acta Arith. 93, 387–399 (2000)
10. Paterson, K.G.: Generalized Reed-Muller codes and power control in OFDM modulation. IEEE Trans. Inform. Theory 46(1), 104–120 (2000)

On the Average Distribution of Power Residues and Primitive Elements in Inversive and Nonlinear Recurring Sequences

Ayça Çeşmelioğlu[1] and Arne Winterhof[2]

[1] Sabancı University, Orhanlı, Tuzla, 34956 Istanbul, Turkey
cesmelioglu@su.sabanciuniv.edu
[2] Johann Radon Institute for Computational and Applied Mathematics (RICAM)
Austrian Academy of Sciences, Altenbergerstr. 69, 4040 Linz, Austria
arne.winterhof@oeaw.ac.at

Abstract. We estimate character sums with inversive and nonlinear recurring sequences 'on average' over all initial values and obtain much stronger bounds than known for 'individual' sequences. As a consequence, we present results 'on average' about the distribution of power residues and primitive elements in such sequences.

On the one hand our bounds can be regarded as results on the pseudorandomness of inversive and nonlinear recurring sequences. On the other hand they shall provide a further step to efficient deterministic algorithms for finding non-powers and primitive elements in a finite field.

1 Introduction

Let q be a prime power and $\mathbb{F}_q$ the finite field of q elements. For $q \geq 4$ and given $a \in \mathbb{F}_q^*$, $b \in \mathbb{F}_q$, let ψ be the permutation of $\mathbb{F}_q$ defined by $\psi(w) = aw^{q-2} + b$, $w \in \mathbb{F}_q$. Let $(u_n) = u_0, u_1, \ldots$ be the sequence of elements of $\mathbb{F}_q$ obtained by the recurrence relation

$$u_{n+1} = \psi(u_n), \quad n \geq 0,$$

with some initial value $u_0 = \vartheta \in \mathbb{F}_q$. Because of the property $w^{q-2} = w^{-1}$ for $w \neq 0$ the sequence (u_n) is called *inversive recurring sequence*. It is obvious that the sequence (u_n) is purely periodic with least period $t_\vartheta \leq q$.

The inversive recurring sequence (u_n) belongs to the class of *nonlinear recurring sequences* $(x_n) = x_0, x_1, \ldots$ obtained by

$$x_{n+1} = f(x_n), \quad n \geq 0,$$

with some initial value $x_0 = \vartheta \in \mathbb{F}_q$ and a polynomial $f(X) \in \mathbb{F}_q[X]$ of degree at least two. Note that, if $f(X)$ is a permutation polynomial, the sequence (x_n) is purely periodic. We restrict ourselves to this case and denote the least period again by $t_\vartheta \leq q$.

We study character sums over the sequences (u_n) and (x_n) 'on average' over all initial values ϑ.

S.W. Golomb et al. (Eds.): SETA 2008, LNCS 5203, pp. 60–70, 2008.
© Springer-Verlag Berlin Heidelberg 2008

Let the sequence $R_n(X)$ of rational functions over $\mathbb{F}_q$ be defined by

$$R_0(X) = X, \quad R_n(X) = R_{n-1}(aX^{-1} + b), \quad n \geq 1.$$

Obviously this sequence is purely periodic and denote by $T_{a,b}$ its least period. From [8, Lemma 6] it follows that there are elements $\varepsilon_1, \ldots, \varepsilon_{T_{a,b}-1} \in \mathbb{F}_q$, such that

$$R_n(X) = \frac{(b - \varepsilon_n)X + a}{X - \varepsilon_n}, \quad 1 \leq n \leq T_{a,b} - 1.$$

This implies that for $1 \leq n \leq T_{a,b} - 1$ we have

$$\psi^n(\gamma) = \frac{(b - \varepsilon_n)\gamma + a}{\gamma - \varepsilon_n}, \quad \gamma \in \mathbb{F}_q \setminus \{\varepsilon_1, \ldots, \varepsilon_n\}, \tag{1}$$

where ψ^n denotes the nth iterate of ψ.

For a permutation polynomial $f(X) \in \mathbb{F}_q[X]$ we define the sequence $f_n(X)$ of polynomials over $\mathbb{F}_q$ by

$$f_0(X) = X, \quad f_n(X) = f(f_{n-1}(X)), \quad n \geq 1.$$

The sequence $f_n(X) \bmod (X^q - X)$ is purely periodic and we denote by T_f its period.

We put

$$S_{\chi,a,b}(N, \vartheta) = \sum_{n=0}^{N-1} \chi(\psi^n(\vartheta)), \quad 1 \leq N \leq T_{a,b},$$

and

$$S_{\chi,f}(N, \vartheta) = \sum_{n=0}^{N-1} \chi(f_n(\vartheta)), \quad 1 \leq N \leq T_f,$$

where χ is a nontrivial multiplicative character of $\mathbb{F}_q$.

For 'individual' sequences it was proved in [10,11] that

$$\max_{\vartheta \in \mathbb{F}_q} |S_{\chi,a,b}(N, \vartheta)| = O(N^{1/2}q^{1/4}), \quad 1 \leq N \leq t_\vartheta, \tag{2}$$

where the implied constant is absolute, and under some restrictions

$$\max_{\vartheta} |S_{\chi,f}(N, \vartheta)| = O(N^{1/2}q^{1/2}(\log q)^{-1/2}), \quad 1 \leq N \leq t_\vartheta, \tag{3}$$

where the maximum is taken over all ϑ such that (x_n) is purely periodic and the implied constant depends only on the degree of $f(X)$.

In this paper we prove that for any $0 < \varepsilon < 1$ and all initial values $\vartheta \in \mathbb{F}_q$, except at most $O(\varepsilon^2 q)$ of them, we have

$$S_{\chi,a,b}(N, \vartheta) = O(\varepsilon^{-1} \max\{Nq^{-1/4}, N^{1/2}\}) \tag{4}$$

and under the same restrictions as needed for (3)

$$S_{\chi,f}(N, \vartheta) = O(\varepsilon^{-1}N(\log q)^{-1/2}). \tag{5}$$

Note that the expected value of $\sum_{n=0}^{N-1} \chi(y_n)$ for a 'truly' random sequence (y_n) over $\mathbb{F}_q$ and a nontrivial multiplicative character of $\mathbb{F}_q$ is $\theta(N^{1/2})$, see Section 5, and (4) and (5) can be regarded as results on the pseudorandomness of the sequences (u_n) and (x_n).

Using a standard technique, from (4) and (5) we derive results 'on average' on the distribution of powers and primitive elements in the sequences (u_n) and (x_n). These results shall provide a further step to efficient deterministic algorithms for finding non-powers and primitive elements in a finite field where the construction of short sequences containing non-powers and primitive elements is crucial.

2   Bounds on Character Sums of Inversive Sequences

In this section we estimate the sums

$$S_{\chi,a,b}(N) = \sum_{\vartheta \in \mathbb{F}_q} |S_{\chi,a,b}(N,\vartheta)|^2, \quad 1 \le N \le T_{a,b}.$$

Theorem 1. *Let χ be a nontrivial multiplicative character of $\mathbb{F}_q$, then for $1 \le N \le T_{a,b}$ we have*

$$S_{\chi,a,b}(N) = O\left(\max\left\{N^2 q^{1/2}, Nq\right\}\right).$$

Proof. We have

$$S_{\chi,a,b}(N) = \sum_{n,k=0}^{N-1} \sum_{\vartheta \in \mathbb{F}_q} \chi(\psi^n(\vartheta))\overline{\chi(\psi^k(\vartheta))}$$

$$\le \sum_{n,k=0}^{N-1} \left| \sum_{\vartheta \in \mathbb{F}_q} \chi(\psi^n(\vartheta))\overline{\chi(\psi^k(\vartheta))} \right|.$$

For $n = k$, the sum over ϑ is equal to $q-1$ and thus

$$S_{\chi,a,b}(N) \le N(q-1) + 2 \sum_{\substack{n,k=0 \\ n>k}}^{N-1} \left| \sum_{\vartheta \in \mathbb{F}_q} \chi(\psi^n(\vartheta))\overline{\chi(\psi^k(\vartheta))} \right|.$$

Since ψ is a permutation, we can write the previous sum as

$$S_{\chi,a,b}(N) \le N(q-1) + 2 \sum_{\substack{n,k=0 \\ n>k}}^{N-1} \left| \sum_{\vartheta \in \mathbb{F}_q} \chi(\psi^{n-k}(\vartheta))\overline{\chi(\vartheta)} \right|$$

$$= N(q-1) + 2 \sum_{\substack{n,k=0 \\ n>k}}^{N-1} \left| \sum_{\vartheta \in \mathbb{F}_q} \chi(\psi^{n-k}(\vartheta)\vartheta^{q-2}) \right|.$$

We put $d = n - k$ and get

$$S_{\chi,a,b}(N) \leq N(q-1) + 2 \sum_{d=1}^{N-1} (N-d) \left| \sum_{\vartheta \in \mathbb{F}_q} \chi(\psi^d(\vartheta)\vartheta^{q-2}) \right|. \tag{6}$$

Put $F_d(X) = ((b - \varepsilon_d)X + a)(X - \varepsilon_d)^{q-2}X^{q-2}$. Then we have by (1)

$$\left| \sum_{\vartheta \in \mathbb{F}_q} \chi(\psi^d(\vartheta)\vartheta^{q-2}) - \sum_{\vartheta \in \mathbb{F}_q} \chi(F_d(\vartheta)) \right| \leq 2(d-1)$$

since for $\vartheta = \varepsilon_1 = 0$ the two sums agree.

Now, we need to verify that the polynomial $F_d(X)$ is not, up to a multiplicative constant, an sth power where $s|q-1$ is the order of the multiplicative character χ. In the case $b = \varepsilon_d$, the polynomial $F_d(X)$ equals $a(X-b)^{q-2}X^{q-2}$. It can only be an sth power, up to the constant a, if $b = \varepsilon_d = 0$ but this implies $T_{a,b} = 2$. In the case $b \neq \varepsilon_d$, $F_d(X)$ is equal to $(b - \varepsilon_d)(X + \frac{a}{b-\varepsilon_d})((X - \varepsilon_d)X)^{q-2}$. It is easily seen that this polynomial can not be an sth power when we consider the cases $\varepsilon_d = 0$ and $\varepsilon_d \neq 0$ separately. Therefore our sum satisfies all the necessary conditions for the Weil bound, see [6, Theorem 5.41], and we get

$$\left| \sum_{\vartheta \in \mathbb{F}_q} \chi(F_d(\vartheta)) \right| \leq 2q^{1/2}.$$

Hence

$$\left| \sum_{\vartheta \in \mathbb{F}_q} \chi(\psi^d(\vartheta)\vartheta^{q-2}) \right| \leq 2q^{1/2} + 2d - 2$$

and with Equation (6) we have

$$S_{\chi,a,b}(N) \leq N(q-1) + 2 \sum_{d=1}^{N-1} (N-d)(2q^{1/2} + 2d - 2)$$
$$= O(Nq + N^2 q^{1/2} + N^3).$$

The term $N^2 q^{1/2}$ never dominates so we have

$$S_{\chi,a,b}(N) = O(Nq + N^3).$$

If we divide the inner sum of $S_{\chi,a,b}(N)$ into at most $N/K + 1$ subsums of length K, except possibly the last sum, and use the triangle inequality we have

$$S_{\chi,a,b}(N) = \sum_{\vartheta \in \mathbb{F}_q} \left| \sum_{n=0}^{N-1} \chi(\psi^n(\vartheta)) \right|^2$$

$$\leq \sum_{\vartheta \in \mathbb{F}_q} \left(\left| \sum_{n=0}^{K-1} \chi(\psi^n(\vartheta)) \right| + \ldots + \left| \sum_{n=0}^{K-1} \chi(\psi^n(\psi^{\ell K}(\vartheta))) \right| \right)^2$$

$$\ll \frac{N^2}{K^2} \sum_{\vartheta \in \mathbb{F}_q} \left| \sum_{n=0}^{K-1} \chi(\psi^n(\vartheta)) \right|^2$$

where $\ell = \lfloor \frac{N}{K} \rfloor$. Therefore

$$S_{\chi,a,b}(N) = O\left(\frac{N^2}{K^2}(Kq + K^3) \right).$$

By choosing $K = \min\{N, \lfloor q^{1/2} \rfloor\}$ we get the result. □

Corollary 1. *For any $0 < \varepsilon < 1$ and all initial values ϑ, except at most $O(\varepsilon^2 q)$ of them, we have*

$$S_{\chi,a,b}(N, \vartheta) = O\left(\varepsilon^{-1} \max\{Nq^{-1/4}, N^{1/2}\} \right).$$

Proof. Let A be the number of exceptional ϑ. Then we have

$$S_{\chi,a,b}(N) = \Omega(A\varepsilon^{-2} \max\{N^2 q^{-1/2}, N\})$$

and Theorem 1 implies the result. □

3   Bounds on Character Sums of Nonlinear Sequences

In this section we estimate the sums

$$S_{\chi,f}(N) = \sum_{\vartheta \in \mathbb{F}_q} |S_{\chi,f}(N, \vartheta)|^2, \quad 1 \leq N \leq T_f.$$

Let t_0 be the least period of the sequence (x_n) with initial value $x_0 = 0$.

Theorem 2. *Let χ be a nontrivial multiplicative character of $\mathbb{F}_q$ of order $s > 1$ and $f(X) \in \mathbb{F}_q[X]$ a permutation polynomial with $d = \deg(f) \geq 2$. Then for $1 \leq N \leq T_f$ we have*

$$S_{\chi,f}(N) = O\left(\frac{N^2 q}{\min\{\log q / \log d, N, t_0\}} \right),$$

where the implied constant is absolute.

Proof. We have

$$S_{\chi,f}(N) = \sum_{\vartheta \in \mathbb{F}_q} \left| \sum_{n=0}^{N-1} \chi(f_n(\vartheta)) \right|^2 = \sum_{n,k=0}^{N-1} \sum_{\vartheta \in \mathbb{F}_q} \chi(f_n(\vartheta)) \overline{\chi(f_k(\vartheta))}$$

$$\leq \sum_{n,k=0}^{N-1} \left| \sum_{\vartheta \in \mathbb{F}_q} \chi(f_n(\vartheta)) \overline{\chi(f_k(\vartheta))} \right| = \sum_{n,k=0}^{N-1} \left| \sum_{\vartheta \in \mathbb{F}_q} \chi(f_n(\vartheta)(f_k(\vartheta))^{q-2}) \right|$$

$$\leq Nq + 2 \sum_{\substack{n,k=0 \\ n>k}}^{N-1} \left| \sum_{\vartheta \in \mathbb{F}_q} \chi(f_n(\vartheta)(f_k(\vartheta))^{q-2}) \right|.$$

Let $G(X) = \gcd(f_n(X), f_k(X))$. Suppose that $G(X)$ is a polynomial of degree at least 1 and let α be a root of $G(X)$ in some extension field of $\mathbb{F}_q$. Then we have

$$f_{n-k}(0) = f_{n-k}(f_k(\alpha)) = f_n(\alpha) = 0, \quad n > k,$$

and $n \equiv k \pmod{t_0}$. Now, we suppose $n \not\equiv k \pmod{t_0}$ and hence $G(X) = 1$. If $f_n(X)(f_k(X))^{q-2}$ is an sth power then $f_n(X)$ and $f_k(X)$ are also sth powers. But this contradicts our assumption that $f(X)$ is a permutation polynomial and thus $f_n(X)$ and $f_k(X)$ are not sth powers. The number of pairs $(n,k) \in \mathbb{Z}^2$ with $0 \leq k < n \leq N-1$ and $n \equiv k \pmod{t_0}$ is at most $N^2/2t_0$. For such pairs, we can bound the inner sum in $S_{\chi,f}(N)$ trivially by q and for the remaining pairs we use the Weil bound.

$$S_{\chi,f}(N) < Nq + N^2 \left(\frac{q}{t_0} + 2d^{N-1}q^{1/2} \right).$$

Now, we divide the inner sum in $S_{\chi,f}(N)$ into at most $N/K + 1$ sums of length at most K and by using the triangle inequality we obtain

$$S_{\chi,f}(N) \leq \sum_{\vartheta \in \mathbb{F}_q} \left(\left| \sum_{n=0}^{K-1} \chi(f_n(\vartheta)) \right| + \ldots + \left| \sum_{n=0}^{K-1} \chi(f_n(f_{\ell K}(\vartheta))) \right| \right)^2$$

$$= O\left(\frac{N^2 q}{K} + \frac{N^2 q}{t_0} + N^2 q^{1/2} d^{K-1} \right),$$

where again $\ell = \lfloor N/K \rfloor$. Here we needed that $f(X)$ is a permutation polynomial again. Choosing

$$K = \min\left\{ \left\lceil 0.4 \frac{\log q}{\log d} \right\rceil, N \right\}$$

completes the proof. $\square$

Corollary 2. *Under the restrictions of Theorem 2, for any $0 < \varepsilon < 1$ and all initial values ϑ, except at most $O(\varepsilon^2 q)$ of them, we have*

$$S_{\chi,f}(N, \vartheta) = O\left(\varepsilon^{-1} \frac{N}{\min\{\log q/\log d, N, t_0\}^{1/2}} \right).$$

4   Distribution of Powers and Primitive Elements

For a positive divisor s of $q - 1$, an element $w \in \mathbb{F}_q^*$ is called an *sth power* if the equation $w = z^s$ has a solution in $\mathbb{F}_q$.

Let $R_{s,a,b}(N, \vartheta)$ and $R_{s,f}(N, \vartheta)$ be the number of sth powers among u_0, $u_1, \ldots, u_{N-1}$ and $x_0, x_1, \ldots, x_{N-1}$, respectively, with initial value ϑ.

Theorem 3. *Let q be a prime power and $s > 1$ be a divisor of $q - 1$. We have*

$$\sum_{\vartheta \in \mathbb{F}_q} \left| R_{s,a,b}(N, \vartheta) - \frac{N}{s} \right| = O\left(\max\{Nq^{3/4}, N^{1/2}q\} \right) \quad \text{for } 1 \le N \le T_{a,b}.$$

Proof. Let X_s denote the set of multiplicative characters χ for which $\chi(w) = 1$ for any sth power $w \in \mathbb{F}_q^*$. By [6, Theorem 5.4] we obtain

$$\frac{1}{s} \sum_{\chi \in X_s} \chi(w) = \begin{cases} 1, \text{ if } w \in \mathbb{F}_q^* \text{ is an } s\text{th power,} \\ 0, \text{ otherwise,} \end{cases}$$

where we used the convention $\chi_0(0) = 0$ for the trivial character χ_0 of $\mathbb{F}_q$. Therefore

$$R_{s,a,b}(N, \vartheta) = \frac{1}{s} \sum_{\chi \in X_s} S_{\chi,a,b}(N, \vartheta).$$

The contribution to $R_{s,a,b}(N)$ of the sum corresponding to the trivial character is either $(N - 1)/s$ or N/s. Therefore

$$\left| R_{s,a,b}(N, \vartheta) - \frac{N}{s} \right| \le \frac{1}{s} + \frac{1}{s} \sum_{\chi \in X_s \setminus \{\chi_0\}} |S_{\chi,a,b}(N, \vartheta)|.$$

Summing over ϑ and applying the Cauchy-Schwarz inequality to

$$\sum_{\vartheta \in \mathbb{F}_q} |S_{\chi,a,b}(N, \vartheta)| \le q^{1/2} S_{\chi,a,b}(N)^{1/2},$$

we get the result by Theorem 1.   □

With the notation of Theorem 2, we have the following result.

Theorem 4. *Let q be a prime power and $s > 1$ be a divisor of $q - 1$. Then*

$$\sum_{\vartheta \in \mathbb{F}_q} \left| R_{s,f}(N, \vartheta) - \frac{N}{s} \right| = O\left(Nq \left(\min\left\{ \frac{\log q}{\log d}, N, t_0 \right\} \right)^{-1/2} \right)$$

for $1 \le N \le T_f$.

Proof. The result is proved analogously to the previous theorem using Theorem 2.   □

Corollary 3. *For any $0 < \varepsilon < 1$ and all initial values ϑ, except at most $O(\varepsilon q)$ of them, we have*

$$\left| R_{s,a,b}(N, \vartheta) - \frac{N}{s} \right| = O\left(\varepsilon^{-1} \max\{ Nq^{-1/4}, N^{1/2} \} \right), \quad 1 \le N \le T_{a,b},$$

and under the conditions of Theorem 2

$$\left| R_{s,f}(N, \vartheta) - \frac{N}{s} \right| = O\left(\varepsilon^{-1} N \left(\min\left\{ \frac{\log q}{\log d}, N, t_0 \right\} \right)^{-1/2} \right) \quad \text{for } 1 \le N \le T_f.$$

We recall that $w \in \mathbb{F}_q^*$ is a *primitive element* of $\mathbb{F}_q$ if it is not an sth power for any divisor $s > 1$ of $q - 1$. For an integer $m \ge 1$ we denote by $\nu(m)$ the number of distinct prime divisors of m and by $\varphi(m)$ Euler's totient function.

Let $Q_{a,b}(N, \vartheta)$ and $Q_f(N, \vartheta)$ be the number of primitive elements of $\mathbb{F}_q$ among $u_0, u_1, \ldots, u_{N-1}$ and $x_0, x_1, \ldots, x_{N-1}$, respectively, with initial value ϑ.

Theorem 5. *For $1 \le N \le t$ we have*

$$\sum_{\vartheta \in \mathbb{F}_q} \left| Q_{a,b}(N, \vartheta) - \frac{\varphi(q-1)}{q-1} N \right| = O\left(2^{\nu(q-1)} \max\{ Nq^{3/4}, N^{1/2}q \} \right).$$

Proof. From Vinogradov's formula (see [5, Lemma 7.5.3], [6, Exercise 5.14]) we obtain

$$Q_{a,b}(N, \vartheta) = \frac{\varphi(q-1)}{q-1} \sum_{s|(q-1)} \left(\frac{\mu(s)}{\varphi(s)} \sum_{\chi \in Y_s} S_{\chi,a,b}(N, \vartheta) \right),$$

where μ denotes the Möbius function and Y_s the set of multiplicative characters of $\mathbb{F}_q$ of order s. The rest follows using the Cauchy-Schwarz inequality from Theorem 1. $\qquad\square$

With the notation in Theorem 2, we have the following result.

Theorem 6. *For $1 \le N \le T_f$ we have*

$$\sum_{\vartheta \in \mathbb{F}_q} \left| Q_f(N, \vartheta) - \frac{\varphi(q-1)}{q-1} N \right| = O\left(2^{\nu(q-1)} Nq \left(\min\left\{ \frac{\log q}{\log d}, N, t_0 \right\} \right)^{-1/2} \right).$$

Proof. We get the result from Theorem 2 in the same way. $\qquad\square$

Corollary 4. *For any $0 < \varepsilon < 1$ and all initial values ϑ, except at most $O(\varepsilon q)$ of them, we have*

$$\left| Q_{s,a,b}(N, \vartheta) - \frac{\varphi(q-1)}{q-1} N \right| = O\left(\varepsilon^{-1} 2^{\nu(q-1)} \max\{ Nq^{-1/4}, N^{1/2} \} \right)$$

for $1 \le N \le T_{a,b}$, and under the conditions of Theorem 2

$$\left| Q_{s,f}(N, \vartheta) - \frac{\varphi(q-1)}{q-1} N \right| = O\left(\varepsilon^{-1} 2^{\nu(q-1)} N \left(\min\left\{ \frac{\log q}{\log d}, N, t_0 \right\} \right)^{-1/2} \right)$$

for $1 \le N \le T_f$.

5   Final Remarks

The bounds (2) and (3) for all ϑ are only nontrivial if N is at least of the order of magnitude $\Omega(q^{1/2})$ and $\Omega(q/\log q)$, respectively. However, the bounds (4) and (5) for almost all ϑ are nontrivial for all N larger than a constant which doesn't depend on q.

Corollary 3 provides the existence of an sth power in the sequence $u_0, u_1, \ldots, u_{N-1}$ or $x_0, x_1, \ldots, x_{N-1}$ for almost all initial values if

$$s = O\left(\varepsilon \min\{q^{1/4}, N^{1/2}\}\right)$$

or

$$s = O\left(\varepsilon \min\left\{\frac{\log q}{\log d}, N, t_0\right\}^{1/2}\right),$$

respectively. It also provides the existence of an sth non-power in these sequences for almost all ϑ if N is larger than a constant depending only on ε provided that t_0 is large enough in the second case.

Corollary 4 implies the existence of a primitive element in $u_0, u_1, \ldots, u_{N-1}$ or $x_0, x_1, \ldots, x_{N-1}$ for almost ϑ if

$$2^{\nu(q-1)} = O\left(\varepsilon \frac{\varphi(q-1)}{q-1} \min\{q^{1/4}, N^{1/2}\}\right)$$

or

$$2^{\nu(q-1)} = O\left(\varepsilon \frac{\varphi(q-1)}{q-1} \min\left\{\frac{\log q}{\log d}, N, t_0\right\}^{1/2}\right),$$

respectively.

It is clear that the maximal value $t_\vartheta = q$ of the least period of (x_n) with initial value $x_0 = \vartheta$ is obtained if and only if $f(X)$ is a permutation polynomial of $\mathbb{F}_q$ representing a permutation which is a cycle of length q. In this case we have $t_\vartheta = t_0$.

Let $f(X) = X^d$ with $d \geq 2$, $x_0 \neq 0$, and s be a divisor of $d - 1$. Then it is clear that for a character χ of order s we have $\chi(x_n) = \chi(x_0)$ for all $n \geq 0$. This example provides some evidence that the dependence of the character sum bound on t_0 is natural.

Let $f(X) = (X + a)^d - a$ with $d \geq 2$, $a \in \mathbb{F}_q^*$, $x_0 = 0$. The sequence (x_n) generated by this polynomial can be obtained by subtracting a from a sequence as in the previous remark. Hence, both sequences have the same least period. For example, if q is even, $q - 1$ a Mersenne prime, d the least primitive root modulo $q - 1$, i.e., $d = O(\log^6(q - 1))$ under ERH (see [12, Theorem 1.3]), and a is a primitive element of $\mathbb{F}_q$, then we have $t_0 = q - 2$.

This shows that examples for which Theorem 2 gives a nontrivial bound can be easily constructed.

Besides the class of inversive recurring sequences there are two other classes of nonlinear recurring sequences for which we know much better worst case bounds than for a general nonlinear recurring sequence where $f(X)$ is either a Dickson polynomial, see [2], or a Rédei function, see [3].

Since for a nontrivial multiplicative character of $\mathbb{F}_q$

$$\sum_{(y_0,\ldots,y_{N-1})\in\mathbb{F}_q^N} \left|\sum_{n=0}^{N-1} \chi(y_n)\right|^2 = Nq^{N-1}(q-1),$$

the expected value of $\left|\sum_{n=0}^{N-1} \chi(y_n)\right|$ of a 'truly' random sequence (y_n) is

$$\left(1-\frac{1}{q}\right)^{1/2} N^{1/2}.$$

Besides the rather small character sum bounds inversive recurring sequences have several other nice pseudorandomness properties as low linear complexity [4] and small discrepancy [8,9]. Similar but weaker results are also known for general nonlinear recurring sequences [4,7].

Finding an sth non-power is the crucial step in the design of deterministic algorithms for solving equations over finite fields, see e.g. [1, Chapter 7]. Algorithms for finding primitive elements in a finite field normally consist of two parts: 1. Finding a short sequence containing at least one primitive element; 2. Testing primitiveness of the sequence elements, see e.g. [13, Chapter 2]. Our results show that inversive recurring sequences are suitable for both algorithmic problems for almost all initial values. However, it is still a problem to detect bad initial values.

Acknowledgment

This paper was written during a pleasant stay of A.Ç. to Linz. She wishes to thank the Radon Institute for its hospitality. A.Ç. was supported by Yousef Jameel scholarship and in part by Tübitak and A.W. was supported in part by the Austrian Science Fund (FWF) Grant P19004-N18.

References

1. Bach, E., Shallit, J.: Algorithmic number theory. Efficient algorithms. Foundations of Computing Series, vol. 1. MIT Press, Cambridge (1996)
2. Gomez, D., Winterhof, A.: Character sums for sequences of iterations of Dickson polynomials. In: Mullen, G., et al. (eds.) Finite Fields and Applications Fq8, Contemporary Mathematics (to appear)
3. Gomez, D., Winterhof, A.: Multiplicative Character Sums of Recurring Sequences with Rédei Functions. In: Proceedings SETA 2008 (to appear, 2008)

4. Gutierrez, J., Shparlinski, I.E.: Winterhof, Arne On the linear and nonlinear complexity profile of nonlinear pseudorandom number-generators. IEEE Trans. Inform. Theory 49(1), 60–64 (2003)

5. Jungnickel, D.: Finite fields. Structure and arithmetics. Bibliographisches Institut, Mannheim (1993)

6. Lidl, R., Niederreiter, H.: Finite fields. Encyclopedia of Mathematics and its Applications, vol. 20. Cambridge University Press, Cambridge (1997)

7. Niederreiter, H., Shparlinski, I.E.: On the distribution and lattice structure of nonlinear congruential pseudorandom numbers. Finite Fields Appl. 5(3), 246–253 (1999)

8. Niederreiter, H., Shparlinski, I.E.: On the distribution of pseudorandom numbers and vectors generated by inversive methods. Appl. Algebra Engrg. Comm. Comput. 10(3), 189–202 (2000)

9. Niederreiter, H., Shparlinski, I.E.: On the average distribution of inversive pseudorandom numbers. Finite Fields Appl. 8(4), 491–503 (2002)

10. Niederreiter, H., Shparlinski, I.E.: On the distribution of power residues and primitive elements in some nonlinear recurring sequences. Bull. London Math. Soc. 35(4), 522–528 (2003)

11. Niederreiter, H., Winterhof, A.: Multiplicative character sums for nonlinear recurring sequences. Acta Arith. 111(3), 299–305 (2004)

12. Shoup, V.: Searching for primitive roots in finite fields. Math. Comp. 58, 369–380 (1992)

13. Shparlinski, I.E.: Finite fields: theory and computation. The meeting point of number theory, computer science, coding theory and cryptography. Mathematics and its Applications, vol. 477. Kluwer Academic Publishers, Dordrecht (1999)

Some Results on the Arithmetic Correlation of Sequences

(Extended Abstract)

Mark Goresky[1],[*] and Andrew Klapper[2],[**]

[1] Department of Mathematics, Institute for Advanced Study
Princeton, N.J. 08540, USA
`http://www.math.ias.edu/~goresky/`
[2] Department of Computer Science, University of Kentucky,
Lexington, KY 40506-0046, USA
`http://www.cs.uky.edu/~klapper/`

Abstract. In this paper we study various properties of arithmetic correlations of sequences. Arithmetic correlations are the with-carry analogs of classical correlations. Here we analyze the arithmetic autocorrelations of non-binary ℓ-sequences, showing that they are nearly optimal. We analyze the expected auto- and cross-correlations of sequences with fixed shift. We study sequences with the arithmetic shift and add property, showing that they are exactly the ℓ-sequences with prime connection element.

Keywords: feedback with carry shift register, pseudo-randomness, correlation, sequences.

1 Introduction

Sequences with good correlation properties are essential ingredients in a wide range of applications including CDMA systems and radar ranging. On the plus side, a great deal is known about the design and generation of sequences with good correlation properties. Unfortunately, we also know that there are fundamental limits on the sizes of families of sequences with such properties.

The purpose of this paper is to study properties of an arithmetic or "with-carry" analog of the classical correlation function. This notion of correlation is interesting in part because it is known (in the binary case) that they do not suffer from some of the constraints on families of sequences with good classical

[*] Research partially supported by DARPA grant no. HR0011-04-1-0031.
[**] Parts of this work were carried out while the author visited the Fields Inst. at the University of Toronto; The Inst. for Advanced Study; and Princeton Universiy. This material is based upon work supported by the National Science Foundation under Grant No. CCF-0514660. Any opinions, findings, and conclusions or recommendations expressed in this material are those of the author and do not necessarily reflect the views of the National Science Foundation.

correlations. However, we do not as yet know of any significant applications of arithmetic correlations. Nonetheless, it is worthwhile studying properties of arithmetic correlations in hope that applications will come.

In previous work we have studied arithmetic auto- and cross-correlations (defined below) of a class of binary sequences called ℓ-sequences [4,6,7]. The arithmetic autocorrelations of these sequences were previously studied in the context of arithmetic coding [9,10]. It is known that the shifted arithmetic autocorrelations of binary ℓ-sequences are identically zero and that the arithmetic cross-correlations of any two distinct decimations of a binary ℓ-sequence is identically zero.

In this paper we study the arithmetic correlations of possibly non-binary sequences. We show that the arithmetic autocorrelations of ℓ-sequences are at most one for a prime connection integer and at most two for a prime power connection integer. We also analyze the expected arithmetic auto- and cross-correlations of sequences with fixed shift. Finally, we define a notion of arithmetic shift and add sequence, generalizing the classical notion of shift and add sequence [1,2,3,5], and prove that the arithmetic shift and add sequences are exactly the ℓ-sequences with prime connection integer.

2   Arithmetic Correlations

Let $N \geq 2$ be a natural number. In this section we define a with carry analog of the usual notion of cross-correlations for N-ary sequences.

A fundamental tool we use is the notion of N-adic numbers. An N-adic number is a formal expression

$$a = \sum_{i=0}^{\infty} a_i N^i,$$

where $a_i \in \{0, 1, \cdots, N-1\}$, $i = 0, 1, \cdots$. The set $\hat{\mathbb{Z}}_n$ of N-adic numbers forms an algebraic ring and has been the subject of extensive study for over 100 years [7,8]. The algebra — addition and multiplication — is defined with carries propagated to higher and higher terms, just as it is for ordinary nonnegative integers, but possibly involving infinitely many terms. It is easy to see that $\hat{\mathbb{Z}}_n$ contains all rational numbers u/q, $u, q \in \mathbb{Z}$, with q relatively prime to N and no other rational numbers. There is a one to one correspondence between N-adic numbers and infinite N-ary sequences. Under this correspondence the rational numbers u/q with q relatively prime to N correspond to the eventually periodic sequences. The rational numbers u/q with q relatively prime to N and $-q \leq u \leq 0$ correspond to the (strictly) periodic sequences. If $\mathbf{a}$ is periodic (resp., eventually periodic) then we say that the associated N-adic number is periodic (resp., eventually periodic). Note that, unlike power series, the sum and difference of strictly periodic N-adic numbers are eventually periodic but may not be strictly periodic.

Let $\mathbf{a}$ be an eventually periodic N-ary sequence and let

$$a = \sum_{i=0}^{\infty} a_i N^i$$

be the associated N-adic number. For each $i = 0, 1, \cdots, N - 1$, let μ_i be the number of occurrences of i in one complete period of $\mathbf{a}$. Let

$$\zeta = e^{2\pi i/N}$$

be a complex primitive Nth root of 1. Let

$$Z(a) = Z(\mathbf{a}) = \sum_{i=0}^{N-1} \mu_i \zeta^i,$$

the *imbalance* of a or of $\mathbf{a}$.

The periodic sequence $\mathbf{a}$ is said to be *balanced* if $\mu_i = \mu_j$ for all i, j. It is *weakly balanced* if $Z(\mathbf{a}) = 0$.

For example, let $N = 3$ and $a = 3/5 = 0 + 2 \cdot 3 + 0 \cdot 3^2 + 1 \cdot 3^3 + 2 \cdot 3^4 + 1 \cdot 3^5 + 0 \cdot 3^6 + 1 \cdot 3^7 + \cdots$. This sequence is periodic with period 4 from the 3^2 term on. Thus $\mu_0 = 1$, $\mu_1 = 2$, and $\mu_2 = 1$. We have $Z(\mathbf{a}) = 1 + 2\zeta + \zeta^2 = \zeta$. The sequence is not weakly balanced.

Lemma 1. *If the N-ary sequence $\mathbf{a}$ is balanced, then it is weakly balanced. If N is prime, then $\mathbf{a}$ is balanced if and only if it is weakly balanced.*

For any N-ary sequence $\mathbf{b}$, let $\mathbf{b}^\tau$ be the sequence formed by shifting $\mathbf{b}$ by τ positions, $b_i^\tau = b_{i+\tau}$. The ordinary cross-correlation with shift τ of two N-ary sequences $\mathbf{a}$ and $\mathbf{b}$ of period T is the imbalance of the term by term difference of $\mathbf{a}$ and $\mathbf{b}^\tau$, or equivalently, of the coefficient sequence of the difference between the power series associated with $\mathbf{a}$ and the power series associated with $\mathbf{b}^\tau$. In the binary case this is the number of zeros minus the number of ones in one period of the bitwise exclusive-or of $\mathbf{a}$ and the τ shift of $\mathbf{b}$ [2]. The arithmetic cross-correlation is the with-carry analog of this [6].

Definition 1. *Let $\mathbf{a}$ and $\mathbf{b}$ be two eventually periodic sequences with period T and let $0 \leq \tau < T$. Let a and $b^{(\tau)}$ be the N-adic numbers whose coefficients are given by $\mathbf{a}$ and $\mathbf{b}^\tau$, respectively. Then the sequence of coefficients associated with $a - b^{(\tau)}$ is eventually periodic and its period divides T. The shifted arithmetic cross-correlation of $\mathbf{a}$ and $\mathbf{b}$ is*

$$\mathcal{C}_{\mathbf{a},\mathbf{b}}^A(\tau) = Z(a - b^{(\tau)}), \tag{1}$$

where the imbalance is taken over a full period of length T. When $\mathbf{a} = \mathbf{b}$, the arithmetic cross-correlation is called the arithmetic autocorrelation *of $\mathbf{a}$ and is denoted $\mathcal{A}_{\mathbf{a}}^A(\tau)$.*

If for all τ such that $\mathbf{a}$ and $\mathbf{b}^\tau$ are distinct we have $\mathcal{C}_{\mathbf{a},\mathbf{b}}^A(\tau) = 0$, then $\mathbf{a}$ and $\mathbf{b}$ are said to have *ideal arithmetic correlations*. A family of sequences is said to have ideal arithmetic correlations if every pair of sequences in the family has ideal arithmetic correlations.

3   ℓ-Sequences

In this section we consider the arithmetic autocorrelations of ℓ-sequences. These are the arithmetic analogs of m-sequences, a class of sequences that have been used in many applications. Recall that an m-sequence over a finite field F is the coefficient sequence of the power series expansion of a rational function $f(x)/q(x)$ such that the degree of f is less than the degree of q, q is irreducible, and x is a primitive element in the multiplicative group of $F[x]/(q)$. It is well known that the classical shifted autocorrelations of an m-sequence all equal -1. However, the cross-correlations of m-sequences are only known in a few special cases.

An N-ary ℓ-sequence **a** is the N-adic expansion of a rational number f/q where $\gcd(q, N) = 1$, $-q < f < 0$ (so that **a** is strictly periodic), and N is a primitive element in the multiplicative group of integers modulo q. This last condition means that the multiplicative order of N modulo q, $\mathrm{ord}_q(N)$, equals $\phi(q)$ (Euler's function). In particular it implies that q is a power of a prime number, $q = p^t$. For the remainder of this section we assume that N, **a**, q, p, t and f satisfy all these conditions.

Quite a lot is known about ℓ-sequences, especially in the binary ($N = 2$) case. For example, we have the following remarkable fact about binary ℓ-sequences [6].

Theorem 1. *Suppose the* **a** *is a binary ℓ-sequence. If* **c** *and* **b** *are decimations of* **a**, *then the arithmetic cross-correlation of* **c** *and* **b** *with shift* τ *is zero unless* $\tau = 0$ *and* **b** $=$ **c**.

Our goal here is to determine the arithmetic autocorrelations of not necessarily binary ℓ-sequences. First we look at their imbalances.

Theorem 2. *Let* **a** *be an N-ary ℓ-sequence based on a connection integer $q = p^e$, p prime, $e \geq 1$. Then*

$$|Z(\mathbf{a})| \begin{cases} \leq 2 \text{ for all } q \\ \leq 1 \text{ if } q \text{ is prime} \\ \leq 1 \text{ if } e \geq 2 \text{ and either } q \equiv 1 \mod N \text{ or } p^{e-1} \equiv 1 \mod N \\ = 0 \text{ if } q \text{ is prime and } q \equiv 1 \mod N \\ = 0 \text{ if } e \geq 2, q \equiv 1 \mod N, \text{ and } p^{e-1} \equiv 1 \mod N. \end{cases}$$

One of the last two cases always holds when $N = 2$.

We can apply this result to estimate the autocorrelations of ℓ-sequences.

Theorem 3. *Let* **a** *be an N-ary ℓ-sequence with period T based on a prime connection integer q. Let τ be an integer that is not a multiple of T. Then $|\mathcal{A}_{\mathbf{a}}^A(\tau)| \leq 1$. If $q \equiv 1 \mod N$, then $\mathcal{A}_{\mathbf{a}}^A(\tau) = 0$. This last statement holds when $N = 2$.*

Proof. The N-adic number associated with **a** is a fraction $-f/q$ as above. By an argument similar to the one in Section 3.1, the arithmetic autocorrelation of **a** with shift τ is the imbalance of the rational number

$$\frac{(N^{T-\tau} - 1)f \mod q}{q},$$

where the reduction modulo q is taken in the range $[-(q-1), 0]$. Since q is prime, this is again the rational number corresponding to an ℓ-sequence. The theorem then follows from Theorem 2.   □

Note that this argument does not apply to ℓ-sequences with prime power connection integer since the numerator $(N^{T-\tau} - 1)f$ may not be relatively prime to q.

3.1   Expected Arithmetic Correlations

In this section we investigate the expected values of the arithmetic autocorrelations and cross-correlations and the second moments and variances of the cross-correlations for a fixed shift. We leave the problem of computing second moments and variances of the arithmetic autocorrelations as open problems.

We need some initial analysis for general N-ary sequences. Fix a period T. As we have seen, the N-ary sequences of period T are the coefficient sequences **a** of rational numbers of the form

$$a = \frac{-f}{N^T - 1}$$

with $0 \leq f \leq N^T - 1$.

Lemma 2. *If a and b are distinct N-adic numbers whose coefficient sequences are periodic with period T, and $a - b \in \mathbb{Z}$, then $\{a, b\} = \{0, -1\}$.*

Next fix a shift τ. Then the τ shift of **a** corresponds to a rational number

$$a^{(\tau)} = c_{f,\tau} + \frac{-N^{T-\tau}f}{N^T - 1},$$

where $0 \leq c_{f,\tau} < N^{T-\tau}$ is an integer.

Now let **b** be another periodic N-ary sequence corresponding to the rational number

$$b = \frac{-g}{N^T - 1}.$$

Then the arithmetic cross-correlation between **a** and **b** with shift τ is

$$\mathcal{C}_{\mathbf{a},\mathbf{b}}^A(\tau) = Z\left(\frac{-f}{N^T - 1} - \left(c_{g,\tau} + \frac{-N^{T-\tau}g}{N^T - 1}\right)\right)$$

$$= Z\left(\frac{N^{T-\tau}g - f}{N^T - 1} - c_{g,\tau}\right). \tag{2}$$

Theorem 4. *For any τ, the expected arithmetic autocorrelation, averaged over all sequences **a** of period T, is*

$$E[\mathcal{A}_{\mathbf{a}}^A(\tau)] = \frac{T}{N^{T-\gcd(\tau,T)}}.$$

*The expected cross-correlation, averaged over all pairs of sequences **a** and **b** is*

$$E[\mathcal{C}_{\mathbf{a},\mathbf{b}}^A(\tau)] = \frac{T}{N^T}.$$

Proof. If the τ shift of **b** equals **a**, then $C^A_{\mathbf{a},\mathbf{b}}(\tau) = T$. Otherwise a and $b^{(\tau)}$ are distinct periodic sequences. In particular, by Lemma 2 $a - b^{(\tau)}$ is an integer only if $\{a, b^{(\tau)}\} = \{0, -1\}$.

First we consider the autocorrelation. Let

$$S = \sum_{f=0}^{N^T-1} Z\left(\frac{(N^{T-\tau} - 1)f}{N^T - 1} - c_{f,\tau}\right).$$

It follows from equation (2) that the expected arithmetic autocorrelation is $E[\mathcal{A}^A_{\mathbf{a}}(\tau)] = S/N^T$.

By the first paragraph of this proof $a - a^{(\tau)}$ is an integer only if $a^{(\tau)} = a$. When it is not an integer, the periodic part of

$$\frac{(N^{T-\tau} - 1)f}{N^T - 1} - c_{f,\tau}$$

is the same as the periodic part of

$$\frac{(N^{T-\tau} - 1)f \mod N^T - 1}{N^T - 1},$$

where we take the reduction modulo $N^T - 1$ in the set of residues $\{-(N^T - 2), -(N^T - 3), \cdots, -1, 0\}$. In particular, this latter rational number has a strictly periodic N-adic expansion, so we can compute its contribution to S by considering the first T coefficients.

Let $d = \gcd(T, T - \tau) = \gcd(T, \tau)$. Then $\gcd(N^T - 1, N^{T-\tau} - 1) = N^d - 1$. Then the set of elements of the form $(N^{T-\tau} - 1)f \mod N^T - 1$ is the same as the set of elements of the form $(N^d - 1)f \mod N^T - 1$. Thus

$$S = \sum_{f=0}^{N^T-1} Z\left(\frac{(N^d - 1)f \mod N^T - 1}{N^T - 1}\right).$$

Now consider the contribution to S from the ith term in the expansion in each element in the sum, say corresponding to an integer f. If we multiply f by N^{T-i} modulo $N^T - 1$, this corresponds to cyclically permuting the corresponding sequence to the right by $T - i$ places. This is equivalent to permuting to the left by i positions, so the elements in the ith place become the elements in the 0th place. Moreover, multiplying by N^{T-i} is a permutation modulo $N^T - 1$, so the distribution of values contributing to S from the ith terms is identical to the distribution of values from the 0th term.

To count the contribution from the 0th position, let

$$D = \frac{N^T - 1}{N^d - 1}$$

and $f = u + vD$ with $0 < u < D$ and $0 \le v < N^d - 1$. Then $(N^d - 1)f \mod N^T - 1 = (N^d - 1)u \mod N^T - 1 = (N^d - 1)u - (N^T - 1)$. Thus

$$\frac{(N^d - 1)f \mod N^T - 1}{N^T - 1} = \frac{(N^d - 1)u}{N^T - 1} - 1. \tag{3}$$

In particular, the contribution to S from the 0th position depends only on u. Thus we can count the contributions over all g with $0 < u < D$, and then multiply by $N^d - 1$. The contribution from the 0th position for a particular u is given by reducing the right hand side of equation (3) modulo N. We have

$$\frac{(N^d - 1)u}{N^T - 1} - 1 = (1 + N^T + N^{2T} + \cdots)(N^d - 1)u - 1$$

$$\equiv -u - 1 \mod N.$$

Since $-(1 + N^d + N^{2d} + \cdots + N^{T-d}) \le -u - 1 \le -2$, as u varies its reduction modulo N takes each value in $\{0, 1, \cdots, N - 1\}$ exactly $N^{d-1} + N^{2d-1} + \cdots + N^{T-d-1}$ times.

It follows that the contribution to S from the sequences that are not equal to their τ shifts is a multiple of $1 + \zeta + \cdots + \zeta^{N-1} = 0$.

Thus we need to count the number of sequences that are equal to their τ shifts. These are the sequences whose minimal periods are divisors of τ. Of course the minimal periods of such sequences are also divisors of T, so it is equivalent to counting the sequences whose minimal period divides d. The number of such sequences is exactly N^d. Thus the expected autocorrelation is

$$E[\mathcal{A}_{\mathbf{a}}^A(\tau)] = \frac{N^d T}{N^T}.$$

The derivation of the expected arithmetic cross-correlation uses similar methods. □

Theorem 5. *For any shift τ, the second moment of the arithmetic cross-correlation, averaged over all pairs of sequences $\mathbf{a}$ and $\mathbf{b}$ is*

$$E[\mathcal{C}_{\mathbf{a},\mathbf{b}}^A(\tau)^2] = T\frac{N^T + 1 - T}{N^T}.$$

The variance is

$$V[\mathcal{C}_{\mathbf{a},\mathbf{b}}^A(\tau)] = T\frac{(N^T + 1)(N^T - T)}{N^{2T}}.$$

Proof. The proof uses methods which are similar to those used in the proof of Theorem 4. □

4   Computing Arithmetic Cross-Correlations

If $\mathbf{b}$ and $\mathbf{c}$ are two periodic sequences with associated N-adic numbers b and c, respectively, then the sequence associated with the difference $b - c$ may not be strictly periodic (although it must be eventually periodic). Thus at first glance computing the arithmetic cross-correlation of two sequences is problematic. How many symbols of the difference must be computed before we reach the periodic part? As it turns out, however, the number of symbols needed is well bounded.

Theorem 6. *Let* **b** *and* **c** *be periodic sequences with period* T. *Let* b *and* c *be the* N-*adic numbers associated with* **b** *and* **c**. *Let* $\mathbf{d} = d_0, d_1, \cdots$ *be the sequence associated with* $b - c$. *Then* **d** *is strictly periodic from at least* d_T *on.*

Proof. As noted earlier, the strict periodicity of **b** and **c** implies that there are integers g and h so that $b = -g/(N^T - 1)$, $c = -h/(N^T - 1)$, $0 \leq g \leq N^T - 1$, and $0 \leq h \leq N^T - 1$. Thus

$$b - c = \frac{h - g}{N^T - 1}.$$

We either have $-(N^T - 1) \leq h - g \leq 0$, in which case $b - c$ is periodic, or $-(N^T - 1) < h - g - 1 \leq 0$. In the latter case we just show that adding 1 to a period T sequence results in a sequence that is periodic from position T on. $\square$

Consequently, the arithmetic cross-correlation of **b** and **c** can be computed by computing the first $2T$ bits of the difference $b - c$, and finding the imbalance of the last T of these $2T$ bits. This is a linear time computation in T (although not easily parallelizable as is the case with standard cross-correlations).

5 Arithmetic Shift and Add Sequences

It is natural to consider an arithmetic analog of the shift and add property. In this section we give some basic definitions and characterize the sequences satisfying the arithmetic shift and add property.

Let $N \geq 2$ be a natural number and let $\mathbf{a} = a_0, a_1, \cdots$ be an infinite N-ary sequence. Let

$$a = \sum_{i=0}^{\infty} a_i N^i$$

be the N-adic number associated with **a**. As above, for any integer τ let $\mathbf{a}_\tau = a_\tau, a_{\tau+1}, \cdots$ be the left shift of **a** by τ positions and let $a^{(\tau)}$ be the N-adic number associated with $\mathbf{a}_\tau$. We shall sometimes refer to the left shift of an N-adic number a, meaning the N-adic number associated with the left shift of the coefficient sequence of **a**

A first attempt would be to ask that the set of shifts of **a** be closed under N-adic addition. This is impossible for eventually periodic sequences — multiples of an N-adic number by distinct positive integers are distinct, so there are infinitely many such multiples. If the set of shifts were closed under addition, all these multiples of **a** would be shifts of **a**. But there are only finitely many distinct shifts of an eventually periodic sequence.

The solution is to concern ourselves only with the periodic part of the sum of two sequences, as we have done in defining arithmetic correlations.

Definition 2. *The sequence* **a** *is said to have the* arithmetic shift and add property *if for any shift* $\tau \geq 0$, *either (1) some left shift of* $a + a^{(\tau)}$ *is zero or (2) some left shift of* $a + a^{(\tau)}$ *equals* **a**. *That is, there is a* τ' *so that* $(a + a^{(\tau)})^{(\tau')} = a$.

We may similarly define the shift and subtract property. It is helpful here to use rational representations of sequences. If $\mathbf{a}$ is periodic with minimal period T then for some q and f we have $\mathbf{a} = f/q$ with $-q \leq f \leq 0$ and $\gcd(q, f) = 1$. Then a shift $\mathbf{a}_\tau$ of $\mathbf{a}$ also corresponds to a rational number of this form, say $\mathbf{a}^{(\tau)} = f_\tau/q$. Moreover, there is an integer c_τ so that

$$\mathbf{a}^{(\tau)} = c_\tau + N^{T-\tau}\mathbf{a}.$$

Thus $f_\tau = c_\tau q + N^{T-\tau}f$. It follows that

$$f_\tau \equiv N^{T-\tau}f \mod q.$$

The set of integers

$$C_f = \{f, Nf \mod q, N^2 f \mod q, \cdots\}$$

is called the fth *cyclotomic coset modulo q relative to N* (the terms "modulo q" and "relative to N" may be omitted if q and/or N are understood). It follows that the set of numerators f_τ of the N-adic numbers associated with the cyclic shifts of $\mathbf{a}$ is the fth cyclotomic coset modulo q relative to N.

Now suppose that $(\mathbf{a} + \mathbf{a}_\tau)_{\tau'} = \mathbf{a}$. Let g/r be the rational representation of the N-adic number associated with $\mathbf{a} + \mathbf{a}_\tau$. Then

$$\frac{g}{r} = d + N^{\tau'}\frac{f}{q}$$

for some integer d. In particular, we can take $r = q$. Then $g = qd + N^{\tau'}f$. Thus $g \equiv N^{\tau'}f \mod q$, so that $-g$ is in the cyclotomic coset of f.

Theorem 7. *The periodic N-ary sequence $\mathbf{a}$ with associated N-adic number f/q with $\gcd(q, f) = 1$ has the arithmetic shift and add property if and only if $G = C_f \cup \{0\}$ is an additive subgroup of $\mathbb{Z}/(q)$.*

Proof (Sketch). The proof amounts to showing that addition in G corresponds mod q to addition of the associated numerators of fractions, and that the integer multiple of q difference does not affect the periodic part. $\qquad\square$

Corollary 1. *A sequence has the arithmetic shift and add property if and only if it has the arithmetic shift and subtract property.*

How can it be that $C_f \cup \{0\}$ is a subgroup of $\mathbb{Z}/(q)$? It implies, in particular, that $C_f \cup \{0\}$ is closed under multiplication by integers modulo q. If we take q and f so that $\gcd(f, q) = 1$, then for any integer g, $C_f \cup \{0\}$ is closed under multiplication by gf^{-1} modulo q, so that $g \in C_f \cup \{0\}$. That is, $C_f \cup \{0\} = \mathbb{Z}/(q)$, and so $C_f = \mathbb{Z}/(q) - \{0\}$. In particular, every nonzero element of $\mathbb{Z}/(q)$ is of the form $N^i f$, hence is a unit. Thus q is prime. Moreover, $1 \in C_f$, so that $(\mathbb{Z}/(q))^* = C_1 = \{N^i \mod q, i = 0, 1, \cdots, q - 2\}$. That is, N is a primitive element modulo q.

Theorem 8. *A non-zero N-ary sequence has the arithmetic shift and add property if and only if it is an ℓ-sequence.*

6 Conclusions

We have analyzed the autocorrelations of ℓ-sequences and the expected arithmetic auto- and cross-correlations for fixed shift, and we have characterized arithmetic shift and add sequences. In all these cases the answers are similar to the answers in the classical (no carry) case.

Two problems are left open here. First, we have not computed the arithmetic cross-correlations of non-binary sequences. In the binary case, this is one instance where the answer in the arithmetic realm is radically different from the answer in the classical realm.

Second, we have not computed the second moment and variance of the arithmetic autocorrelation. This appears to be a much harder problem than computing the expected arithmetic auto- and cross-correlations or the second moment of the arithmetic cross-correlation.

References

1. Blackburn, S.: A Note on Sequences with the Shift and Add Property. Designs, Codes, and Crypt. 9, 251–256 (1996)
2. Golomb, S.: Shift Register Sequences. Aegean Park Press, Laguna Hills (1982)
3. Gong, G., Di Porto, A., Wolfowicz, W.: Galois linear group sequences. La Comm., Note Rec. Not.f XLII, 83–89 (1993)
4. Goresky, M., Klapper, A.: Periodicity and Correlations of of d-FCSR Sequences. Designs, Codes, and Crypt. 33, 123–148 (2004)
5. Goresky, M., Klapper, A.: Polynomial pseudo-noise sequences based on algebraic shift register sequences. IEEE Trans. Info. Theory 53, 1649–1662 (2006)
6. Klapper, A., Goresky, M.: Arithmetic cross-correlation of FCSR sequences. IEEE Trans. Info. Theory 43, 1342–1346 (1997)
7. Klapper, A., Goresky, M.: Feedback Shift Registers, Combiners with Memory, and 2-Adic Span. J. Crypt. 10, 111–147 (1997)
8. Koblitz, N.: p-Adic Numbers, p-Adic Analysis, and Zeta Functions. Graduate Texts in Mathematics, vol. 58. Springer, Heidelberg (1984)
9. Mandelbaum, D.: Arithmetic codes with large distance. IEEE Trans. Info. Theory IT-13, 237–242 (1967)
10. Rao, T.R.N.: Error Coding For Arithmetic Processors. Academic Press, New York (1974)

A Class of Nonbinary Codes and Sequence Families

Xiangyong Zeng[1,*], Nian Li[1], and Lei Hu[2]

[1] The Faculty of Mathematics and Computer Science, Hubei University,
Xueyuan Road 11, Wuhan 430062, P.R. China
xzeng@hubu.edu.cn
[2] The State Key Laboratory of Information Security,
Graduate School of Chinese Academy of Sciences,
Beijing 10049, P.R. China
hu@is.ac.cn

Abstract. In this paper, for an even integer $n \geq 4$ and any positive integer k with $\gcd(n/2, k) = \gcd(n/2 - k, 2k) = d$ being odd, a class of p-ary codes $\mathcal{C}^k$ is defined and the weight distribution is completely determined, where p is an odd prime. A class of nonbinary sequence families is constructed from these codes, and the correlation distribution is also determined.

Keywords: Linear code, weight distribution, exponential sum, quadratic form, correlation distribution.

1 Introduction

Nonlinear functions have important applications in coding theory and cryptography [7,17]. Linear codes constructed from functions with high nonlinearity [2,6,11] can be good and have useful applications in communications [9,10,19] or cryptography [3,4,5,24].

Throughout this paper, let $\mathbb{F}_q$ be the finite field with $q = p^n$ elements for a prime p and a positive integer n, and let $\mathbb{F}_q^* = \mathbb{F}_q \setminus \{0\}$. For an even integer $n \geq 4$, let $\mathcal{C}^k$ denote the $[p^n - 1, 5n/2]$ cyclic code given by $\mathcal{C}^k = \{c(\gamma, \delta, \epsilon) = \left(\Pi_{\gamma,\delta}(x) + Tr_1^n(\epsilon x) \right)_{x \in \mathbb{F}_{p^n}^*} \mid \gamma \in \mathbb{F}_{p^{n/2}}, \delta, \epsilon \in \mathbb{F}_{p^n} \}$ constructed from the function

$$\Pi_{\gamma,\delta}(x) = Tr_1^{n/2}(\gamma x^{p^{n/2}+1}) + Tr_1^n(\delta x^{p^k+1}), \tag{1}$$

where $1 \leq k < n$ with $k \neq n/2$, and $Tr_1^l(\cdot)$ is the trace function from $\mathbb{F}_{p^l}$ to $\mathbb{F}_p$.

Several classes of binary codes $\mathcal{C}^k$ have been extensively studied for various values of the parameter k. The binary code $\mathcal{C}^{n/2 \pm 1}$ is exactly the Kasami code

---

* The work of Zeng and Li was supported in part by the National Science Foundation of China (NSFC) under Grant 60603012 and Chenguang plan of Wuhan City under Grant 200850731340. Hu's work was supported in part by the NSFC under Grant 60573053.

S.W. Golomb et al. (Eds.): SETA 2008, LNCS 5203, pp. 81–94, 2008.

in Theorem 14 of [9]. By choosing cyclicly inequivalent codewords from the Kasami code, the large set of binary Kasami sequences was obtained in [19]. The minimum distance bound of $\mathcal{C}^1$ was established by evaluating the exponential sums $\sum_{x \in \mathbb{F}_{2^n}} (-1)^{\Pi_{\gamma,\delta}(x) + Tr_1^n(\epsilon x)}$ in [12] and [16], and the weight distribution and then the minimum distance were completely determined in [22]. For even $n/2$, the binary code $\mathcal{C}^1$ has the same weight distribution as the Kasami code [9,22]. Furthermore, for any k with $\gcd(k,n) = 2$ if $n/2$ is odd or 1 if $n/2$ is even, the weight distribution of binary codes $\mathcal{C}^k$ was also determined, and these codes were used to construct families of generalized Kasami sequences, which have the same correlation distribution and family size as the large set of Kasami sequences [26].

This paper discusses the code $\mathcal{C}^k$ in the nonbinary case, namely we assume p is odd, for a wide range of k that satisfies

$$\gcd(n/2, k) = \gcd(n/2 - k, 2k) = d \text{ being odd.} \qquad (2)$$

Applying the techniques developed in [1], we describe some properties of the roots to the equation $\delta^{p^{n-k}} y^{p^{n/2-k}+1} + \gamma y + \delta = 0$ with $\gamma\delta \neq 0$. Based on these properties and the theory of quadratic form over finite fields of odd characteristic, we completely determine the weight distribution. These codes are also used to construct a class of nonbinary sequence families $\mathcal{F}^k$. The correlation function of sequences in $\mathcal{F}^k$ takes $5p + 2$ values. Let $k = n/2 - t$ for any odd integer t relatively prime to $n/2$, then $d = 1$ and the families $\mathcal{F}^k$ have the maximum magnitude $p^{\frac{n}{2}+1} + 1$. Some families of p-ary sequences of period $p^n - 1$ are listed in Table 1. The maximum magnitude of the proposed family deviates from the optimal correlation value [23], however, it has a large family size.

Table 1. Families of p-ary sequences for odd prime p

Family	n	Period	Family size	Maximum magnitude
Kumar, Scholtz, Welch [11]	even	$p^n - 1$	$p^{\frac{n}{2}}$	$p^{\frac{n}{2}} + 1$
Liu, Komo[13]	even	$p^n - 1$	$p^{\frac{n}{2}}$	$p^{\frac{n}{2}} + 1$
Moriuchi, Imamura [15]	even	$p^n - 1$	$p^{\frac{n}{2}}$	$p^{\frac{n}{2}} + 1$
Sidelnikov [18]	even or odd	$p^n - 1$	p^n	$p^{\frac{n}{2}} + 1$
Kumar, Moreno [10]	$(2m+1)k$	$p^n - 1$	$p^n + 1$	$p^{\frac{n}{2}} + 1$
Trachtenberg [20]	odd	$p^n - 1$	$p^n + 1$	$p^{\frac{n+1}{2}} + 1$
Tang, Udaya, Fan [21]	$(2m+1)k$	$p^n - 1$	$p^n + 1$	$p^{\frac{n+k}{2}} + 1$
The proposed family	even	$p^n - 1$	$p^{\frac{3n}{2}}$	$p^{\frac{n}{2}+1} + 1$

2   Preliminaries and Main Result

The field $\mathbb{F}_q$ is an n-dimensional vector space over $\mathbb{F}_p$. For any given basis $\{\alpha_1, \alpha_2, \cdots, \alpha_n\}$ of $\mathbb{F}_q$ over $\mathbb{F}_p$, each element $x \in \mathbb{F}_q$ can be uniquely represented as $x = \sum_{i=1}^{n} x_i \alpha_i$ with $x_i \in \mathbb{F}_p$. Under this representation, the field $\mathbb{F}_q$ is identical to the $\mathbb{F}_p$-vector space $\mathbb{F}_p^n$. A function $f(x)$ on $\mathbb{F}_{p^n}$ is a *quadratic form* if it can be written as a homogeneous polynomial of degree 2 on $\mathbb{F}_p^n$, namely of the form $f(x_1, \cdots, x_n) = \sum\limits_{1 \leq i \leq j \leq n} a_{ij} x_i x_j, \; a_{ij} \in \mathbb{F}_p$. The *rank* of $f(x)$ is defined as the codimension of the $\mathbb{F}_p$-vector space $V_f = \{z \in \mathbb{F}_{p^n} \mid f(x + z) = f(x) \text{ for all } x \in \mathbb{F}_{p^n}\}$, denoted by $\text{rank}(f)$. Then $|V_f| = p^{n-\text{rank}(f)}$.

For the quadratic form $f(x)$, there exists a symmetric matrix A such that $f(x) = X^T A X$, where $X^T = (x_1, x_2, \cdots, x_n) \in \mathbb{F}_p^n$ denotes the transpose of a column vector X. The *determinant* $\det(f)$ of $f(x)$ is defined to be the determinant of A, and $f(x)$ is *nondegenerate* if $\det(f) \neq 0$. By Theorem 6.21 of [14], there exists a nonsingular matrix B such that $B^T A B$ is a diagonal matrix. Making a nonsingular substitution $X = BY$ with $Y^T = (y_1, y_2, \cdots, y_n)$, one has

$$f(x) = Y^T B^T A B Y = \sum_{i=1}^{n} a_i y_i^2, \; a_i \in \mathbb{F}_p. \tag{3}$$

The *quadratic character* of $\mathbb{F}_q$ is defined by $\eta(x) = 1$ if x is a square element in $\mathbb{F}_q^*$, -1 if x is a non-square element in $\mathbb{F}_q^*$, and 0 if $x = 0$.

Lemma 1 (Theorems 6.26 and 6.27 of [14]). *For odd q, let f be a nondegenerate quadratic form over $\mathbb{F}_q$ in l indeterminates, and a function $v(x)$ over $\mathbb{F}_q$ is defined by $v(0) = q-1$ and $v(x) = -1$ for $x \in \mathbb{F}_q^*$. Then for $\rho \in \mathbb{F}_q$ the number of solutions to the equation $f(x_1, \cdots, x_l) = \rho$ is $q^{l-1} + q^{\frac{l-1}{2}} \eta \left((-1)^{\frac{l-1}{2}} \rho \cdot \det(f) \right)$ for odd l, and $q^{l-1} + v(\rho) q^{\frac{l-2}{2}} \eta \left((-1)^{\frac{l}{2}} \det(f) \right)$ for even l.*

Lemma 2 (Theorem 5.15 of [14]). *Let ω be a complex primitive p-th root of unity. Then $\sum\limits_{k=1}^{p-1} \eta(k) \omega^k = \sqrt{(-1)^{\frac{p-1}{2}} p}$.*

A *p-ary $[m, l]$ linear code* C is a linear subspace of $\mathbb{F}_p^m$ with dimension l. The *Hamming weight* of a codeword $c_1 c_2 \cdots c_m$ of C is the number of nonzero c_i for $1 \leq i \leq m$.

Let $\mathcal{F} = \left\{ \{s_i(t)\}_{t=0}^{p^n-2} \mid 0 \leq i \leq M - 1 \right\}$ be a family of M p-ary sequences of period $p^n - 1$. The *periodic correlation function* of the sequences $\{s_i(t)\}$ and $\{s_j(t)\}$ in $\mathcal{F}$ is $C_{i,j}(\tau) = \sum\limits_{t=0}^{p^n-2} \omega^{s_i(t)-s_j(t+\tau)}, \; 0 \leq \tau \leq p^n - 2$, and $\{s_i(t)\}$ and $\{s_j(t)\}$ are *cyclicly inequivalent* if $|C_{i,j}(\tau)| < p^n - 1$ for any τ. The *maximum magnitude* of the correlation values is $C_{\max} = \max\{|C_{i,j}(\tau)| : i \neq j \text{ or } \tau \neq 0\}$.

From now on, we always assume that the prime p is odd, and $n = 2m \geq 4$.

Without loss of generality, the integer k in the definition of code C^k can be assumed to satisfy $1 \leq k < n/2$. For an odd integer t relatively prime to m, the integer $k = m - t$ satisfies Equation (2), and $d = 1$. In particular, for $t = 1$, $\gcd(m, k) = \gcd(m - k, 2k) = 1$. The parameter $k = m - 1$ corresponds to the binary Kasami code [9], and for this reason, we call these p-ary $[p^n - 1, 5m]$ linear codes C^k with k satisfying Equation (2) *the nonbinary Kasami codes*.

The main result of this paper is stated as the following theorem.

Theorem 3. *For $n = 2m \geq 4$ and any positive integer k satisfying Equation (2), the weight distribution of the nonbinary Kasami codes C^k is given as follows:*

Table 2. The weight distribution of code C^k

Weight	Frequency
0	1
$(p-1)p^{n-1}$	$(p^n - 1)(1 + p^{n-2d}(p^{m+d} - p^m + p^{m-1} + p^{m-d} - 1))$
$(p-1)(p^{n-1} - p^{\frac{n-2}{2}})$	$p^d(p^m + 1)(p^n - 1)(p^{n-1} + (p-1)p^{\frac{n-2}{2}})/(2p^d + 2)$
$(p-1)(p^{n-1} + p^{\frac{n-2}{2}})$	$\frac{p^{n+d} - 2p^n + p^d}{2(p^d - 1)}(p^m - 1)(p^{n-1} - (p-1)p^{\frac{n-2}{2}})$
$(p-1)p^{n-1} + p^{\frac{n-2}{2}}$	$p^d(p^m + 1)(p^n - 1)(p-1)(p^{n-1} - p^{\frac{n-2}{2}})/(2p^d + 2)$
$(p-1)p^{n-1} - p^{\frac{n-2}{2}}$	$\frac{p^{n+d} - 2p^n + p^d}{2(p^d - 1)}(p^m - 1)(p-1)(p^{n-1} + p^{\frac{n-2}{2}})$
$(p-1)p^{n-1} - p^{\frac{n+d-1}{2}}$	$p^{m-d}(p^n - 1)(p-1)(p^{n-d-1} + p^{\frac{n-d-1}{2}})/2$
$(p-1)p^{n-1} + p^{\frac{n+d-1}{2}}$	$p^{m-d}(p^n - 1)(p-1)(p^{n-d-1} - p^{\frac{n-d-1}{2}})/2$
$(p-1)(p^{n-1} + p^{\frac{n+2d-2}{2}})$	$\frac{(p^{m-d} - 1)(p^n - 1)}{p^{2d} - 1}(p^{n-2d-1} - (p-1)p^{\frac{n-2d-2}{2}})$
$(p-1)p^{n-1} - p^{\frac{n+2d-2}{2}}$	$\frac{(p^{m-d} - 1)(p^n - 1)}{p^{2d} - 1}(p-1)(p^{n-2d-1} + p^{\frac{n-2d-2}{2}})$

It will be proven by the techniques developed in the next two sections.

3   Rank Distribution of Quadratic Form $\Pi_{\gamma,\delta}(x)$

This section investigates the rank distribution of the quadratic form $\Pi_{\gamma,\delta}(x)$ defined by (1) for either $\gamma \neq 0$ or $\delta \neq 0$.

Lemma 4 (Theorems 5.4 and 5.6 of [1]). *Let $h_c(x) = x^{p^s + 1} - cx + c$, $c \in \mathbb{F}_{p^l}^*$. Then $h_c(x) = 0$ has either 0, 1, 2, or $p^{\gcd(s,l)} + 1$ roots in $\mathbb{F}_{p^l}^*$. Further, let N_1 denote the number of $c \in \mathbb{F}_{p^l}^*$ such that $h_c(x) = 0$ has exactly one solution in $\mathbb{F}_{p^l}^*$, then $N_1 = p^{l - \gcd(s,l)}$ and if $x_0 \in \mathbb{F}_{p^l}^*$ is the unique solution of the equation, then $(x_0 - 1)^{\frac{p^l - 1}{p^{\gcd(s,l)} - 1}} = 1$.*

Proposition 5. *Let $g_{\delta,\gamma}(y) = \delta^{p^{n-k}} y^{p^{m-k} + 1} + \gamma y + \delta$ with $\gamma\delta \neq 0$, and d be defined as in (2). Then*

(1) The equation $g_{\delta,\gamma}(y) = 0$ has either 0, 1, 2, or $p^d + 1$ roots in $\mathbb{F}_{p^n}$;

(2) If $y_1, y_2 \in \mathbb{F}_{p^n}$ are different solutions of $g_{\delta,\gamma}(y) = 0$, then $(y_1 y_2)^{\frac{p^n - 1}{p^d - 1}} = 1$;

(3) If $g_{\delta,\gamma}(y) = 0$ has exactly one solution $y_0 \in \mathbb{F}_{p^n}$, then $y_0^{\frac{p^n - 1}{p^d - 1}} = 1$.

Proof. (1) Let $y = -\frac{\delta}{\gamma}x$ and $c = \frac{\gamma^{p^{m-k} + 1}}{\delta^{p^{m-k}}(p^m + 1)}$. Then $g_{\delta,\gamma}(y) = 0$ becomes

$$x^{p^{m-k} + 1} - cx + c = 0. \tag{4}$$

Since $c \in \mathbb{F}_{p^m}^* \subseteq \mathbb{F}_{p^n}^*$ and $\gcd(m - k, n) = d$ by Equation (2), by Lemma 4, Equation (4) has either 0, 1, 2, or $p^d + 1$ roots in $\mathbb{F}_{p^n}$. Thus, so does $g_{\delta,\gamma}(y) = 0$.

(2) Let y_1, y_2 be two different solutions of $g_{\delta,\gamma}(y) = 0$ in $\mathbb{F}_{p^n}$. Then $y_1 y_2 (y_1 - y_2)^{p^{m-k}} = (-\frac{(\gamma y_1 + \delta) y_2}{\delta p^{n-k}}) - (-\frac{(\gamma y_2 + \delta) y_1}{\delta p^{n-k}}) = \frac{\delta (y_1 - y_2)}{\delta p^{n-k}}$, i.e., $y_1 y_2 = \delta^{1-p^{n-k}} (y_1 - y_2)^{1-p^{m-k}}$. This together with Equation (2) imply $(y_1 y_2)^{\frac{p^n-1}{p^d-1}} = 1$.

(3) Suppose that y_0 is the unique solution of $g_{\delta,\gamma}(y) = 0$. Thus $x_0 = -\frac{\gamma y_0}{\delta}$ is the unique solution of Equation (4) in $\mathbb{F}_{p^n}$. Since $c \in \mathbb{F}_{p^m}$, $x_0^{p^m}$ is also a solution of Equation (4). Then $x_0 = x_0^{p^m}$, i.e., $x_0 \in \mathbb{F}_{p^m}$. By Lemma 4, one has $1 = (x_0 - 1)^{\frac{p^m-1}{p^d-1}} = (-\frac{\gamma y_0}{\delta} - 1)^{\frac{p^m-1}{p^d-1}} = (\delta^{p^{n-k}-1} y_0^{p^{m-k}+1})^{\frac{p^m-1}{p^d-1}}$. Similarly, if y is a solution of $g_{\delta,\gamma}(y) = 0$, one can verify that $y^{p^m} \delta^{1-p^m}$ is also a solution of $g_{\delta,\gamma}(y) = 0$. Thus, $y_0 = y_0^{p^m} \delta^{1-p^m}$, i.e., $y_0^{p^m-1} = \delta^{p^m-1}$. By Equation (2), we have $1 = \delta^{\frac{(p^{n-k}-1)(p^m-1)}{p^d-1}} y_0^{\frac{(p^{m-k}+1)(p^m-1)}{p^d-1}} = y_0^{\frac{p^{m-k}}{p^d-1}(p^n-1)}$. This leads to $y_0^{\frac{p^n-1}{p^d-1}} = (y_0^{\frac{p^{m-k}}{p^d-1}(p^n-1)})^{p^{m+k}} = 1$. $\square$

Remark 6. (1) For given γ and δ with $\gamma \delta \neq 0$, by Proposition 5(3), if $g_{\delta,\gamma}(y) = 0$ has exactly one solution in $\mathbb{F}_{p^n}$, then the unique solution is $(p^d - 1)$-th power in $\mathbb{F}_{p^n}$.

(2) By Proposition 5(2), if $g_{\delta,\gamma}(y) = 0$ has at least two different solutions in $\mathbb{F}_{p^n}$, then either all these solutions or none of them are $(p^d - 1)$-th powers in $\mathbb{F}_{p^n}$. Over the finite field of characteristic 2, an analogy of Proposition 5(2) was obtained in Proposition 2 of [8]. But the analogy of Proposition 5(3) does not exist in [8].

With Proposition 5, $\mathrm{rank}(\Pi_{\gamma,\delta})$ can be determined as follows.

Proposition 7. *For $\gamma \neq 0$ or $\delta \neq 0$, the rank of $\Pi_{\gamma,\delta}(x)$ is n, $n - d$, or $n - 2d$.*

Proof. The integer $p^{n-\mathrm{rank}(\Pi_{\gamma,\delta})}$ is equal to the number of $z \in \mathbb{F}_{p^n}$ such that $\Pi_{\gamma,\delta}(x + z) = \Pi_{\gamma,\delta}(x)$ holds for all $x \in \mathbb{F}_{p^n}$. This equation holds if and only if

$$\gamma z^{p^m} + \delta z^{p^k} + (\delta z)^{p^{n-k}} = 0 \tag{5}$$

since Equation (5) implies $Tr_1^m(\gamma z^{p^m+1}) + Tr_1^n(\delta z^{p^k+1}) = 0$.

When $\delta = 0$, one has $\gamma \neq 0$ and then Equation (5) has only zero solution. In the sequel, we only consider the case $\delta \neq 0$. If $\gamma = 0$, Equation (5) is equivalent to $z(z^{p^{2k}-1} + \delta^{1-p^k}) = 0$. The number of all solutions to this equation is $p^{\gcd(2k,n)} = p^{2d}$ or 1 depending on whether $-\delta^{1-p^k}$ is a $(p^{2d} - 1)$-th power in $\mathbb{F}_{p^n}$ or not.

For $\gamma \neq 0$, $\gamma z^{p^m} + \delta z^{p^k} + (\delta z)^{p^{n-k}} = z^{p^k}(\delta + \gamma z^{p^m-p^k} + \delta^{p^{n-k}} z^{p^{n-k}-p^k}) = 0$. Thus, we only need consider the number of nonzero solutions to the equation

$$\delta + \gamma z^{p^m-p^k} + \delta^{p^{n-k}} z^{p^{n-k}-p^k} = 0, \tag{6}$$

which becomes $g_{\delta,\gamma}(y) = \delta^{p^{n-k}} y^{p^{m-k}+1} + \gamma y + \delta = 0$ if let $y = z^{p^k(p^{m-k}-1)}$. By Proposition 5(1), $g_{\delta,\gamma}(y) = 0$ has either 0, 1, 2, or $p^d + 1$ roots in $\mathbb{F}_{p^n}$. Remark 6 and the fact $\gcd(p^k(p^{m-k}-1), p^n-1) = p^d-1$ show that Equation (6) has 0, p^d-1, $2(p^d-1)$, or $(p^d-1)(p^d+1) = p^{2d}-1$ nonzero solutions in $\mathbb{F}_{p^n}$. Then, Equation (5) has 1, p^d, $2p^d-1$, or p^{2d} solutions. Since $2p^d-1$ is not a power of p and hence is not the number of solutions of an $\mathbb{F}_p$-linearization polynomial, the number of solutions to Equation (5) is equal to 1, p^d, or p^{2d}.

Therefore, the rank of $\Pi_{\gamma,\delta}(x)$ is either n, $n-d$, or $n-2d$. $\square$

In order to further determine the rank distribution of $\Pi_{\gamma,\delta}(x)$, we define

$$R_i = \Big\{ (\gamma,\delta) \mid \mathrm{rank}(\Pi_{\gamma,\delta}) = n-i, (\gamma,\delta) \in \mathbb{F}_{p^m} \times \mathbb{F}_{p^n} \setminus \{(0,0)\} \Big\} \qquad (7)$$

for $i = 0, d$, and $2d$. To achieve this goal, we need to evaluate the sum

$$S(\gamma,\delta,\epsilon) = \sum_{x \in \mathbb{F}_{p^n}} \omega^{\Pi_{\gamma,\delta}(x)+Tr_1^n(\epsilon x)}, \qquad (8)$$

where $\gamma \in \mathbb{F}_{p^m}, \delta, \epsilon \in \mathbb{F}_{p^n}$. For $\rho \in \mathbb{F}_p$, let $N_{\gamma,\delta,\epsilon}(\rho)$ denote the number of solutions to $\Pi_{\gamma,\delta}(x) + Tr_1^n(\epsilon x) = \rho$. Then, (8) can be expressed as

$$S(\gamma,\delta,\epsilon) = \sum_{\rho=0}^{p-1} N_{\gamma,\delta,\epsilon}(\rho)\omega^\rho. \qquad (9)$$

Let $f(x) = \Pi_{\gamma,\delta}(x)$ be as in (1). For convenience, for $i \in \{0, d, 2d\}$, we define $\Delta_i = (-1)^{\lfloor \frac{n-i}{2} \rfloor} \prod_{j=1}^{n-i} a_j$, where $\lfloor \frac{n-i}{2} \rfloor$ denotes the largest integer not exceeding $\frac{n-i}{2}$, and the coefficients a_j are defined by (3).

In what follows, we will study the values of $S(\gamma,\delta,0)$ according to the rank of $\Pi_{\gamma,\delta}(x)$, and then use them to determine the rank distribution.

Case 1. $(\gamma,\delta) \in R_0$: in this case, $\mathrm{rank}(\Pi_{\gamma,\delta}) = n$ and every a_i in (3) is nonzero. Since $\det(\Pi_{\gamma,\delta})(\det(B))^2 = \prod_{i=1}^{n} a_i$, one has $\eta(\det(\Pi_{\gamma,\delta})) = \eta(\prod_{i=1}^{n} a_i)$. Then by Lemma 1, $N_{\gamma,\delta,0}(\rho) = p^{n-1} + v(\rho)p^{\frac{n-2}{2}}\eta(\Delta_0)$ and then by (9), we have $S(\gamma,\delta,0) = \eta(\Delta_0)p^{\frac{n}{2}}$ since $\sum_{\rho=0}^{p-1} v(\rho)\omega^\rho = p$.

Case 2. $(\gamma,\delta) \in R_d$: since $\mathrm{rank}(\Pi_{\gamma,\delta}) = n-d$, without loss of generality, we assume $\prod_{i=1}^{n-d} a_i \neq 0$ and $a_i = 0$ for $n-d < i \leq n$. Then, $\Pi_{\gamma,\delta}(x) = \sum_{i=1}^{n-d} a_i y_i^2$, and by Lemma 1, for odd d, one has $N_{\gamma,\delta,0}(\rho) = p^d \left(p^{n-d-1} + p^{\frac{n-d-1}{2}}\eta(\rho\Delta_d) \right) = p^{n-1} + p^{\frac{n+d-1}{2}}\eta(\rho\Delta_d)$. By (9) and Lemma 2, $S(\gamma,\delta,0) = \eta(\Delta_d)\sqrt{(-1)^{\frac{p-1}{2}}}p^{\frac{n+d}{2}}$.

Case 3. $(\gamma,\delta) \in R_{2d}$: similarly, let $\prod_{i=1}^{n-2d} a_i \neq 0$ and $a_i = 0$ for $n-2d < i \leq n$. Then, $N_{\gamma,\delta,0}(\rho) = p^{n-1} + v(\rho)p^{\frac{n+2d-2}{2}}\eta(\Delta_{2d})$ and $S(\gamma,\delta,0) = \eta(\Delta_{2d})p^{\frac{n}{2}+d}$.

For each $i \in \{0, d, 2d\}$, we define $R_{i,j} = \{(\gamma,\delta) \in R_i \mid \eta(\Delta_i) = j\}$ for $j = \pm 1$. Since d is odd, we have

Lemma 8. $|R_{d,1}| = |R_{d,-1}|$.

Proof. Let $(\gamma, \delta) \in R_d$ and let $u \in \mathbb{F}_p^*$ such that its inverse element satisfies $\eta(u^{-1}) = -1$. By (8) and the analysis in Case 2, one has $S(u\gamma, u\delta, 0) =$

$$\sum_{x \in \mathbb{F}_{p^n}} \omega^{u\Pi_{\gamma,\delta}(x)} = \sum_{\rho \in \mathbb{F}_p} N_{\gamma,\delta,0}(u^{-1}\rho)\omega^\rho = p^{\frac{n+d-1}{2}} \sum_{\rho=0}^{p-1} \eta(u^{-1}\rho\Delta_d)\omega^\rho = -S(\gamma, \delta, 0).$$

The above equality shows that for $j \in \{1, -1\}$, if $(\gamma, \delta) \in R_{d,j}$, then $(u\gamma, u\delta) \in R_{d,-j}$. Thus, one has $|R_{d,1}| = |R_{d,-1}|$. □

Applying Proposition 5 and Lemma 8, we have

Proposition 9. $|R_d| = p^{m-d}(p^n - 1)$ and $|R_{d,\pm 1}| = p^{m-d}(p^n - 1)/2$.

Proof. By (7), it is sufficient to determine the number of $(\gamma, \delta) \in \mathbb{F}_{p^m} \times \mathbb{F}_{p^n} \setminus \{(0,0)\}$ such that Equation (5) has exactly p^d solutions in $\mathbb{F}_{p^n}$. By the proof of Proposition 7, this case can occur only if $\gamma\delta \neq 0$.

Let $W = \{x^{p^d-1} \mid x \in \mathbb{F}_{p^n}^*\}$ be the set of nonzero $(p^d - 1)$-th powers in $\mathbb{F}_{p^n}$. By Proposition 5(1), $g_{\delta,\gamma}(y) = 0$ has either 0, 1, 2, or $p^d + 1$ roots in $\mathbb{F}_{p^n}$.

When $g_{\delta,\gamma}(y) = 0$ has at least two different roots in $\mathbb{F}_{p^n}^*$, by Proposition 5(2) and Remark 6, if one of the solutions belongs to W, then all these solutions are also in W and Equation (5) has at least $2(p^d - 1) + 1 = 2p^d - 1$ solutions. If none of these solutions belong to W, then Equation (5) has only zero solution. Thus, in this case the number of solutions to Equation (5) can never be p^d.

If $g_{\delta,\gamma}(y) = 0$ has exactly one solution in $\mathbb{F}_{p^n}$, by Proposition 5(3) and Remark 6, Equation (5) has $1 \times (p^d - 1) + 1 = p^d$ solutions. Since $y = -\frac{\delta}{\gamma}x$, $g_{\delta,\gamma}(y) = 0$ has exactly one solution in $\mathbb{F}_{p^n}^*$ if and only if Equation (4) has exactly one solution in $\mathbb{F}_{p^n}^*$. Furthermore, in this case the unique solution to Equation (4) belongs to $\mathbb{F}_{p^m}^*$ by the analysis in the proof of Proposition 5(3). When Equation (4) has exactly one solution in $\mathbb{F}_{p^m}^*$, it has exactly one solution in $\mathbb{F}_{p^n}^*$ since this equation has either 0, 1, 2, or $p^d + 1$ roots in $\mathbb{F}_{p^n}^*$ and its solutions in $\mathbb{F}_{p^n} \setminus \mathbb{F}_{p^m}$ occur in pairs. Therefore, $g_{\delta,\gamma}(y) = 0$ has exactly one solution in $\mathbb{F}_{p^n}^*$ if and only if Equation (4) has exactly one solution in $\mathbb{F}_{p^m}^*$.

By Lemma 4, the number of $c \in \mathbb{F}_{p^m}^*$ such that Equation (4) has exactly one solution in $\mathbb{F}_{p^m}$ is p^{m-d}. For any fixed $\gamma \in \mathbb{F}_{p^m}^*$, $c = \frac{\gamma^{p^{m-k}+1}}{\delta^{p^{m-k}(p^m+1)}}$ runs through $\mathbb{F}_{p^m}^*$ exactly (p^m+1) times when δ runs through $\mathbb{F}_{p^n}^*$. Therefore, when γ and δ run throughout $\mathbb{F}_{p^m}^*$ and $\mathbb{F}_{p^n}^*$, respectively, there are exactly $(p^m - 1)p^{m-d}(p^m + 1) = p^{m-d}(p^n - 1)$ elements in R_d. This together with Lemma 8 finish the proof. □

Proposition 10 describes the sums of i-th powers of $S(\gamma, \delta, 0)$ for $1 \leq i \leq 3$.

Proposition 10. (i) $\displaystyle\sum_{\gamma \in \mathbb{F}_{p^m}} \sum_{\delta \in \mathbb{F}_{p^n}} S(\gamma, \delta, 0) = p^{n+m}$.

(ii) $\displaystyle\sum_{\gamma \in \mathbb{F}_{p^m}} \sum_{\delta \in \mathbb{F}_{p^n}} S(\gamma, \delta, 0)^2 = \begin{cases} p^{n+m}, & p \equiv 3 \bmod 4, \\ (2p^n - 1)p^{n+m}, & p \equiv 1 \bmod 4. \end{cases}$

(iii) $\displaystyle\sum_{\gamma \in \mathbb{F}_{p^m}} \sum_{\delta \in \mathbb{F}_{p^n}} S(\gamma, \delta, 0)^3 = p^{n+m}(p^{n+d} + p^n - p^d)$.

Proof. The proof of (i) is trivial, and we only give the proof of (ii) and (iii).

(ii) It is true that

$$\sum_{\gamma\in\mathbb{F}_{p^m},\delta\in\mathbb{F}_{p^n}} S(\gamma,\delta,0)^2$$

$$= \sum_{x,y\in\mathbb{F}_{p^n}} \sum_{\gamma\in\mathbb{F}_{p^m}} \omega^{Tr_1^m\left(\gamma(x^{p^m+1}+y^{p^m+1})\right)} \sum_{\delta\in\mathbb{F}_{p^n}} \omega^{Tr_1^n\left(\delta(x^{p^k+1}+y^{p^k+1})\right)}$$

$$= p^{n+m}|T_2|$$

where T_2 consists of all solutions $(x,y) \in \mathbb{F}_{p^n} \times \mathbb{F}_{p^n}$ to the system of equations

$$\begin{cases} x^{p^m+1} + y^{p^m+1} = 0, \\ x^{p^k+1} + y^{p^k+1} = 0. \end{cases} \tag{10}$$

If $xy = 0$, then $x = y = 0$ by (10). If $xy \neq 0$, again by (10), one has $(\frac{x}{y})^{p^k(p^{m-k}-1)} = 1$ which implies $(\frac{x}{y})^{p^k(p^{m-k}-1)} = \left((\frac{x}{y})^{p^k(p^{m-k}-1)}\right)^{p^{n-k}} = 1$, then $\frac{x}{y} \in \mathbb{F}_{p^n}^* \cap \mathbb{F}_{p^{m-k}}^* = \mathbb{F}_{p^d}^*$ since $\gcd(m-k,n) = d$ by (2). Let $x = ty$ for some $t \in \mathbb{F}_{p^d}^*$, then Equation (10) becomes $t^2 + 1 = 0$ since $\gcd(m,k) = d$ and $xy \neq 0$.

For $p \equiv 3 \bmod 4$, -1 is a non-square element in $\mathbb{F}_{p^d}$ since d is odd. Thus, $t^2 + 1 = 0$ has no solutions and $|T_2| = 1$. For $p \equiv 1 \bmod 4$, -1 is a square element in $\mathbb{F}_{p^d}$ and $t^2 + 1 = 0$ has 2 solutions in $\mathbb{F}_{p^d}^*$. Then, $|T_2| = 1 + 2(p^n - 1) = 2p^n - 1$.

(iii) Similar analysis as in (ii) shows $\sum_{\gamma\in\mathbb{F}_{p^m},\delta\in\mathbb{F}_{p^n}} S(\gamma,\delta,0)^3 = p^{n+m}|T_3|$, where T_3 consists of all solutions $(x,y,z) \in \mathbb{F}_{p^n} \times \mathbb{F}_{p^n} \times \mathbb{F}_{p^n}$ to the system of equations

$$\begin{cases} x^{p^m+1} + y^{p^m+1} + z^{p^m+1} = 0, \\ x^{p^k+1} + y^{p^k+1} + z^{p^k+1} = 0. \end{cases} \tag{11}$$

For $xyz = 0$, then $x = y = z = 0$, or there are exactly two nonzero elements in $\{x,y,z\}$. Thus, by (10), in this case the number of solutions to Equation (11) is equal to $3(|T_2| - 1) + 1 = 3|T_2| - 2$. For $xyz \neq 0$, the number of solutions to Equation (11) is $(p^n - 1)$ multiples of that to

$$\begin{cases} x^{p^m+1} + y^{p^m+1} + 1 = 0, \\ x^{p^k+1} + y^{p^k+1} + 1 = 0 \end{cases} \tag{12}$$

where $x,y \in \mathbb{F}_{p^n}^*$. Equation (12) implies $y^{(p^k+1)p^m} + y^{p^k+1} - y^{(p^m+1)p^k} - y^{p^m+1} = (y^{p^{m+k}} - y)(y^{p^m} - y^{p^k}) = 0$. By (2), one has $\gcd(m+k,n) = d$. Then $y \in \mathbb{F}_{p^{m+k}} \cap \mathbb{F}_{p^n} = \mathbb{F}_{p^d}$, or $y \in \mathbb{F}_{p^m} \cap \mathbb{F}_{p^k} = \mathbb{F}_{p^d}$, i.e., $y \in \mathbb{F}_{p^d}$. Similarly, one has $x \in \mathbb{F}_{p^d}$. By the fact $\gcd(m,k) = d$, Equation (12) is equivalent to

$$x^2 + y^2 + 1 = 0, \ x, y \in \mathbb{F}_{p^d}^*. \tag{13}$$

By Lemma 1, the number of solutions $(x,y) \in \mathbb{F}_{p^d} \times \mathbb{F}_{p^d}$ to $x^2 + y^2 + 1 = 0$ is $p^d + v(-1)\eta(-1)$. Notice that the number of solutions to $xy = 0$ and $x^2 + y^2 + 1 = 0$ in $\mathbb{F}_{p^d}$ is 0 for $p \equiv 3 \bmod 4$, and 4 for $p \equiv 1 \bmod 4$. Then, the number of solutions to Equation (13) is equal to $p^d + 1$ for $p \equiv 3 \bmod 4$, and $(p^d - 1) - 4$ for $p \equiv 1 \bmod 4$. Therefore, one has $|T_3| = p^{n+d} + p^n - p^d$. $\qquad\square$

With the above preparations, the rank distribution of $\Pi_{\delta,\gamma}(x)$ can be determined as follows. Since $S(0,0,0) = p^n$, by Proposition 10 and the values of $S(\gamma, \delta, 0)$ corresponding to the rank of $\Pi_{\gamma,\delta}(x)$, one has

$$p^{\frac{n}{2}}(|R_{0,1}| - |R_{0,-1}|) + p^{\frac{n}{2}+d}(|R_{2d,1}| - |R_{2d,-1}|) + p^n = \sum_{\gamma \in \mathbb{F}_{p^m}} \sum_{\delta \in \mathbb{F}_{p^n}} S(\gamma, \delta, 0),$$

$$p^n|R_0| + (-1)^{\frac{p-1}{2}} p^{n+d}|R_d| + p^{n+2d}|R_{2d}| + p^{2n} = \sum_{\gamma \in \mathbb{F}_{p^m}} \sum_{\delta \in \mathbb{F}_{p^n}} S(\gamma, \delta, 0)^2,$$

$$p^{\frac{3n}{2}}(|R_{0,1}| - |R_{0,-1}|) + p^{\frac{3n}{2}+3d}(|R_{2d,1}| - |R_{2d,-1}|) + p^{3n} = \sum_{\gamma \in \mathbb{F}_{p^m}} \sum_{\delta \in \mathbb{F}_{p^n}} S(\gamma, \delta, 0)^3.$$

This together with the equalities $|R_{0,1}| + |R_{0,-1}| + |R_d| + |R_{2d,1}| + |R_{2d,-1}| = p^{n+m} - 1$, $|R_i| = |R_{i,1}| + |R_{i,-1}|(i = 0, 2d)$ as well as Proposition 9 give

$$|R_{0,1}| = \frac{p^d(p^m+1)(p^n-1)}{2(p^d+1)}, \; |R_{0,-1}| = \frac{(p^{n+d} - 2p^n + p^d)(p^m-1)}{2(p^d-1)},$$

$$|R_{2d,1}| = 0, \qquad\qquad |R_{2d,-1}| = \frac{(p^{m-d}-1)(p^n-1)}{p^{2d}-1}. \tag{14}$$

Therefore, we have the following result.

Proposition 11. *When (γ, δ) runs through $\mathbb{F}_{p^m} \times \mathbb{F}_{p^n} \setminus \{(0,0)\}$, the rank distribution of the quadratic form $\Pi_{\gamma,\delta}(x)$ is given as follows:*

$$\begin{cases} n, & (p^{m+n+2d} + p^n + p^{m+d} - p^{m+n} - p^{m+n+d} - p^{2d})/(p^{2d} - 1) \text{ times,} \\ n - d, & p^{m-d}(p^n - 1) \text{ times,} \\ n - 2d, & (p^{m-d} - 1)(p^n - 1)/(p^{2d} - 1) \text{ times.} \end{cases}$$

4   Weight Distribution of the Nonbinary Kasami Codes

This section determines the weight distribution of the nonbinary Kasami codes C^k. Further, we also give the distribution of $S(\gamma, \delta, \epsilon)$, which will be used to derive the correlation distribution of the sequence families proposed in next section.

Since the weight of the codeword $c(\gamma, \delta, \epsilon)$ is equal to $p^n - 1 - (N_{\gamma,\delta,\epsilon}(0) - 1) = p^n - N_{\gamma,\delta,\epsilon}(0)$, it is sufficient to find $N_{\gamma,\delta,\epsilon}(0)$ for any given γ, δ, ϵ.

Under the basis $\{\alpha_1, \alpha_2, \cdots, \alpha_n\}$ of $\mathbb{F}_{p^n}$ over $\mathbb{F}_p$, let $\epsilon = \sum_{i=1}^{n} \epsilon_i \alpha_i$ with $\epsilon_i \in \mathbb{F}_p$.

Then, $Tr_1^n(\epsilon x) = \Lambda^T C X$ where $\Lambda^T = (\epsilon_1, \epsilon_2, \cdots, \epsilon_n) \in \mathbb{F}_p^n$ and the matrix $C = (Tr_1^n(\alpha_i \alpha_j))_{1 \le i,j \le n}$, which is nonsingular since $\{\alpha_1, \alpha_2, \cdots, \alpha_n\}$ is a basis of $\mathbb{F}_{p^n}$ over $\mathbb{F}_p$. Then let $X = BY$ as in Section 2 and $\Lambda^T CB = (b_1, b_2, \cdots, b_n)$, one has

$$\Pi_{\gamma,\delta}(x) + Tr_1^n(\epsilon x) = Y^T B^T ABY + \Lambda^T CBY = \sum_{i=1}^{n} a_i y_i^2 + \sum_{i=1}^{n} b_i y_i.$$

We calculate $N_{\gamma,\delta,\epsilon}(\rho)$ $(\rho \in \mathbb{F}_p)$ and $S(\gamma, \delta, \epsilon)$ as follows:

Case 1. $(\gamma, \delta) = (0, 0)$: the weight of $c(\gamma, \delta, \epsilon)$ is $(p - 1)p^{n-1}$ for $\epsilon \neq 0$, and 0 for $\epsilon = 0$. Then $S(0, 0, \epsilon) = 0$ for $\epsilon \neq 0$, and p^n for $\epsilon = 0$.

Case 2. $(\gamma, \delta) \neq (0, 0)$:

Case 2.1. $(\gamma, \delta) \in R_0$: a substitution $y_i = z_i - \frac{b_i}{2a_i}$ for $1 \leq i \leq n$ leads to

$$\sum_{i=1}^{n}(a_i y_i^2 + b_i y_i) = \rho \iff \sum_{i=1}^{n} a_i z_i^2 = \lambda_{\gamma,\delta,\epsilon} + \rho, \text{ where } \lambda_{\gamma,\delta,\epsilon} = \sum_{i=1}^{n} \frac{b_i^2}{4a_i}. \text{ Then, for}$$

any $\rho \in \mathbb{F}_p$ and given $(\gamma, \delta) \in R_0$, by Lemma 1, one has

$$N_{\gamma,\delta,\epsilon}(\rho) = p^{n-1} + v(\lambda_{\gamma,\delta,\epsilon} + \rho)p^{\frac{n-2}{2}}\eta(\Delta_0). \tag{15}$$

When ϵ runs through $\mathbb{F}_{p^n}$, $(b_1, b_2, \cdots, b_n)$ runs through $\mathbb{F}_p^n$ since CB is non-singular. Notice that $\lambda_{\gamma,\delta,\epsilon}$ is a quadratic form with n variables b_i for $1 \leq i \leq n$. Then, for any given $(\gamma, \delta) \in R_0$, by Lemma 1, when ϵ runs through $\mathbb{F}_{p^n}$, one has

$$\lambda_{\gamma,\delta,\epsilon} = \sum_{i=1}^{n} \frac{b_i^2}{4a_i} = \rho' \text{ occurring } p^{n-1} + v(\rho')p^{\frac{n-2}{2}}\eta(\Delta_0) \text{ times} \tag{16}$$

for each $\rho' \in \mathbb{F}_p$ since $\eta(\frac{1}{4^n \prod\limits_{i=1}^{n} a_i}) = \eta(\prod\limits_{i=1}^{n} a_i)$. Thus, when ϵ runs through

$\mathbb{F}_{p^n}$, by (15) and (16), $N_{\gamma,\delta,\epsilon}(0) = p^{n-1} + (p-1)p^{\frac{n-2}{2}}\eta(\Delta_0)$ occurs $p^{n-1} + (p-1)p^{\frac{n-2}{2}}\eta(\Delta_0)$ times, and $N_{\gamma,\delta,\epsilon}(0) = p^{n-1} - p^{\frac{n-2}{2}}\eta(\Delta_0)$ occurs $(p-1)(p^{n-1} - p^{\frac{n-2}{2}}\eta(\Delta_0))$ times.

By (9) and (15), $S(\gamma, \delta, \epsilon) = \eta(\Delta_0)p^{\frac{n}{2}}\omega^{-\lambda_{\gamma,\delta,\epsilon}}$ since $\sum\limits_{\rho \in \mathbb{F}_p} v(\lambda_{\gamma,\delta,\epsilon} + \rho)\omega^{\rho+\lambda_{\gamma,\delta,\epsilon}} = p$. Notice that $v(-\lambda_{\gamma,\delta,\epsilon}) = v(\lambda_{\gamma,\delta,\epsilon})$. By (16), for given $(\gamma, \delta) \in R_0$, when ϵ runs through $\mathbb{F}_{p^n}$, $S(\gamma, \delta, \epsilon) = \eta(\Delta_0)p^{\frac{n}{2}}\omega^{\rho}$ occurs $p^{n-1} + v(\rho)p^{\frac{n-2}{2}}\eta(\Delta_0)$ times for each $\rho \in \mathbb{F}_p$.

Case 2.2. $(\gamma, \delta) \in R_d$: in this case, $\Pi_{\gamma,\delta}(x) + Tr_1^n(\epsilon x) = \sum\limits_{i=1}^{n-d} a_i y_i^2 + \sum\limits_{i=1}^{n} b_i y_i$. If there exists some $b_i \neq 0$ for $n - d < i \leq n$, then for any $\rho \in \mathbb{F}_p$, $N_{\gamma,\delta,\epsilon}(\rho) = p^{n-1}$ and by (9), $S(\gamma, \delta, \epsilon) = 0$. Further, for given $(\gamma, \delta) \in R_d$, when ϵ runs through $\mathbb{F}_{p^n}$, there are exactly $p^n - p^{n-d}$ choices for ϵ such that there is at least one $b_i \neq 0$ with $n - d < i \leq n$ since CB is nonsingular.

If $b_i = 0$ for all $n - d < i \leq n$, a similar analysis as in Case 2.1 shows that $\sum\limits_{i=1}^{n-d}(a_i y_i^2 + b_i y_i) = \rho \iff \sum\limits_{i=1}^{n-d} a_i z_i^2 = \lambda_{\gamma,\delta,\epsilon} + \rho$, where $\lambda_{\gamma,\delta,\epsilon} = \sum\limits_{i=1}^{n-d} \frac{b_i^2}{4a_i}$ and $z_i = y_i + \frac{b_i}{2a_i}$ for $1 \leq i \leq n - d$. Then, for any $\rho \in \mathbb{F}_p$ and given $(\gamma, \delta) \in R_d$, by Lemma 1, for odd $n - d$, $N_{\gamma,\delta,\epsilon}(\rho) = p^d(p^{n-d-1} + p^{\frac{n-d-1}{2}}\eta((\lambda_{\gamma,\delta,\epsilon} + \rho)\Delta_d))$, i.e.,

$$N_{\gamma,\delta,\epsilon}(\rho) = p^{n-1} + p^{\frac{n+d-1}{2}}\eta((\lambda_{\gamma,\delta,\epsilon} + \rho)\Delta_d). \tag{17}$$

For given $(\gamma, \delta) \in R_d$, by Lemma 1, when $(b_1, b_2, \cdots, b_{n-d})$ runs through $\mathbb{F}_p^{n-d}$,

$$\lambda_{\gamma,\delta,\epsilon} = \sum_{i=1}^{n-d} \frac{b_i^2}{4a_i} = \rho' \text{ occurring } p^{n-d-1} + \eta(\rho')p^{\frac{n-d-1}{2}}\eta(\Delta_d) \text{ times} \tag{18}$$

for each $\rho' \in \mathbb{F}_p$. Thus, $\eta(\lambda_{\gamma,\delta,\epsilon}) = 0$ occurs p^{n-d-1} times, and ± 1 occur $\frac{p-1}{2}(p^{n-d-1} \pm p^{\frac{n-d-1}{2}}\eta(\Delta_d))$ times, respectively. Therefore, when $(b_1, b_2, \cdots, b_{n-d})$

runs through $\mathbb{F}_p^{n-d}$, $N_{\gamma,\delta,\epsilon}(0) = p^{n-1}$ occurs p^{n-d-1} times, and $N_{\gamma,\delta,\epsilon}(0) = p^{n-1} \pm p^{\frac{n+d-1}{2}}\eta(\Delta_d)$ occurs $\frac{p-1}{2}(p^{n-d-1} \pm p^{\frac{n-d-1}{2}}\eta(\Delta_d))$ times.

By (9) and (17), $S(\gamma,\delta,\epsilon) = \eta(\Delta_d)p^{\frac{n+d}{2}}\sqrt{(-1)^{\frac{p-1}{2}}}\omega^{-\lambda_{\gamma,\delta,\epsilon}}$. By (18), for given $(\gamma,\delta) \in R_d$, when $(b_1,b_2,\cdots,b_{n-d})$ runs through $\mathbb{F}_p^{n-d}$, $S(\gamma,\delta,\epsilon) = \eta(\Delta_d)p^{\frac{n+d}{2}}\sqrt{(-1)^{\frac{p-1}{2}}}\omega^\rho$ occurs $p^{n-d-1} + \eta(-\rho)p^{\frac{n-d-1}{2}}\eta(\Delta_d)$ times for each $\rho \in \mathbb{F}_p$.

Case 2.3. $(\gamma,\delta) \in R_{2d}$: in this case, $\Pi_{\gamma,\delta}(x) + Tr_1^n(\epsilon x) = \sum_{i=1}^{n-2d} a_i y_i^2 + \sum_{i=1}^{n} b_i y_i$. Similarly as in Case 2.2, if there exists some $b_i \neq 0$ with $n - 2d < i \leq n$, then $N_{\gamma,\delta,\epsilon}(\rho) = p^{n-1}$ for any $\rho \in \mathbb{F}_p$, and $S(\gamma,\delta,\epsilon) = 0$. Further, for given $(\gamma,\delta) \in R_{2d}$, when ϵ runs through $\mathbb{F}_{p^n}$, there are $p^n - p^{n-2d}$ choices for ϵ such that there is at least one $b_i \neq 0$ with $n - 2d < i \leq n$.

If $b_i = 0$ for all $n - 2d < i \leq n$, a similar analysis shows that for any given $(\gamma,\delta) \in R_{2d}$, when $(b_1,b_2,\cdots,b_{n-2d})$ runs through $\mathbb{F}_p^{n-2d}$, $N_{\gamma,\delta,\epsilon}(0) = p^{n-1} + (p-1)p^{\frac{n+2d-2}{2}}\eta(\Delta_{2d})$ occurs $p^{n-2d-1} + (p-1)p^{\frac{n-2d-2}{2}}\eta(\Delta_{2d})$ times, $N_{\gamma,\delta,\epsilon}(0) = p^{n-1} - p^{\frac{n+2d-2}{2}}\eta(\Delta_{2d})$ occurs $(p-1)(p^{n-2d-1} - p^{\frac{n-2d-2}{2}}\eta(\Delta_{2d}))$ times, and $S(\gamma,\delta,\epsilon) = \eta(\Delta_{2d})p^{\frac{n}{2}+d}\omega^\rho$ occurs $p^{n-2d-1} + v(\rho)p^{\frac{n-2d-2}{2}}\eta(\Delta_{2d})$ times for each $\rho \in \mathbb{F}_p$.

For $i \in \{0,d,2d\}$ and $j \in \{1,-1\}$, since $\eta(\Delta_i) = j$ for $(\gamma,\delta) \in R_{i,j}$, Theorem 3 can be proven by the above analysis, Equation (14), and Proposition 9.

Proof (of Theorem 3). We only give the frequency of the codewords with weight $(p-1)p^{n-1}$. The other cases can be proven in a similar way. The weight of $c(\gamma,\delta,\epsilon)$ is equal to $(p-1)p^{n-1}$ if and only if $N_{\gamma,\delta,\epsilon}(0) = p^{n-1}$, which occurs only in Cases 1, 2.2 and 2.3. The frequency is equal to $p^n - 1 + ((p^n - p^{n-d}) + p^{n-d-1})|R_d| + (p^n - p^{n-2d})|R_{2d}| = (p^n - 1)(1 + p^{n-2d}(p^{m+d} - p^m + p^{m-1} + p^{m-d} - 1))$.   □

Remark 12. By Theorem 3, the code C^k has 9 different weights for $d = 1$, and 10 different weights for $d > 1$. The codewords with weight $(p-1)(p^{n-1} - p^{\frac{n+2d-2}{2}})$ or $(p-1)p^{n-1} + p^{\frac{n+2d-2}{2}}$ do not exist in C^k since $|R_{2d,1}| = 0$.

The following result can also be similarly proven and we omit its proof here.

Proposition 13. *For $n = 2m \geq 4$, let $\varphi = \sqrt{(-1)^{\frac{p-1}{2}}}$, when (γ,δ,ϵ) runs through $\mathbb{F}_{p^m} \times \mathbb{F}_{p^n} \times \mathbb{F}_{p^n}$, the exponential sum $S(\gamma,\delta,\epsilon)$ defined in (8) has the following distribution, where $\rho = 0,1,\cdots,p-1$.*

$$\begin{cases} p^n, & 1 \text{ time,} \\ 0, & (p^n - 1)(1 + p^{m+n-d} - p^{m+n-2d} + p^{m+n-3d} - p^{n-2d}) \text{ times,} \\ p^{\frac{n}{2}}\omega^\rho, & p^d(p^m + 1)(p^n - 1)(p^{n-1} + v(\rho)p^{\frac{n-2}{2}})/(2(p^d + 1)) \text{ times,} \\ -p^{\frac{n}{2}}\omega^\rho, & (p^{n+d} - 2p^n + p^d)(p^m - 1)(p^{n-1} - v(\rho)p^{\frac{n-2}{2}})/(2(p^d - 1)) \text{ times,} \\ p^{\frac{n+d}{2}}\varphi\omega^\rho, & p^{m-d}(p^n - 1)(p^{n-d-1} + \eta(-\rho)p^{\frac{n-d-1}{2}})/2 \text{ times,} \\ -p^{\frac{n+d}{2}}\varphi\omega^\rho, & p^{m-d}(p^n - 1)(p^{n-d-1} - \eta(-\rho)p^{\frac{n-d-1}{2}})/2 \text{ times,} \\ -p^{\frac{n}{2}+d}\omega^\rho, & (p^{m-d} - 1)(p^n - 1)(p^{n-2d-1} - v(\rho)p^{\frac{n-2d-2}{2}})/(p^{2d} - 1) \text{ times.} \end{cases}$$

5   A Class of Nonbinary Sequence Families

By choosing cyclicly inequivalent codewords from $\mathcal{C}^k$, a family of nonbinary sequences is defined by

$$\mathcal{F}^k = \left\{ \{s_{a,b}(\alpha^t)\}_{0 \le t \le p^n - 2} \mid a \in \mathbb{F}_{p^m}, b \in \mathbb{F}_{p^n} \right\}, \tag{19}$$

where $s_{a,b}(\alpha^t) = Tr_1^m(a\alpha^{(p^m+1)t}) + Tr_1^n(b\alpha^{(p^k+1)t} + \alpha^t)$, and α is a primitive element of $\mathbb{F}_{p^n}$. The possible correlation values of $\mathcal{F}^{m+1}$ were discussed in [25], but the correlation distribution remains unsolved.

It is easy to verify that for two sequences s_{a_1,b_1} and s_{a_2,b_2}, $C_{a_1 b_1, a_2 b_2}(\tau) = -1 + S(\lambda_1, \lambda_2, \lambda_3)$, where $\lambda_1 = a_1 - a_2 \alpha^{(p^m+1)\tau}$, $\lambda_2 = b_1 - b_2 \alpha^{(p^k+1)\tau}$, $\lambda_3 = 1 - \alpha^\tau$. With this, the correlation distribution of $\mathcal{F}^k$ can be described in terms of the exponential sum $S(\lambda_1, \lambda_2, \lambda_3)$.

Theorem 14. *Let $\mathcal{F}^k$ be the family of sequences defined in (19) and $\varphi = \sqrt{(-1)^{\frac{p-1}{2}}}$. Then, $\mathcal{F}^k$ is a family of $p^{\frac{3n}{2}}$ nonbinary sequences with period $p^n - 1$, and its maximum correlation magnitude is equal to $p^{\frac{n}{2}+d} + 1$. Further, the correlation distribution is given as follows, where $\rho = 1, 2, \cdots, p-1$:*

Correlation value	Frequency
$p^n - 1$	$p^{\frac{3n}{2}}$
-1	$p^{\frac{3n}{2}}(p^n - 2)(1 + p^{\frac{3n}{2}-d} - p^{\frac{3n}{2}-2d} + p^{\frac{3n}{2}-3d} - p^{n-2d})$
$p^{\frac{n}{2}} - 1$	$\frac{p^{\frac{3n}{2}+d}(p^{\frac{n}{2}}+1)}{2(p^d+1)}((p^n - 2)(p^{n-1} + p^{\frac{n}{2}} - p^{\frac{n-2}{2}}) + 1)$
$p^{\frac{n}{2}}\omega^\rho - 1$	$p^{\frac{3n}{2}+d}(p^{\frac{n}{2}} + 1)(p^n - 2)(p^{n-1} - p^{\frac{n-2}{2}})/(2(p^d + 1))$
$-p^{\frac{n}{2}} - 1$	$\frac{p^{\frac{3n}{2}}(p^{n+d}-2p^n+p^d)}{2(p^{\frac{n}{2}}+1)(p^d-1)}((p^n - 2)(p^{n-1} - p^{\frac{n}{2}} + p^{\frac{n-2}{2}}) + 1)$
$-p^{\frac{n}{2}}\omega^\rho - 1$	$p^{2n-1}(p^{n+d} - 2p^n + p^d)(p^n - 2)/(2(p^d - 1))$
$\pm p^{\frac{n+d}{2}}\varphi - 1$	$p^{2n-d}((p^n - 2)p^{n-d-1} + 1)/2$
$p^{\frac{n+d}{2}}\varphi\omega^\rho - 1$	$p^{2n-d}(p^n - 2)(p^{n-d-1} + \eta(-\rho)p^{\frac{n-d-1}{2}})/2$
$-p^{\frac{n+d}{2}}\varphi\omega^\rho - 1$	$p^{2n-d}(p^n - 2)(p^{n-d-1} - \eta(-\rho)p^{\frac{n-d-1}{2}})/2$
$-p^{\frac{n}{2}+d} - 1$	$\frac{p^{2n-2d}-p^{\frac{3n}{2}-d}}{p^{2d}-1}((p^n - 2)(p^{n-d-1} - p^{\frac{n}{2}} + p^{\frac{n-2}{2}}) + p^d)$
$-p^{\frac{n}{2}+d}\omega^\rho - 1$	$(p^{2n-d} - p^{\frac{3n}{2}})(p^n - 2)(p^{n-2d-1} + p^{\frac{n}{2}-d-1})/(p^{2d} - 1)$

Proof. For any fixed $(a_2, b_2) \in \mathbb{F}_{p^m} \times \mathbb{F}_{p^n}$, when (a_1, b_1) runs through $\mathbb{F}_{p^m} \times \mathbb{F}_{p^n}$ and τ varies from 0 to $p^n - 2$, $(\lambda_1, \lambda_2, \lambda_3)$ runs through $\mathbb{F}_{p^m} \times \mathbb{F}_{p^n} \times \{\mathbb{F}_{p^n} \setminus \{1\}\}$ one time. Thus, the correlation distribution of $\mathcal{F}^k$ is $p^{\frac{3n}{2}}$ times as that of $S(\gamma, \delta, \epsilon) - 1$ when $(\gamma, \delta, \epsilon)$ runs through $\mathbb{F}_{p^m} \times \mathbb{F}_{p^n} \times \{\mathbb{F}_{p^n} \setminus \{1\}\}$. By Proposition 9, Equation (14) and the possible values of $S(\gamma, \delta, 0)$ corresponding to (γ, δ), the distribution of $S(\gamma, \delta, 0) - 1$ is obtained when (γ, δ) runs through $\mathbb{F}_{p^m} \times \mathbb{F}_{p^n}$. This together with Proposition 13 give the distribution of $S(\gamma, \delta, \epsilon) - 1$ as $(\gamma, \delta, \epsilon)$ runs through $\mathbb{F}_{p^m} \times \mathbb{F}_{p^n} \times \mathbb{F}_{p^n}^*$. Notice that $S(\gamma, \delta, \epsilon) = S(\gamma\epsilon^{-(p^m+1)}, \delta\epsilon^{-(p^k+1)}, 1)$ for any fixed $\epsilon \in \mathbb{F}_{p^n}^*$, and then for any given $\epsilon \in \mathbb{F}_{p^n}^*$, when (γ, δ) runs through $\mathbb{F}_{p^m} \times \mathbb{F}_{p^n}$, the distribution of $S(\gamma, \delta, \epsilon)$ is the same as that of $S(\gamma, \delta, 1)$. Thus, the distribution of

$S(\gamma, \delta, \epsilon) - 1$ can be determined when $(\gamma, \delta, \epsilon)$ runs through $\mathbb{F}_{p^m} \times \mathbb{F}_{p^n} \times \{\mathbb{F}_{p^m}^* \setminus \{1\}\}$. Together with the distribution of $S(\gamma, \delta, 0) - 1$ as (γ, δ) runs through $\mathbb{F}_{p^m} \times \mathbb{F}_{p^n}$, the correlation distribution is obtained.

By the definition of $\{s_{a,b}(\alpha^t)\}_{0 \leq t \leq p^n - 2}$, its period is $p^n - 1$. From the correlation distribution given in Theorem 14, one easily knows there are exactly $p^{\frac{3n}{2}}$ sequences in the family $\mathcal{F}^k$ and the maximum magnitude is $p^{\frac{n}{2}+d} + 1$.   □

Acknowledgment. The authors thank anonymous reviewers for their useful comments.

References

1. Bluher, A.W.: On $x^{q+1} + ax + b$. Finite Fields and Their Applications 10, 285–305 (2004)
2. Carlet, C., Ding, C.: Highly Nonlinear Functions. J. Complexity 20, 205–244 (2004)
3. Carlet, C., Ding, C., Niederreiter, H.: Authentication Schemes from Highly Nonlinear Functions. Des. Codes Cryptography 40, 71–79 (2006)
4. Carlet, C., Ding, C., Yuan, J.: Linear Codes from Perfect Nonlinear Mappings and Their Secret Sharing Schemes. IEEE Trans. Inform. Theory 51, 2089–2102 (2005)
5. Ding, C., Niederreiter, H.: Systematic Authentication Codes from Highly Nonlinear Functions. IEEE Trans. Inform. Theory 50, 2421–2428 (2004)
6. Ding, C., Yuan, J.: A Family of Skew Hadamard Difference Sets. J. Combin. Theory, series A 113, 1526–1535 (2006)
7. Golomb, S.W., Gong, G.: Signal Design for Good Correlation–For Wireless Communication, Cryptography and Radar. Cambridge Univ. Press, New York (2005)
8. Hu, L., Zeng, X., Li, N., Jiang, W.: Period-different m-sequences with at Most a Four-valued Cross Correlation, http://arxiv.org/abs/0801.0857
9. Kasami, T.: Weight Distribution of Bose-Chaudhuri-Hocquenghem Codes. In: Bose, R.C., Dowling, T.A. (eds.) Combinatorial Mathematics and Its Applications, pp. 335–357. University of North Carolina Press, Chapel Hill (1969)
10. Kumar, P.V., Moreno, O.: Prime-phase Sequences with Periodic Correlation Properties better than Binary Sequences. IEEE Trans. Inform. Theory 37, 603–616 (1991)
11. Kumar, P.V., Scholtz, R.A., Welch, L.R.: Generalized Bent Functions and Their Properties. J. Combin. Theory, series A 40, 90–107 (1985)
12. Lahtonen, J.: Two Remarks on a Paper by Moreno and Kumar. IEEE Trans. Inform. Theory 41, 859–861 (1995)
13. Liu, S.-C., Komo, J.J.: Nonbinary Kasami Sequences over GF(p). IEEE Trans. Inform. Theory 38, 1409–1412 (1992)
14. Lidl, R., Niederreiter, H.: Finite Fields. Encyclopedia of Mathematics and Its Applications. Addison-Wesley, Reading (1983)
15. Moriuchi, T., Imamura, K.: Balanced Nonbinary Sequences With Good Periodic Correlation Properties Obtained From Modified Kumar–Moreno Sequences. IEEE Trans. Inform. Theory 41, 572–576 (1995)
16. Moreno, O., Kumar, P.V.: Minimum Distance Bounds for Cyclic Codes and Deligne's Theorem. IEEE Trans. Inform. Theory 39, 1524–1534 (1993)
17. MacWilliams, F.J., Sloane, N.J.: The Theory of Error-Correcting Codes. North-Holland, Amsterdam (1977)

18. Sidelnikov, V.M.: On Mutual Correlation of Sequences. Soviet Math. Dokl. 12, 197–201 (1971)
19. Sarwate, D.V., Pursley, M.B.: Crosscorrelation Properties of Pseudorandom and Related Sequences. Proc. IEEE 68, 593–619 (1980)
20. Trachtenberg, H.M.: On the Cross-correlation Functions of Maximal Recurring Sequences. Ph.D. dissertation. Univ. of Southern California, Los Angeles, CA (1970)
21. Tang, X.H., Udaya, P., Fan, P.Z.: A New Family of Nonbinary Sequences With Three-Level Correlation Property and Large Linear Span. IEEE Trans. Inform. Theory 51, 2906–2914 (2005)
22. van der Vlugt, M.: Surfaces and the Weight Distribution of a Family of Codes. IEEE Trans. Inform. Theory 43, 1354–1360 (1997)
23. Welch, L.R.: Lower Bounds on the Maximum Cross Correlation of Signals. IEEE Trans. Inform. Theory 20, 397–399 (1974)
24. Yuan, J., Carlet, C., Ding, C.: The Weight Distribution of a Class of Linear Codes from Perfect Nonlinear Functions. IEEE Trans. Inform. Theory 52, 712–717 (2006)
25. Xia, Y., Zeng, X., Hu, L.: The Large Set of p-ary Kasami Sequences (preprint)
26. Zeng, X., Liu, J.Q., Hu, L.: Generalized Kasami Sequences: The Large Set. IEEE Trans. Inform. Theory 53, 2587–2598 (2007)

Results on the Crosscorrelation and Autocorrelation of Sequences

Faruk Göloğlu* and Alexander Pott

Institute for Algebra and Geometry
Otto-von-Guericke-University
39016 Magdeburg, Germany
goeloglu@student.uni-magdeburg.de, alexander.pott@ovgu.de

Abstract. In this paper we investigate some properties of the cross-correlation spectrum of an m-sequence $\mathbf{a}$ with period $2^m - 1$ and a d-decimation $\mathbf{b}$. Recently, Lahtonen et. al. [1] calculated the crosscorrelation value $\Theta_d(1)$ for specific exponents (i.e. for Gold and Kasami type). In this paper we generalize this result to all known almost bent exponents. In [1], the authors also prove that Gold functions are bent on some hyperplanes with respect to the base field $\mathbb{F}_2$. We also generalize this result and show that Gold functions are bent on all hyperplanes with respect to any subfield $\mathbb{F}_{2^k}$. We also show that $\Theta_d(1) \neq 0$ for many exponents d, and conclude that many sequences of type $\mathbf{a} + \mathbf{b}$ (including m-sequences added to an almost bent decimation) do not have perfect autocorrelation.

1 Introduction

Consider binary sequences $\mathbf{a} := (a_i), \mathbf{b} := (b_i)$ with period v, i.e. $a_{i+v} = a_i, b_{i+v} = b_i$. Quantifying the similarity between two periodic binary sequences is important in many applications, and a lot of research has been done on the analysis of the correlation of sequences. The **crosscorrelation** of a sequence $\mathbf{a}$ with a sequence $\mathbf{b}$ at shift τ is defined as $C_{\mathbf{a},\mathbf{b}}(\tau) := \sum_{i:=1}^{v}(-1)^{a_i+b_{i+\tau}}$, where v is the period. When the period $v = 2^m - 1$ for some integer m, one can use several algebraic tools. Let $\mathbb{F} = \mathbb{F}_q$ be the finite field with $q = 2^m$ elements. The multiplicative group $\mathbb{F}^*$ of $\mathbb{F}$ is cyclic and has cardinality $2^m - 1$, and any binary sequence of period $2^m - 1$ can be generated by a polynomial in $\mathbb{F}[x]$. The polynomial is called the **trace representation** of the sequence, and has the form $a_i := \mathsf{Tr}_1^m \left(\beta_1 \alpha^{d_1 i} + \cdots + \beta_s \alpha^{d_s i} \right)$, where $\beta_j \in \mathbb{F}^*$, α is a generator of $\mathbb{F}^*$, $d_j \in \{1, \ldots, q - 1\}$, and the trace map $\mathsf{Tr}_k^m : \mathbb{F}_{2^m} \to \mathbb{F}_{2^k}$ defined as $\mathsf{Tr}_k^m(\alpha) := \sum_{i=0}^{\frac{m}{k}-1} \alpha^{2^{ki}}$. When we omit the sub- and superscripts it should be understood that $k = 1$, and we have the so-called absolute trace. Note that the trace representation is not unique, however one can choose the exponents from the so-called cyclotomic coset leaders to make the representation unique (see for instance [2]).

---

* Research supported by Deutscher Akademischer Austausch Dienst (DAAD).

S.W. Golomb et al. (Eds.): SETA 2008, LNCS 5203, pp. 95–105, 2008.

One can also consider the correlation of a sequence with itself. The **auto-correlation** of a sequence is the crosscorrelation of the sequence with itself, and defined as follows. $A_{\mathbf{a}}(\tau) := \sum_{i:=1}^{v}(-1)^{a_i+a_{i+\tau}}$. Note that $A_{\mathbf{a}}(0) = v$. If the autocorrelation spectrum $\mathcal{A}_{\mathbf{a}} := \{A_{\mathbf{a}}(\tau) \ : \ \tau \in \{0,\ldots,v-1\}\}$, consists of -1 and v, then the sequence **a** is said to be a sequence with **perfect autocorrelation**. For instance m-sequences, which are defined to be $a_i := \mathsf{Tr}\left(\beta\alpha^i\right)$, with $\beta \in \mathbb{F}^*$, and α a generator of $\mathbb{F}^*$, have perfect autocorrelation. There are many known types of perfect autocorrelation sequences when the period is $2^m - 1$. It is not known whether the following list of perfect autocorrelation sequences is complete: m-sequences, Dillon-Dobbertin-sequences [3], Legendre-sequences [4], twin-prime-sequences [4], GMW-construction [2].

Let us return to the discussion of the crosscorrelation. The crosscorrelation of an m-sequence **a** with a decimation **b**, (i.e. $b_i := a_{di}$) is well studied [5,6]. Note that neither the crosscorrelation spectrum nor $C_{\mathbf{a},\mathbf{b}}(1)$ depends on the choice of the primitive element α. When i runs through $\{1,\ldots,q-1\}$, then α^i runs through all nonzero elements $\mathbb{F}^*$ of $\mathbb{F}$: $C_{\mathbf{a},\mathbf{b}}(\tau) = \sum_{i:=1}^{q-1}(-1)^{\mathsf{Tr}(\alpha^{di}+\alpha^{i+\tau})} = -1 + \sum_{x\in\mathbb{F}}(-1)^{\mathsf{Tr}\left(x^d+\beta x\right)}$, where $\beta = \alpha^\tau$. In this paper we will be interested in $\Theta_d(\beta) = \sum_{x\in\mathbb{F}}(-1)^{\mathsf{Tr}\left(x^d+\beta x\right)}$, which satisfy $C_{\mathbf{a},\mathbf{b}}(\tau) + 1 = \Theta_d(\alpha^\tau)$,[1] for the chosen α. When d has some specific values it was proved that (for odd m) the crosscorrelation function is optimal (i.e. $\mathsf{max}_{\alpha\in\mathbb{F}}\{\Theta_d(\alpha)\}$ is minimal), and it has a 3-valued spectrum. These pairs of sequences are called **preferred pairs** and they satisfy: $\Theta_d(\alpha) \in \{0,\pm 2^{(m+1)/2}\}$ for all $\alpha \in \mathbb{F}$.

Note also that the crosscorrelation spectrum of d-th decimation is basically the Walsh spectrum of the function $f \ : \ \mathbb{F} \ \to \ \mathbb{F}, x \ \mapsto \ x^d$. We denote the **Walsh transform** of a function at (β,γ) by $W_f(\beta,\gamma) := \sum_{x\in\mathbb{F}}(-1)^{\mathsf{Tr}(\beta f(x)+\gamma x)}$. When we are dealing with monomials we will abuse the notation and write $W_f(\gamma)$ instead of $W_f(1,\gamma)$. The functions above with optimal 3-valued Walsh spectra are called **almost bent (AB)**. The functions $f : \mathbb{F} \to \mathbb{F}$ for which the derivatives $\{f(x) + f(x+\alpha) \ : \ x \in \mathbb{F}\}$ have cardinality 2^{m-1} are called **almost perfect nonlinear (APN)**. AB functions are necessarily APN. Known AB exponents are:

- **Gold:** $2^e + 1$, where $\gcd(e,m) = 1$,
- **Kasami:** $2^{2e} - 2^e + 1$, where $\gcd(e,m) = 1$,
- **Welch:** $2^t + 3$, where $m = 2t + 1$,
- **Niho:**
  - $2^t + 2^{t/2} - 1$, if t even and $m = 2t + 1$,
  - $2^t + 2^{(3t+1)/2} - 1$, if t odd and $m = 2t + 1$.

In this paper the sets Gold, Kasami, Welch and Niho will denote the sets of exponents which are AB in $\mathbb{F}$ of the mentioned type. The reader can consult [7] for more on AB and APN functions.

We need some elementary results on finite fields. For more on finite fields, the reader can consult [8,9]. The **cyclotomic coset** containing $\alpha \in \mathbb{F}$ is defined as:

---

[1] In the literature, generally, $\theta_d(\beta) = \sum_{x\in\mathbb{F}^*}(-1)^{\mathsf{Tr}\left(x^d+\beta x\right)}$. In this paper we have $\Theta_d(\beta) = \theta_d(\beta) + 1$.

$C_\alpha := \{\alpha^{2^i} \ : \ i \in \{0,\ldots,m-1\}\}$. We can view $\mathbb{F} = \mathbb{F}_{2^m}$ as a vector space of dimension m over $\mathbb{F}_2$. A **hyperplane** is an $(m-1)$-dimensional linear subspace of this vector space, and can be defined as one of the subsets $H_\alpha := \{x \in \mathbb{F} : \mathsf{Tr}(\alpha x) = 0\}$ of $\mathbb{F}$, for all nonzero $\alpha \in \mathbb{F}$. The Trace-0 hyperplane H_1, is denoted simply by H. Obviously for all $\alpha \in \mathbb{F}^*$, $H_\alpha = \alpha^{-1}H$. Note that $\mathbb{F}$ is the union of cyclotomic cosets of length k whenever $k \mid m$. That is: $\mathbb{F}_{2^m} = \bigcup_{k \mid m} C_k$, where C_k is the set of field elements in a cyclotomic coset of length k, i.e. $C_k := \bigcup_{\alpha \in \mathbb{F}, |C_\alpha|=k} C_\alpha$. The cardinalities $c_k := |C_k|$ can be calculated as follows:

$$c_1 = 2, \quad \text{since } C_1 = \{0,1\}, \tag{1}$$

$$c_k = 2^k - \sum_{l \mid k, l \neq k} c_l. \tag{2}$$

When m is odd, the intersections of the Trace-0-hyperplane with the subfields of $\mathbb{F}_{2^m}$ have the following cardinalities:

Lemma 1. *Let m be odd, $k \mid m$, and $H \cap \mathbb{F}_{2^k}$ denote the intersection of the hyperplane H with the subfield $\mathbb{F}_{2^k}$ of $\mathbb{F}_{2^m}$. Then*

$$|H \cap \mathbb{F}_{2^k}| = 2^{k-1}.$$

Moreover, the intersection with the cyclotomic cosets of length k is

$$|H \cap C_k| = \frac{c_k}{2}. \tag{3}$$

Proof. Let $\alpha \in \mathbb{F}_{2^k} \leqslant \mathbb{F}_{2^m}$. Then

$$\mathsf{Tr}_1^m(\alpha) = \mathsf{Tr}_1^k(\mathsf{Tr}_k^m(\alpha)) = \mathsf{Tr}_1^k\left(\frac{m}{k}\alpha\right) = \frac{m}{k}\mathsf{Tr}_1^k(\alpha) = \mathsf{Tr}_1^k(\alpha).$$

Hence $|H \cap \mathbb{F}_{2^k}| = 2^{k-1}$. We prove the second claim by an easy induction on the length $l = e_1 + \cdots + e_r$ of the prime decomposition $m = p_1^{e_1} \cdots p_r^{e_r}$ of m.

If $l = 0$, then $H \cap \mathbb{F}_2 = \{0\}$, and $|H \cap \mathbb{F}_2| = c_1/2$. Also for $l = 1$, we have $c_m = 2^m - 2$ and $|H \cap C_k| = 2^{m-1} - 1$. Now if all $n \mid m$, $n \neq m$ satisfy (3), then we have (using (2))

$$|H \cap C_m| = 2^{m-1} - \sum_{k \mid m, k \neq m} \frac{c_k}{2} = \frac{c_m}{2},$$

which was to be shown. $\qquad\square$

It is well known that the crosscorrelation spectrum of Θ_d is related to the cardinality of intersections $|H^d \cap \alpha H|$ if $\gcd(d, q-1) = 1$,

$$\Theta_d(\alpha) = \sum_{x \in \mathbb{F}}(-1)^{\mathsf{Tr}(x^d + \alpha x)} = \sum_{x \in \alpha^{-1}H}(-1)^{\mathsf{Tr}(x^d)} - \sum_{x \in \overline{\alpha^{-1}H}}(-1)^{\mathsf{Tr}(x^d)}$$

$$= 2\sum_{x \in \alpha^{-1}H}(-1)^{\mathsf{Tr}(x^d)} = 2\left(\left|H^d \cap \alpha^{-1}H\right| - \left|H^d \cap \overline{\alpha^{-1}H}\right|\right)$$

$$= -2^m + 4\left|H^d \cap \alpha^{-1}H\right|,$$

where $\overline{S}$ denote complementation of S in $\mathbb{F}$. We can now characterize the AB exponents simply by these cardinalities: An exponent d is AB if and only if

$$|H^d \cap \alpha H| \in \{2^{m-2}, 2^{m-2} \pm 2^{(m-3)/2}\}, \ \forall \alpha \in \mathbb{F}^*. \tag{4}$$

2 Computing $\Theta_d(1)$

Our main result in this section is the following theorem.

Theorem 1. *Let* $\mathbb{F} = \mathbb{F}_{2^m}$, *$m$ odd, and let d be an AB exponent in $\mathbb{F}$ such that d is also AB in $\mathbb{F}_{2^k}$ for all $k \mid m$, that is the function $f : \mathbb{F}_{2^k} \to \mathbb{F}_{2^k}$, $x \mapsto x^d$ is AB. Then:*

$$\Theta_d(1) = \begin{cases} +2^{(m+1)/2} & \text{if } m \equiv \pm 1 \pmod 8, \\ -2^{(m+1)/2} & \text{if } m \equiv \pm 3 \pmod 8. \end{cases} \tag{5}$$

Recently Lahtonen et. al. [1] proved the following theorem:

Theorem 2. *[1] Let* $\mathbb{F} = \mathbb{F}_{2^m}$, *and* $d \in$ Gold $\cup$ Kasami *be an AB exponent in $\mathbb{F}$. Then $\Theta_d(1)$ satisfies (5).*

Actually the Gold case is due to Dillon and Dobbertin [3]. Also we should mention that when d is Gold, $\Theta_d(1)$ determines all the values in the crosscorrelation spectrum of d, as shown in [1].

   The following theorem proves that (5) is true for all AB exponents d in $\mathbb{F}_{2^p}$, where p is an odd prime. This will also act as a basis of a further inductive generalization.

Theorem 3. *Let* $\mathbb{F} = \mathbb{F}_{2^m}$, *$m = p$ be an odd prime, and let d be an AB exponent in $\mathbb{F}$. Then $\Theta_d(1)$ satisfies (5).*

Proof. Since p is a prime, $c_1 = 2$ and $c_p = 2^p - 2$. Now, H and H^d are combinations of cosets (since $\text{Tr}\,(\alpha) = \text{Tr}\,(\alpha^2)$, and x^d is a permutation which maps cyclotomic cosets to cyclotomic cosets), and consist of $c_p/2p$ cosets of cardinality p and $C_0 = \{0\}$. Then $|H \cap H^d| = lp + 1$, where $0 \leqslant l \leqslant c_p/2p$. Now we have $\frac{c_p}{2} + 1 = 2^{p-1}$, and $lp + 1 \in \{2^{p-2}, 2^{p-2} \pm 2^{(p-3)/2}\}$. But $lp + 1 = 2^{p-2}$ leads to a quick contradiction since

$$2lp + 2 = 2^p - 1 = \frac{c_p}{2} + 1$$

$$1 = p(\frac{c_p}{2p} - 2l) = p(\frac{2^p - 2}{2p})$$

is impossible (note that $\frac{2^p - 2}{2p}$ is an integer). When we analyse the other possibilities we see that

$$lp + 1 = 2^{p-2} \pm 2^{(p-3)/2}$$

$$2lp + 2 = \frac{c_p}{2} + 1 \pm 2^{(p-1)/2}$$

$$1 = p(\frac{c_p}{2p} - 2l) \pm 2^{(p-1)/2} = p(\frac{2^p - 2}{2p}) \pm 2^{(p-1)/2}.$$

Now consider the last equalities modulo p, that is

$$1 \equiv p(\frac{c_p}{2p} - 2l) \pm 2^{(p-1)/2} \pmod{p} \equiv \pm 2^{(p-1)/2} \pmod{p}.$$

We have 2 is a square in $\mathbb{F}_p$ if $p \equiv \pm 1 \pmod 8$, and non-square if $p \equiv \pm 3 \pmod 8$ (see any book on elementary number theory). Now if 2 is a square (resp. non-square) equality with $+$ (resp. $-$) holds. That is, if 2 is a square (resp. non-square) $|H \cap H^d| = 2^{p-2} + 2^{(p-3)/2}$ (resp. $|H \cap H^d| = 2^{p-2} - 2^{(p-3)/2}$), which implies (using (4)) $\Theta_d(1) = +2^{(p+1)/2}$ if $p \equiv \pm 1 \pmod 8$ (resp. $\Theta_d(1) = -2^{(p+1)/2}$ if $p \equiv \pm 3 \pmod 8$). □

Proof of Theorem 1: We will prove the claim by induction on the length $l = e_1 + \cdots + e_r$ of the prime decomposition $m = p_1^{e_1} \cdots p_r^{e_r}$ of m.

Base case $l = 1$: We have m prime. This case is proved by Theorem 3. We also prove for $l = 2$, as an example of the inductive step. Consider $m = pq$, where p, q are odd primes. Recall that the sets $C_k \subset \mathbb{F}$ have cardinalities c_k given by (1) and (2). Also, by Lemma 1, $|H \cap C_k| = \frac{c_k}{2}$ for all $k \mid m$. Since x^d is a permutation not only on $\mathbb{F}$, but also on the restrictions to C_k, the intersections $|H^d \cap C_k|$ satisfy $|H \cap C_k| = |H^d \cap C_k|$, for all $k \mid m$.

Let $h := |H \cap H^d|$ and $h_k := |(H \cap C_k) \cap (H^d \cap C_k)|$. Then we have $H \cap H^d = \bigcup_{k|m} (H \cap C_k) \cap (H^d \cap C_k)$, and $h = \sum_{k|m} h_k$. Note that the numbers h_k satisfy $k|h_k$. Since d is AB on subfields $\mathbb{F}_{2^p}, \mathbb{F}_{2^q}$, we get h_k for all $k \mid m$ and $k \neq m$ by Theorem 3.

Note that $3 \in$ Gold for any finite field $\mathbb{F}_{2^m}$. Let $g := |H \cap H^3|$ and $g_k := |(H \cap C_k) \cap (H^3 \cap C_k)|$. Then we know g by Theorem 2. Note that $g = \sum_{k|m} g_k$, and $g_k = h_k$ for $k \mid m$ and $k \neq m$ by Theorem 3 and Theorem 2. Hence $h_m \in \{g_m, g_m + 2^{(m-3)/2}, g_m + 2 \cdot 2^{(m-3)/2}\}$ (or $h_m \in \{g_m, g_m - 2^{(m-3)/2}, g_m - 2 \cdot 2^{(m-3)/2}\}$, depending on $(\frac{2}{m})$) by (4). Now since $pq \nmid \pm 2^{(m-3)/2}$, then $h_m = g_m$ and therefore $h = g$. Note that $(\frac{2}{m})$ is the so-called Jacobi symbol and is equal to $+1$ (resp. -1) if 2 is a square (resp. non-square) in $\mathbb{Z}_{2^m-1}$.

Inductive step: Now let the theorem be true for all $k \mid m$, but $k \neq m$. Then since $3 \in$ Gold we get $g := |H \cap H^3|$ and $g_k := |(H \cap C_k) \cap (H^3 \cap C_k)|$ for all $k \mid m$. We also get $h_k = g_k$ for $k \mid m$ and $k \neq m$, again inductively (starting from k's with simpler prime decomposition). The divisibility argument above is employed to get $h_m = g_m$ and consequently $h = g$. □

Now we will prove that all known AB exponents, i.e. Gold, Kasami, Welch and Niho satisfy the assumption of the Theorem 1, that is they are AB on all subfields.

Lemma 2. Let $\mathbb{F} = \mathbb{F}_{2^m}$ be a finite field, m odd, and $K = \mathbb{F}_{2^k} \leqslant \mathbb{F}$ be an arbitrary subfield. If d is an AB exponent in $\mathbb{F}$ such that $d \in$ Gold $\cup$ Kasami $\cup$ Niho $\cup$ Welch, then d is also AB in K. Moreover d belongs to the same subclass of AB exponents in the subfield.

Proof. We analyse case by case.

- Gold: Let $d = 2^e + 1$, with $\gcd(e, m) = 1$, and let $e = rk + s$, where $s < k \leqslant 0$. Then

$$2^e + 1 = 2^s(2^{rk} - 1) + 2^s + 1 \equiv 2^s + 1 \pmod{2^k - 1},$$

  with $\gcd(e, m) = 1 \Rightarrow \gcd(e, k) = 1 \Rightarrow \gcd(s, k) = 1$.
- Kasami: Let $d = 2^{2e} - 2^e + 1$, with $\gcd(e, m) = 1$, and let $e = rk + s$, where $s < k \leqslant 0$. Then

$$2^{2e} - 2^e + 1 = 2^{2s}(2^{rk} - 1) + 2^{2s} - 2^s(2^{rk} - 1) - 2^s + 1$$
$$= 2^{2s} - 2^s + 1 \pmod{2^k - 1}.$$

  with $\gcd(e, m) = 1 \Rightarrow \gcd(e, k) = 1 \Rightarrow \gcd(s, k) = 1$.
- Welch and Niho: Let $m = 2t + 1$, and $m = kl$. Then

$$t = \frac{kl - 1}{2} = \frac{kl - k}{2} + \frac{k - 1}{2} = \frac{l - 1}{2}k + \frac{k - 1}{2}.$$

Note that $k > \frac{k-1}{2} > 0$. Therefore, for the Welch exponent $d = 2^t + 3$, we have

$$2^t + 3 = 2^{\frac{k-1}{2}}(2^{\frac{l-1}{2}k} - 1) + 2^{\frac{k-1}{2}} + 3$$
$$\equiv 2^{\frac{k-1}{2}} + 3 \pmod{2^k - 1}.$$

Let $r = \frac{l-1}{2}k$, and $s = \frac{k-1}{2}$. Then $t = rk + s$. For the Niho case we have to consider two cases:

• t **even:** In $t = rk + s$, if s is also even, then $t/2 = rk/2 + s/2$ and:

$$2^t + 2^{t/2} - 1 = 2^s(2^{rk} - 1) + 2^s + 2^{s/2}(2^{rk/2} - 1) + 2^{s/2} - 1$$
$$\equiv 2^s + 2^{s/2} - 1 \pmod{2^k - 1}.$$

If s is odd, then

$$\frac{t}{2} = \frac{rk + s}{2} = \frac{(r-1)k + k + s}{2} = \frac{(r-1)k + 3s + 1}{2} = \frac{(r-1)k}{2} + \frac{3s + 1}{2},$$

and finally $2^t + 2^{t/2} - 1 \equiv 2^s + 2^{(3s+1)/2} - 1 \pmod{2^k - 1}$.
• t **odd:** We have

$$\frac{t + 1}{2} = \frac{rk + s + 1}{2}$$
$$t + \frac{t + 1}{2} = rk + s + \frac{rk + s + 1}{2}$$
$$\frac{3t + 1}{2} = \frac{3rk + 3s + 1}{2}$$

If s is also odd, then

$$\frac{3t+1}{2} = \frac{3rk}{2} + \frac{3s+1}{2},$$

and, $2^t + 2^{(3t+1)/2} - 1 \equiv 2^s + 2^{(3s+1)/2} - 1 \pmod{2^k - 1}$.
If s is even, then

$$\frac{3t+1}{2} = \frac{3rk+3s+1}{2} = \frac{3rk+k+s}{2} = \frac{(3r+1)k+s}{2}$$

$$= \frac{(3r+1)k}{2} + \frac{s}{2},$$

and consequently $2^t + 2^{(3t+1)/2} - 1 \equiv 2^s + 2^{s/2} - 1 \pmod{2^k - 1}$.

$\square$

As a corollary we have:

Corollary 1. *Let* $\mathbb{F} = \mathbb{F}_{2^m}$ *and* $d \in$ Gold $\cup$ Kasami $\cup$ Welch $\cup$ Niho, *then* $\Theta_d(1)$ *satisfies (5).*

3   Decomposition of Gold Exponent

In this section we let $m = kk'$, $K = \mathbb{F}_{2^k}$, d be an AB exponent such that it is AB on all subfields, $f : \mathbb{F}_{2^m} \to \mathbb{F}_{2^m}, x \mapsto x^d$, and $\tilde{f} : K \to K, x \mapsto x^d$. Rewriting Corollary 1, we have $W_f(1) = W_{\tilde{f}}(1) \, 2^{\frac{m-k}{2}} \left(\frac{2}{k'}\right)$. Actually if d is a Gold exponent, we can generalize the above equality. Since, as mentioned before, the Gold spectrum is completely known, the following theorem is not new. But it emphasizes an interesting property of Gold exponents, which is not shared by other exponents, even by Kasami exponents; and our proof has a corollary, which is also a generalization of a result in [1].

Theorem 4. *Let* $d = 2^e + 1 \in$ Gold *and* $K^\perp = \{x \in \mathbb{F} : \mathrm{Tr}\,(xy) = 0, \forall y \in K\}$. *If* $\gamma \in K^\perp$ *and* $a \in K$, *then:*

$$W_f(\alpha + \gamma) = W_{\tilde{f}}(\alpha)\, 2^{\frac{m-k}{2}} \delta(\gamma), \tag{6}$$

and, in particular,

$$W_f(\alpha) = W_{\tilde{f}}(\alpha)\, 2^{\frac{m-k}{2}} \left(\frac{2}{k'}\right), \tag{7}$$

where $\delta(\gamma) := \mathrm{sign}\left(\frac{W_f(1+\gamma)}{W_{\tilde{f}}(1)}\right)$.

Proof. We have

$$W_f(1) = \sum_{x \in \mathbb{F}} (-1)^{\mathrm{Tr}(x^{2^e+1}+x)} = \sum_{u \in K, v \in K^\perp} (-1)^{\mathrm{Tr}((u+v)^{2^e+1}+u+v)}$$

$$= \sum_{u \in K, v \in K^\perp} (-1)^{\mathrm{Tr}(u^{2^e+1}+u^{2^e}v+v^{2^e}u+v^{2^e+1}+u+v)}$$

$$= \sum_{u \in K} (-1)^{\mathrm{Tr}(u^d+u)} \sum_{v \in K^\perp} (-1)^{\mathrm{Tr}(v^d)},$$

since $\text{Tr}\left(u^{2^e}v\right) = \text{Tr}\left(v^{2^e}u\right) = \text{Tr}\left(v\right) = 0$. Now put $s_d(0) := \sum_{v \in K^\perp}(-1)^{\text{Tr}\left(v^d\right)}$. Then $W_f(1) = W_{\tilde{f}}(1)\, s_d(0)$, hence $s_d(0) = 2^{\frac{m-k}{2}}\left(\frac{2}{k'}\right)$, since $\left(\frac{2}{m}\right) = \left(\frac{2}{k}\right)\left(\frac{2}{k'}\right)$. Now for $\alpha \in K$, we get

$$W_f(\alpha) = \sum_{x \in \mathbb{F}}(-1)^{\text{Tr}\left(x^{2^e+1}+\alpha x\right)}$$

$$= \sum_{u \in K}(-1)^{\text{Tr}\left(u^d+\alpha u\right)}\sum_{v \in K^\perp}(-1)^{\text{Tr}\left(v^d\right)}$$

$$= W_{\tilde{f}}(\alpha)\, 2^{\frac{m-k}{2}}\left(\frac{2}{k'}\right).$$

For $\gamma \in K^\perp$ we get:

$$W_f(\alpha+\gamma) = \sum_{x \in \mathbb{F}}(-1)^{\text{Tr}\left(x^{2^e+1}+(\alpha+\gamma)x\right)}$$

$$= \sum_{u \in K}(-1)^{\text{Tr}\left(u^d+\alpha u\right)}\sum_{v \in K^\perp}(-1)^{\text{Tr}\left(v^d+\gamma v\right)}$$

$$= W_{\tilde{f}}(\alpha)\sum_{v \in K^\perp}(-1)^{\text{Tr}\left(v^d+\gamma v\right)}.$$

Putting $\alpha = 1$ we have $s_d(\gamma) := \sum_{v \in K^\perp}(-1)^{\text{Tr}\left(v^d+\gamma v\right)} = \frac{W_f(1+\gamma)}{W_{\tilde{f}}(1)} \in \{\pm 2^{\frac{m-k}{2}}\}$ since $1+\gamma \in \mathbb{F}\setminus H$, where H is the Trace-0 hyperplane of $\mathbb{F}$ (if $d \in$ Gold then we have $W_f(\alpha) = 0 \iff \alpha \in H$ cf. [10]). Letting $\delta(\gamma) := \text{sign}\left(\frac{W_f(1+\gamma)}{W_{\tilde{f}}(1)}\right)$, the proof is finished. $\qquad\square$

Remark 1. Equality (6) is in general not true if $d \in$ Kasami $\cup$ Welch $\cup$ Niho. We expect, however, (7) is a property of AB exponents.

The following corollary shows that a Gold exponent can be decomposed into bent functions. We note here that a **bent function** is a boolean function defined on a vector space V with a 2-valued Walsh spectrum (i.e. $\{\sum_{x \in V}(-1)^{f(x)+u\cdot x} : u \in V\} = \{\pm\sqrt{|V|}\}$).

Corollary 2. *Let $d \in$ Gold and $f|_V$ denote the restriction of f to a subspace V. Then*

1. *$f|_K$ is AB.*
2. *$\text{Tr}(f)|_{K^\perp}$ is bent.*

Proof. The first item is already proved. For the second, note that all the linear functionals on $K^\perp$ are of the form $\text{Tr}(\gamma x)$, where $\gamma \in K^\perp$, as a result of nondegeneracy of the trace bilinear form (cf. [8]). Noting that $s_d(\gamma)$ are Walsh coefficients of $\text{Tr}\left(x^d\right)$ on $K^\perp$ completes the proof. $\qquad\square$

Remark 2. Note that $K^\perp$ is the Trace-0 hyperplane, when $\mathbb{F}$ is considered as an $(m-k)$-dimensional vector space over K. Hence, the above corollary generalizes the result in [1], which states that Gold functions restricted to a hyperplane are bent. Since $\text{Tr}\left((\alpha+\gamma)^d\right) = \text{Tr}\left(\alpha^d + \gamma^d\right)$, where $\alpha \in K, \gamma \in K^\perp, d \in$ Gold, functions $\text{Tr}\left(f|_{\alpha+K^\perp}\right)$ are all bent, hence $\text{Tr}\left(f\right)$ is decomposed into 2^k bent functions. We also expect that similar to Gold exponents, Kasami exponents restricted to $K^\perp$ are bent (see [11] for the restriction of Kasami exponents to H and [12] for functions decomposing to bent functions) not satisfying the decomposition.

4   Sequences from Binomials

The **linear complexity** of a sequence **a** with period $2^m - 1$, is equal to the number of nonzero monomials of the expanded trace representation of **a**, which is a polynomial in $\mathbb{F}[x]/(x^{2^m-1} - 1)$. The m-sequences have linear complexity m, and they have minimal linear complexity among all perfect autocorrelation sequences. We think that the following question is interesting:

Question 1. What is the minimal linear complexity l of a perfect autocorrelation sequence such that $l > m$.

In the case $m = 2k$ even, GMW construction $\text{Tr}_1^k\left(\text{Tr}_k^m\left(\alpha^i\right)^d\right)$, where $\text{wt}(d) = 2$, leads to a sequence with perfect autocorrelation and linear complexity $2m$. Here $\text{wt}(d)$ denotes the number of 1s in the binary expansion of d. In the odd case, however, the GMW construction cannot lead to complexity $2m$ [2,8]. In this section we prove that many sequences whose trace representations are of the form $\text{Tr}\left(\beta_1 x^{d_1} + \beta_2 x^{d_2}\right)$ cannot have perfect autocorrelation. Note that the linear complexity of these sequences are at most $2m$.

In [13], Helleseth et. al. prove that in $\mathbb{F}_{3^m}$, the sequence $\text{Tr}\left(\alpha^{di} + \alpha^i\right)$ gives a sequence with perfect autocorrelation, where $d = 3^{2k} - 3^k + 1$ (a Kasami exponent), and $m = 3k$. This exponent also gives a 3-valued crosscorrelation spectrum in $\mathbb{F}_{3^m}$, as proved by Trachtenberg [6]. In this paper we will give an argument why AB exponents cannot be used as above to produce perfect autocorrelation sequences in the binary case.

First note that if $\gcd(d_1, q-1) = \gcd(d_2, q-1) = 1$, then it is sufficient to study the autocorrelation spectrum of $u_i := \text{Tr}\left(\alpha^{di} + \beta\alpha^i\right)$, where $d = d_1 d_2^{-1}$ and $\beta = \frac{\beta_2}{\beta_1^{d^{-1}}}$. We assume for the following that $d \neq 2^i$ for some $i \in \{0, \dots, m-1\}$, otherwise the linear complexity would be at most m. The corresponding difference set $D_\mathbf{u} \in \mathbb{Z}_{2^m-1}$ is defined as $D_\mathbf{u} := \{i : u_i = 0\}$. A **multiplier** of a difference set in a group G is a number m which satisfies $mD = D + g$ for some $g \in G$. Note that 2 is a multiplier of the difference sets we are interested in (i.e. difference sets with Singer parameters) in $\mathbb{Z}_{2^m-1}$. It is well know that there exists a 'translate' (or 'shift') $E = D + g$ of D, which satisfies $2E = E$ [4,2]. Therefore if **u** has perfect autocorrelation, then an l shift $u_i^l :\doteq \text{Tr}\left(\alpha^{dl}\alpha^{di} + \alpha^l\beta\alpha^i\right)$ of **u** must be constant on cosets, i.e. must be of the form $\text{Tr}\left(\alpha^{di} + \alpha^i\right)$ (cf. [2]). But this is not possible unless $\beta = 1$ and $l = 0$. Also if **u** is a perfect autocorrelation sequence with Singer parameters, then it should be the case that $|D| = 2^{m-1}$, hence $\Theta_d(1) = 0$ should be satisfied.

For the remaining let $s_i = \text{Tr}\left(\alpha^i\right), t_i = \text{Tr}\left(\alpha^{di}\right)$ and $u_i = s_i + t_i = \text{Tr}\left(\alpha^{di} + \alpha^i\right)$, and $\gcd(d, q-1) = 1$. The following relation is also in [13]: $A_u(\tau) = -1 + \sum_{x \in \mathbb{F}}(-1)^{\text{Tr}\left(x^d + \psi_d(\alpha^\tau)x\right)}$, where $\psi_d : \mathbb{F}^* \rightarrow \mathbb{F}, \beta \mapsto (1+\beta)(1+\beta^d)^e$ and $e = q - 1 - d^{-1}$, that is $de \equiv -1 \pmod{q-1}$. Note that the values of the autocorrelation spectrum of u depends on the image of the map ψ_d and the kernel of the crosscorrelation function Θ_d. Note also that ψ_d maps 1 to 0, and $\Theta_d(0) = 0$ for all d. Next two proposition follows from these arguments:

Proposition 1. *[13] Let $\phi_d : \mathbb{F} \backslash \mathbb{F}_2 \rightarrow \mathbb{F} \backslash \mathbb{F}_2$ denote the map $x \mapsto \left[x^d + (x+1)^d\right]^e$, and $\psi_d : \mathbb{F} \backslash \mathbb{F}_2 \rightarrow \mathbb{F} \backslash \mathbb{F}_2$ denote the map $x \mapsto (1+x)(1+x^d)^e$. Then $Im(\phi_d) = Im(\psi_d)$, where Im denotes the image of a map, and $e = q - 1 - d^{-1}$.*

Proposition 2. *The sequence $\text{Tr}\left(\alpha^{di} + \alpha^i\right)$ has perfect autocorrelation if and only if $\Theta_d(1) = 0$ and $Im(\phi_d) \subseteq Ker(\Theta_d)$, where Ker denotes the zeroes of a function.*

The following theorem shows that for many d and m, $\Theta_d(1) \neq 0$.

Theorem 5. *Let $\gcd(d, q-1) = 1$, and $\mathbb{F} = \mathbb{F}_{2^m}$. If m is p^e, $2p^e$, or $3p^e$, where p is an odd prime, $e \geqslant 0$. Then $\Theta_d(1) \neq 0$.*

Proof. For cases $m = p^e$ and $m = 3p^e$, we will make use of Lemma 1. Let a_i and b_i denote the number of intersecting cosets $|H \cap H^d \cap C_k|/k$, where k will be clear from the context. First consider $m = p^e$. We must show $a_e p^e + a_{e-1}p^{e-1} + \cdots + a_1 p + 1 \neq 2^{p^e - 2}$. If the theorem is not true then again considering modulo p we would have $a_e p^e + a_{e-1}p^{e-1} + \cdots + a_1 p + 1 \equiv 2^{p^e - 2} \pmod{p}$ and $1 \equiv 2^{-1} \pmod{p}$, which is a contradiction.

Now let $m = 2p^e$. Note that if $\alpha \in \mathbb{F}_4 \backslash \mathbb{F}_2$, then $\text{Tr}_1^m(\alpha) = p^e \text{Tr}_1^2(\alpha) = 1$, and if $\alpha \in \mathbb{F}_{p^e}$, then $\text{Tr}_1^m(\alpha) = 2\text{Tr}_1^k(\alpha) = 0$. Hence we should show: $a_e 2p^e + \cdots + a_1 2p + 2^{p^e} \neq 2^{2p^e - 2}$. The equation modulo p is $2 \equiv 2^{2(p^e-1)} \equiv 1 \pmod{p}$, which cannot hold.

Finally let $m = 3p^e$. Now we have to show $a_e 3p^e + \cdots + a_1 3p + a_0 3 + b_e p^e + \cdots + b_1 p + 1 \neq 2^{3p^e - 2}$. Note that $|H \cap C_3| = 3$, which means $a_0 \in \{0, 1\}$ and therefore the modulo p reduction $a_0 3 + 1 \equiv 2^{3(p^e - 1)+1} \equiv 2 \pmod{p}$, cannot hold. $\square$

Remark 3. If $m = 4k$, then we have exponents d with $\gcd(d, q-1) = 1$, satisfying $\Theta_d(1) = 0$. The theorem covers all odd m (resp. $m = 2k$, k odd), for $m < 35$ (resp. $m < 30$). We believe that there are exponents d with $\gcd(d, q-1) = 1$ in larger fields like $\mathbb{F}_{2^{30}}$ or $\mathbb{F}_{2^{35}}$, satisfying $\Theta_d(1) = 0$, that is, we believe the above theorem is as general as possible.

Corollary 3. *If*

- *d is AB, or*
- *d satisfies $\gcd(d, q-1) = 1$ and m is p^e, $2p^e$, or $3p^e$, where p is an odd prime and $e \geqslant 0$,*

then $\text{Tr}\left(\beta_1\alpha^{d_1 i} + \beta_2\alpha^{d_2 i}\right)$ where d_1, d_2 satisfy $d_1 d_2^{-1} = d$, does not have perfect autocorrelation.

Proof. The second item follows immediately by Theorem 5 and Proposition 2. If d is AB with $\Theta_d(1) \neq 0$ then the result follows from Proposition 2. Now assume d is AB with $\Theta_d(1) = 0$. Then $|Im(\psi_d)| = |Im(\phi_d)| = 2^{m-1} - 1$, since d is AB, and in turn APN. But $|Ker(\Theta_d) \setminus \{0, 1\}| = 2^{m-1} - 2$, and therefore $Im(\psi_d) \not\subseteq Ker(\Theta_d)$, since $|Ker(\Theta_d)| = 2^{m-1}$ if d is AB (see for instance [2]). □

Question 2. If m is odd, are there sequences $\text{Tr}\left(\beta_1\alpha^{d_1 i} + \beta_2\alpha^{d_2 i}\right)$ with perfect autocorrelation where d_1, d_2 satisfy $d_1 d_2^{-1} = d$, and $\gcd(d, q - 1) = 1$?

References

1. Lahtonen, J., McGuire, G., Ward, H.N.: Gold and Kasami-Welch functions, quadratic forms, and bent functions. Adv. Math. Commun. 1(2), 243–250 (2007)
2. Golomb, S.W., Gong, G.: Signal design for good correlation. For wireless communication, cryptography, and radar, p. 438. Cambridge University Press, Cambridge (2005)
3. Dillon, J.F., Dobbertin, H.: New cyclic difference sets with Singer parameters. Finite Fields Appl. 10(3), 342–389 (2004)
4. Beth, T., Jungnickel, D., Lenz, H.: Design theory, 2nd edn. Encyclopedia of Mathematics and its Applications, vol. 69. Cambridge University Press, Cambridge (1999)
5. Niho, Y.: Multi-valued Cross-Correlation Functions Between Two Maximal Linear Recursive Sequences, PhD thesis, University of Southern California, Los Angeles (1972)
6. Trachtenberg, H.M.: On the Cross-Correlation Functions of Maximal Linear Sequences. PhD thesis, University of Southern California, Los Angeles (1970)
7. Carlet, C.: Boolean Methods and Models. In: Vectorial Boolean functions for cryptography, Cambridge University Press, Cambridge (manuscript) (to appear, 2006)
8. Jungnickel, D.: Finite fields. Bibliographisches Institut, Mannheim, Structure and arithmetics (1993)
9. Lidl, R., Niederreiter, H.: Finite fields and their applications. In: Handbook of algebra, vol. 1. North-Holland, Amsterdam (1996)
10. Gold, R.: Maximal recursive sequences with 3-valued recursive cross-correlation functions. IEEE Trans. Inf. Th. 14, 377–385 (1968)
11. Dillon, J.F., McGuire, G.: Kasami-Welch functions on hyperplane (preprint, 2006)
12. Canteaut, A., Charpin, P.: Decomposing bent functions. IEEE Trans. Inform. Theory 49(8), 2004–2019 (2003)
13. Helleseth, T., Kumar, P.V., Martinsen, H.: A new family of ternary sequences with ideal two-level autocorrelation function. Des. Codes Cryptogr. 23(2), 157–166 (2001)

m-Sequences of Lengths $2^{2k} - 1$ and $2^k - 1$ with at Most Four-Valued Cross Correlation*

Tor Helleseth and Alexander Kholosha

The Selmer Center
Department of Informatics, University of Bergen
P.O. Box 7800, N-5020 Bergen, Norway
{Tor.Helleseth, Alexander.Kholosha}@uib.no

Abstract. Considered is the distribution of the cross correlation between m-sequences of length $2^m - 1$, where m is even, and m-sequences of a shorter length $2^{m/2} - 1$. Pairs of this type with at most four-valued cross correlation are found and the complete correlation distribution is determined. These results cover the two-valued Kasami case and all three-valued decimations found earlier. Conjectured is that there are no other cases leading to at most four-valued cross correlation apart from the ones proven here and except for a single, seemingly degenerate, case.

Keywords: m-sequences, cross correlation, linearized polynomials.

1 Introduction and Preliminaries

Let $\{a_t\}$ and $\{b_t\}$ be two binary sequences of length p. The cross-correlation function between these two sequences at shift τ, where $0 \leq \tau < p$, is defined by

$$C(\tau) = \sum_{t=0}^{p-1} (-1)^{a_t + b_{t+\tau}} \ .$$

Recently, Ness and Helleseth [1] studied the cross correlation between any m-sequence $\{s_t\}$ of length $p = 2^m - 1$ and any m-sequences $\{u_{dt}\}$ of shorter length $2^{m/2} - 1$, where m is even and $\gcd(d, 2^{m/2} - 1) = 1$. For convenience, $\{u_t\}$ is selected to be the m-sequence used in the small Kasami sequence family. The only known families of m-sequences of these periods giving a two-valued cross correlation are related to the Kasami sequences [2] and are obtained taking $d = 1$. Further, families with three-valued cross correlation have been constructed by Ness and Helleseth in [1] and [3]. These results were generalized by Helleseth, Kholosha and Ness [4] who covered all known cases of three-valued cross correlation and conjectured that these were the only existing. The first family with four-valued cross correlation was described in [5].

In this paper, we consider pairs of sequences with at most four-valued cross correlation. We completed a full search for all values of $m \leq 32$ and revealed a

---

* This work was supported by the Norwegian Research Council.

few examples that did not fit into the known families. Except for a single case with $m = 8$ and $d = 7$, all decimations leading to at most four-valued cross correlation are such that $m = 2k$ and $d(2^l + 1) \equiv 2^i \pmod{2^k - 1}$ for some integer l with $0 \le l < k$ and $i \ge 0$. Note that the Kasami case satisfies this general condition on d taking $l = 0$, three-valued cases from [4] are achieved when $\gcd(l, k) = 1$, and the four-valued pairs found in [5] hold with $k = 3t$ and $l = t$. The main result of this paper is finding the cross-correlation distribution for the above decimations. Moreover, for any given $\tau \in \{0, 1, \ldots, 2^k - 2\}$ we exactly compute the corresponding cross-correlation value (see Corollary 3). We also conjecture that we have a characterization of all decimations leading to at most four-valued cross correlation of m-sequences with the described parameters except for a single case.

In the remaining part of this section, we present preliminaries needed to prove our main results. In Sect. 2, we give the distribution of the number of zeros of a particular affine polynomial $A_a(x)$ and find the zeros. This information is useful when obtaining the cross-correlation values and their distribution. Sect. 3 determines the cross-correlation distribution of our family.

Let $\mathrm{GF}(q)$ denote a finite field with q elements and let $\mathrm{GF}(q)^* = \mathrm{GF}(q) \setminus \{0\}$. The finite field $\mathrm{GF}(q^l)$ is a subfield of $\mathrm{GF}(q^m)$ if and only if l divides m. The trace and norm mappings from $\mathrm{GF}(q^m)$ to $\mathrm{GF}(q^l)$ are defined respectively by

$$\mathrm{Tr}_l^m(x) = \sum_{i=0}^{m/l-1} x^{q^{li}} \quad \text{and} \quad \mathrm{N}_l^m(x) = \prod_{i=0}^{m/l-1} x^{q^{li}} .$$

In the case when $l = 1$, we use the notation $\mathrm{Tr}_m(x)$ instead of $\mathrm{Tr}_1^m(x)$.

Let $m = 2k$ and α be an element of order $p = 2^m - 1$ in $\mathrm{GF}(2^m)$. Then the m-sequence $\{s_t\}$ of length $p = 2^m - 1$ can be written in terms of the trace mapping as

$$s_t = \mathrm{Tr}_m(\alpha^t) .$$

Let $\beta = \alpha^{2^k+1}$ be an element of order $2^k - 1$. The sequence $\{u_t\}$ of length $2^k - 1$ (which is used in the construction of the well known Kasami family) is obtained as

$$u_t = \mathrm{Tr}_k(\beta^t) .$$

In this paper, we consider the cross correlation between the m-sequences $\{s_t\}$ and $\{v_t\} = \{u_{dt}\}$ at shift τ defined by

$$C_d(\tau) = \sum_{t=0}^{p-1} (-1)^{s_t + v_{t+\tau}} ,$$

where $\gcd(d, 2^k - 1) = 1$ and $\tau = 0, 1, \ldots, 2^k - 2$.

Take any $a \in \mathrm{GF}(2^k)^*$ and let

$$S(a) = \sum_{x \in \mathrm{GF}(2^m)} (-1)^{\mathrm{Tr}_m(ax) + \mathrm{Tr}_k(x^{d(2^k+1)})} . \tag{1}$$

The following moment identities are known to hold for $S(a)$.

Lemma 1 ([1]). *For any decimation d with $\gcd(d, 2^k - 1) = 1$ the exponential sum $S(a)$ defined in (1) satisfies the following moment identities*

$$\sum_{a \in \mathrm{GF}(2^k)^*} S(a) = 2^k$$

$$\sum_{a \in \mathrm{GF}(2^k)^*} S(a)^2 = 2^{2k}(2^k - 1)$$

$$\sum_{a \in \mathrm{GF}(2^k)^*} S(a)^3 = -2^{4k} + (\lambda + 3)2^{m+k} \ ,$$

where λ is the number of solutions for $x_1, x_2 \in \mathrm{GF}(2^m)^*$ of the equation system

$$1 + x_1 + x_2 = 0$$
$$1 + x_1^{d(2^k+1)} + x_2^{d(2^k+1)} = 0 \ .$$

For the values of d that we consider it is easy to show that $\lambda = 2^{\gcd(l,k)} - 2$.

2   The Affine Polynomial $A_a(x)$

In this section, we consider zeros in $\mathrm{GF}(2^k)$ of the affine polynomial

$$A_a(x) = a^{2^l} x^{2^{2l}} + x^{2^l} + ax + c \ , \tag{2}$$

where $a \in \mathrm{GF}(2^k)$ and $c \in \mathrm{GF}(2^e)$, where $e = \gcd(l, k)$. Some additional conditions on the parameters will be imposed later. The distribution of zeros in $\mathrm{GF}(2^k)$ of (2) will determine to a large extent the distribution of our cross-correlation function. It is clear that $A_a(x)$ does not have multiple roots if $a \neq 0$. Zeros of the linearized homogeneous part of $A_a(x)$ were previously studied in [6,7]. Some results needed here will be cited from [7].

Let $k = ne$ for some n and introduce a particular sequence of polynomials over $\mathrm{GF}(2^k)$ that will play a crucial role when finding zeros of (2). For any $u \in \mathrm{GF}(2^k)$ denote $u_i = u^{2^{il}}$ for $i = 0, \ldots, n-1$ so $A_a(x) = a_1 x_2 + x_1 + a_0 x_0 + c$. Let

$$B_1(x) = 1 \ ,$$
$$B_2(x) = 1 \ ,$$
$$B_{i+2}(x) = B_{i+1}(x) + x_i B_i(x) \quad \text{for} \quad 1 \leq i \leq n-1 \ . \tag{3}$$

Observe the following recursive identity that can be seen as an equivalent definition of $B_i(x)$ and which was proved in [7]

$$B_{i+2}(x) = B_{i+1}^{2^l}(x) + x_1 B_i^{2^{2l}}(x) \quad \text{for} \quad 1 \leq i \leq n-1 \ . \tag{4}$$

We also define polynomials $Z_n(x)$ over $\mathrm{GF}(2^k)$ as $Z_1(x) = 1$ and

$$Z_n(x) = B_{n+1}(x) + x B_{n-1}^{2^l}(x) \tag{5}$$

for $n > 1$. The following lemma describes zeros of $B_n(x)$ in $\mathrm{GF}(2^k)$.

Proposition 1. *Take any $v \in \mathrm{GF}(2^{ne}) \setminus \mathrm{GF}(2^e)$ with $n > 1$ and let*

$$V = \frac{v_0^{2^{2l}+1}}{(v_0 + v_1)^{2^l+1}} \; . \tag{6}$$

Then

$$B_n(V) = \frac{\mathrm{Tr}_e^{ne}(v_0)}{(v_1 + v_2)} \prod_{j=2}^{n-1} \left(\frac{v_0}{v_0 + v_1}\right)^{2^{jl}} .$$

If n is odd (respectively, n is even) then the total number of distinct zeros of $B_n(x)$ in $\mathrm{GF}(2^{ne})$ is equal to $\frac{2^{(n-1)e}-1}{2^{2e}-1}$ (respectively, $\frac{2^{(n-1)e}-2^e}{2^{2e}-1}$). All zeros have the form of (6) with $\mathrm{Tr}_e^{ne}(v_0) = 0$ and occur with multiplicity 2^l. Moreover, polynomial $B_n(x)$ splits in $\mathrm{GF}(2^{ne})$ if and only if $e = l$ or $n < 4$.

Proof. Most of the proof is completed in [7, Lemma 1] and the remaining part follows. In particular, it was shown that

$$B_i(V) = \frac{\sum_{j=1}^{i} v_j}{(v_1 + v_2)} \prod_{j=2}^{i-1} \left(\frac{v_0}{v_0 + v_1}\right)^{2^{jl}} \tag{7}$$

for $2 \le i \le n + 1$ and that $B_n(x)$ splits in $\mathrm{GF}(2^{nl})$ and zeros of $B_n(x)$ in $\mathrm{GF}(2^{ne})$ are exactly the elements obtained by (6) using $w_0 \in \mathrm{GF}(2^{nl}) \setminus \mathrm{GF}(2^l)$ with $\mathrm{Tr}_l^{nl}(w_0) = 0$ that result in $V \in \mathrm{GF}(2^{ne})$.

It also follows from the proof of [7, Proposition 4] that polynomial $f_b(y) = y^{2^l+1} + by + b$ with $b \in \mathrm{GF}(2^{ne})^*$ has exactly $2^e + 1$ zeros in $\mathrm{GF}(2^{ne})$ if and only if b^{-1} has the form of (6) with $\mathrm{Tr}_e^{ne}(v_0) = 0$. Take any $V \in \mathrm{GF}(2^{ne})$ obtained by (6) using $w_0 \in \mathrm{GF}(2^{nl}) \setminus \mathrm{GF}(2^l)$ with $\mathrm{Tr}_l^{nl}(w_0) = 0$. Then $f_{V^{-1}}(y)$ splits in $\mathrm{GF}(2^{nl})$ and, by [6, Corollary 7.2], this is equivalent to $f_{V^{-1}}(y)$ having $2^e + 1$ zeros in $\mathrm{GF}(2^{ne})$. Thus, there exists some $v_0 \in \mathrm{GF}(2^{ne}) \setminus \mathrm{GF}(2^e)$ with $\mathrm{Tr}_e^{ne}(v_0) = 0$ that gives this V using (6). □

Corollary 1. *If n is odd (respectively, n is even) then the total number of distinct zeros of $Z_n(x)$ in $\mathrm{GF}(2^{ne})$ is equal to $\frac{2^{(n+1)e}-2^{2e}}{2^{2e}-1}$ (respectively, $\frac{2^{(n+1)e}-2^e}{2^{2e}-1}$). All zeros have the form of (6) and occur with multiplicity one. Moreover, polynomial $Z_n(x)$ splits in $\mathrm{GF}(2^{ne})$ if and only if $e = l$ or $n = 1$.*

Proof. Most of the proof is completed in [7, Corollary 1] and the remaining part follows.

In particular, it was shown that $Z_n(x)$ splits in $\mathrm{GF}(2^{nl})$ and zeros of $Z_n(x)$ in $\mathrm{GF}(2^{ne})$ are exactly the elements obtained by (6) using $w_0 \in \mathrm{GF}(2^{nl}) \setminus \mathrm{GF}(2^l)$ that result in $V \in \mathrm{GF}(2^{ne})$. It also follows from the proof of [7, Propositions 3, 4] that polynomial $f_b(y) = y^{2^l+1} + by + b$ with $b \in \mathrm{GF}(2^{ne})^*$ has exactly one or $2^e + 1$ zeros in $\mathrm{GF}(2^{ne})$ if and only if b^{-1} has the form of (6). Take any $V \in \mathrm{GF}(2^{ne})$ obtained by (6) using $w_0 \in \mathrm{GF}(2^{nl}) \setminus \mathrm{GF}(2^l)$. Then $f_{V^{-1}}(y)$ has exactly one or $2^l + 1$ zeros $\mathrm{GF}(2^{nl})$ and, by [6, Corollaries 7.2, 7.3], this is equivalent to $f_{V^{-1}}(y)$ having one or $2^e + 1$ zeros in $\mathrm{GF}(2^{ne})$ respectively. Thus, there exists some $v_0 \in \mathrm{GF}(2^{ne}) \setminus \mathrm{GF}(2^e)$ that gives this V using (6). □

Corollary 2. *For any* $V \in \mathrm{GF}(2^{ne})$ *having the form of (6) with* $n > 2$ *and* $\mathrm{Tr}_e^{ne}(v_0) \neq 0$ *we have*

$$\mathrm{Tr}_e^{ne}\left(\frac{B_{n-1}^{2^l}(V)}{B_n^{2^l+1}(V)}\right) = 0 = \mathrm{Tr}_e^{ne}\left(\frac{B_{n-1}^{2^l}(V)B_{n+1}(V)}{B_n^{2^l+1}(V)}\right) ,$$

where the second identity holds if and only if n *is odd.*

Proof. Using (7), it can be verified directly that

$$\frac{B_{n-1}^{2^l}(V)}{B_n^{2^l+1}(V)} = N_e^{ne}\left(1 + \frac{v_1}{v_0}\right)\frac{v_1\sum_{j=2}^{n}v_j}{\mathrm{Tr}_e^{ne}(v_0)^2} \quad \text{and}$$

$$\frac{B_{n-1}^{2^l}(V)B_{n+1}(V)}{B_n^{2^l+1}(V)} = \frac{v_1^2}{\mathrm{Tr}_e^{ne}(v_0)^2} + 1$$

for any $V \in \mathrm{GF}(2^{ne})$ having the form of (6) and $n > 2$. Now note that

$$\mathrm{Tr}_e^{ne}\left(v_1\sum_{j=2}^{n}v_j\right) = \mathrm{Tr}_e^{ne}\left(v_1\mathrm{Tr}_e^{ne}(v_0) + v_1^2\right) = \mathrm{Tr}_e^{ne}(v_0)^2 + \mathrm{Tr}_e^{ne}(v_0^2) = 0 .$$

The second trace is obviously equal 0 if n is odd and is equal 1 if n is even.   □

Lemma 2. *Assume* $A_a(x)$ *has a zero in* $\mathrm{GF}(2^k)$. *Take* v_0 *being any zero of* $A_a(x)$ *in* $\mathrm{GF}(2^k)$. *Then for any zero* v *of* $A_a(x)$ *in* $\mathrm{GF}(2^k)$ *holds*

$$\mathrm{Tr}_k(v) = \mathrm{Tr}_k(v_0) .$$

Proof. The trace identity follows by observing that any zero of $A_a(v)$ is obtained as a sum of v_0 and a zero of its homogeneous part $a^{2^l}x^{2^{2l}} + x^{2^l} + ax$. To prove the identity it therefore suffices to show that $\mathrm{Tr}_k(v_1) = 0$ for any v_1 with $a^{2^l}v_1^{2^{2l}} + v_1^{2^l} + av_1 = 0$. This follows from

$$\mathrm{Tr}_k(v_1) = \mathrm{Tr}_k(v_1^{2^{l+1}}) = \mathrm{Tr}_k(v_1^{2^l+2^l}) = \mathrm{Tr}_k(a^{2^l}v_1^{2^{2l}+2^l} + av_1^{2^l+1}) = 0 ,$$

as claimed.   □

We will need the following result that can be obtained combining Theorems 5.6 and 6.4 in [6].

Theorem 1 ([6]). *Take polynomials over* $\mathrm{GF}(2^k)$

$$f(x) = x^{2^l+1} + b^2x + b^2 \quad \text{and} \quad g(x) = b^{-1}f(bx^{2^l-1}) = b^{2^l}x^{2^{2l}-1} + b^2x^{2^l-1} + b$$

with $b \neq 0$ *and* $\gcd(l,k) = e$. *Then exactly one of the following holds*

(i) $f(x)$ *has none or two zeros in* $\mathrm{GF}(2^k)$ *and* $g(x)$ *has none zeros in* $\mathrm{GF}(2^k)$;
(ii) $f(x)$ *has one zero in* $\mathrm{GF}(2^k)$, $g(x)$ *has* $2^e - 1$ *zeros in* $\mathrm{GF}(2^k)$ *and each rational root* δ *of* $g(x)$ *satisfies* $\mathrm{Tr}_e^k(b^{-1}\delta^{-(2^l+1)}) \neq 0$;

(iii) $f(x)$ has $2^e + 1$ zeros in $\mathrm{GF}(2^k)$ and $g(x)$ has $2^{2e} - 1$ zeros in $\mathrm{GF}(2^k)$.

Let N_i denote the number of $b \in \mathrm{GF}(2^k)^*$ such that $g(x) = 0$ has exactly i roots in $\mathrm{GF}(2^k)$. Then the following distribution holds for k/e odd (respectively, k/e even)

$$
\begin{aligned}
N_0 &= \tfrac{2^{k+2e} - 2^{k+e} - 2^k + 1}{2^{2e} - 1} &\quad &(\text{resp. } \tfrac{2^{k+2e} - 2^{k+e} - 2^k - 2^{2e} + 2^e + 1}{2^{2e} - 1}), \\
N_{2^e - 1} &= 2^{k-e} - 1 &\quad &(\text{resp. } 2^{k-e}), \\
N_{2^{2e} - 1} &= \tfrac{2^{k-e} - 1}{2^{2e} - 1} &\quad &(\text{resp. } \tfrac{2^{k-e} - 2^e}{2^{2e} - 1}).
\end{aligned}
$$

Note 1. Let

$$
M_i = \{a \mid a \neq 0, A_a(x) \text{ has exactly } i \text{ zeros in } \mathrm{GF}(2^k)\} .
$$

Obviously, either $A_a(x)$ has no zeros in $\mathrm{GF}(2^k)$ or it has exactly the same number of zeros as its linearized homogeneous part that is

$$
l_a(x) = a_1 x^{2^{2l}} + x^{2^l} + a_0 x . \tag{8}
$$

The zeros in $\mathrm{GF}(2^k)$ of $l_a(x)$ form a vector subspace over $\mathrm{GF}(2^e)$ and thus, the number of zeros can be equal to $1, 2^e, 2^{2e}, \ldots, 2^{2l}$ (we will see that, in fact, $l_a(x)$ can not have more than 2^{2e} zeros). Assume $a \neq 0$, then dividing $l_a(x)$ by $a_0 a_1 x$ (we remove one zero $x = 0$) and then substituting x with $a_0^{-1} x$ leads to $a_1^{-2^l} x^{2^{2l} - 1} + a_1^{-2} x^{2^l - 1} + a_1^{-1}$ which has the form of polynomial $g(x)$ from Theorem 1 taking $b = a_1^{-1}$ (note a 1-to-1 correspondence between a and b). Thus, $l_a(x)$ has either $1, 2^e$ or 2^{2e} zeros in $\mathrm{GF}(2^k)$ and $|M_i| \leq N_{i-1}$ for $i \in \{1, 2^e, 2^{2e}\}$. The equality would hold if we prove that $A_a(x)$ always has a zero in $\mathrm{GF}(2^k)$.

Proposition 2. *For any $a \in \mathrm{GF}(2^{ne})^*$, polynomial $A_a(x)$ has exactly one zero in $\mathrm{GF}(2^{ne})$ if and only if $Z_n(a) \neq 0$. Moreover, this zero is equal to $V_a = cB_n(a)/Z_n(a)$ and $\mathrm{Tr}_{ne}(V_a) = \mathrm{Tr}_e(nc)$. Also if n is odd (resp. n is even) then*

$$
|M_1| = \frac{2^{k+2e} - 2^{k+e} - 2^k + 1}{2^{2e} - 1} \quad (\text{resp. } \frac{2^{k+2e} - 2^{k+e} - 2^k - 2^{2e} + 2^e + 1}{2^{2e} - 1}) .
$$

Proof. If $n = 1$ then the proof is obvious, so we further assume $n > 1$ and start with proving that $cB_n(a)/Z_n(a)$ indeed is a zero of $A_a(x)$ if $Z_n(a) \neq 0$. First, for any $v \in \mathrm{GF}(2^{ne})$, using both recursive definitions of $B_n(x)$

$$
\begin{aligned}
Z_n^{2^l}(v) &\overset{(5)}{=} B_{n+1}^{2^l}(v) + v_1 B_{n-1}^{2^{2l}}(v) \\
&\overset{(3)}{=} B_n^{2^l}(v) + v_0 B_{n-1}^{2^l}(v) + v_1 B_{n-1}^{2^{2l}}(v) \\
&\overset{(4)}{=} B_{n+1}(v) + v_0 B_{n-1}^{2^l}(v) \\
&\overset{(5)}{=} Z_n(v)
\end{aligned}
$$

and thus, $Z_n(v) \in \mathrm{GF}(2^l)$. Since $\mathrm{GF}(2^{ne}) \cap \mathrm{GF}(2^l) = \mathrm{GF}(2^e)$, we have $Z_n(v) \in \mathrm{GF}(2^e)$. Therefore,

$$A_a(\mathcal{V}_a) = \frac{c}{Z_n(a)} \left(a_1 B_n^{2^{2l}}(a) + B_n^{2^l}(a) + a_0 B_n(a) + Z_n(a) \right) \tag{9}$$

$$\overset{(3)}{=} \frac{c}{Z_n(a)} \left(a_1 B_{n-1}^{2^{2l}}(a) + a_1 a_0 B_{n-2}^{2^{2l}}(a) + B_n^{2^l}(a) + a_0 B_n(a) + Z_n(a) \right)$$

$$\overset{(4)}{=} \frac{c}{Z_n(a)} \left(B_{n+1}(a) + a_0 B_{n-1}^{2^l}(a) + Z_n(a) \right) = 0 .$$

Showing that in our case $\mathcal{V}_a$ is the only zero of $A_a(x)$ is done exactly the way presented in [7, Proposition 2].

Using Corollary 1, we can obtain the number of $a \in \mathrm{GF}(2^{ne})^*$ such that $Z_n(a) \neq 0$ (note that $Z_n(0) = 1$). Observe that this number is identical to N_0 from Theorem 1 that is equal to the number of $a \in \mathrm{GF}(2^{ne})^*$ such that $l_a(x) = 0$ has exactly one root in $\mathrm{GF}(2^{ne})$ (see explanations following Theorem 1). Therefore, if $A_a(x)$ has exactly one zero in $\mathrm{GF}(2^{ne})$ then its homogeneous part $l_a(x)$ has the same number of zeros and so a is necessarily such that $Z_n(a) \neq 0$.

Finally, to prove the trace identity for $\mathcal{V}_a$ first note that for any $v \in \mathrm{GF}(2^{ne})$

$$\mathrm{Tr}_e^{ne}(B_n(v) + Z_n(v)) \overset{(5)}{=} \mathrm{Tr}_e^{ne} \left(B_n(v) + B_{n+1}(v) + v_0 B_{n-1}^{2^l}(v) \right) \tag{10}$$

$$\overset{(3)}{=} \mathrm{Tr}_e^{ne} \left(B_n(v) + B_n(v) + v_{n-1} B_{n-1}(v) + v_0 B_{n-1}^{2^l}(v) \right)$$

$$= \mathrm{Tr}_e^{ne} \left(v_{n-1} B_{n-1}(v) + (v_{n-1} B_{n-1}(v))^{2^l} \right) = 0 .$$

Therefore, since c and $Z_n(v)$ are both in $\mathrm{GF}(2^e)$, then

$$\mathrm{Tr}_{ne}(\mathcal{V}_v) = \mathrm{Tr}_{ne} \left(c + c \frac{B_n(v) + Z_n(v)}{Z_n(v)} \right)$$

$$= \mathrm{Tr}_e(nc) + \mathrm{Tr}_e \left(\frac{c}{Z_n(v)} \mathrm{Tr}_e^{ne}(B_n(v) + Z_n(v)) \right)$$

$$= \mathrm{Tr}_e(nc) .$$

This completes the proof.   □

Proposition 3. *Let n be odd and take any $a \in \mathrm{GF}(2^{ne})^*$. Then polynomial $A_a(x)$ has exactly 2^e zeros in $\mathrm{GF}(2^{ne})$ if and only if $Z_n(a) = 0$ and $B_n(a) \neq 0$. Moreover, these zeros are the following*

$$v_\mu = c \sum_{i=0}^{\frac{n-1}{2}} \frac{B_{n-1}^{2^{(2i+1)l}}(a)}{B_n^{2^{(2i+1)l} + 2^{2il} - 1}(a)} + \mu B_n(a)$$

with $\mu \in \mathrm{GF}(2^e)$ and for each zero of this type $\mathrm{Tr}_{ne}(v_\mu) = 0$. Also $|M_{2^e}| = 2^{k-e} - 1$.

Proof. First, we consider $l_a(x)$ defined in (8) that is the linearized homogeneous part of $A_a(x)$, and prove that it has exactly 2^e zeros in $GF(2^{ne})$ if and only if $Z_n(a) = 0$ and $B_n(a) \neq 0$. We quote the following fact that can be found in the proof of [7, Proposition 3]. Namely, polynomial $f_b(y) = y^{2^l+1} + by + b$ with $b \in GF(2^{ne})^*$ has exactly one zero in $GF(2^{ne})$ if and only if b^{-1} has the form of (6) with $\text{Tr}_e^{ne}(v_0) \neq 0$.

By Note 1 and Theorem 1 (ii), $l_a(x)$ has exactly 2^e zeros in $GF(2^{ne})$ if and only if $f_{a^{-1}}(y)$ (as well as $f_{a^{-2^l+1}}(y)$) has exactly one zero in $GF(2^{ne})$ and, by the fact cited above, this is equivalent to a having the form of (6) with $\text{Tr}_e^{ne}(v_0) \neq 0$. Then it remains to apply Proposition 1 and Corollary 1.

If $A_a(x)$ has exactly 2^e zeros in $GF(2^{ne})$ then the same holds for its homogeneous part $l_a(x)$ and we already proved that in this case, $Z_n(a) = 0$ and $B_n(a) \neq 0$. Now we have to find a particular solution of $A_a(x) = 0$ assuming $Z_n(a) = 0$ and $B_n(a) \neq 0$. Then, by (9),

$$a_1 B_n^{2^{2l}}(a) + B_n^{2^l}(a) + a_0 B_n(a) = 0 \tag{11}$$

which means that all 2^e distinct elements $\mu B_n(a) \in GF(2^{ne})$ for $\mu \in GF(2^e)$ are zeros of $l_a(x)$ (since $GF(2^{ne}) \cap GF(2^l) = GF(2^e)$).

The remaining calculations are technical and are placed in Appendix A. The identity for $|M_{2^e}|$ follows from Theorem 1. □

Proposition 4. *Take any $a \in GF(2^{ne})^*$. Then polynomial $l_a(x)$ from (8) has exactly 2^{2e} zeros in $GF(2^{ne})$ if and only if $B_n(a) = 0$.*

Proof. We quote the following fact that can be found in the proof of [7, Proposition 4]. Namely, polynomial $f_b(y) = y^{2^l+1} + by + b$ with $b \in GF(2^{ne})^*$ has exactly $2^e + 1$ zeros in $GF(2^{ne})$ if and only if b^{-1} has the form of (6) with $\text{Tr}_e^{ne}(v_0) = 0$.

By Note 1 and Theorem 1 (iii), $l_a(x)$ has exactly 2^{2e} zeros in $GF(2^{ne})$ if and only if $f_{a^{-1}}(y)$ (as well as $f_{a^{-2^l+1}}(y)$) has exactly $2^e + 1$ zeros in $GF(2^{ne})$ and, by the fact cited above, this is equivalent to a having the form of (6) with $\text{Tr}_e^{ne}(v_0) = 0$. Then it remains to apply Proposition 1. □

3   Three- and Four-Valued Cross Correlation

In this section, we prove our main result. First, we consider the exponential sum denoted $S_0(a)$ that to some extent is determined by the following proposition.

Proposition 5. *Take integers l and k with $0 \leq l < k$ such that k/e is odd, where $e = \gcd(l, k)$. For any $a \in GF(2^k)$ define*

$$S_0(a) = \sum_{y \in GF(2^{2k})} (-1)^{\text{Tr}_{2k}(ay^{2^l+1}) + \text{Tr}_k(y^{2^k+1})} ,$$

Then

$$S_0(a) = 2^k \sum_{v \in GF(2^k), A_a(v) = 0} (-1)^{\text{Tr}_k \left(a(l/e+1)c^{-2}v^{2^l+1} + v \right)} , \tag{12}$$

where $A_a(x)$ comes from (2) with $c^{-1} = \delta + \delta^{-1} \in \mathrm{GF}(2^e)$ for δ being a primitive $(2^e + 1)^{\mathrm{th}}$ root of unity over $\mathrm{GF}(2)$, and $\mathrm{Tr}_e(c) = 1$. Moreover, $S_0(a)^2$ taken for all $a \in \mathrm{GF}(2^k)^*$ has the following distribution for l/e even:

0	*occurs*	$2^{k-e} - 1$	*times*
2^{2k}	*occurs*	$\frac{2^{k+2e} - 2^{k+e} - 2^k + 1}{2^{2e} - 1}$	*times*
$2^{2(k+e)}$	*occurs*	$\frac{2^{k-e} - 1}{2^{2e} - 1}$	*times* .

Proof. Let δ be a primitive $(2^e + 1)^{\mathrm{th}}$ root of unity over $\mathrm{GF}(2)$. Then any element in $\mathrm{GF}(2^{2k})$ can be written uniquely as $y = u + \delta v$ with $u, v \in \mathrm{GF}(2^k)$. This easily follows from the fact that $\delta \in \mathrm{GF}(2^{2e}) \setminus \mathrm{GF}(2^e)$ and noting that $\mathrm{GF}(2^k) \cap \mathrm{GF}(2^{2e}) = \mathrm{GF}(2^e)$ since k/e is odd.

Denote $\bar{y} = y^{2^k}$ and $c^{-1} = \delta + \delta^{-1} \in \mathrm{GF}(2^e) \subset \mathrm{GF}(2^l)$, then we obtain

$$y^{2^l+1} + \bar{y}^{2^l+1} = (u + \delta v)^{2^l+1} + (u + \delta^{2^e} v)^{2^l+1}$$
$$= (\delta^{2^l+1} + \delta^{2^e(2^l+1)})v^{2^l+1} + (\delta^{2^l} + \delta^{2^{e+l}})uv^{2^l} + (\delta + \delta^{2^e})u^{2^l}v$$
$$= (\delta + \delta^{-1})(uv^{2^l} + u^{2^l}v) + (l/e + 1)(\delta^2 + \delta^{-2})v^{2^l+1}$$

and further

$$y^{2^k+1} = (u + \delta v)^{2^k+1}$$
$$= u^{2^k+1} + u^{2^k}v\delta + uv^{2^k}\delta^{2^k} + v^{2^k+1}$$
$$= u^2 + (\delta + \delta^{-1})uv + v^2 .$$

Hence, we get

$$S_0(a) = \sum_{y \in \mathrm{GF}(2^{2k})} (-1)^{\mathrm{Tr}_k\left(a(y^{2^l+1} + \bar{y}^{2^l+1}) + y^{2^k+1}\right)}$$

$$= \sum_{u,v \in \mathrm{GF}(2^k)} (-1)^{\mathrm{Tr}_k\left(a\left(c^{-1}(uv^{2^l} + u^{2^l}v) + (l/e+1)c^{-2}v^{2^l+1}\right) + u^2 + c^{-1}uv + v^2\right)}$$

$$= \sum_{v \in \mathrm{GF}(2^k)} (-1)^{\mathrm{Tr}_k\left(a(l/e+1)c^{-2}v^{2^l+1} + v\right)}$$

$$\times \sum_{u \in \mathrm{GF}(2^k)} (-1)^{\mathrm{Tr}_k\left(u^{2^l}c^{-1}(a^{2^l}v^{2^{2l}} + v^{2^l} + av + c)\right)}$$

$$= 2^k \sum_{v \in \mathrm{GF}(2^k),\, A_a(v)=0} (-1)^{\mathrm{Tr}_k\left(a(l/e+1)c^{-2}v^{2^l+1} + v\right)} ,$$

where $A_a(x) = a^{2^l}x^{2^{2l}} + x^{2^l} + ax + c$ and $c^{-1} = \delta + \delta^{-1}$.

Consider equation $x^2 + c^{-1}x = 1$ that has two roots δ and δ^{-1} which are elements in $\mathrm{GF}(2^{2e})$ but not in $\mathrm{GF}(2^e)$. Letting $x = c^{-1}y$ we get $y^2 + y = c^2$

that has two solutions $c\delta$ and $c\delta^{-1}$ which do not belong to $\mathrm{GF}(2^e)$. Thus, $\mathrm{Tr}_e(c^2) = \mathrm{Tr}_e(c) = 1$.

Define function $\chi(x) = (-1)^{\mathrm{Tr}_{2k}(ax^{2^l+1})+\mathrm{Tr}_k(x^{2^k+1})}$ on $\mathrm{GF}(2^{2k})$ and linearized polynomial $L_a(x) = a^{2^l}x^{2^{2l}} + x^{2^{k+l}} + ax$. Note that for any $u, v \in \mathrm{GF}(2^{2k})$ with $L_a(u) = L_a(v) = 0$ we have

$$\chi(u+v) = \chi(u)\chi(v)(-1)^{\mathrm{Tr}_{2k}\left(a(uv^{2^l}+u^{2^l}v)\right)+\mathrm{Tr}_k(uv^{2^k}+u^{2^k}v)}$$

$$= \chi(u)\chi(v)(-1)^{\mathrm{Tr}_{2k}\left(u^{2^l}L_a(v)\right)} = \chi(u)\chi(v) \ .$$

Therefore, $\chi(x)$ defines homomorphism on the set of zeros of $L_a(x)$ and, thus, since $L_a(x)$ is a linearized polynomial, $\chi(x)$ is either identically 1 or is balanced on this set. Now we can compute

$$S_0(a)^2 = \sum_{x,y\in\mathrm{GF}(2^{2k})} (-1)^{\mathrm{Tr}_{2k}(a(x^{2^l+1}+y^{2^l+1}))+\mathrm{Tr}_k(x^{2^k+1}+y^{2^k+1})}$$

$$= \sum_{y,v\in\mathrm{GF}(2^{2k})} (-1)^{\mathrm{Tr}_{2k}\left(a((v+y)^{2^l+1}+y^{2^l+1})\right)+\mathrm{Tr}_k\left((v+y)^{2^k+1}+y^{2^k+1}\right)}$$

$$= \sum_{y,v\in\mathrm{GF}(2^{2k})} (-1)^{\mathrm{Tr}_{2k}\left(a(vy^{2^l}+v^{2^l}y+v^{2^l+1})+yv^{2^k}\right)+\mathrm{Tr}_k(v^{2^k+1})}$$

$$= \sum_{v\in\mathrm{GF}(2^{2k})} (-1)^{\mathrm{Tr}_{2k}(av^{2^l+1})+\mathrm{Tr}_k(v^{2^k+1})} \sum_{y\in\mathrm{GF}(2^{2k})} (-1)^{\mathrm{Tr}_{2k}\left(y^{2^l}L_a(v)\right)}$$

$$= 2^{2k} \sum_{v\in\mathrm{GF}(2^{2k}),\, L_a(v)=0} \chi(v)$$

$$= 2^{2k}\#\{v \in \mathrm{GF}(2^{2k}) \mid L_a(v) = 0\} \quad \text{or} \quad 0 \ .$$

Now recall that $a \in \mathrm{GF}(2^k)$ and, thus, $L_a(x) = a^{2^{k+l}}x^{2^{2(k+l)}} + x^{2^{k+l}} + ax$ is similar to $l_a(x)$ from (8). By Note 1, $L_a(x)$ has either 1, 2^e or 2^{2e} zeros in $\mathrm{GF}(2^{2k})$ if l/e is even because

$$\gcd(k+l, 2k) = e\gcd(k/e+l/e, 2k/e) = e\gcd(k/e+l/e, k/e) = e$$

since k/e is odd.

Now we show that $L_a(x)$ can not have 2^e zeros in $\mathrm{GF}(2^{2k})$ if l/e is even. This obviously holds for $a = 0$. Take $a \neq 0$ and assume the opposite. Then there exists some $\mathcal{V} \in \mathrm{GF}(2^{2k})^*$ with $L_a(\mathcal{V}) = 0$ and all zeros of $L_a(x)$ are exactly $\{\mu\mathcal{V} \mid \mu \in \mathrm{GF}(2^e)\}$. Note that $L_a(\mathcal{V}^{2^k}) = L_a(\mathcal{V})^{2^k} = 0$ since $a \in \mathrm{GF}(2^k)$ and, thus, $\mathcal{V}^{2^k-1} \in \mathrm{GF}(2^e)$. Take ξ being a primitive element of $\mathrm{GF}(2^{2k})$ and assume $\mathcal{V} = \xi^i$. Then $\mathcal{V}^{2^k-1} \in \mathrm{GF}(2^e)$ if and only if $2^{2k}-1$ divides $i(2^k-1)(2^e-1)$ which is equivalent to 2^k+1 divide $i(2^e-1)$ and, further, to 2^k+1 divide i since $\gcd(2^k+1, 2^e-1) = 1$ if k/e is odd. Therefore, $\mathcal{V} \in \mathrm{GF}(2^k)^*$. Taking any $\delta \neq 0$ it can be checked directly that $L_a(a^{-1}\delta) = 0$ if and only if $G(\delta) = 0$, where

$G(x) = b^{2^{k+l}} x^{2^{2(k+l)}-1} + b^2 x^{2^{k+l}-1} + b$ and $b = a^{-2^{k+l}}$. Note that $G(x)$ has the form of polynomial $g(x)$ from Theorem 1. Thus, $G(x)$ has 2^e zeros in $\mathrm{GF}(2^{2k})$ with $a\mathcal{V}$ being one of them and, by the trace condition from Theorem 1 (ii), $\mathrm{Tr}_e^{2k}(a^{2^{k+l}}(a\mathcal{V})^{-(2^{k+l}+1)}) \neq 0$ which is wrong.

Finally, we prove the value distribution of $S_0(a)^2$. By (12), $S_0(a)^2 = 2^{2k}$ if and only if $A_a(x)$ has just one zero in $\mathrm{GF}(2^k)$. Therefore, by Proposition 2, $S_0(a)^2$ is equal to 2^{2k} for $|M_1| = \frac{2^{k+2e} - 2^{k+e} - 2^k + 1}{2^{2e}-1}$ values of a. If l/e is even then suppose $S_0(a)^2$ takes on value zero r times and the value $2^{2(k+e)}$ occurs t times. Taking the sum-of-squares identity from Lemma 1 giving $2^{2k}|M_1| + 2^{2(k+e)}t = 2^{2k}(2^k - 1)$ and knowing also that $r + |M_1| + t = 2^k - 1$, we are directly led to the claimed distribution.   □

Corollary 3. *Under the conditions of Proposition 5, let $n = k/e$. Then the distribution of $S_0(a)$ for l/e being even is as follows:*

$$
\begin{array}{lll}
- 2^k(-1)^{\mathrm{Tr}_k(aB_n(a)^{2^l+1}/Z_n(a)^2)} & if & Z_n(a) \neq 0 \\
0 & if & Z_n(a) = 0 \text{ and } B_n(a) \neq 0 \\
\pm 2^{k+e} & if & B_n(a) = 0 \ .
\end{array}
$$

Proof. If $Z_n(a) \neq 0$ then, by Proposition 2, $A_a(x)$ has exactly one zero in $\mathrm{GF}(2^k)$ that is equal to $\mathcal{V}_a = cB_n(a)/Z_n(a)$ and $\mathrm{Tr}_k(\mathcal{V}_a) = \mathrm{Tr}_e(c) = 1$ since n is odd. Further,

$$
\mathrm{Tr}_k\left(ac^{-2}\mathcal{V}_a^{2^l+1}\right) = \mathrm{Tr}_k\left(\frac{aB_n(a)^{2^l+1}}{Z_n(a)^2}\right)
$$

since $c, Z_n(a) \in \mathrm{GF}(2^e)$ as proved in Proposition 2.

Further, assume $Z_n(a) = 0$ and $B_n(a) \neq 0$. By Proposition 1 and Corollary 1, there exists some $v \in \mathrm{GF}(2^{ne}) \setminus \mathrm{GF}(2^e)$ with $\mathrm{Tr}_e^{ne}(v) \neq 0$ such that $a = v_0^{2^{2l}+1}/(v_0 + v_1)^{2^l+1}$. Using Proposition 1 we obtain

$$
aB_n(a)^{2^l+1} = N_e^{ne}\left(\frac{v_0}{v_0+v_1}\right)^2 \frac{\mathrm{Tr}_e^{ne}(v_0)^2}{v_1^2} \ .
$$

Note that $\mathrm{Tr}_e^{ne}(v) = \mathrm{Tr}_l^{nl}(v)$ and $N_e^{ne}(v) = N_l^{nl}(v)$ if $v \in \mathrm{GF}(2^{ne})$.

By Proposition 3, $A_a(x)$ has exactly 2^e zeros in $\mathrm{GF}(2^k)$ that have particular form v_μ corresponding to every $\mu \in \mathrm{GF}(2^e)$ and $\mathrm{Tr}_k(v_\mu) = 0$. Checking the trace calculations in the proof of Proposition 3 (see Appendix A), it is easy to see that

$$
\mathrm{Tr}_e^{ne}(\mathcal{V}B_n(a)) = \mathrm{Tr}_e^{ne}\left(\frac{B_{n-1}^{2^l}(a)B_{n+1}^2(a)}{B_n^{2^l+1}(a)}\right)
$$

$$
\overset{(7)}{=} N_e^{ne}\left(\frac{v_0}{v_0+v_1}\right)\mathrm{Tr}_e^{ne}\left(\frac{(v_1 + \mathrm{Tr}_e^{ne}(v_0))^3}{\mathrm{Tr}_e^{ne}(v_0)^2 v_1}\right)
$$

$$
= N_e^{ne}\left(\frac{v_0}{v_0+v_1}\right)\left(1 + \mathrm{Tr}_e^{ne}(v_0)\mathrm{Tr}_e^{ne}(v_0^{-1})\right) \ ,
$$

where $\mathcal{V} = c^{-1}(v_\mu + \mu B_n(a))$. Now it is quite technical to show that

$$\sum_{\mu \in \mathrm{GF}(2^e)} (-1)^{\mathrm{Tr}_k \left(ac^{-2} v_\mu^{2^l+1} + v_\mu \right)} = 0$$

(see Appendix B for the details).

From Proposition 5 we know that the number of $a \in \mathrm{GF}(2^k)^*$ for which $S_0(a) = 0$ is equal to $2^{k-e} - 1$ which is equal to $|M_{2^e}|$ from Proposition 3. Thus, $S_0(a) = 0$ only if $Z_n(a) = 0$ and $B_n(a) \neq 0$.

Finally, if $B_n(a) = 0$ then, by Proposition 4, $A_a(x)$ has either none or 2^{2e} zeros in $\mathrm{GF}(2^k)$ and, thus, $S_0(a) = \pm 2^{k+e}$ since zero is impossible and $\pm 2^k$ occurs if and only if $A_a(x)$ has exactly one zero.   □

Using the trace representation, Ness and Helleseth [1] showed that the set of values of $C_d(\tau) + 1$ for $\tau = 0, 1, \ldots, 2^k - 2$ is equal to the set of values of $S(a)$ defined in (1) taking $a \in \mathrm{GF}(2^k)^*$. Lemma 1 and Proposition 5 allow to determine the distribution of $S(a)$ completely.

Theorem 2. *Let $m = 2k$ and $d(2^l + 1) \equiv 2^i \pmod{2^k - 1}$ for some integer l with $0 \leq l < k$ and $i \geq 0$. Then the exponential sum $S(a)$ defined in (1) for $a \in \mathrm{GF}(2^k)^*$ (and $C_d(\tau) + 1$ for $\tau = 0, 1, \ldots, 2^k - 2$) have the following distribution:*

-2^{k+e}	*occurs*	$\frac{2^{k-e}-1}{2^{2e}-1}$	*times*
-2^k	*occurs*	$\frac{(2^k-1)(2^{e-1}-1)}{2^e-1}$	*times*
0	*occurs*	$2^{k-e} - 1$	*times*
2^k	*occurs*	$\frac{(2^k+1)2^{e-1}}{2^e+1}$	*times* ,

where $e = \gcd(l, k)$.

Proof. If $l = 0$ we assume $e = k$. Now note that conditions of the theorem allow k/e to be odd only. Indeed, suppose k/e is even then $\gcd(2^l + 1, 2^k - 1) = 2^e + 1$ and, thus, $2^e + 1$ divides 2^i which is impossible. Further, note that $d(2^l + 1) \equiv 2^i \pmod{2^k - 1}$ conditioned by the theorem, holds if and only if $d(2^{k-l} + 1) \equiv 2^{i+k-l} \pmod{2^k - 1}$ meaning that l can be substituted by $k - l$. Now suppose that $\gcd(2^l + 1, 2^{2k} - 1) \neq 1$ which holds if and only if $v_2(l) = v_2(k)$ (since we know already that $v_2(l) \geq v_2(k)$). Obviously, $\max\{v_2(l), v_2(k-l)\} > v_2(k)$ since if $v_2(l) = v_2(k)$ then $v_2(k-l) \geq v_2(k) + 1$. Thus, we can assume that $\gcd(2^l + 1, 2^{2k} - 1) = 1$. The latter is equivalent to l/e being even.

Finding the distribution of the cross-correlation function $C_d(\tau) + 1$ is equivalent to computing the distribution of $S(a)$ defined in (1) for $a \in \mathrm{GF}(2^k)^*$. Since $\gcd(2^l + 1, 2^{2k} - 1) = 1$, substituting $x = y^{2^l+1}$ in the expression for $S(a)$ and since $d(2^l + 1)(2^k + 1) \equiv 2^i(2^k + 1) \pmod{2^m - 1}$, we are led to

$$S(a) = \sum_{y \in \mathrm{GF}(2^m)} (-1)^{\mathrm{Tr}_m(ay^{2^l+1}) + \mathrm{Tr}_k(y^{2^k+1})} = S_0(a) ,$$

where $S_0(a)$ comes from Proposition 5.

Suppose $S_0(a)$ takes on value 2^k totally r times, the value -2^k is taken on s times, the value 2^{k+e} occurs t times and -2^{k+e} occurs z times. From the distribution of $S_0(a)^2$ proven in Proposition 5 we get $r + s = \frac{2^{k+2e} - 2^{k+e} - 2^k + 1}{2^{2e} - 1}$ and $t + z = \frac{2^{k-e} - 1}{2^{2e} - 1}$. Taking the simple sum and the sum-of-cubes identities from Lemma 1 we get

$$2^k(r - s) + 2^{k+e}(t - z) = 2^k$$
$$2^{3k}(r - s) + 2^{3(k+e)}(t - z) = -2^{4k} + (2^e + 1)2^{3k} \ .$$

Solving this system of four equations for four unknowns gives the claimed distribution.   □

As noted above in the proof, from the conditions of Theorem 2 it follows that k/e is odd. Thus, the condition for k to be odd can safely be removed from Theorem 2 in [4], where particular case of $e = 1$ is considered. Also note that if $e = 1$ then the cross correlation is three-valued since the value -2^k is not taken on. If $l = 0$ then the cross correlation is two-valued since the values 0 and -2^{k+e} are not taken on, which gives the Kasami case where $d = 1$. Also note that the value of $C_d(\tau)$ and $S(a)$ can be found directly using Corollary 3 which is usually a much more difficult task than finding the value distribution.

Conjecture 1. Except for the case when $m = 8$ and $d = 7$, all decimations leading to at most four-valued cross correlation between two m-sequences of different lengths $2^m - 1$ and $2^k - 1$, where $m = 2k$, are described in Theorem 2.

4   Conclusions

We have identified pairs of m-sequences having different lengths $2^{2k} - 1$ and $2^k - 1$ with at most four-valued cross correlation and we have completely determined the cross-correlation distribution. In these pairs, decimation d is taken such that $d(2^l + 1) \equiv 2^i \pmod{2^k - 1}$ for some integer l with $0 \leq l < k$ and $i \geq 0$. Our results cover the two-valued Kasami case where $d = 1$ and all three-valued decimations found in [4]. Conjectured is that we have a characterization of all decimations leading to at most four-valued cross correlation of m-sequences with the described parameters except for a single, seemingly degenerate, case.

References

1. Ness, G.J., Helleseth, T.: Cross correlation of m-sequences of different lengths. IEEE Trans. Inf. Theory 52(4), 1637–1648 (2006)
2. Kasami, T.: Weight distribution formula for some classes of cyclic codes. Technical Report R-285 (AD 637524), Coordinated Science Laboratory, University of Illinois, Urbana (April 1966)
3. Ness, G.J., Helleseth, T.: A new three-valued cross correlation between m-sequences of different lengths. IEEE Trans. Inf. Theory 52(10), 4695–4701 (2006)
4. Helleseth, T., Kholosha, A., Ness, G.J.: Characterization of m-sequences of lengths $2^{2k} - 1$ and $2^k - 1$ with three-valued crosscorrelation. IEEE Trans. Inf. Theory 53(6), 2236–2245 (2007)

5. Ness, G.J., Helleseth, T.: A new family of four-valued cross correlation between *m*-sequences of different lengths. IEEE Trans. Inf. Theory 53(11), 4308–4313 (2007)
6. Bluher, A.W.: On $x^{q+1} + ax + b$. Finite Fields and Their Applications 10(3), 285–305 (2004)
7. Helleseth, T., Kholosha, A.: On the equation $x^{2^l+1} + x + a = 0$ over GF(2^k). Finite Fields and Their Applications 14(1), 159–176 (2008)

A   Remaining Proof of Proposition 3

By Corollary 1, a has the form of (6) and, by Proposition 1, $\mathrm{Tr}_e^{ne}(v_0) \neq 0$. Using these facts and assuming that n is odd (note that the latter assumption is involved only at this stage), we compute

$$
A_a \left(c \sum_{i=0}^{\frac{n-1}{2}} \frac{B_{n-1}^{2^{(2i+1)l}}(a)}{B_n^{2^{(2i+1)l}+2^{2il}-1}(a)} \right) = ca_1 B_n^{2^{2l}}(a) \sum_{i=1}^{\frac{n-1}{2}} \frac{B_{n-1}^{2^{(2i+1)l}}(a)}{B_n^{2^{(2i+1)l}+2^{2il}}(a)}
$$

$$
+ ca_1 B_n^{2^{2l}}(a) \frac{B_{n-1}^{2^{2l}}(a)}{B_n^{2^{2l}+2^l}(a)} + c B_n^{2^l}(a) \sum_{i=1}^{\frac{n-1}{2}} \frac{B_{n-1}^{2^{2il}}(a)}{B_n^{2^{2il}+2^{(2i-1)l}}(a)}
$$

$$
+ c B_n^{2^l}(a) \frac{B_{n-1}^{2^l}(a)}{B_n^{2^l+1}(a)} + ca_0 B_n(a) \sum_{i=0}^{\frac{n-1}{2}} \frac{B_{n-1}^{2^{(2i+1)l}}(a)}{B_n^{2^{(2i+1)l}+2^{2il}}(a)} + c
$$

$$
\overset{(11)}{=} c B_n^{2^l}(a) \sum_{i=0}^{\frac{n-1}{2}} \frac{B_{n-1}^{2^{(2i+1)l}}(a)}{B_n^{2^{(2i+1)l}+2^{2il}}(a)} + c B_n^{2^l}(a) \sum_{i=1}^{\frac{n-1}{2}} \frac{B_{n-1}^{2^{2il}}(a)}{B_n^{2^{2il}+2^{(2i-1)l}}(a)}
$$

$$
+ ca_0 B_n(a) \frac{B_{n-1}^{2^l}(a)}{B_n^{2^l+1}(a)} + ca_1 B_n^{2^{2l}}(a) \frac{B_{n-1}^{2^{2l}}(a)}{B_n^{2^{2l}+2^l}(a)} + c
$$

$$
= c B_n^{2^l}(a) \mathrm{Tr}_l^{nl} \left(\frac{B_{n-1}^{2^l}(a)}{B_n^{2^l+1}(a)} \right) + c \frac{a_0 B_{n-1}^{2^l}(a) + (a_0 B_{n-1}^{2^l}(a))^{2^l}}{B_n^{2^l}(a)} + c
$$

$$
\overset{(*)}{=} c \frac{(a_{n-1} B_{n-1}(a))^{2^l} + B_{n+1}^{2^l}(a)}{B_n^{2^l}(a)} + c \overset{(3)}{=} 0 \ ,
$$

where $(*)$ holds by Corollary 2 (since the value under the trace function is an element of GF(2^{ne})) and since $B_{n+1}(a) = a_0 B_{n-1}^{2^l}(a)$ resulting from (5) if $Z_n(a) = 0$.

Finally, to prove the trace identity for v_μ first note that, by (10), $\mathrm{Tr}_e^{ne}(B_n(a) + Z_n(a)) = \mathrm{Tr}_e^{ne}(B_n(a)) = 0$ if $Z_n(a) = 0$. Further,

$$
\mathrm{Tr}_e^{ne} \left(\sum_{i=0}^{\frac{n-1}{2}} \frac{B_{n-1}^{2^{(2i+1)l}}(a)}{B_n^{2^{(2i+1)l}+2^{2il}-1}(a)} \right) = \sum_{j=0}^{n-1} \sum_{i=0}^{\frac{n-1}{2}} \frac{B_n^{2^{jl}}(a) B_{n-1}^{2^{(2i+j+1)l}}(a)}{B_n^{2^{(2i+j+1)l}+2^{(2i+j)l}}(a)}
$$

$$= \sum_{j=0}^{n-1} \frac{B_{n-1}^{2^{(j+1)l}}(a) \sum_{i=0}^{\frac{n-1}{2}} B_n^{2^{(j-2i)l}}(a)}{B_n^{2^{(j+1)l}+2^{jl}}(a)} = \text{Tr}_e^{ne} \left(\frac{B_{n-1}^{2^l}(a) \sum_{i=0}^{\frac{n-1}{2}} B_n^{2^{(2i+1)l}}(a)}{B_n^{2^l+1}(a)} \right)$$

$$\stackrel{(4)}{=} \text{Tr}_e^{ne} \left(\frac{B_{n-1}^{2^l}(a) \sum_{i=0}^{\frac{n-1}{2}} \left(B_{n+1}(a) + B_{n+1}^{2^l}(a) \right)^{2^{2il}}}{B_n^{2^l+1}(a)} \right)$$

$$= \text{Tr}_e^{ne} \left(\frac{B_{n-1}^{2^l}(a) \left(\text{Tr}_e^{ne}(B_{n+1}(a)) + B_{n+1}(a) \right)}{B_n^{2^l+1}(a)} \right)$$

$$= \text{Tr}_e^{ne}(B_{n+1}(a)) \text{Tr}_e^{ne} \left(\frac{B_{n-1}^{2^l}(a)}{B_n^{2^l+1}(a)} \right) + \text{Tr}_e^{ne} \left(\frac{B_{n-1}^{2^l}(a) B_{n+1}(a)}{B_n^{2^l+1}(a)} \right) = 0 \; ,$$

where the latest identity follows by Corollary 2 (note that $\text{Tr}_e^{ne}(v) = \text{Tr}_l^{nl}(v)$ if $v \in \text{GF}(2^{ne})$).

B   Remaining Proof of Corollary 3

We can compute

$$\text{Tr}_k \left(ac^{-2} v_\mu^{2^l+1} \right)$$

$$= \text{Tr}_k \left(a\mathcal{V}^{2^l+1} + ac^{-1}\mu\mathcal{V}^{2^l} B_n(a) + ac^{-1}\mu\mathcal{V}B_n^{2^l}(a) + ac^{-2}\mu^2 B_n^{2^l+1}(a) \right)$$

$$\stackrel{(11)}{=} \text{Tr}_k \left(a\mathcal{V}^{2^l+1} + c^{-1}\mu\mathcal{V}^{2^l} B_n^{2^l}(a) + ac^{-2}\mu^2 B_n^{2^l+1}(a) \right)$$

$$= \text{Tr}_k \left(a\mathcal{V}^{2^l+1} + c^{-1}\mu\mathcal{V}B_n(a) + c^{-2}\mu^2 \text{N}_e^{ne} \left(\frac{v_0}{v_0 + v_1} \right)^2 \text{Tr}_e^{ne}(v_0)^2 v_1^{-2} \right)$$

$$= \text{Tr}_k \left(a\mathcal{V}^{2^l+1} \right) + \text{Tr}_e \left(c^{-1}\mu \text{N}_e^{ne} \left(\frac{v_0}{v_0 + v_1} \right) \left(1 + \text{Tr}_e^{ne}(v_0) \text{Tr}_e^{ne}(v_0^{-1}) \right) \right.$$

$$\left. + c^{-1}\mu \text{N}_e^{ne} \left(\frac{v_0}{v_0 + v_1} \right) \text{Tr}_e^{ne} \left(\text{Tr}_e^{ne}(v_0) v_1^{-1} \right) \right)$$

$$= \text{Tr}_k \left(a\mathcal{V}^{2^l+1} \right) + \text{Tr}_e \left(c^{-1}\mu \text{N}_e^{ne} \left(\frac{v_0}{v_0 + v_1} \right) \right) \; .$$

Thus,

$$\sum_{\mu \in \text{GF}(2^e)} (-1)^{\text{Tr}_k \left(ac^{-2} v_\mu^{2^l+1} + v_\mu \right)}$$

$$= (-1)^{\text{Tr}_k \left(a\mathcal{V}^{2^l+1} \right)} \sum_{\mu \in \text{GF}(2^e)} (-1)^{\text{Tr}_e \left(c^{-1}\mu \text{N}_e^{ne} \left(\frac{v_0}{v_0 + v_1} \right) \right)} = 0 \; .$$

On the Correlation Distribution of Kerdock Sequences[*]

Xiaohu Tang[1], Tor Helleseth[2], and Aina Johansen[2]

[1] The Provincial Key Lab of Information Coding and Transmission, Institute of Mobile Communications
Southwest Jiaotong University, Chengdu 610031, Sichuan, China
xhutang@ieee.org
[2] The Selmer Center
Department of Informatics, University of Bergen
P.O. Box 7800, N-5020 Bergen, Norway
tor.helleseth@ii.uib.no, Aina.Johansen@ii.uib.no

Abstract. For any even integer n, the binary Kerdock sequences of period $2(2^n - 1)$ are optimal with respect to the well-known Welch bound. Until now the correlation distribution of this family has not been known. In this paper we completely determine its correlation distribution using connections between the correlation properties of binary sequences and quaternary sequences under the Gray map.

Keywords: Binary sequences, quaternary sequences, Gray map, correlation distribution.

1 Introduction

CDMA systems provide several users with simultaneous access to the full channel bandwidth by assigning a unique sequence from a sequence family to each user [2]. In order to distinguish each user and minimize interference due to competition and simultaneous traffic across the channel, the sequence family must have as low maximal correlation as possible. Nevertheless, the maximal correlation is subjected to the limitation of some theoretical bounds, for example the Welch and Sidelnikov bounds. The Welch bound states that for a family of sequences of period N, the maximal value of cross correlations and out-of-phase autocorrelations (usually refer to maximal nontrivial correlation value R_{max}), is lower bounded by $\sqrt{N}$ approximately.

The well-known Kasami sequence is the first optimal family of binary sequence with respect to the Welch bound, consisting of $2^{n/2}$ sequences of period $2^n - 1$, whose $R_{\max}$ is lower bounded by $2^{n/2} + 1$, for any even integer n. In 1996, by employing interleaved maximal length sequences over $\mathbf{Z}_4$, Udaya and Siddiqi

---

[*] This work of Xiaohu Tang was supported Humboldt Research Fellowship 2007, and the Teacher Research Projects of Southwest Jiaotong University. The research of T. Helleseth and A. Johansen was supported by the Norwegian Research Council.

S.W. Golomb et al. (Eds.): SETA 2008, LNCS 5203, pp. 121–129, 2008.

constructed a family of 2^{n-1} binary sequences of period $2(2^n - 1)$ satisfying the Welch bound on maximum out of phase correlations [7]. Compared with the Kasami sequences, US sequences have almost the same maximal nontrivial correlation value, more precisely $R_{\max} = 2^{n/2}+2$, but offer much more sequences.

Later in [4], by the Gray map of the optimal Family $\mathcal{A}$ of maximal length sequences over $\mathbf{Z}_4$, Helleseth and Kumar defined Kerdock sequences $Q(2)$, comprised of 2^n sequences of period $2(2^n - 1)$ having $R_{\max} = 2^{n/2} + 2$. In fact, Kerdock sequences $Q(2)$ turn out to be the Kerdock code punctured in two coordinates in cyclic form given by Nechaev [6], which includes US sequences as a subset. Recently aiming at doubling the size of US sequences, Tang, Udaya, and Fan obtained Kerdock sequences from a distinct technique [8]. However, the correlation distribution of Kerdock sequences is still open problem.

Most recently, Johansen, Helleseth, and Tang studied the correlation distribution of sequences of period $2(2^n - 1)$ over $\mathbf{Z}_4$ [5]. During this study, the authors developed some results for determining the correlation distribution of sequences in Family $\mathcal{A}$. Thanks to one result (c.f. Lemma 2 in this paper), we are able to completely determine the correlation distribution of the binary Kerdock sequences using connections between the correlation of binary and quaternary sequences under the Gray map.

2   Preliminaries

The Galois ring $\mathbf{R} = \mathbf{GR}(4, n)$ with 4^n elements is the Galois extension of degree n over $\mathbf{Z}_4$. $\mathbf{R}$ is a commutative ring having the maximal ideal $2\mathbf{R}$. Let $\mu : \mathbf{R} \leftarrow \mathbf{R}/(2\mathbf{R})$ be the mod-2 reduction map given by

$$\mu(x) = x + 2\mathbf{R}, \ x \in \mathbf{R}.$$

Clearly, $\mu(\mathbf{R}) = \mathbf{R}/(2\mathbf{R}) \cong \mathbf{F}_{2^n}$, where $\mathbf{F}_{2^n}$ is the finite field with 2^n elements.

As a multiplicative group, the units $\mathbf{R}^*$ in $\mathbf{R}$ has a cyclic group $\mathbf{G}_C$ of order $2^n - 1$. Let β be a generator of the cyclic group $\mathbf{G}_C$, i.e.,

$$\mathbf{G}_C = \{1, \beta, \beta^2, \cdots, \beta^{2^n-2}\}.$$

Then $\alpha = \mu(\beta)$ is a primitive root of $\mathbf{F}_{2^n}$. The set $\mathcal{T} = \mathbf{G}_C \cup \{0\}$ is called the Teichmuller set, which is isomorphic to the finite field $\mathbf{F}_2^n$.

The trace function $Tr_1^n(\cdot)$ maps elements of $\mathbf{R}$ to $\mathbf{Z}_4$, defined as

$$Tr_1^n(x) = \sum_{i=0}^{n}(\sigma(x))^i,$$

where $\sigma(\cdot)$ is the automorphisms of $\mathbf{R}$ given by:

$$\sigma(a + 2b) = a^2 + 2b^2 \ for \ a, b \in \mathbf{R}.$$

Let $tr(\cdot)$ denote the analogous trace function over $\mathbf{F}_{2^n}$, defined by:

$$tr_1^n(x) = \sum_{i=0}^{n-1} x^{2^i}.$$

A. Correlation Function

Let $a = (a(t), 0 \leq t < L)$ and $b = (b(t), 0 \leq t < L)$ be two sequences over $\mathbf{Z}_N$. The correlation function between the sequences a and b at shift $0 \leq \tau < L$ is defined by

$$R_{a,b}(\tau) = \sum_{t=0}^{L-1} \omega^{a(t+\tau)-b(t)}, 0 \leq \tau < N \qquad (1)$$

where ω is a primitive complex Nth root of unity.

Specifically, in this paper there are two types of sequences: binary sequences and quaternary sequences. They are corresponding to different ω in (1), i.e., $\omega = -1$ for binary sequences, and $\omega = \sqrt{-1}$ for quaternary sequences.

B. Gray Map

For arbitrary element $x = 2x_1 + x_2$ in $\mathbf{Z}_4$, $x_1, x_2 \in \mathbf{Z}_2$, the Gray map $\phi : \mathbf{Z}_4 \rightarrow \mathbf{Z}_2 \times \mathbf{Z}_2$ is given by $\phi(x) = (x_1, x_1 + x_2 (\text{mod } 2))$, i.e.,

$$0 \rightarrow 00, \ 1 \rightarrow 01, \ 2 \rightarrow 11, \ 3 \rightarrow 10. \qquad (2)$$

If defining $\pi(x) = x_1$ and $\nu(x) = x_1 + x_2 (\text{mod } 2)$, the Gray map can be described by the two maps π and ν, i.e., $\phi(x) = (\pi(x), \nu(x))$. In particular, it is easy to see that

$$\pi(x + 1) = x_1 + x_2 \,(\text{mod } 2), \ \nu(x + 1) = x_1 + 1 \,(\text{mod } 2), \qquad (3)$$
$$\pi(x + 3) = x_1 + x_2 + 1 \,(\text{mod } 2), \ \nu(x + 3) = x_1 \,(\text{mod } 2). \qquad (4)$$

Applying the Gray map to all entries of the sequence $(a(0), \cdots, a(L - 1))$, we naturally obtain the Gray map sequence $\phi_a = (\phi_a(t), 0 \leq t < 2L)$ of length $2L$ respectively, where

$$\phi_a(t) = \begin{cases} \pi(a(t_1)), \ t = 2t_1 \\ \nu(a(t_1)), \ t = 2t_1 + 1. \end{cases}$$

When L is odd, it is more convenient to use the modified Gray map sequence $\varphi_a = (\varphi_a(t), 0 \leq t < 2L)$ proposed by Nechaev [6], which is equivalent to the Gray map sequence via a permutation of coordinates, and is defined by

$$\varphi_a(t) = \begin{cases} \pi(a(t_1)), & t = 2t_1 \\ \nu(a(t_1 + \frac{L+1}{2})), & t = 2t_1 + 1. \end{cases}$$

The following lemma establishes the connection between the correlation of quaternary sequences and the correlation of their (modified) Gray map sequence.

Lemma 1 ([4]). *Let* $(a(0), \cdots, a(L - 1))$ *and* $(b(0), \cdots, b(L - 1))$ *be two quaternary sequences of length* L. *Then,*

$$\sum_{t=0}^{2L-1} (-1)^{\phi(a(t))+\phi(b(t))} = 2 \cdot \Re(\sum_{t=0}^{L-1} \omega^{a(t)-b(t)}),$$

where $\Re(x)$ denotes the real part of complex-valued variable x. Specifically, when L is odd,

$$\sum_{t=0}^{2L-1}(-1)^{\varphi(a(t))+\varphi(b(t))} = 2 \cdot \Re(\sum_{t=0}^{L-1}\omega^{a(t)-b(t)}).$$

3   Correlation Distribution of Family $\mathcal{A}$

Throughout this paper let n be an odd integer and $\{\eta_0, \eta_1, \cdots, \eta_{2^n-1}\}$ an enumeration of the elements in $\mathcal{T}$, satisfying $\eta_{i+2^{n-1}} = 1+\eta_i$ (mod 2), $0 \leq i < 2^{n-1}$, and

$$Tr_1^n(\eta_i) = \begin{cases} 0 \ (\text{mod } 2), 0 \leq i < 2^{n-1} \\ 1 \ (\text{mod } 2), 2^{n-1} \leq i < 2^n. \end{cases}$$

The well-known sequence Family $\mathcal{A}$ is defined in [1].

Definition 1 ([1]). *The sequence Family* $\mathcal{A} = \{a_i, i = 0, 1, \cdots, 2^n\}$ *of length* $2^n - 1$ *is defined by*

$$a_i(t) = \begin{cases} Tr_1^n((1+2\eta_i)\beta^t), 0 \leq i < 2^n \\ 2Tr_1^n(\beta^t), \qquad i = 2^n. \end{cases}$$

In this paper, we will focus on a smaller set comprised of the first 2^n sequences. For convenience, we still call it Family $\mathcal{A}$, i.e., $\mathcal{A} = \{a_i, i = 0, 1, \cdots, 2^n - 1\}$.

Let $\mathcal{A}_1 = \{a_i, i = 0, \cdots, 2^{n-1} - 1\}$ and $\mathcal{A}_2 = \{a_i, i = 2^{n-1}, \cdots, 2^n - 1\}$. Specifically, we are interested in the correlation distribution between $\mathcal{A}_i$ and $\mathcal{A}_j$, $1 \leq i, j \leq 2$, which is crucial to derive our main result on the correlation distribution of the family of Kerdock sequences. First of all, consider the correlation function between $a_i \in \mathcal{A}$ and $a_j \in \mathcal{A}$ at the shift τ as

$$R_{i,j}(\tau) = \sum_{t=0}^{2^n-2}\omega^{Tr_1^n((1+2\eta_i)\beta^{t+\tau}-(1+2\eta_j)\beta^t)}.$$

Define $\theta_\tau = (1+2\eta_i)\beta^\tau - (1+2\eta_j) = b_\tau + 2c_\tau, b_\tau, c_\tau \in \mathcal{T}$. If $b_\tau = 0$, which implies $\tau = 0$, then $R_{i,j}(\tau) = -1$ when $c_\tau \neq 0$ (that is $i \neq j$) and $R_{i,j}(\tau) = 2^n - 1$ when $c_\tau = 0$ (that is $i = j$). Otherwise let $U_\tau = \frac{c_\tau}{b_\tau} \in \mathcal{T}$, the correlation function then becomes

$$R_{i,j}(\tau) = \sum_{t=0}^{2^n-2}\omega^{Tr_1^n((1+2U_\tau)\beta^t)}.$$

Next, the following lemma is important to determine the distribution of U_τ when a_i, a_j range over $\mathcal{A}_1$ or $\mathcal{A}_2$.

Lemma 2 ([5]). *Let* $0 < \tau < 2^n - 1$. *Then for any* $c \in \mathcal{T}$ *the number of solutions* η_i, η_j *satisfying* $U_\tau = c$ *is* 2^{n-2} *if* i *runs from* $2^{n-1}\delta_1$ *to* $2^{n-1}(1 + \delta_1) - 1$, *and* j *runs from* $2^{n-1}\delta_2$ *to* $2^{n-1}(1 + \delta_2) - 1$, *respectively, where* $\delta_1, \delta_2 \in \{0, 1\}$.

On the other hand, the correlation sum $\Delta(c) = \sum_{t=0}^{2^n-2} \omega^{Tr_1^n((1+2c)\beta^t)}$ has the following weight distribution.

Lemma 3 ([1], Theorem 5). *As c is varying over T, the correlation sum $\Delta(c)$ assumes values as*

$$\Delta(c) = \begin{cases} -1 + 2^{\frac{n-1}{2}} + 2^{\frac{n-1}{2}}\omega, \ 2^{n-2} + 2^{\frac{n-3}{2}} \ times \\ -1 + 2^{\frac{n-1}{2}} - 2^{\frac{n-1}{2}}\omega, \ 2^{n-2} + 2^{\frac{n-3}{2}} \ times \\ -1 - 2^{\frac{n-1}{2}} + 2^{\frac{n-1}{2}}\omega, \ 2^{n-2} - 2^{\frac{n-3}{2}} \ times \\ -1 - 2^{\frac{n-1}{2}} - 2^{\frac{n-1}{2}}\omega, \ 2^{n-2} - 2^{\frac{n-3}{2}} \ times. \end{cases}$$

Combining Lemma 2 and 3, we therefore have the following correlation distribution between the sequences in A_1 and A_2.

Theorem 4. *The correlation distribution of Family A_1 and A_2 are as follows:*

1. *for $0 \le i, j < 2^{n-1}$ or $2^{n-1} \le i, j < 2^n$,*

$$R_{i,j}(\tau) = \begin{cases} -1 + 2^n, & 2^{n-1} \ times \\ -1, & 2^{n-1}(2^{n-1} - 1) \ times \\ -1 + 2^{\frac{n-1}{2}} + 2^{\frac{n-1}{2}}\omega, \ 2^{n-2}(2^n - 2)(2^{n-2} + 2^{\frac{n-3}{2}}) \ times \\ -1 + 2^{\frac{n-1}{2}} - 2^{\frac{n-1}{2}}\omega, \ 2^{n-2}(2^n - 2)(2^{n-2} + 2^{\frac{n-3}{2}}) \ times \\ -1 - 2^{\frac{n-1}{2}} + 2^{\frac{n-1}{2}}\omega, \ 2^{n-2}(2^n - 2)(2^{n-2} - 2^{\frac{n-3}{2}}) \ times \\ -1 - 2^{\frac{n-1}{2}} - 2^{\frac{n-1}{2}}\omega, \ 2^{n-2}(2^n - 2)(2^{n-2} - 2^{\frac{n-3}{2}}) \ times; \end{cases}$$

2. *for $0 \le i < 2^{n-1} \le j < 2^n$ or $0 \le j < 2^{n-1} \le i < 2^n$,*

$$R_{i,j}(\tau) = \begin{cases} -1, & 2^{2n-2} \ times \\ -1 + 2^{\frac{n-1}{2}} + 2^{\frac{n-1}{2}}\omega, \ 2^{n-2}(2^n - 2)(2^{n-2} + 2^{\frac{n-3}{2}}) \ times \\ -1 + 2^{\frac{n-1}{2}} - 2^{\frac{n-1}{2}}\omega, \ 2^{n-2}(2^n - 2)(2^{n-2} + 2^{\frac{n-3}{2}}) \ times \\ -1 - 2^{\frac{n-1}{2}} + 2^{\frac{n-1}{2}}\omega, \ 2^{n-2}(2^n - 2)(2^{n-2} - 2^{\frac{n-3}{2}}) \ times \\ -1 - 2^{\frac{n-1}{2}} - 2^{\frac{n-1}{2}}\omega, \ 2^{n-2}(2^n - 2)(2^{n-2} - 2^{\frac{n-3}{2}}) \ times. \end{cases}$$

In particular, $R_{i,j}(\tau) = 2^n - 1$ if and only if $\tau = 0$ and $i = j$, and $R_{i,j}(\tau) = -1$ if and only if $\tau = 0$ and $i \ne j$.

4   Correlation Distribution of Kerdock Sequences

The following 2-adic expression of the quaternary sequences in trace form is helpful to obtain the counterparts of the sequences in Family A under the Gray maps.

Lemma 5 ([3]). *Assuming that $\eta \in T$, the 2-adic representation of $a(t) = Tr_1^n((1+2\eta)\beta^t)$, i.e., $a(t) = 2b(t) + c(t)$, is given by*

$$b(t) = tr_1^n(\alpha^t), \tag{5}$$
$$c(t) = tr_1^n(\zeta\alpha^t) + p(\alpha^t) \tag{6}$$

where $\zeta = \mu(\eta)$ and $p(x) = \sum_{l=1}^{\frac{n-1}{2}} tr_1^n(x^{2^l+1})$.

Let $\zeta_i = \mu(\eta_i)$, $0 \le i < 2^n$. Obviously, $\zeta_{i+2^{n-1}} = 1 + \zeta_i$, $0 \le i < 2^{n-1}$. Applying the modified Gray map to the sequences in $\{a_0, \cdots, a_{2^{n-1}-1}\}$, and doubling the size by the technique proposed in [4] and [8], we obtain the binary family of Kerdock sequences $\mathcal{V}$.

Definition 2. *The binary Family $\mathcal{V}$ of sequences $\{v_i, i = 0, 1, \cdots, 2^n - 1\}$ of length $2(2^n - 1)$ is defined as*

− $0 \le i < 2^{n-1}$:

$$v_i(t) = \begin{cases} tr_1^n(\zeta_i \alpha^{t_1}) + p(\alpha^{t_1}), & t = 2t_1 \\ tr_1^n((1+\zeta_i)\alpha^{t_1+2^{n-1}}) + p(\alpha^{t_1+2^{n-1}}), & t = 2t_1 + 1 \end{cases}, \quad (7)$$

− $2^{n-1} \le i < 2^n$:

$$v_i(t) = \begin{cases} tr_1^n(\zeta_{i-2^{n-1}} \alpha^{t_1}) + p(\alpha^{t_1}) + 1, & t = 2t_1 \\ tr_1^n((1+\zeta_{i-2^{n-1}})\alpha^{t_1+2^{n-1}}) + p(\alpha^{t_1+2^{n-1}}), & t = 2t_1 + 1. \end{cases} \quad (8)$$

Remark 1. *When $2^{n-1} \le i < 2^n$, the sequence v_i could be viewed as the modified Gray map sequence of $a_i + 3 = (a_i(t) + 3 = Tr_1^n((1+2\eta_i)\beta^t) + 3, 0 \le t < 2^n)$. From (4), the Gray map of $a_i(t) + 3 = (\pi(a_i(t) + 3), \nu(a_i(t) + 3))$ where*

$$\pi(a_i(t) + 3) = tr_1^n(\zeta_i \alpha^{t_1}) + p(\alpha^{t_1}) + tr_1^n(\alpha^{t_1}) + 1,$$
$$\nu(a_i(t) + 3) = tr_1^n(\zeta_i \alpha^{t_1}) + p(\alpha^{t_1}).$$

Hence, the expression of the modified Gray map sequence in (8) follows from the fact that $\zeta_i = 1 + \zeta_{i-2^{n-1}}$, $2^{n-1} \le i < 2^n$.
This connection was firstly observed in [4].

We have the following main result on the correlation distribution of Kerdock sequences family.

Theorem 6. *Family $\mathcal{V}$ has the following correlation distribution:*

$$R_{i,j}(\tau) = \begin{cases} -2 + 2^{n+1}, & 2^n \ times \\ 0, & 2^{2n}\ times \\ -2, & 3 \cdot 2^{2n-2} - 2^n\ times \\ 2, & 2^{2n-2}\ times \\ -2 + 2^{\frac{n+1}{2}}, & 3 \cdot 2^{n-1}(2^n - 2)(2^{n-2} + 2^{\frac{n-3}{2}})\ times \\ -2 - 2^{\frac{n-1}{2}}, & 3 \cdot 2^{n-1}(2^n - 2)(2^{n-2} - 2^{\frac{n-3}{2}})\ times \\ 2 - 2^{\frac{n+1}{2}}, & 2^{n-1}(2^n - 2)(2^{n-2} + 2^{\frac{n-3}{2}})\ times \\ 2 + 2^{\frac{n+1}{2}}, & 2^{n-1}(2^n - 2)(2^{n-2} - 2^{\frac{n-3}{2}})\ times \\ 2^{\frac{n+1}{2}}, & 2^{2n-1}(2^n - 2)\ times \\ -2^{\frac{n+1}{2}}, & 2^{2n-1}(2^n - 2)\ times. \end{cases}$$

To prove Theorem 6, we need the following lemma.

Lemma 7. *Let $0 \le \tau = 2\tau_1 + \tau_2 < 2(2^n - 1)$ where $0 \le \tau_1 < 2^n - 1$ and $\tau_2 = 0, 1$. Then*

1. *for $\tau = 2\tau_1$,*

$$v_i(t+\tau) = \begin{cases} \varphi(a_i(t+\tau_1)), & 0 \le i < 2^{n-1} \\ \varphi(a_i(t+\tau_1)+3), & 2^{n-1} \le i < 2^n \end{cases};$$

2. *for $\tau = 2\tau_1 + 1$,*

$$v_i(t+\tau) = \begin{cases} \varphi(a_{i+2^{n-1}}(t+\tau_1+2^{n-1})), & 0 \le i < 2^{n-1} \\ \varphi(a_{i-2^{n-1}}(t+\tau_1+2^{n-1})+1), & 2^{n-1} \le i < 2^n. \end{cases}$$

Proof: According to Lemma 5, Remark 1, and the definition of the modified Gray map, the case of $\tau = 2\tau_1$ is straightforward. Hereafter we only prove the result for $\tau = 2\tau_1 + 1$.

When $0 \le i < 2^{n-1}$, $a_{i+2^{n-1}}(t+\tau_1+2^{n-1}) = Tr_1^n((1+2\eta_{i+2^{n-1}})\beta^{t+\tau_1+2^{n-1}})$. Applying Lemma 5, we have

$$\varphi(a_{i+2^{n-1}}(t+\tau_1+2^{n-1}))$$
$$= \begin{cases} tr_1^n(\zeta_{i+2^{n-1}}\alpha^{t_1+\tau_1+2^{n-1}}) + p(\alpha^{t_1+\tau_1+2^{n-1}}), & t = 2t_1 \\ tr_1^n((1+\zeta_{i+2^{n-1}})\alpha^{t_1+\tau_1+2^n}) + p(\alpha^{t_1+\tau_1+2^n}), & t = 2t_1+1 \end{cases}$$
$$= \begin{cases} tr_1^n((1+\zeta_i)\alpha^{t_1+\tau_1+2^{n-1}}) + p(\alpha^{t_1+\tau_1+2^{n-1}}), & t = 2t_1 \\ tr_1^n(\zeta_i\alpha^{t_1+\tau_1+1}) + p(\alpha^{t_1+\tau_1+1}), & t = 2t_1+1 \end{cases}$$
$$= v_i(t+\tau),$$

where we make use of the fact that $\zeta_{i+2^{n-1}} = 1 + \zeta_i$, $0 \le i < 2^{n-1}$.

When $2^{n-1} \le i < 2^n$, then

$$a_{i-2^{n-1}}(t+\tau_1+2^{n-1}) + 1 = Tr_1^n((1+2\eta_{i-2^{n-1}})\beta^{t+\tau_1+2^{n-1}}) + 1.$$

Applying Lemma 5 and (4) to $a_{i-2^{n-1}} + 1$, we now have

$$\varphi(a_{i-2^{n-1}}(t+\tau_1+2^{n-1})+1)$$
$$= \begin{cases} tr_1^n(\zeta_{i-2^{n-1}}\alpha^{t_1+\tau_1+2^{n-1}}) + p(\alpha^{t_1+\tau_1+2^{n-1}}) + tr_1^n(\alpha^{t_1+\tau_1+2^{n-1}}), & t = 2t_1 \\ tr_1^n(\zeta_{i-2^{n-1}}\alpha^{t_1+\tau_1+2^n}) + p(\alpha^{t_1+\tau_1+2^n}) + 1, & t = 2t_1+1 \end{cases}$$
$$= \begin{cases} tr_1^n((1+\zeta_{i-2^{n-1}})\alpha^{t_1+\tau_1+2^{n-1}}) + p(\alpha^{t_1+\tau_1+2^{n-1}}), & t = 2t_1 \\ tr_1^n(\zeta_{i-2^{n-1}}\alpha^{t_1+\tau_1+1}) + p(\alpha^{t_1+\tau_1+1}) + 1, & t = 2t_1+1 \end{cases}$$
$$= v_i(t+\tau).$$

$\square$

Proof of Theorem 6: We investigate the following 7 cases for computing the correlation function:

Case 1. $0 \le i = j < 2^n$ and $\tau = 0$.
This is trivial case, $R_{i,i}(0) = 2(2^n - 1)$.
Case 2. ($0 \le i \ne j < 2^{n-1}$ or $2^{n-1} \le i \ne j < 2^n$) and $\tau = 0$.
In this case,

$$R_{i,j}(\tau) = 2\left(\sum_{x \in \mathbf{F}_{2^n}} (-1)^{tr_1^n(\zeta_i+\zeta_j)x} - 1 \right)$$
$$= -2.$$

where $\zeta_{i-2^{n-1}} + \zeta_{j-2^{n-1}} = \zeta_i + \zeta_j$, $2^{n-1} \le i, j < 2^n$, come from the fact that $\zeta_i = 1 + \zeta_{i-2^{n-1}}$ and $\zeta_j = 1 + \zeta_{j-2^{n-1}}$.

Case 3. $(0 \le i < 2^{n-1} \le j < 2^n$ or $0 \le i < 2^{n-1} \le j < 2^n)$ and $\tau = 0$.

For this case,

$$R_{i,j}(\tau) = -(\sum_{x \in \mathbf{F}_{2^n}} (-1)^{tr_1^n((\zeta_i + \zeta_j)x)} - 1) + \sum_{x \in \mathbf{F}_{2^n}} (-1)^{tr_1^n((\zeta_i + \zeta_j)x)} - 1$$

$$= 0,$$

where we also use the fact that $\zeta_i = 1 + \zeta_{i-2^{n-1}}$ for $2^{n-1} \le i < 2^n$, or $\zeta_j = 1 + \zeta_{j-2^{n-1}}$ for $2^{n-1} \le j < 2^n$.

Case 4. $0 \le i, j < 2^{n-1}$ and $\tau \ne 0$.

From Lemma 1 and Lemma 7 it holds that

$$R_{i,j}(\tau) = \sum_{t=0}^{2^{n+1}-3} (-1)^{v_i(t+\tau)+v_j(t)}$$

$$= \begin{cases} 2 \cdot \Re(\sum_{t=0}^{2^n-2} \omega^{a_i(t+\tau_1)-a_j(t)}), & \tau = 2\tau_1 \\ 2 \cdot \Re(\sum_{t=0}^{2^n-2} \omega^{a_{i+2^{n-1}}(t+\tau_1+2^{n-1})-a_j(t)}), & \tau = 2\tau_1 + 1. \end{cases}$$

As a direct consequence of Theorem 4, when $\tau = 2\tau_1$ we have

$$R_{i,j}(\tau) = \begin{cases} -2 + 2^{\frac{n+1}{2}}, & 2^{n-1}(2^n - 2)(2^{n-2} + 2^{\frac{n-3}{2}}) \ times \\ -2 - 2^{\frac{n+1}{2}}, & 2^{n-1}(2^n - 2)(2^{n-2} - 2^{\frac{n-3}{2}}) \ times \end{cases}$$

as τ_1 ranges over $0 < \tau_1 < 2^n - 1$ and i, j vary from 0 to $2^{n-1} - 1$, respectively.

If $\tau = 2\tau_1 + 1 = 2^n - 1$, then $R_{i,j}(\tau) = -2$, which occurs 2^{2n-2} times. Otherwise from Theorem 4, the other correlation distribution is

$$R_{i,j}(\tau) = \begin{cases} -2 + 2^{\frac{n+1}{2}}, & 2^{n-1}(2^n - 2)(2^{n-2} + 2^{\frac{n-3}{2}}) \ times \\ -2 - 2^{\frac{n+1}{2}}, & 2^{n-1}(2^n - 2)(2^{n-2} - 2^{\frac{n-3}{2}}) \ times \end{cases}$$

as $\tau_1 \ne 2^{n-1} - 1$ varies from 0 to $2^n - 2$, and i, j ranges from 0 to 2^{n-1}, respectively.

Case 5. $2^{n-1} \le i, j < 2^n$ and $\tau \ne 0$.

Then,

$$R_{i,j}(\tau) = \begin{cases} R_{i-2^{n-1},j-2^{n-1}}(\tau), & \tau = 2\tau_1 \\ -R_{i-2^{n-1},j-2^{n-1}}(\tau), & \tau = 2\tau_1 + 1. \end{cases}$$

Hence, it is immediate from case 4 that the correlation distribution is

$$R_{i,j}(\tau) = \begin{cases} 2, & 2^{2n-2} \ times \\ -2 + 2^{\frac{n+1}{2}}, & 2^{n-1}(2^n - 2)(2^{n-2} + 2^{\frac{n-3}{2}}) \ times \\ 2 - 2^{\frac{n+1}{2}}, & 2^{n-1}(2^n - 2)(2^{n-2} + 2^{\frac{n-3}{2}}) \ times \\ -2 - 2^{\frac{n+1}{2}}, & 2^{n-1}(2^n - 2)(2^{n-2} - 2^{\frac{n-3}{2}}) \ times \\ 2 + 2^{\frac{n+1}{2}}, & 2^{n-1}(2^n - 2)(2^{n-2} - 2^{\frac{n-3}{2}}) \ times \end{cases}$$

where τ runs through $0 < \tau < 2(2^n - 1)$ and i, j vary from 2^{n-1} to $2^n - 1$, respectively.

Case 6. $0 \leq i < 2^{n-1} \leq j < 2^n$ and $\tau \neq 0$.
It follows from Lemma 1 and Lemma 7 that

$$R_{i,j}(\tau) = \begin{cases} 2 \cdot \Re(\sum_{t=0}^{2^n-2} \omega^{a_i(t+\tau_1)-a_j(t)-3}), & \tau = 2\tau_1 \\ 2 \cdot \Re(\sum_{t=0}^{2^n-2} \omega^{a_{i+2^{n-1}}(t+\tau_1+2^{n-1})-a_j(t)-3}), & \tau = 2\tau_1 + 1. \end{cases}$$

In a similar manner to Case 4, we arrive at the following distribution

$$R_{i,j}(\tau) = \begin{cases} 0, & 2^{2n-2} \text{ times} \\ 2^{\frac{n+1}{2}}, & 2^{2n-2}(2^n - 2) \text{ times} \\ -2^{\frac{n+1}{2}}, & 2^{2n-2}(2^n - 2) \text{ times} \end{cases} \qquad (9)$$

with τ ranging through $0 < \tau < 2(2^n - 1)$, i and j varying from 0 to $2^{n-1} - 1$ and 2^{n-1} to $2^n - 1$, respectively.

Case 7. $0 \leq j < 2^{n-1} \leq i < 2^n$ and $\tau \neq 0$.
The correlation function is

$$R_{i,j}(\tau) = \begin{cases} 2 \cdot \Re(\sum_{t=0}^{2^n-2} \omega^{a_i(t+\tau_1)+3-a_j(t)}), & \tau = 2\tau_1 \\ 2 \cdot \Re(\sum_{t=0}^{2^n-2} \omega^{a_{i-2^{n-1}}(t+\tau_1+2^{n-1})+1-a_j(t)}), & \tau = 2\tau_1 + 1 \end{cases},$$

which has the same correlation distribution as (9).

References

1. Boztas, S., Hammons, R., Kumar, P.V.: 4-phase sequences with near-optimum correlation properties. IEEE Trans. Inform. Theory 38, 1101–1113 (1992)
2. Fan, P.Z., Darnell, M.: Sequence Design for Communications Applications. John Wiley, Chichester (1996)
3. Hammons, R., Kumar, P.V., Calderbank, A.N., Sloane, N.J.A., Solé, P.: The Z_4-Linearity of Kerdock, Preparata, Goethals and Related Codes. IEEE Trans. Inform. Theory 40, 301–319 (1994)
4. Helleseth, T., Kumar, P.V.: Sequences with low correlation. In: Pless, V., Huffman, C. (eds.) Handbook of Coding Theory, Elsevier, Amsterdam (1998)
5. Johansen, A., Helleseth, T., Tang, X.H.: The correlation distribution of sequences of period $2(2^n - 1)$. IEEE Trans. Inform. Theory (to appear)
6. Nechaev, A.A.: Kerdock code in a cyclic form. Discrete Mathematics Appl. 1, 365–384 (1991)
7. Udaya, P., Siddiqi, M.U.: Optimal biphase sequences with large linear complexity derived from sequences over Z4. IEEE Trans. Inform. Theory 42, 206–216 (1996)
8. Tang, X.H., Udaya, P., Fan, P.Z.: Generalized binary Udaya-Siddiqi sequences. IEEE Transactions on Information Theory 53, 1225–1230 (2007)

Two New Families of Low-Correlation Interleaved QAM Sequences

Gagan Garg[1], P. Vijay Kumar[2,*], and C.E. Veni Madhavan[1]

[1] Department of Computer Science and Automation,
Indian Institute of Science, Bangalore 560012, India
[2] Department of Electrical Communication Engineering,
Indian Institute of Science, Bangalore 560012, India
gagan.garg@gmail.com, vijayk@usc.edu, cevm@csa.iisc.ernet.in

Abstract. Two families of low correlation QAM sequences are presented here. In a CDMA setting, these sequences have the ability to transport a large amount of data as well as enable variable-rate signaling on the reverse link.

The first family $\mathcal{I}^2 \mathcal{S} \mathcal{Q} - \mathcal{B}$ is constructed by interleaving 2 selected QAM sequences. This family is defined over M^2-QAM, where $M = 2^m, m \geq 2$. Over 16-QAM, the normalized maximum correlation $\bar{\theta}_{\max}$ is bounded above by $\lesssim 1.17 \sqrt{N}$, where N is the period of the sequences in the family. This upper bound on $\bar{\theta}_{\max}$ is the lowest among all known sequence families over 16-QAM.

The second family $\mathcal{I}^4 \mathcal{S} \mathcal{Q}$ is constructed by interleaving 4 selected QAM sequences. This family is defined over M^2-QAM, where $M = 2^m, m \geq 3$, i.e., 64-QAM and beyond. The $\bar{\theta}_{\max}$ for sequences in this family over 64-QAM is upper bounded by $\lesssim 1.60 \sqrt{N}$. For large M, $\bar{\theta}_{\max} \lesssim 1.64 \sqrt{N}$. These upper bounds on $\bar{\theta}_{\max}$ are the lowest among all known sequence families over M^2-QAM, $M = 2^m, m \geq 3$.

1 Introduction

Low-correlation sequences over QAM constellations are of interest partly on account of the increasingly common usage of the QAM alphabet as a signaling alphabet and also on account of their ability to carry a larger number of data bits per sequence period in comparison with the one or the two bits traditionally associated with BPSK or QPSK sequence families.

1.1 Results in Perspective

In this subsection, we summarize the results of this paper with respect to the currently known QAM sequence families.

* P. Vijay Kumar is on leave of absence from the Department of EE-Systems, University of Southern California, Los Angeles, CA 90089 USA. This research is supported in part by NSF-ITR CCR-0326628 and in part by the DRDO-IISc Program on Advanced Research in Mathematical Engineering.

Table 1. Families of sequences over M^2-QAM and $2M$-ary Q-PAM, $M = 2^m, m \geq 2$

Family	Constellation	Period N	Data rate	$\bar{\theta}_{max}$
$\mathcal{CQ}$ [1]	16-QAM to M^2-QAM	$2^r - 1$	$2m$	$1.80\sqrt{N}$ to $3.00\sqrt{N}$
$\mathcal{I}^2\mathcal{CQ}$ [5]	16-QAM to M^2-QAM	$2(2^r - 1)$	$2m$	$1.41\sqrt{N}$ to $2.12\sqrt{N}$
$\mathcal{SQ}$ [1]	16-QAM to M^2-QAM	$2^r - 1$	$m + 1$	$1.61\sqrt{N}$ to $2.76\sqrt{N}$
$\mathcal{I}^2\mathcal{SQ} - \mathcal{A}$ [1,5]	16-QAM to M^2-QAM	$2(2^r - 1)$	$m + 1$	$1.41\sqrt{N}$ to $1.99\sqrt{N}$
$\mathcal{I}^2\mathcal{SQ} - \mathcal{B}$ (new)	16-QAM to M^2-QAM	$2(2^r - 1)$	$m + 1$	$1.17\sqrt{N}$ to $1.99\sqrt{N}$
$\mathcal{I}^4\mathcal{SQ}$ (new)	64-QAM to M^2-QAM	$4(2^r - 1)$	$m + 1$	$1.60\sqrt{N}$ to $1.64\sqrt{N}$
$\mathcal{P}$ [1]	8-ary Q-PAM to $2M$-ary Q-PAM	$2^r - 1$	$m + 1$	$1.34\sqrt{N}$ to $2.24\sqrt{N}$
$\mathcal{I}^2\mathcal{P}$ [1,5]	8-ary Q-PAM to $2M$-ary Q-PAM	$2(2^r - 1)$	$m + 1$	$\sqrt{N}$ to $1.72\sqrt{N}$

In 1996, Boztaş [2] proposed a 16-QAM CDMA sequence family built from quaternary sequences drawn from family $\mathcal{A}$ [4]. The first concerted look at low correlation sequence families over QAM constellations was presented by Anand and Kumar (AK) [1] recently. Extending the concept of interleaving proposed by AK [1], Garg, Kumar, and Madhavan (GKM) [5] proposed three interleaved sequence families with improved correlation properties.

The main properties of the five sequence families proposed in [1], the three sequence families proposed in [5], and the two sequence families proposed in this paper are summarized in Table 1. In the present paper, we are interested in the middle block of rows of the table, i.e., the families of selected sequences viz. $\mathcal{SQ}$, $\mathcal{I}^2\mathcal{SQ} - \mathcal{A}$, $\mathcal{I}^2\mathcal{SQ} - \mathcal{B}$ and $\mathcal{I}^4\mathcal{SQ}$.

Family $\mathcal{SQ}$ was proposed by AK in [1]. Over M^2-QAM, it permits $(m+1)$-bit data modulation. As can be seen from the table, the upper bound on $\bar{\theta}_{max}$ for sequences in this family ranges from $1.61\sqrt{N}$ in the case of 16-QAM to $2.76\sqrt{N}$ for large M.

In the same paper, AK propose the interleaved selected family $\mathcal{I}^2\mathcal{SQ}_{16} - \mathcal{A}$ over 16-QAM. This family has $\bar{\theta}_{max} \lesssim 1.41\sqrt{N}$. Motivated by this and building upon the techniques introduced in [1], GKM [5] extend the interleaved construction of family $\mathcal{I}^2\mathcal{SQ}_{16} - \mathcal{A}$ to 64-QAM and beyond. The upper bound on $\bar{\theta}_{max}$ for family $\mathcal{I}^2\mathcal{SQ}_{M^2} - \mathcal{A}$ ranges from $1.89\sqrt{N}$ for 64-QAM to $1.99\sqrt{N}$ for large M. This implies that family $\mathcal{I}^2\mathcal{SQ} - \mathcal{A}$ has better correlation properties as compared to family $\mathcal{SQ}$. This can also be seen from the table.

In the present paper, we propose two new interleaved families. Both these families have better correlation properties than family $\mathcal{I}^2\mathcal{SQ} - \mathcal{A}$.

We first construct family $\mathcal{I}^2\mathcal{SQ}_{M^2} - \mathcal{B}$, with $M = 2^m, m \geq 2$. This family is obtained by interleaving 2 selected sequences. Over 16-QAM, the $\bar{\theta}_{max}$ for sequences in this family is upper bounded by $1.17\sqrt{N}$, which is the lowest over all known 16-QAM sequence families.

We then construct family $\mathcal{I}^4\mathcal{SQ}_{M^2}$, with $M = 2^m, m \geq 3$. This family is obtained by interleaving 4 selected sequences. The upper bound on $\bar{\theta}_{max}$ for sequences in this family ranges from $1.60\sqrt{N}$ for 64-QAM to $1.64\sqrt{N}$ for large M. From the table, we can see that this value of $\bar{\theta}_{max}$ for large M is substantially lower than the previous upper bound of $1.99\sqrt{N}$.

1.2 Preliminaries

The M^2-QAM constellation is the set

$$\{a + ib \mid -M + 1 \le a, b \le M - 1, \ a, b \text{ odd}\}.$$

When $M = 2^m$, this constellation can alternately be described as

$$\left\{ \sqrt{2i} \left(\sum_{k=0}^{m-1} 2^k i^{a_k} \right) \,\middle|\, a_i \in \mathbb{Z}_4 \right\},$$

where $\sqrt{2i}$ denotes the element $(1+i) = \sqrt{2}\exp(\frac{i2\pi}{8})$. The 16-QAM constellation is shown in Fig. 1. This representation suggests that quaternary sequences can be used in the construction of low correlation sequences over these constellations.

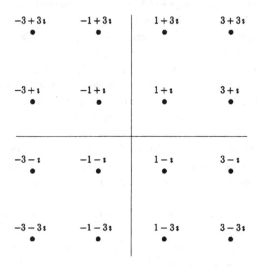

Fig. 1. 16-QAM Constellation

Galois rings are Galois extensions of the prime ring $\mathbb{Z}_p$. Let $R = GR(4, r)$ denote a Galois extension of $\mathbb{Z}_4$ of degree r. R is a commutative ring with identity and contains a unique maximal ideal $M = 2R$ generated by the element 2. The quotient R/M is isomorphic to $\mathbb{F}_q$, the finite field with $q = 2^r$ elements. As a multiplicative group, the set R^* of units of R has the following structure:

$$R^* \cong \mathbb{Z}_{2^r - 1} \times \underbrace{\mathbb{F}_2 \times \mathbb{F}_2 \dots \times \mathbb{F}_2}_{r \text{ times}}.$$

Let ξ be a generator for the multiplicative cyclic subgroup isomorphic to $\mathbb{Z}_{2^r - 1}$ contained within R^*. Let $\mathcal{T}$ denote the set $\mathcal{T} = \{0, 1, \xi, \dots, \xi^{2^r - 2}\}$. $\mathcal{T}$ is called the set of Teichmueller representatives (of $\mathbb{F}_q$ in R). More details on Galois rings can be found in [6,7,8].

As explained in [1], the maximum non-trivial correlation magnitude of a sequence family is given the modified definition:

$$\theta_{\max} := \max \left\{ |\theta_{s(j),s(k)}(\tau)| \;\middle|\; \begin{array}{l} \text{either } s(j,t), s(k,t) \\ \text{have been assigned} \\ \text{to distinct users or} \\ \tau \neq 0 \end{array} \right\}$$

We use $\bar{\theta}_{\max}$ to denote the maximum correlation magnitude after energy normalization and this is used as a basis for comparison across signal constellations.

Family $\mathcal{A}$ is an asymptotically optimal family of quaternary sequences (i.e., over $\mathbb{Z}_4$) discovered independently by Solé [10] and Boztaş, Hammons and Kumar [4]. Let $\{\gamma_i\}_{i=1}^{2^r}$ denote 2^r distinct elements in $\mathcal{T}$, i.e., we have the alternate expression $\mathcal{T} = \{\gamma_1, \gamma_2, \ldots, \gamma_{2^r}\}$. There are $2^r + 1$ cyclically distinct sequences in Family $\mathcal{A}$, each of period $2^r - 1$. For details on family $\mathcal{A}$, the reader is referred to [4,10,14].

There are some more related papers: Boztaş [3] gives new lower bounds on the periodic crosscorrelation; Udaya and Siddiqi [12] give families of biphase sequences derived from families of interleaved maximal-length sequences over $\mathbb{Z}_4$; Tang and Udaya [11] modify families $\mathcal{B}$ and $\mathcal{C}$ to obtain family $\mathcal{D}$, which is a larger family of optimal quadriphase sequences over $\mathbb{Z}_4$.

2   Family $\mathcal{I}^2\mathcal{SQ} - \mathcal{B}$

Family $\mathcal{I}^2\mathcal{SQ} - \mathcal{B}$ is constructed by interleaving 2 selected QAM sequences. It is defined over the M^2-QAM constellation, with $M = 2^m, m \geq 2$. Compared to family $\mathcal{SQ}$ and family $\mathcal{I}^2\mathcal{SQ} - \mathcal{A}$, this family has a lower value of normalized maximum correlation $\bar{\theta}_{\max}$. The data rate for all these families is the same - $(m + 1)$ bits of data per sequence period.

Let $m \geq 2$. Let $\{\delta_0 = 0, \delta_1, \delta_2, \ldots, \delta_{m-1}\}$ be elements from $\mathbb{F}_q$ such that $tr(\delta_k) = 1, \forall\, k \geq 1$. Set $H = \{\delta_0, \delta_1 \cdots, \delta_{m-1}\}$. Let $G = \{g_k\}$ be the largest subset of $\mathbb{F}_q$ having the property that

$$g_k + \delta_p \neq g_l + \delta_q\,, \quad g_k, g_l \in G\,, \quad \delta_p, \delta_q \in H\,,$$

unless $g_k = g_l$ and $\delta_p = \delta_q$.

Then the corresponding Gilbert-Varshamov and Hamming bounds [9] on the size of G are given by

$$\frac{2^r}{1 + \binom{m-1}{1} + \binom{m-1}{2}} \leq |G| \leq \frac{2^r}{1 + \binom{m-1}{1}} \tag{1}$$

2.1   Subspace-Based Construction for G and H [1]

Given constellation parameter m, let l denote the smallest power of 2 greater than $(m - 1)$, i.e., l is defined by

$$2^{l-1} < (m - 1) \leq 2^l. \tag{2}$$

We refer to the integer l as the *subspace-size exponent* (sse) associated with the *constellation parameter* (c-p) m. Thus l lies in the range $0 \le l \le (r-1)$. Let μ denote the function that, given c-p m in the range $1 \le m \le 2^{r-1} + 1$, maps m to the corresponding sse l given above, i.e., $\mu(m) = l$.

Treating $\mathbb{F}_q$ as a vector space over $\mathbb{F}_2$ of dimension r, let W_{r-1} denote the subspace of $\mathbb{F}_q$ of dimension $(r-1)$ corresponding to the elements of trace $= 0$. Let W_l denote a subspace of W_{r-1} having dimension l. Let ζ be an element in $\mathbb{F}_q$ having trace 1 and let V_l denote the subspace $V_l = W_l \cup \{W_l + \zeta\}$ of size 2^{l+1}.

Noting that every element in the coset $W_l + \zeta$ of W_l has trace 1, we select as the elements $\{\delta_k\}_{k=1}^{m-1}$ to be used in the construction of family $\mathcal{I}^2 \mathcal{SQ} - \mathcal{B}$, an arbitrary collection of $(m-1) \le 2^l$ elements selected from the set $W_l + \zeta$.

Next, we partition W_{r-1} into the 2^{r-l-1} cosets $W_l + g$ of W_l. With each coset, we associate a distinct user. To this user, we assign the coefficient set $\{g, g + \delta_1, g + \delta_2, \dots, g + \delta_l\}$. The coefficients $\{g + \delta_k\}_{k=1}^{m-1}$ belong to the coset $W_l + (g + \zeta)$ of W_l. Thus, in general, each user is assigned m coefficients, with one coefficient g belonging to the coset $W_l + g$ of W_l lying in W_{r-1} and the remaining drawn from the coset $W_l + g + \zeta$ of W_l. Since $V_l = W_l \cup (W_l + \zeta)$, all m coefficients taken together belong to the coset $V_l + g$ of V_l. Note that $V_l + g = V_l + g'$ implies

$$\{W_l + g\} \cup \{W_l + \zeta + g\} = \{W_l + g'\} \cup \{W_l + \zeta + g'\}.$$

But this is impossible since g, g' belong to different cosets of W_l and g, g' have trace zero, whereas, $tr(\zeta) = 1$. It follows that the coefficient sets of distinct users belong to different cosets of V_l and are hence distinct.

Let G be the set of all such coset representatives of W_l in W_{r-1}. Since each user is associated to a unique coset representative, the number of users is given by $G = 2^{r-l-1}$. When combined with (2), we obtain

$$\frac{2^r}{4(m-1)} < |G| \le \frac{2^r}{2(m-1)}. \tag{3}$$

Thus, the size of G is at most a factor of 4 smaller than the best possible suggested by the Hamming bound (1). The reader is referred to [1] for more details on the construction of G and H.

2.2 Sequence Definition

We describe our interleaved construction for sequences in the family $\mathcal{I}^2 \mathcal{SQ}_{M^2} - \mathcal{B}$.

Let $\{\tau_1, \tau_2, \dots, \tau_{m-1}\}$ be a set of non-zero, distinct time-shifts with where the set $\{1, \alpha^{\tau_1}, \alpha^{\tau_2}, \dots, \alpha^{\tau_{m-1}}\}$ being a linearly independent set. Let $\kappa = (\kappa_0, \kappa_1, \dots, \kappa_{m-1}) \in \mathbb{Z}_4 \times \mathbb{F}_2^{m-1}$.

Family $\mathcal{I}^2 \mathcal{SQ}_{M^2} - \mathcal{B}$ is then defined as

$$\mathcal{I}^2 \mathcal{SQ} - \mathcal{B} = \left\{ \left\{ s(g, \kappa, t) \mid \kappa \in \mathbb{Z}_4 \times \mathbb{F}_2^{m-1} \right\} \mid g \in G \right\}$$

so that each user is identified by an element of G.

Each user is assigned the collection

$$\left\{ s(g, \kappa, t) \mid \kappa \in \mathbb{Z}_4 \times \mathbb{F}_2^{m-1} \right\}$$

of sequences. The κ-th sequence $s(g, \kappa, t)$ for $m = 2$, $m = 3$ and $m \geq 4$ is given by (4), (5) and (6) respectively,

$$s(g, \kappa, t) = \begin{cases} \sqrt{2\imath} \left(\imath^{u_1(t)}(-1)^{\kappa_1} + 2\,\imath^{u_0(t)} \right) \imath^{\kappa_0} & , t \text{ even} \\ \sqrt{2\imath}\,\imath \left(\imath^{u_1(t)}(-1)^{\kappa_1} - 2\,\imath^{u_0(t)} \right) \imath^{\kappa_0} & , t \text{ odd} \end{cases} \tag{4}$$

$$s(g, \kappa, t) = \begin{cases} \sqrt{2\imath} \left(\imath^{u_2(t)}(-1)^{\kappa_2} + 2\,\imath^{u_1(t)}(-1)^{\kappa_1} + 4\,\imath^{u_0(t)} \right) \imath^{\kappa_0} & , t \text{ even} \\ \sqrt{2\imath}\,\imath \left(\imath^{u_2(t)}(-1)^{\kappa_2} + 2\,\imath^{u_1(t)}(-1)^{\kappa_1} - 4\,\imath^{u_0(t)} \right) \imath^{\kappa_0} & , t \text{ odd} \end{cases} \tag{5}$$

$$s(g, \kappa, t) = \begin{cases} \sqrt{2\imath} \left(\sum_{k=1}^{m-1} 2^{m-k-1}\,\imath^{u_k(t)}(-1)^{\kappa_k} + 2^{m-1}\,\imath^{u_0(t)} \right) \imath^{\kappa_0} & , t \text{ even} \\ \sqrt{2\imath}\,\imath \left(\sum_{k=3}^{m-1} 2^{m-k-1}\,\imath^{u_k(t)}(-1)^{\kappa_k} - 2^{m-3}\,\imath^{u_2(t)}(-1)^{\kappa_2} + \right. & \\ \left. 2^{m-2}\,\imath^{u_1(t)}(-1)^{\kappa_1} - 2^{m-1}\,\imath^{u_0(t)} \right) \imath^{\kappa_0} & , t \text{ odd} \end{cases} \tag{6}$$

where

$$u_0(t) = T([1 + 2g]\xi^t)$$
$$u_k(t) = T([1 + 2(g + \delta_k)]\xi^{t+\tau_k}),$$
$$k = 1, 2, \ldots, m - 1.$$

We refer to the element g as the *ground coefficient*. Note that given the ground coefficient and the set $\{\delta_1, \cdots, \delta_{m-1}\}$, the set of coefficients used by a user are uniquely determined. The elements $\{\delta_k\}$ provide a selection of the component sequences that leads to low correlation values.

2.3 Properties

Let $m \geq 2$ be a positive integer and let $\mathcal{I}^2 S\mathcal{Q}_{M^2} - \mathcal{B}$ be the family of sequences over M^2-QAM, $M = 2^m$, defined in the previous subsection. Then,

1. All sequences in the family $\mathcal{I}^2 S\mathcal{Q}_{M^2} - \mathcal{B}$ have period $N = 2(2^r - 1)$.
2. For large values of N, the energy of the sequences in the family is given by

$$\mathcal{E} \approx \frac{2}{3}\left(M^2 - 1\right) N.$$

3. For large values of m and N, the maximum normalized correlation $\overline{\theta}_{\max}$ of family $\mathcal{I}^2 S\mathcal{Q}_{M^2} - \mathcal{B}$ is bounded as

$$\overline{\theta}_{\max} \lesssim \frac{45\sqrt{2}}{32}\sqrt{N} \approx 1.99\sqrt{N}.$$

4. The family size is given by (3). Note from (1) that this can potentially be improved by a different construction of the set G.
5. Each user in the family can transmit $m + 1$ bits of information per sequence period.

6. The normalized minimum squared Euclidean distance between all sequences assigned to a user is given by

$$\overline{d}_{\min}^2 \approx \frac{12}{M^2 - 1} N.$$

7. The number $\mathbf{N}$ of times an element from the M^2-QAM constellation occurs in sequences of large period can be bounded as:

$$\left| \mathbf{N} - \frac{N+2}{M^2} \right| \leq \sqrt{2} \frac{M^2 - 1}{M^2} \sqrt{N+2} \, .$$

This implies that the sequences in family $\mathcal{I}^2 \mathcal{SQ}_{M^2}-\mathcal{B}$ are approximately balanced, i.e., all points from the M^2-QAM constellation occur approximately equally often in sequences of long period.

The period is clearly $N = 2(2^r - 1)$ since the period of each of the component sequences is $2^r - 1$. The correlation properties of family $\mathcal{I}^2 \mathcal{SQ}_{16}-\mathcal{B}$ are computed in the next subsection. For lack of space, we omit the rest of the proofs.

2.4   Correlation Properties of Family $\mathcal{I}^2 \mathcal{SQ}_{16}-\mathcal{B}$

We define basic sequences in family $\mathcal{I}^2 \mathcal{SQ}_{16}-\mathcal{B}$ as sequences corresponding to $\kappa_0 = \kappa_1 = 0$ in (4). We analyze the correlation between two basic sequences from family $\mathcal{I}^2 \mathcal{SQ}_{16}-\mathcal{B}$; it is straightforward to extend the results to the case of modulated sequences.

Let $\{s(g_1, 0, t)\}$ and $\{s(g_2, 0, t)\}$ be two basic sequences belonging to family $\mathcal{I}^2 \mathcal{SQ}_{16}-\mathcal{B}$, i.e.,

$$s(g_1, 0, t) = \begin{cases} \sqrt{2i} \, \left(\imath^{u_1(t)} + 2 \, \imath^{u_0(t)} \right) , & t \text{ even} \\ \sqrt{2i} \, \imath \left(\imath^{u_1(t)} - 2 \, \imath^{u_0(t)} \right) , & t \text{ odd} \end{cases}$$

$$s(g_2, 0, t) = \begin{cases} \sqrt{2i} \, \left(\imath^{v_1(t)} + 2 \, \imath^{v_0(t)} \right) , & t \text{ even} \\ \sqrt{2i} \, \imath \left(\imath^{v_1(t)} - 2 \, \imath^{v_0(t)} \right) , & t \text{ odd} \end{cases}$$

where

$$u_0(t) = T([1 + 2\,g_1]\xi^t),$$
$$u_1(t) = T([1 + 2\,(g_1 + \delta_1)]\xi^{t+\tau_1}),$$
$$v_0(t) = T([1 + 2\,g_2]\xi^t) \text{ and}$$
$$v_1(t) = T([1 + 2\,(g_2 + \delta_1)]\xi^{t+\tau_1}).$$

The expression for the correlation between the two QAM sequences would take one of the two forms depending on whether τ is even or odd.

Assume that $\tau \equiv 0 \pmod 2$. The correlation between the two sequences can be written as:

$$\theta_{s(g_1),s(g_2)}(\tau)$$

$$= \sum_{t=0}^{N-1} s(g_1,0,t+\tau)\,\overline{s(g_2,0,t)}$$

$$= 2 \sum_{t \text{ even}} \left(\imath^{u_1(t+\tau)} + 2\,\imath^{u_0(t+\tau)} \right) \left(\imath^{-v_1(t)} + 2\,\imath^{-v_0(t)} \right) +$$

$$\quad 2 \sum_{t \text{ odd}} \left(\imath^{u_1(t+\tau)} - 2\,\imath^{u_0(t+\tau)} \right) \left(\imath^{-v_1(t)} - 2\,\imath^{-v_0(t)} \right)$$

$$= 2 \left(\theta_{u_1,v_1}(\tau) + 2\,\theta_{u_1,v_0}(\tau) + 2\,\theta_{u_0,v_1}(\tau) + 4\,\theta_{u_0,v_0}(\tau) \right) +$$

$$\quad 2 \left(\theta_{u_1,v_1}(\tau) - 2\,\theta_{u_1,v_0}(\tau) - 2\,\theta_{u_0,v_1}(\tau) + 4\,\theta_{u_0,v_0}(\tau) \right)$$

$$= 4 \left(\theta_{u_1,v_1}(\tau) + 4\,\theta_{u_0,v_0}(\tau) \right)$$

Since $\theta_{u_0,v_0}(\tau)$ is orthogonal to $\theta_{u_1,v_1}(\tau)$ [1], we bound the magnitude of the number $\theta_{s(g_1),s(g_2)}(\tau)$ in the above expression as

$$\left| \theta_{s(g_1),s(g_2)}(\tau) \right| \le 4\,|\Gamma(1)(1+4\imath)|$$

$$\lesssim 16.49\,\sqrt{N/2}$$

$$\lesssim 11.66\,\sqrt{N}. \tag{7}$$

Now, if $\tau \equiv 1 \pmod 2$, we get

$$\theta_{s(g_1),s(g_2)}(\tau)$$

$$= \sum_{t=0}^{N-1} s(g_1,0,t+\tau)\,\overline{s(g_2,0,t)}$$

$$= -2\imath \sum_{t \text{ even}} \left(\imath^{u_1(t+\tau)} + 2\,\imath^{u_0(t+\tau)} \right) \left(\imath^{-v_1(t)} - 2\,\imath^{-v_0(t)} \right) +$$

$$\quad 2\imath \sum_{t \text{ odd}} \left(\imath^{u_1(t+\tau)} - 2\,\imath^{u_0(t+\tau)} \right) \left(\imath^{-v_1(t)} + 2\,\imath^{-v_0(t)} \right)$$

$$= -2\imath \left(\theta_{u_1,v_1}(\tau) - 2\theta_{u_1,v_0}(\tau) + 2\theta_{u_0,v_1}(\tau) - 4\theta_{u_0,v_0}(\tau) \right) +$$

$$\quad 2\imath \left(\theta_{u_1,v_1}(\tau) + 2\,\theta_{u_1,v_0}(\tau) - 2\,\theta_{u_0,v_1}(\tau) - 4\,\theta_{u_0,v_0}(\tau) \right)$$

$$= 8\imath \left(\theta_{u_1,v_0}(\tau) - \theta_{u_0,v_1}(\tau) \right).$$

Since we do not have any orthogonality relation between $\theta_{u_1,v_0}(\tau)$ and $\theta_{u_0,v_1}(\tau)$, we assume the worst case and bound the magnitude of $\theta_{s(g_1),s(g_2)}(\tau)$ in the above expression as

$$\left| \theta_{s(g_1),s(g_2)}(\tau) \right| \le 8\,|\Gamma(1)(1+1)|$$

$$\lesssim 16\,\sqrt{N/2}$$

$$\lesssim 11.31\,\sqrt{N}. \tag{8}$$

Comparing (7) and (8), we note that the maximum correlation given by (7) is higher. Dividing (7) by the energy, we bound the normalized maximum correlation of family $\mathcal{I}^2 S Q_{16} - \mathcal{B}$ as

$$\bar{\theta}_{s(g_1), s(g_2)}(\tau) \lesssim 1.166\sqrt{N} .$$

3 Family $\mathcal{I}^4 S Q$

Family $\mathcal{I}^4 S Q$ is constructed by interleaving 4 selected QAM sequences. It is defined over the M^2-QAM constellation, with $M = 2^m, m \geq 3$. Compared to families $S Q$, $\mathcal{I}^2 S Q - \mathcal{A}$ and $\mathcal{I}^2 S Q - \mathcal{B}$, this family has a lower value of normalized correlation parameter $\bar{\theta}_{\max}$ for 64-QAM and beyond. The data rate for all these families is the same - $(m + 1)$ bits of data per sequence period.

3.1 Sequence Definition

Let $m \geq 3$. Let $\{\delta_0 = 0, \delta_1, \delta_2, \ldots, \delta_{m-1}\}$ be elements from $\mathbb{F}_q$ such that $tr(\delta_k) = 1, \forall\, k \geq 1$. Set $H = \{\delta_0, \delta_1 \cdots, \delta_{m-1}\}$. Let $G = \{g_k\}$ be the largest subset of $\mathbb{F}_q$ having the property that

$$g_k + \delta_p \neq g_l + \delta_q , \quad g_k, g_l \in G , \quad \delta_p, \delta_q \in H ,$$

unless $g_k = g_l$ and $\delta_p = \delta_q$. A subspace-based construction for G and H has been described in subsection 2.1; we refer the reader to [1] for details.

Let $\{\tau_1, \tau_2, \ldots, \tau_{m-1}\}$ be a set of non-zero, distinct time-shifts with $\{1, \alpha^{\tau_1}, \alpha^{\tau_2}, \ldots, \alpha^{\tau_{m-1}}\}$ being a linearly independent set. Let $\kappa = (\kappa_0, \kappa_1, \ldots, \kappa_{m-1}) \in \mathbb{Z}_4 \times \mathbb{F}_2^{m-1}$.

Family $\mathcal{I}^4 S Q_{M^2}$ is then defined as

$$\mathcal{I}^4 S Q = \big\{ \{ s(g, \kappa, t) \,|\, \kappa \in \mathbb{Z}_4 \times \mathbb{F}_2^{m-1} \} \,\big|\, g \in G \big\}$$

so that each user is identified by an element of G.

Each user is assigned the collection

$$\big\{ s(g, \kappa, t) \mid \kappa \in \mathbb{Z}_4 \times \mathbb{F}_2^{m-1} \big\}$$

of sequences. The κ-th sequence $s(g, \kappa, t)$ is given by

$s(g, \kappa, t) =$

$$= \begin{cases} \sqrt{2}\imath \left(\sum_{k=1}^{m-1} 2^{m-k-1} \imath^{u_k(t)} (-1)^{\kappa_k} + 2^{m-1} \imath^{u_0(t)} \right) \imath^{\kappa_0} & , t \equiv 0 \ (\mathrm{mod}\ 4) \\[2mm] \sqrt{2}\imath \left(-\sum_{k=3}^{m-1} 2^{m-k-1} \imath^{u_k(t)} (-1)^{\kappa_k} - 2^{m-3} \imath^{u_2(t)} (-1)^{\kappa_2} + \right. \\ \left. 2^{m-2} \imath^{u_1(t)} (-1)^{\kappa_1} + 2^{m-1} \imath^{u_0(t)} \right) \imath^{\kappa_0} & , t \equiv 1 \ (\mathrm{mod}\ 4) \\[2mm] \sqrt{2}\imath \left(-\sum_{k=3}^{m-1} 2^{m-k-1} \imath^{u_k(t)} (-1)^{\kappa_k} + 2^{m-3} \imath^{u_2(t)} (-1)^{\kappa_2} - \right. \\ \left. 2^{m-2} \imath^{u_1(t)} (-1)^{\kappa_1} + 2^{m-1} \imath^{u_0(t)} \right) \imath^{\kappa_0} & , t \equiv 2 \ (\mathrm{mod}\ 4) \\[2mm] \sqrt{2}\imath \left(-\sum_{k=3}^{m-1} 2^{m-k-1} \imath^{u_k(t)} (-1)^{\kappa_k} + 2^{m-3} \imath^{u_2(t)} (-1)^{\kappa_2} + \right. \\ \left. 2^{m-2} \imath^{u_1(t)} (-1)^{\kappa_1} - 2^{m-1} \imath^{u_0(t)} \right) \imath^{\kappa_0} & , t \equiv 3\,(\mathrm{mod}\ 4) \end{cases}$$

where,

$$u_0(t) = T([1 + 2g]\xi^t)$$
$$u_k(t) = T([1 + 2(g + \delta_k)]\xi^{t+\tau_k}),$$
$$k = 1, 2, \ldots, m - 1.$$

3.2   Properties

Let $m \geq 3$ be a positive integer and let $\mathcal{I}^4 S \mathcal{Q}_{M^2}$ be the family of sequences over M^2-QAM, $M = 2^m$, defined in the previous subsection. Then,

1. All sequences in the family $\mathcal{I}^4 S \mathcal{Q}_{M^2}$ have period $N = 4(2^r - 1)$.
2. For large values of N, the energy of the sequences in the family is given by

$$\mathcal{E} \approx \frac{2}{3}(M^2 - 1) N.$$

3. For large values of m and N, the maximum normalized correlation $\overline{\theta}_{max}$ of family $\mathcal{I}^4 S \mathcal{Q}_{M^2}$ is bounded as

$$\overline{\theta}_{max} \lesssim \frac{\sqrt{689}}{16} \sqrt{N} \approx 1.64\sqrt{N}.$$

4. The family size is given by (3). Note from (1) that this can potentially be improved by a different construction of the set G.
5. Each user in the family can transmit $m + 1$ bits of information per sequence period.
6. The normalized minimum squared Euclidean distance between all sequences assigned to a user is given by

$$\overline{d}_{min}^2 \approx \frac{12}{M^2 - 1} N.$$

7. The number $\mathbf{N}$ of times an element from the M^2-QAM constellation occurs in sequences of large period can be bounded as:

$$\left| \mathbf{N} - \frac{N + 4}{2 M^2} \right| \leq \frac{M^2 - 1}{M^2} \sqrt{N + 4} .$$

This implies that the sequences in family $\mathcal{I}^4 S \mathcal{Q}_{M^2}$ are approximately balanced, i.e., all points from the M^2-QAM constellation occur approximately equally often in sequences of long period.

4   Summary and Further Work

Various properties of the four sequence families discussed in this paper ($S\mathcal{Q}$, $\mathcal{I}^2 S\mathcal{Q} - \mathcal{A}$, $\mathcal{I}^2 S\mathcal{Q} - \mathcal{B}$ and $\mathcal{I}^4 S\mathcal{Q}$) have been summarized in Table 2. From this table, we can see that family $\mathcal{I}^2 S\mathcal{Q} - \mathcal{B}$ has the lowest correlation value over

Table 2. Properties of families $\mathcal{SQ}$, $\mathcal{I}^2\mathcal{SQ} - \mathcal{A}$, $\mathcal{I}^2\mathcal{SQ} - \mathcal{B}$ and $\mathcal{I}^4\mathcal{SQ}$

Family	Constellation	Period N	Euclidean distance	$\bar{\theta}_{max}$
$\mathcal{SQ}_{16}$ [1]	16-QAM	$2^r - 1$	$0.8N$	$1.61\sqrt{N}$
$\mathcal{I}^2\mathcal{SQ}_{16}-\mathcal{A}$ [1]	16-QAM	$2(2^r - 1)$	$2.0N$	$1.41\sqrt{N}$
$\mathcal{I}^2\mathcal{SQ}_{16}-\mathcal{B}$ (new)	16-QAM	$2(2^r - 1)$	$0.8N$	$1.17\sqrt{N}$
$\mathcal{SQ}_{64}$ [1]	64-QAM	$2^r - 1$	$0.19N$	$2.11\sqrt{N}$
$\mathcal{I}^2\mathcal{SQ}_{64}-\mathcal{A}$ [5]	64-QAM	$2(2^r - 1)$	$0.19N$	$1.89\sqrt{N}$
$\mathcal{I}^2\mathcal{SQ}_{64}-\mathcal{B}$ (new)	64-QAM	$2(2^r - 1)$	$0.19N$	$1.62\sqrt{N}$
$\mathcal{I}^4\mathcal{SQ}_{64}$ (new)	64-QAM	$4(2^r - 1)$	$0.19N$	$1.60\sqrt{N}$
$\mathcal{SQ}_{256}$ [1]	256-QAM	$2^r - 1$	$0.047N$	$2.41\sqrt{N}$
$\mathcal{I}^2\mathcal{SQ}_{256}-\mathcal{A}$ [5]	256-QAM	$2(2^r - 1)$	$0.12N$	$1.95\sqrt{N}$
$\mathcal{I}^2\mathcal{SQ}_{256}-\mathcal{B}$ (new)	256-QAM	$2(2^r - 1)$	$0.047N$	$1.77\sqrt{N}$
$\mathcal{I}^4\mathcal{SQ}_{256}$ (new)	256-QAM	$4(2^r - 1)$	$0.047N$	$1.58\sqrt{N}$
$\mathcal{SQ}_{M^2}$ [1]	M^2-QAM	$2^r - 1$	$12N/(M^2 - 1)$	$2.76\sqrt{N}$
$\mathcal{I}^2\mathcal{SQ}_{M^2}-\mathcal{A}$ [5]	M^2-QAM	$2(2^r - 1)$	$30N/(M^2 - 1)$	$1.99\sqrt{N}$
$\mathcal{I}^2\mathcal{SQ}_{M^2}-\mathcal{B}$ (new)	M^2-QAM	$2(2^r - 1)$	$12N/(M^2 - 1)$	$1.99\sqrt{N}$
$\mathcal{I}^4\mathcal{SQ}_{M^2}$ (new)	M^2-QAM	$4(2^r - 1)$	$12N/(M^2 - 1)$	$1.64\sqrt{N}$

16-QAM. Over 64-QAM, 256-QAM and large M^2-QAM constellations, family $\mathcal{I}^4\mathcal{SQ}$ has the lowest correlation value.

However, family $\mathcal{I}^2\mathcal{SQ} - \mathcal{A}$ has a higher value of normalized minimum squared Euclidean distance as compared to the new families $\mathcal{I}^2\mathcal{SQ} - \mathcal{B}$ and $\mathcal{I}^4\mathcal{SQ}$.

From Table 1 (of subsection 1.1), we note that although the selected families have a lower value of correlation, they also have a lower data rate as compared to the canonical families $\mathcal{CQ}$ and $\mathcal{I}^2\mathcal{CQ}$. It would be interesting to reduce the correlation bounds of these canonical families without changing their data rate.

In all the interleaved sequence families proposed in [1] and [5], and in the interleaved sequence families proposed in this paper, we note that interleaving lowers the correlation parameter without effecting the data rate. It also increases the normalized minimum squared-Euclidean distance in most cases.

From these tables, we can also see that the Welch lower bound on sequence correlation [13] has not been achieved for any sequence family over M^2-QAM, $M = 2^m, m \geq 2$.

References

1. Anand, M., Kumar, P.V.: Low Correlation Sequences over QAM and AM-PSK Constellations. IEEE Trans. Inform. Theory 54(2), 791–810 (2008)
2. Boztaş, S.: CDMA over QAM and other Arbitrary Energy Constellations. Proc. IEEE Int. Conf. on Comm. Systems 2, 21.7.1–21.7.5 (1996)
3. Boztaş, S.: New Lower Bounds on the Periodic Crosscorrelation of QAM Codes with Arbitrary Energy. In: Fossorier, M.P.C., Imai, H., Lin, S., Poli, A. (eds.) AAECC 1999. LNCS, vol. 1719, pp. 410–419. Springer, Heidelberg (1999)

4. Boztaş, S., Hammons, R., Kumar, P.V.: 4-phase sequences with near-optimum correlation properties. IEEE Trans. Inform. Theory 38(3), 1101–1113 (1992)
5. Garg, G., Kumar, P.V., Veni Madhavan, C.E.: Low correlation interleaved QAM sequences. In: Int. Symp. on Inf. Theory (ISIT), Toronto, Canada, July 6–11 (2008)
6. Hammons Jr., A.R., Kumar, P.V., Calderbank, A.R., Sloane, N.J.A., Solé, P.: The $\mathbb{Z}_4$-linearity of Kerdock, Preparata, Goethals, and related codes. IEEE Trans. Inform. Theory 40(2), 301–319 (1994)
7. Helleseth, T., Kumar, P.V.: Sequences with low correlation. In: Pless, V.S., Huffman, W.C. (eds.) Handbook of Coding Theory, Elsevier, Amsterdam (1998)
8. Kumar, P.V., Helleseth, T., Calderbank, A.R.: An upper bound for Weil exponential sums over Galois rings and applications. IEEE Trans. Inform. Theory 41(2), 456–468 (1995)
9. MacWilliams, F.J., Sloane, N.J.A.: The Theory of Error-Correcting Codes. North-Holland, Amsterdam (1977)
10. Solé, P.: A quaternary cyclic code and a family of quadriphase sequences with low correlation properties. In: Coding Theory and Applications. LNCS, vol. 388. Springer, Heidelberg (1989)
11. Tang, X., Udaya, P.: A Note on the Optimal Quadriphase Sequences Families. IEEE Trans. Inform. Theory 53(1), 433–436 (2007)
12. Udaya, P., Siddiqi, M.U.: Optimal Biphase Sequences with Large Linear Complexity Derived from Sequences over Z4. IEEE Trans. on Inform. Theory 42(1), 206–216 (1996)
13. Welch, L.R.: Lower bounds on the maximum cross correlation of signals. IEEE Trans. Inform. Theory 20(3), 397–399 (1974)
14. Yang, K., Helleseth, T., Kumar, P.V., Shanbhag, A.: The weight hierarchy of Kerdock codes over $\mathbb{Z}_4$. IEEE Trans. Inform. Theory 42(5), 1587–1593 (1996)

The Combinatorics of Differentiation

Anna S. Bertiger, Robert J. McEliece, and Sarah Sweatlock

Cornell University
California Institute of Technology

Notation: We will try to adhere to the notation established by the seminal works of Stanley [5, 6]. But for quick reference:

$|S|$:	Number of elements in the finite set S
$[n]$:	The finite set $\{1, \ldots, n\}$
$\mathbb{Z}$:	The integers
$\mathbb{P}$:	The positive integers.

Abstract. Let $S_1, S_2, \ldots$ be a sequence of finite sets, and suppose we are asked to find the sequence of cardinalities $s[1], s[2], \ldots$. We are usually satisfied to find a closed-form expression for the a-generating function $F_S(z) = \sum_{n \geq 0} s[n]a[n]z^n$, where $a[n]$ is a fixed positive causal sequence. But extracting $s[n]$ from $F_S(z)$ is often itself a challenging problem, because of the unavoidable link to calculus $s[n] = \frac{a[n]}{n!} D^n[F(z)]_{z=0}$. In this paper we will consider the case $a[n] = 1/n!$, (exponential generating functions), and find many links between combinatorics and calculus.

1   Introduction

In a typical problem of enumerative combinatorics, one is given a sequence $S_0, S_1, \ldots$, of finite sets and asked to determine the corresponding sequence of cardinalities[1] $f[0], f[1], \ldots$, where $f[n] = |S_n|$. One approach to this problem is to show that the exponential generating function (a.k.a. Taylor series)[2]

$$F(z) = \sum_{n \geq 0} f[n] \frac{z^n}{n!} \qquad (1)$$

has a closed-form expression, say $F(z)$. The sequence $f[n]$ can then be obtained by successive differentiation of $F(z)$:

$$f[n] = [D^n F(z)]_{z=0} \quad \text{for } n = 1, 2, \ldots,$$

In this way the *combinatorial* problem of determining the sequence $f[n]$ has been converted to the routine *calculus* problem of determining the Taylor expansion of $F(z)$, and for this and other reasons it is usual to accept $F(z)$ as a solution to the combinatorial problem [5].

---

[1] We use the notation $|S|$ to denote the number of elements in the finite set S.

[2] Of course an equivalent approach is to use the ordinary generating function (a.k.a. z-transform).

S.W. Golomb et al. (Eds.): SETA 2008, LNCS 5203, pp. 142–152, 2008.

However, finding the nth derivative of even the simplest $F(z)$ may not be easy (try $F(z) = e^{e^z}$). Still, if $F(z)$ is a hybrid of two or more simpler functions, a divide and conquer approach can be helpful. For example, if $F(z) = F_1(z)F_2(z)$, the well-known rule of Leibniz shows how the Taylor coefficients of $F(z)$ can be obtained from those of $F_1(z)$ and $F_2(z)$:

$$f[n] = \sum_{k=0}^{n} \binom{n}{k} f_1[k] f_2[n-k] \quad \text{for } n \geq 0. \quad . \tag{2}$$

This formula requires n additions and n multiplications to compute $f[n]$. Another expression for $f[n]$ is

$$f[n] = \sum_{S \subseteq [n]} f_1[|S|] f_2[|S^c|], \tag{3}$$

where the summation in 3 is over all 2^n subsets of $[n] = \{1, \ldots, n\}$, thus requiring 2^n additions and 2^n multiplications to compute $f[n]$. To distinguish the two expressions, let us call them $f_A[n]$ and $f_B[n]$. $f_A[n]$ simpler but redundant; $f_B[n]$ is more complicated but compact. One suspects that expression A will be useful for proving theorems but expression B will be preferable for computations. Note also that both formulas seem sommehow combinatorial.

What is true for the Leibniz rule is true in considerable generality. We have found many other "(A, B)," pairs, and in every case, the resulting formulae are essentially combinatorial; thus the derived *calculus* problem becomes again a *combinatorial* problem. In this paper, which is mostlly, but not entirely, tutorial, we will discuss a number of instances of this phenomenon (see Table 2).

2   Compositions and Partitions of Integers and Finite Sets

In this section we will define the combinatorial objects we need to describe our formulae for inverse Taylor series. The impatient reader is invited to skip ahead to Table 1 where this material is summarized.

Let n and k be positive integers. A **composition of n** is an **ordered** list of **positive** integers whose sum is n. A **weak composition of n** is an **ordered** list of **nonnegative** integers whose sum is n. A **k-composition of n** is a composition of n with exactly k parts A **weak k-composition of n** is an ordered list of k nonnegative integers whose sum is n.

Example 1. The weak 2-compositions of 4 are:

$$C_{4,2}^{\star} = (4,0), (3,1), (2,2), (1,3), (0,4).$$

A **partition of n** is an **unordered** list (multiset) of **positive** integers whose sum is n. A **weak partition** of n is an **unordered** list of **nonnegative** integers whose sum us n. A **k-partition** of n is an **unordered** list of k **positive** integers whose sum is n. A **weak k-partition** of n is an **unordered** list of k nonnegative integers whose sum is n.

Example 2. The 2-partitions of 4:

$$P_{4,2} = \{\{3,1\}\}, \{\{2,2\}\}.$$

A **composition** of $[n] = \{1,\ldots,n\}$ is an **ordered** list of pairwise disjoint **nonempty** subsets of $[n]$ whose union is n. A **weak composition** of $[n]$ is an **ordered** list of pairwise disjoint subsets of $[n]$ whose union is $[n]$. A k-composition of $[n]$ is an ordered list of k nonempty subsets of $[n]$ whose union is $[n]$. A **weak** k-composition of $[n]$ is an ordered list of k disjoint subsets of $[n]$ whose union is n.

Example 3. The weak 2-compositions of $[2]$:

$$\mathfrak{C}^*_{2,2} = (\{1,2\}, \emptyset), (\{1\}, \{2\}), (\{2\}, \{1\}), (\emptyset, \{1,2\}).$$

Table 1. Twelve Combinatorial Structures

	1	2	3		4	5	6
S:	$\mathfrak{C}_n$	$\mathfrak{C}_{n,k}$	$\mathfrak{C}^*_{n,k}$	S:	$\mathfrak{P}_n$	$\mathfrak{P}_{n,k}$	$\mathfrak{P}^*_{n,k}$
$|S|$:	O_n	$\sum_j (-1)^{k-j}\binom{k}{j}j^n$	k^n	$|S|$:	B_n	$S_{n,k}$	$\sum_{j=1}^k S_{n,j}$

	7	8	9		10	11	12
S:	C_n	$C_{n,k}$	$C^*_{n,k}$	S:	P_n	$P_{n,k}$	$P^*_{n,k}$
$|S|$:	2^{n-1}	$\binom{n-1}{k-1}$	$\binom{n+k-1}{k-1}$	$|S|$:	$p(n)$	$p_k(n)$	$\sum_{j=1}^k p_j(n)$

Legend:

$\mathfrak{C}_n$	compositions of $[n]$	$\mathfrak{P}_n$	partitions of $[n]$
$\mathfrak{C}_{n,k}$	k-compositions of $[n]$	$\mathfrak{P}_{n,k}$	k-partitions of $[n]$
$\mathfrak{C}^*_{n,k}$	weak k-compositions of $[n]$	$\mathfrak{P}^*_{n,k}$	weak k-partitions of $[n]$
C_n	compositions of n	P_n	partitions of n
$C_{n,k}$	k-compositions of n	$P_{n,k}$	k-partitions of n
$C^*_{n,k}$	weak k-compositions of n	$P^*_{n,k}$	weak k-partitions of n
$p(n)$	number of partitions of n	$p_k(n)$	number of k-partitions of n
B_n	Number of partitions of $[n]$ (Bell number)	$S_{n,k}$	number of k-partitions of $[n]$ (Stirling Number second kind)
O_n	ordered Bell number	$O_{n,k}$	$\sum_j (-1)^{k-j}\binom{k}{j}j^n$

Similarly, a *partition* of $[n]$ is an **unordered** list of *nonempty* disjoint subsets of $[n]$ whose union is $[n]$, and a *weak partition* of $[n]$ is an *unordered* list of disjoibt subsets of $[n]$ whose union is $[n]$. k-partitions and k-compositions are defined as expected.

Example 4. The weak 3-partitions of $[2]$:

$$\mathfrak{P}^*_{2,3} = \{\{1,2\}, \emptyset, \emptyset\}, \{\{1\}, \{2\}, \emptyset\}.$$

Counting the (weak/strong)(compositions/partitions) of (integers/sets) is a charming problem in elementary enumerative combinatorics, reminiscent of Rota's "Twelvefold Way" [5]. Some of the solutions are given in Table 1.

3   A Curious Class of Partitions

Finally, we define a curious class of weak integer partitions that arises in the study of the Taylor series of inverse functions (the fourth and fifth lines of Table 2). It can be defined recursively, as follows:

$$V_0 := \emptyset$$
$$V_n := P_n \cup (V_{n-1} * 0) \quad \text{for } n \geq 1$$

And nonrecursively:

$$V_0 = \emptyset$$
$$V_1 = \{1, 0\}$$
$$V_2 = \{2, 11, 10, 00\}$$
$$V_3 = \{3, 21, 111, 20, 110, 100, 000\}$$
$$V_4 = \{4, 31, 22, 211, 1111, 30, 210, 1110, 200, 1100, 1000, 0000\}$$
$$\vdots$$
$$V_n = \bigcup_{j=0}^{n} P_{n-j} * 0^j$$

4   The D Operator

Definition. If $\lambda = \{\{\lambda_1, \ldots, \lambda_k\}\}$ is an integer partition, then

$$D\lambda = \{\{\lambda_1 + 1, \ldots, \lambda_k + 1\}\}$$
$$DV_0 = \emptyset$$
$$DV_1 = \{2, 1\}$$
$$DV_2 = \{3, 22, 21, 11\}$$
$$DV_3 = \{4, 32, 222, 31, 221, 211, 111\}$$
$$\vdots$$
$$DV_n = DP_n \cup DV_{n-1} * 1 + \cdots$$

5   More Definitions and Notation

Let $\lambda = \{\{\lambda_1, \ldots, \lambda_k\}\} \in P^*_{n,k}$, and let m be its multiplicity function:

$$m_p(\lambda) = |\{i : 1 \leq i \leq k, \lambda_i = p\}| \quad \text{for } p \in \mathbb{Z} = 0, \ldots, n \quad .$$

We now associate several coefficients with $\lambda = \{\{\lambda_1, \ldots, \lambda_k\}\} = (0^{m_0} 1^{m_1} \ldots n^{m_n})$:

$$A(\lambda) = \binom{n}{\lambda_1, \ldots, \lambda_k} \qquad B(\lambda) = \binom{k}{m_0, m_1, \ldots, m_n}$$

$$C(\lambda) = A(\lambda) B(\lambda) \frac{m_0!}{k!} \qquad D(\lambda) = \frac{k!}{m_0!}$$

Now let us define the L (loosen) and R (roughen) operators:

$$L : (x_1, \ldots, x_k) \rightarrow \{\{x_1, \ldots, x_k\}\}$$

$$R : (y_1, \ldots, y_k) \rightarrow (|y_1|, \ldots, |y_k|)$$

Theorem. As summarized in the diagram below,

(a) For all $c \in C^*$, $|R^{\langle -1\rangle}(c)| = A(c)$.
(b) For all $\lambda \in P^*$, $|L^{\langle -1\rangle}(\lambda)| = B(\lambda)$.
(c) For all $\lambda \in P^*$, $|R^{\langle -1\rangle}(\lambda)| = C(\lambda)$.
(d) For all $\pi \in \mathfrak{P}^*$, $|L^{\langle -1\rangle}(\pi)| = D(\pi)$.

$$
\begin{array}{ccc}
\mathfrak{C}^* & \xrightarrow{L} & \mathfrak{P}^* \\
\downarrow{R} & & \downarrow{R} \\
C^* & \xrightarrow{L} & P^*
\end{array}
$$

Corollaries

$$\sum_{\pi \in \mathfrak{P}_n} f[\pi] = \sum_{\lambda \in P_n} C(\lambda) f[\lambda]$$

$$\sum_{\alpha \in \mathfrak{C}_n^*} f[\alpha] = \sum_{c \in C_n^*} A(c) f[c] = \sum_{\lambda \in P_n^*} A(\lambda) B(\lambda) f[\lambda]$$

$$\vdots$$

6   Illustrative Example

In this section we will illustrate in some detail one of the entries given in Table 2. Specifically, given a Taylor series $F(z)$ and assuming $F(0) = 1$, we seek the Taylor coefficients for $G(z) = F(z)^{-2}$, i.e.,

$$g[n] := \left[\frac{x^n}{n!}\right] F(z)^{-2}, \quad \text{for } n \geq 0.$$

According to Table 2A,

$$g[n] = \sum_{\alpha \in \mathfrak{C}_n} \binom{-2}{|\alpha|} f[|\alpha_1|] \cdots f[|\alpha_k|], \tag{4}$$

where the sum is over all compositions (ordered partitions) of the set $[n] = \{1, \ldots, n\}$, i.e., $\alpha = (\alpha_1, \ldots, \alpha_k)$, where $\emptyset \subset \alpha_i \subseteq [n]$, $\alpha_i \cap \alpha_j = \emptyset$. However, $|\mathfrak{C}_n|$ is enormous — for $n = 10$ it is 102247563 — so we seek elsewhere (Table 2.2B) for a computationally efficient formula.

$$g[n] = \sum_{\lambda \in P_n} (-1)^{|\lambda|}(|\lambda| + 1)A(\lambda)B(\lambda)f[\lambda_1] \cdots f[\lambda_k] \tag{5}$$

This formula is more manageable – exactly 42 terms for $n = 10$. We present a Tableau for computing $g[1], \ldots g[4]$ below.

	λ	$(-1)^{|\lambda|}$	$|\lambda| + 1$	$A(\lambda)$	$B(\lambda)$		Coeff.
$n = 1$	(1)	$-$	2	1	1	-2	$f[1]$
$n = 2$	(2)	$-$	2	1	1	-2	$f[2]$
	$(1,1)$	$+$	3	2	1	$+6$	$f[1]^2$
$n = 3$	(3)	$-$	2	1	1	-2	$f[3]$
	$(2,1)$	$+$	3	3	2	$+18$	$f[2]f[1]$
	$(1,1,1)$	$-$	4	6	1	-24	$f[1^3$
$n = 4$	(4)	$-$	2	1	1	-2	$f[4]$
	$(3,1)$	$+$	3	4	2	24	$f[3]f[1]$
	$(2,2)$	$+$	3	6	1	18	$f[2]^2$
	$(2,1,1)$	$-$	4	12	3	-144	$f[2]f[1]^2$
	$(1,1,1,1)$	$+$	5	24	1	120	$f[1]^4$

Thus

$$g[0] = 1$$

$$g[1] = -2f[2] + 6f[1,1]$$

$$\vdots$$

$$g[4] = -2f[4] + 24f[3,1] + 18f[2,2] - 144f[2,1,1] + 120f[1,1,1,1]$$

If more generally, we asked for the Taylor series for $F(x)^x$ where x is an arbitrary real number, we obtain

$$g[0] = \binom{x}{0}$$

$$g[1] = \binom{x}{1}f[1]$$

$$g[2] = \binom{x}{1}f[2] + 2\binom{x}{2}f[1,1]$$

$$g[3] = \binom{x}{1} f[3] + 6 \binom{x}{2} f[2]f[1] + 6 \binom{x}{3} f[1]^3$$

$$g[4] = \binom{x}{1} f[4] + 8 \binom{x}{2} f[3]f[1] + 6 \binom{x}{2} f[2]^2 + 36 \binom{x}{3} f[2]f[1]^2 + 24 \binom{x}{4} f[1]^4 \blacksquare$$

That concludes the illustrative example. In Table 2, which follows, we list a few of many possible variations on the theme. For the record, entry 1 is Leibniz' rule, entry 5 is Faa di Bruno, and entries 6 and 7 are essentialy due to Lagrange.

Table 2. Some $(H(z) \leftrightarrow h[n])$ pairs

	$H(z)$	$h_A[n]$	$h_B[n]$
1.	$F_1(z) \cdots F_k(z)$	$\displaystyle\sum_{\alpha \in \mathfrak{C}_{n,k}^*} f_1[\alpha_1] \cdots f_k[\alpha_k]$	$\displaystyle\sum_{c \in C_{n,k}^*} A(c) f_1[c_1] \cdots f_k[c_k]$
2.[a,d]	$F(z)^x$	$\displaystyle\sum_{\alpha \in \mathfrak{C}_n} \binom{x}{|\alpha|} f[\alpha]$	$\displaystyle\sum_{\lambda \in P_n} \binom{x}{|\lambda|} A(\lambda) B(\lambda) f[\lambda]$
3.[b,c]	$G(F(z))$	$\displaystyle\sum_{\pi \in \mathfrak{P}_n} f \wr g[\pi]$	$\displaystyle\sum_{\lambda \in P_n} C(\lambda) f \wr g[\lambda]$
	$H'(z) = G'(F(z))F'(z)$		
4.[b]	$F^{\langle -1 \rangle}(z)$	$\displaystyle\sum_{R(\pi) \in DP_{n-1}} (-1)^{|\pi|} f[\pi]$	$\displaystyle\sum_{\lambda \in DP_{n-1}} (-1)^{|\lambda|} C(\lambda) f[\lambda]$
	$F'(z) = \frac{1}{G'(F(z))}$		
5.[b,c]	$G(F^{\langle -1 \rangle}(z))$	$\displaystyle\sum_{R(\pi) \in DV_{n-1}} (-1)^{|\pi|} f \wr g[\pi]$	$\displaystyle\sum_{\lambda \in DV_{n-1}} (-1)^{|\lambda|} C(\lambda) f \wr g[\lambda]$
6.	$F(H, z) = 0$	(See [4])	
7.[b]	$H = G(w), z = F(w)$	(see no. 5)	
8.	$H(z) = G(F_1(z), \ldots F_q(z))$	(requires q-chromatic partitions)	

[a] $F(0) = 1$, [b] $F(0) = 0, F'(0) = 1$, [c] $f \wr g[\alpha] \triangleq f[\alpha]g[|\alpha|]$, [d] $x \in \mathbb{R}$

c	an integer composition	λ	an integer partition
C_n	a set of integer compositions	P_n	a set of integer partitions
α	a set composition	π	a set partition
$\mathfrak{C}_n$	a set of set compositions	$\mathfrak{P}_n$	a set of set partitions
n	sum of parts	k	number of parts
j	number of null parts	$*$	null parts ok

7   The Lagrange Inversion Formula

The following is required for some of our proofs.

Lagrange Inversion Formula [2] [6]. Suppose $F(0) = 0$, and $F_0(z) = F(z)/z$. Then

$$\left[\frac{z^n}{n!} \right] F^{\langle -1 \rangle}(z) = \left[\frac{z^{n-1}}{(n-1)!} \right] F_0(z)^{-n}.$$

More generally,

$$\left[\frac{z^n}{n!}\right] G(F^{\langle -1\rangle}(z)) = \left[\frac{z^{n-1}}{(n-1)!}\right] G'(z) F_0(z)^{-n}.$$

8 Proof of Entry 1 in Table 2

Let $F_j(z) := \sum_{j\geq 0} f_j[n_j]\frac{z^{n_j}}{n_j!}$, for $j = 1,\ldots,k$. Then

$$\prod_{j=1}^k F_j(z) = \prod_{j=1}^k \sum_{n_j\geq 0} f_j[n_j]\frac{z^{n_j}}{n_j!}$$

$$= \sum_{n_1,\ldots,n_k\in\mathbb{N}} f_1[n_1]\cdots f_k[n_k]\frac{z^{n_1+\cdots+n_k}}{n_1!\cdots n_k!} \qquad \text{(Distributive Law.)}$$

$$\left[\frac{z^n}{n!}\right] \prod_{j=1}^k F_j(z) = \sum_{c\in C^*_{n,k}} A(c) f_1[c_1]\cdots f_k[c_k]$$

$$= \sum_{\alpha\in\mathfrak{C}^*_{n,k}} f_1[\alpha_1]\cdots f_k[\alpha_k].$$

9 Proof of Entry 2 in Table 2

Since $F(0) = 1$, we may write $F(z) = 1 + G(z)$ with $G(0) = 0$ and use the binomial theorem:

$$F(z)^x = (1 + G(z))^x$$

$$= \sum_{j\geq 0} \binom{x}{j} G(z)^j$$

But by One, since $g[0] = 0$,

$$\left[\frac{z^n}{n!}\right] G(z)^j = \sum_{\alpha\in\mathfrak{C}_{n,j}} g[\alpha] = \sum_{\alpha\in\mathfrak{C}_{n,j}} f[\alpha]$$

so

$$h[n] = \sum_{j\geq 0} \binom{x}{j} \sum_{\alpha\in\mathfrak{C}_{n,j}} f[\alpha]$$

$$= \sum_{\alpha\in\mathfrak{C}_n} \binom{x}{|\alpha|} f[\alpha]$$

$$= \sum_{c\in C_n} A(c) \binom{x}{|c|} f[c]$$

$$= \sum_{k=1}^{n} \binom{x}{k} \sum_{\lambda \in P_{n,k}} A(\lambda)B(\lambda)f[\lambda]$$

$$= \sum_{\lambda \in P_n} \binom{x}{|\lambda|} A(\lambda)B(\lambda)f[\lambda]$$

10   Proof of Entry 3 in Table 2

By definition,

$$G(F(z)) = \sum_{k \geq 0} \frac{g[k]}{k!} F(z)^k.$$

But by 1, since $F(0) = 0$,

$$\left[\frac{z^n}{n!}\right] F(z)^k = \sum_{\alpha \in \mathfrak{C}_{n,k}} f[\alpha]$$

Thus

$$h[n] = \sum_{\alpha \in \mathfrak{C}_n} \frac{1}{|\alpha|!} f[\alpha]g[|\alpha|]$$

$$= \sum_{\pi \in \mathfrak{P}_n} f[\pi]g[|\pi|]$$

$$= \sum_{\lambda \in P_n} C(\lambda)f \wr g[\lambda]$$

11   Proof of Entry 4 in Table 2

By Lagrange's theorem, if $F(0) = 0$, and $F_0(z) := F(z)/z$ (whence $f_0[n] = f[n+1]/(n+1)$),
then

$$\left[\frac{w^n}{n!}\right] F^{\langle -1 \rangle}(w) = \left[\frac{z^{n-1}}{(n-1)!}\right] F_0(z)^{-n}$$

$$= \sum_{\lambda \in P_{n-1}} C(\lambda) \binom{-n}{|\lambda|} f_0[\lambda]$$

$$= \sum_{\lambda \in P_{n-1}} (-1)^{|\lambda|+1} C(\lambda+1)f[\lambda+1].$$

$$= \sum_{\lambda \in DP_{n-1}} (-1)^{|\lambda|} C(\lambda)f[\lambda]$$

$$= \sum_{R(\pi) \in DP_{n-1}} (-1)^{|\pi|} f[\pi]$$

12   Proof of Entry 5 in Table 2

Follows from Three and Four. Details omitted.

13   Proof of Entry 8 in Table 2

The folowing theorem, which is a straightforward generalization of Stanley's Thorem 5.1.4 [6], is nevertheless useful as a startign point for inverse multivariable multivariale Taylor series.

We begin by defining a q-chromatic partition of n to be a pair (α, c) consisting of partition $\alpha = \{\{\alpha_1, \ldots, \alpha_k\}\}$ of n together with a coloring function $c : \{\{\alpha_1, \ldots, \alpha_k\}\} \to [q]$.

Note that repeated parts of the partition need not have the same color, but that, for example the partition $\{\{1, 1\}\}$ of 2, could be 2-colored with one red 1 and one blue 1, both 1s red or both 1s blue, but that changing which 1 is red and which 1 is blue does not make any distinction for a total of three possible colorings.

Theorem. Multi-variable Compositional Formula

Let $f_1, \ldots, f_q : \mathbb{P} \to \mathbb{C}$ be functions with exponential generating functions $F_1, \ldots, F_q$ respectively. Assume that $f_1[0] = \cdots = f_q[0] = 0$, and let $g : \mathbb{N}^q \to K$ be such that $g[0, \ldots, 0] = 0$. Let $h : \mathbb{N} \to K$ have $h[0] = 0$ and for all $n \geq 0$,

$$h[n] = \sum \prod_{i=1}^{l(\alpha)} f_{c(\alpha_i)}(|\alpha_i|) g(\mathbf{k}),$$

where the sum is over all colored set partitions (α, c) of $[n]$ and $\mathbf{k} = (k_1, \ldots, k_q)$ is such that k_j is the number of parts of (α, c) of color j. Then,

$$H(z) = G(F_1(x), \ldots, F_q(x)).$$

Proof: Fix $\mathbf{k} \in \mathbb{N}^q$, and assume (without loss of generality) that any colored set partition is such that we have all parts of color 1 listed first, followed by all parts of color 2 and so forth, ending with all parts of color q. We can then use Table 2, entry 1, with $k_1 + \cdots + k_q$ functions to prove the theorem. In table 2, entry 1, let the functions that tell us the coefficients of the exponential generating functions be $a_1, \ldots, a_{k_1 + \cdots + k_q}$

$$a_1 = \cdots = a_{k_1} = f_1,$$

$$a_{k_1+1} = \cdots = a_{k_1+k_2} = f_2,$$

and so on, up to

$$a_{k_1+\cdots+k_{q-1}} = \cdots = a_{k_1+\cdots+k_q} = f_q.$$

Then, there are $k_1! k_2! \ldots k_q!$ ways to arrange our ordered blocks so that we have not changed the product. Thus,

$$E_{h_k}(x) = g(k) \prod_{j=1}^{q} \frac{E_{f_j}(x)^{k_j}}{k_j!},$$

and summing over all k gives the desired result.   ∎

References

1. Apostol, T.M.: Calculus, 2nd edn., vol. 2. John Wiley and sons, Chichester (1969)
2. Copson, E.T.: An Introduction to the Theory of Functions of a Complex Variable. The Clarendon Press, Oxford (1935)
3. Knuth, D.E.: The Art of Computer Programming, 2nd edn. Fundamental Algorithms, vol. 1. Addison-Wesley, Reading (1973)
4. Sweatlock, S.L.: Caltech Ph.D. thesis (2008)
5. Stanley, R.P.: Enumerative Combinatorics, vol. 1
6. Stanley, R.P.: Enumerative Combinatorics, vol. 2

Group Representation Design of Digital Signals and Sequences

Shamgar Gurevich[1], Ronny Hadani[2], and Nir Sochen[3]

[1] Department of Mathematics, University of California, Berkeley, CA 94720, USA
shamgar@math.berkeley.edu
[2] Department of Mathematics, University of Chicago, IL 60637, USA
hadani@math.uchicago.edu
[3] School of Mathematical Sciences, Tel Aviv University, Tel Aviv 69978, Israel
sochen@post.tau.ac.il

Abstract. In this survey a novel system, called the *oscillator system*, consisting of order of p^3 functions (signals) on the finite field $\mathbb{F}_p$, is described and studied. The new functions are proved to satisfy good auto-correlation, cross-correlation and low peak-to-average power ratio properties. Moreover, the oscillator system is closed under the operation of discrete Fourier transform. Applications of the oscillator system for discrete radar and digital communication theory are explained. Finally, an explicit algorithm to construct the oscillator system is presented.

Keywords: Weil representation, commutative subgroups, eigenfunctions, good correlations, low supremum, Fourier invariance, explicit algorithm.

1 Introduction

One-dimensional *analog signals* are complex valued functions on the real line $\mathbb{R}$. In the same spirit, one-dimensional *digital signals*, also called *sequences*, might be considered as complex valued functions on the finite line $\mathbb{F}_p$, i.e., the finite field with p elements, where p is an odd prime. In both situations the parameter of the line is denoted by t and is referred to as *time*. In this survey, we will consider digital signals only, which will be simply referred to as signals. The space of signals $\mathcal{H} = \mathbb{C}(\mathbb{F}_p)$ is a Hilbert space with the Hermitian product given by

$$\langle \phi, \varphi \rangle = \sum_{t \in \mathbb{F}_p} \phi(t)\overline{\varphi(t)}.$$

A central problem is to construct interesting and useful systems of signals. Given a system $\mathfrak{S}$, there are various desired properties which appear in the engineering wish list. For example, in various situations [1,2] one requires that the signals will be weakly correlated, i.e., that for every $\phi \neq \varphi \in \mathfrak{S}$

$$|\langle \phi, \varphi \rangle| \ll 1.$$

S.W. Golomb et al. (Eds.): SETA 2008, LNCS 5203, pp. 153–166, 2008.

This property is trivially satisfied if $\mathfrak{S}$ is an orthonormal basis. Such a system cannot consist of more than $\dim \mathcal{H}$ signals, however, for certain applications, e.g., CDMA (Code Division Multiple Access) [3] a larger number of signals is desired, in that case the orthogonality condition is relaxed.

During the transmission process, a signal φ might be distorted in various ways. Two basic types of distortions are *time shift* $\varphi(t) \mapsto L_\tau \varphi(t) = \varphi(t + \tau)$ and *phase shift* $\varphi(t) \mapsto M_w \varphi(t) = e^{\frac{2\pi i}{p} wt} \varphi(t)$, where $\tau, w \in \mathbb{F}_p$. The first type appears in asynchronous communication and the second type is a Doppler effect due to relative velocity between the transmitting and receiving antennas. In conclusion, a general distortion is of the type $\varphi \mapsto M_w L_\tau \varphi$, suggesting that for every $\varphi \neq \phi \in \mathfrak{S}$ it is natural to require [2] the following stronger condition

$$|\langle \phi, M_w L_\tau \varphi \rangle| \ll 1.$$

Due to technical restrictions in the transmission process, signals are sometimes required to admit low peak-to-average power ratio [4], i.e., that for every $\varphi \in \mathfrak{S}$ with $\|\varphi\|_2 = 1$

$$\max \{ |\varphi(t)| : t \in \mathbb{F}_p \} \ll 1.$$

Finally, several schemes for digital communication require that the above properties will continue to hold also if we replace signals from $\mathfrak{S}$ by their Fourier transform.

In this survey we demonstrate a construction of a novel system of (unit) signals $\mathfrak{S}_O$, consisting of order of p^3 signals, called the *oscillator system*. These signals constitute, in an appropriate formal sense, a finite analogue for the eigenfunctions of the harmonic oscillator in the real setting and, in accordance, they share many of the nice properties of the latter class. In particular, the system $\mathfrak{S}_O$ satisfies the following properties

1. *Auto-correlation (ambiguity function)*. For every $\varphi \in \mathfrak{S}_O$ we have

$$|\langle \varphi, M_w L_\tau \varphi \rangle| = \begin{cases} 1 & \text{if } (\tau, w) = 0, \\ \leq \frac{2}{\sqrt{p}} & \text{if } (\tau, w) \neq 0. \end{cases} \tag{1}$$

2. *Cross-correlation (cross-ambiguity function)*. For every $\phi \neq \varphi \in \mathfrak{S}_O$ we have

$$|\langle \phi, M_w L_\tau \varphi \rangle| \leq \frac{4}{\sqrt{p}}, \tag{2}$$

for every $\tau, w \in \mathbb{F}_p$.

3. *Supremum*. For every signal $\varphi \in \mathfrak{S}_O$ we have

$$\max \{ |\varphi(t)| : t \in \mathbb{F}_p \} \leq \frac{2}{\sqrt{p}}.$$

4. *Fourier invariance*. For every signal $\varphi \in \mathfrak{S}_O$ its Fourier transform is $\widehat{\varphi}$ (up to multiplication by a unitary scalar) also in $\mathfrak{S}_O$.

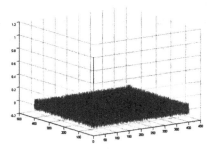

Fig. 1. Ambiguity function of an "oscillator" signal

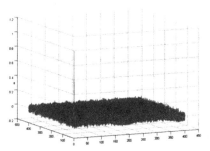

Fig. 2. Ambiguity function of a random signal

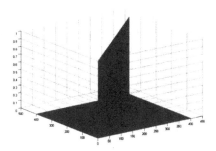

Fig. 3. Ambiguity function of a chirp signal

In Figures 1, 2, and 3, the ambiguity function of a signal from the oscillator system is compared with that of a random signal and a typical chirp.

The oscillator system can be extended to a much larger system $\mathfrak{S}_E$, consisting of order of p^5 signals if one is willing to compromise Properties 1 and 2 for a weaker condition. The extended system consists of all signals of the form $M_w L_\tau \varphi$ for $\tau, w \in \mathbb{F}_p$ and $\varphi \in \mathfrak{S}_O$. It is not hard to show that $\# (\mathfrak{S}_E) = p^2 \cdot \# (\mathfrak{S}_O) \approx p^5$. As a consequence of (1) and (2) for every $\varphi \neq \phi \in \mathfrak{S}_E$ we have

$$|\langle \varphi, \phi \rangle| \leq \frac{4}{\sqrt{p}}.$$

The characterization and construction of the oscillator system is representation theoretic and we devote the rest of the survey to an intuitive explanation of the main underlying ideas. As a suggestive model example we explain first the construction of the well known system of chirp (Heisenberg) signals, deliberately taking a representation theoretic point of view (see [2,5] for a more comprehensive treatment).

2   Model Example (Heisenberg System)

Let us denote by $\psi : \mathbb{F}_p \to \mathbb{C}^\times$ the character $\psi(t) = e^{\frac{2\pi i}{p}t}$. We consider the pair of orthonormal bases $\Delta = \{\delta_a : a \in \mathbb{F}_p\}$ and $\Delta^\vee = \{\psi_a : a \in \mathbb{F}_p\}$, where $\psi_a(t) = \frac{1}{\sqrt{p}}\psi(at)$.

2.1   Characterization of the Bases Δ and $\Delta^\vee$

Let $\mathsf{L} : \mathcal{H} \to \mathcal{H}$ be the time shift operator $\mathsf{L}\varphi(t) = \varphi(t + 1)$. This operator is unitary and it induces a homomorphism of groups $\mathsf{L} : \mathbb{F}_p \to U(\mathcal{H})$ given by $\mathsf{L}_\tau \varphi(t) = \varphi(t + \tau)$ for any $\tau \in \mathbb{F}_p$.

Elements of the basis $\Delta^\vee$ are character vectors with respect to the action L, i.e., $\mathsf{L}_\tau \psi_a = \psi(a\tau)\psi_a$ for any $\tau \in \mathbb{F}_p$. In the same fashion, the basis Δ consists of character vectors with respect to the homomorphism $\mathsf{M} : \mathbb{F}_p \to U(\mathcal{H})$ generated by the phase shift operator $\mathsf{M}\varphi(t) = \psi(t)\varphi(t)$.

2.2   The Heisenberg Representation

The homomorphisms L and M can be combined into a single map $\widetilde{\pi} : \mathbb{F}_p \times \mathbb{F}_p \to U(\mathcal{H})$ which sends a pair (τ, w) to the unitary operator $\widetilde{\pi}(\tau, \omega) = \psi\left(-\frac{1}{2}\tau w\right)\mathsf{M}_w \circ \mathsf{L}_\tau$. The plane $\mathbb{F}_p \times \mathbb{F}_p$ is called the *time-frequency plane* and will be denoted by V. The map $\widetilde{\pi}$ is not an homomorphism since, in general, the operators L_τ and M_w do not commute. This deficiency can be corrected if we consider the group $H = V \times \mathbb{F}_p$ with multiplication given by

$$(\tau, w, z) \cdot (\tau', w', z') = (\tau + \tau', w + w', z + z' + \frac{1}{2}(\tau w' - \tau' w)).$$

The map $\widetilde{\pi}$ extends to a homomorphism $\pi : H \to U(\mathcal{H})$ given by

$$\pi(\tau, w, z) = \psi\left(-\frac{1}{2}\tau w + z\right)\mathsf{M}_w \circ \mathsf{L}_\tau.$$

The group H is called the *Heisenberg* group and the homomorphism π is called the *Heisenberg representation*.

2.3   Maximal Commutative Subgroups

The Heisenberg group is no longer commutative, however, it contains various commutative subgroups which can be easily described. To every line $L \subset V$, which pass through the origin, one can associate a maximal commutative subgroup $A_L = \{(l, 0) \in V \times \mathbb{F}_p : l \in L\}$. It will be convenient to identify the subgroup A_L with the line L.

2.4 Bases Associated with Lines

Restricting the Heisenberg representation π to a subgroup L yields a decomposition of the Hilbert space $\mathcal{H}$ into a direct sum of one-dimensional subspaces $\mathcal{H} = \bigoplus_{\chi} \mathcal{H}_\chi$, where χ runs in the set $L^\vee$ of (complex valued) characters of the group L. The subspace $\mathcal{H}_\chi$ consists of vectors $\varphi \in \mathcal{H}$ such that $\pi(l)\varphi = \chi(l)\varphi$. In other words, the space $\mathcal{H}_\chi$ consists of common eigenvectors with respect to the commutative system of unitary operators $\{\pi(l)\}_{l \in L}$ such that the operator $\pi(l)$ has eigenvalue $\chi(l)$.

Choosing a unit vector $\varphi_\chi \in \mathcal{H}_\chi$ for every $\chi \in L^\vee$ we obtain an orthonormal basis $\mathcal{B}_L = \{\varphi_\chi : \chi \in L^\vee\}$. In particular, $\Delta^\vee$ and Δ are recovered as the bases associated with the lines $T = \{(\tau, 0) : \tau \in \mathbb{F}_p\}$ and $W = \{(0, w) : w \in \mathbb{F}_p\}$ respectively. For a general L the signals in $\mathcal{B}_L$ are certain kinds of chirps. Concluding, we associated with every line $L \subset V$ an orthonormal basis $\mathcal{B}_L$, and overall we constructed a system of signals consisting of a union of orthonormal bases

$$\mathfrak{S}_H = \{\varphi \in \mathcal{B}_L : L \subset V\}.$$

For obvious reasons, the system $\mathfrak{S}_H$ will be called the *Heisenberg* system.

2.5 Properties of the Heisenberg System

It will be convenient to introduce the following general notion. Given two signals $\phi, \varphi \in \mathcal{H}$, their matrix coefficient is the function $m_{\phi,\varphi} : H \to \mathbb{C}$ given by $m_{\phi,\varphi}(h) = \langle \phi, \pi(h)\varphi \rangle$. In coordinates, if we write $h = (\tau, w, z)$ then $m_{\phi,\varphi}(h) = \psi\left(-\frac{1}{2}\tau w + z\right)\langle \phi, \mathsf{M}_w \circ \mathsf{L}_\tau \varphi \rangle$. When $\phi = \varphi$ the function $m_{\varphi,\varphi}$ is called the *ambiguity* function of the vector φ and is denoted by $A_\varphi = m_{\varphi,\varphi}$.

The system $\mathfrak{S}_H$ consists of $p + 1$ orthonormal bases[1], altogether $p(p+1)$ signals and it satisfies the following properties [2,5]

1. *Auto-correlation.* For every signal $\varphi \in \mathcal{B}_L$ the function $|A_\varphi|$ is the characteristic function of the line L, i.e.,

$$|A_\varphi(v)| = \begin{cases} 0, & v \notin L, \\ 1, & v \in L. \end{cases}$$

2. *Cross-correlation.* For every $\phi \in \mathcal{B}_L$ and $\varphi \in \mathcal{B}_M$ where $L \neq M$ we have

$$|m_{\varphi,\phi}(v)| \leq \frac{1}{\sqrt{p}},$$

for every $v \in V$. If $L = M$ then $|m_{\varphi,\phi}|$ is the characteristic function of some translation of the line L.

3. *Supremum.* A signal $\varphi \in \mathfrak{S}_H$ is a unimodular function, i.e., $|\varphi(t)| = \frac{1}{\sqrt{p}}$ for every $t \in \mathbb{F}_p$, in particular we have

$$\max\{|\varphi(t)| : t \in \mathbb{F}_p\} = \frac{1}{\sqrt{p}} \ll 1.$$

---

[1] Note that $p + 1$ is the number of lines in V.

Remark 1. Note the main differences between the Heisenberg and the oscillator systems. The oscillator system consists of order of p^3 signals, while the Heisenberg system consists of order of p^2 signals. Signals in the oscillator system admit an ambiguity function concentrated at $0 \in V$ (thumbtack pattern) while signals in the Heisenberg system admit ambiguity function concentrated on a line (see Figures 1, 3).

3   The Oscillator System

Reflecting back on the Heisenberg system we see that each vector $\varphi \in \mathfrak{S}_H$ is characterized in terms of action of the additive group $G_a = \mathbb{F}_p$. Roughly, in comparison, each vector in the oscillator system is characterized in terms of action of the multiplicative group $G_m = \mathbb{F}_p^\times$. Our next goal is to explain the last assertion. We begin by giving a model example.

Given a multiplicative character[2] $\chi : G_m \to \mathbb{C}^\times$, we define a vector $\underline{\chi} \in \mathcal{H}$ by

$$\underline{\chi}(t) = \begin{cases} \frac{1}{\sqrt{p-1}}\chi(t), & t \neq 0, \\ 0, & t = 0. \end{cases}$$

We consider the system $\mathcal{B}_{std} = \{\underline{\chi} : \chi \in G_m^\vee,\ \chi \neq 1\}$, where $G_m^\vee$ is the dual group of characters.

3.1   Characterizing the System $\mathcal{B}_{std}$

For each element $a \in G_m$ let $\rho_a : \mathcal{H} \to \mathcal{H}$ be the unitary operator acting by scaling $\rho_a\varphi(t) = \varphi(at)$. This collection of operators form a homomorphism $\rho : G_m \to U(\mathcal{H})$.

Elements of $\mathcal{B}_{std}$ are character vectors with respect to ρ, i.e., the vector $\underline{\chi}$ satisfies $\rho_a\left(\underline{\chi}\right) = \chi(a)\underline{\chi}$ for every $a \in G_m$. In more conceptual terms, the action ρ yields a decomposition of the Hilbert space $\mathcal{H}$ into character spaces $\mathcal{H} = \bigoplus \mathcal{H}_\chi$, where χ runs in the group $G_m^\vee$. The system $\mathcal{B}_{std}$ consists of a representative unit vector for each space $\mathcal{H}_\chi$, $\chi \neq 1$.

3.2   The Weil Representation

We would like to generalize the system $\mathcal{B}_{std}$ in a similar fashion to the way we generalized the bases Δ and $\Delta^\vee$ in the Heisenberg setting. In order to do this we need to introduce several auxiliary operators.

Let $\rho_a : \mathcal{H} \to \mathcal{H}$, $a \in \mathbb{F}_p^\times$, be the operators acting by $\rho_a\varphi(t) = \sigma(a)\varphi(a^{-1}t)$ (scaling), where σ is the unique quadratic character of $\mathbb{F}_p^\times$, let $\rho_T : \mathcal{H} \to \mathcal{H}$ be the operator acting by $\rho_T\varphi(t) = \psi(t^2)\varphi(t)$ (quadratic modulation), and finally let $\rho_S : \mathcal{H} \to \mathcal{H}$ be the operator of Fourier transform

$$\rho_S\varphi(t) = \frac{\nu}{\sqrt{p}} \sum_{s \in \mathbb{F}_p} \psi(ts)\varphi(s),$$

---

[2] A multiplicative character is a function $\chi : G_m \to \mathbb{C}^\times$ which satisfies $\chi(xy) = \chi(x)\chi(y)$ for every $x, y \in G_m$.

where ν is a normalization constant [6]. The operators ρ_a, ρ_T and ρ_S are unitary. Let us consider the subgroup of unitary operators generated by ρ_a, ρ_S and ρ_T. This group turns out to be isomorphic to the finite group $Sp = SL_2(\mathbb{F}_p)$, therefore we obtained a homomorphism $\rho : Sp \to U(\mathcal{H})$. The representation ρ is called the *Weil representation* [7] and it will play a prominent role in this survey.

3.3   Systems Associated with Maximal (Split) Tori

The group Sp consists of various types of commutative subgroups. We will be interested in maximal *diagonalizable* commutative subgroups. A subgroup of this type is called maximal *split torus*. The standard example is the subgroup consisting of all diagonal matrices

$$A = \left\{ \begin{pmatrix} a & 0 \\ 0 & a^{-1} \end{pmatrix} : a \in G_m \right\},$$

which is called the *standard torus*. The restriction of the Weil representation to a split torus $T \subset Sp$ yields a decomposition of the Hilbert space $\mathcal{H}$ into a direct sum of character spaces $\mathcal{H} = \bigoplus \mathcal{H}_\chi$, where χ runs in the set of characters $T^\vee$. Choosing a unit vector $\varphi_\chi \in \mathcal{H}_\chi$ for every χ we obtain a collection of orthonormal vectors $\mathcal{B}_T = \{\varphi_\chi : \chi \in T^\vee, \chi \neq \sigma\}$. Overall, we constructed a system

$$\mathfrak{S}_O^s = \{\varphi \in \mathcal{B}_T : T \subset Sp \text{ split}\},$$

which will be referred to as the *split oscillator system*. We note that our initial system $\mathcal{B}_{std}$ is recovered as $\mathcal{B}_{std} = \mathcal{B}_A$.

3.4   Systems Associated with Maximal (Non-split) Tori

From the point of view of this survey, the most interesting maximal commutative subgroups in Sp are those which are diagonalizable over an extension field rather than over the base field $\mathbb{F}_p$. A subgroup of this type is called maximal *non-split torus*. It might be suggestive to first explain the analogue notion in the more familiar setting of the field $\mathbb{R}$. Here, the standard example of a maximal non-split torus is the circle group $SO(2) \subset SL_2(\mathbb{R})$. Indeed, it is a maximal commutative subgroup which becomes diagonalizable when considered over the extension field $\mathbb{C}$ of complex numbers.

The above analogy suggests a way to construct examples of maximal non-split tori in the finite field setting as well. Let us assume for simplicity that -1 does not admit a square root in $\mathbb{F}_p$. The group Sp acts naturally on the plane $V = \mathbb{F}_p \times \mathbb{F}_p$. Consider the symmetric bilinear form B on V given by

$$B((t, w), (t', w')) = tt' + ww'.$$

An example of maximal non-split torus is the subgroup $T_{ns} \subset Sp$ consisting of all elements $g \in Sp$ preserving the form B, i.e., $g \in T_{ns}$ if and only if $B(gu, gv) = B(u, v)$ for every $u, v \in V$.

In the same fashion like in the split case, restricting the Weil representation to a non-split torus T yields a decomposition into character spaces $\mathcal{H} = \bigoplus \mathcal{H}_\chi$. Choosing a unit vector $\varphi_\chi \in \mathcal{H}_\chi$ for every $\chi \in T^\vee$ we obtain an orthonormal basis $\mathcal{B}_T$. Overall, we constructed a system of signals

$$\mathfrak{S}_O^{ns} = \{\varphi \in \mathcal{B}_T : T \subset Sp \text{ non-split}\}.$$

The system $\mathfrak{S}_O^{ns}$ will be referred to as the *non-split oscillator* system. The construction of the system $\mathfrak{S}_O = \mathfrak{S}_O^s \cup \mathfrak{S}_O^{ns}$ together with the formulation of some of its properties are the main contribution of this survey.

3.5   Behavior under Fourier Transform

The oscillator system is closed under the operation of Fourier transform, i.e., for every $\varphi \in \mathfrak{S}_O$ we have $\widehat{\varphi} \in \mathfrak{S}_O$. The Fourier transform on the space $\mathbb{C}(\mathbb{F}_p)$ appears as a specific operator $\rho(\mathrm{w})$ in the Weil representation, where

$$\mathrm{w} = \begin{pmatrix} 0 & 1 \\ -1 & 0 \end{pmatrix} \in Sp.$$

Given a signal $\varphi \in \mathcal{B}_T \subset \mathfrak{S}_O$, its Fourier transform $\widehat{\varphi} = \rho(\mathrm{w})\varphi$ is, up to a unitary scalar, a signal in $\mathcal{B}_{T'}$ where $T' = \mathrm{w}T\mathrm{w}^{-1}$. In fact, $\mathfrak{S}_O$ is closed under all the operators in the Weil representation! Given an element $g \in Sp$ and a signal $\varphi \in \mathcal{B}_T$ we have, up to a unitary scalar, that $\rho(g)\varphi \in \mathcal{B}_{T'}$, where $T' = gTg^{-1}$.

In addition, the Weyl element w is an element in some maximal torus T_w (the split type of T_w depends on the characteristic p of the field) and as a result signals $\varphi \in \mathcal{B}_{T_\mathrm{w}}$ are, in particular, eigenvectors of the Fourier transform. As a consequences a signal $\varphi \in \mathcal{B}_{T_\mathrm{w}}$ and its Fourier transform $\widehat{\varphi}$ differ by a unitary constant, therefore are practically the "same" for all essential matters.

These properties might be relevant for applications to OFDM (Orthogonal Frequency Division Multiplexing) [8] where one requires good properties both from the signal and its Fourier transform.

3.6   Relation to the Harmonic Oscillator

Here we give the explanation why functions in the non-split oscillator system $\mathfrak{S}_O^{ns}$ constitute a finite analogue of the eigenfunctions of the harmonic oscillator in the real setting. The Weil representation establishes the dictionary between these two, seemingly, unrelated objects. The argument works as follows.

The one-dimensional harmonic oscillator is given by the differential operator $D = \partial^2 - t^2$. The operator D can be exponentiated to give a unitary representation of the circle group $\rho : SO(2, \mathbb{R}) \longrightarrow U(L^2(\mathbb{R}))$ where $\rho(\theta) = e^{i\theta D}$. Eigenfunctions of D are naturally identified with character vectors with respect to ρ. The crucial point is that ρ is the restriction of the Weil representation of $SL_2(\mathbb{R})$ to the maximal non-split torus $SO(2, \mathbb{R}) \subset SL_2(\mathbb{R})$.

Summarizing, the eigenfunctions of the harmonic oscillator and functions in $\mathfrak{S}_O^{ns}$ are governed by the same mechanism, namely both are character vectors with respect to the restriction of the Weil representation to a maximal non-split torus in SL_2. The only difference appears to be the field of definition, which for the harmonic oscillator is the reals and for the oscillator functions is the finite field.

4   Applications

Two applications of the oscillator system will be described. The first application is to the theory of discrete radar. The second application is to CDMA systems. We will give a brief explanation of these problems, while emphasizing the relation to the Heisenberg representation.

4.1   Discrete Radar

The theory of discrete radar is closely related [2] to the finite Heisenberg group H. A radar sends a signal $\varphi(t)$ and obtains an echo $e(t)$. The goal [9] is to reconstruct, in maximal accuracy, the target range and velocity. The signal $\varphi(t)$ and the echo $e(t)$ are, principally, related by the transformation

$$e(t) = e^{2\pi i w t}\varphi(t + \tau) = \mathsf{M}_w \mathsf{L}_\tau \varphi(t),$$

where the time shift τ encodes the distance of the target from the radar and the phase shift encodes the velocity of the target. Equivalently, the transmitted signal φ and the received echo e are related by an action of an element $h_0 \in H$, i.e., $e = \pi(h_0)\varphi$. The problem of discrete radar can be described as follows. Given a signal φ and an echo $e = \pi(h_0)\varphi$ extract the value of h_0.

It is easy to show that $|m_{\varphi,e}(h)| = |A_\varphi(h \cdot h_0)|$ and it obtains its maximum at h_0^{-1}. This suggests that a desired signal φ for discrete radar should admit an ambiguity function A_φ which is highly concentrated around $0 \in H$, which is a property satisfied by signals in the oscillator system (Property 2).

Remark 2. It should be noted that the system $\mathfrak{S}_O$ is "large" consisting of order of p^3 signals. This property becomes important in a *jamming* scenario.

4.2   Code Division Multiple Access (CDMA)

We are considering the following setting.

- There exists a collection of users $i \in I$, each holding a *bit* of information $b_i \in \mathbb{C}$ (usually b_i is taken to be an N'th root of unity).
- Each user transmits his bit of information, say, to a central antenna. In order to do that, he multiplies his bit b_i by a private signal $\varphi_i \in \mathcal{H}$ and forms a message $u_i = b_i\varphi_i$.

- The transmission is carried through a single channel (for example in the case of cellular communication the channel is the atmosphere), therefore the message received by the antenna is the sum

$$u = \sum_i u_i.$$

The main problem [3] is to extract the individual bits b_i from the message u. The bit b_i can be estimated by calculating the inner product

$$\langle \varphi_i, u \rangle = \sum_i \langle \varphi_i, u_j \rangle = \sum_j b_j \langle \varphi_i, \varphi_j \rangle = b_i + \sum_{j \neq i} b_j \langle \varphi_i, \varphi_j \rangle.$$

The last expression above should be considered as a sum of the information bit b_i and an additional noise caused by the interference of the other messages. This is the standard scenario also called the *Synchronous* scenario. In practice, more complicated scenarios appear, e.g., *asynchronous scenario* - in which each message u_i is allowed to acquire an arbitrary time shift $u_i(t) \mapsto u_i(t + \tau_i)$, *phase shift scenario* - in which each message u_i is allowed to acquire an arbitrary phase shift $u_i(t) \mapsto e^{\frac{2\pi i}{p} w_i t} u_i(t)$ and probably also a combination of the two where each message u_i is allowed to acquire an arbitrary distortion of the form $u_i(t) \mapsto e^{\frac{2\pi i}{p} w_i t} u_i(t + \tau_i)$.

The previous discussion suggests that what we are seeking for is a large system $\mathfrak{S}$ of signals which will enable a reliable extraction of each bit b_i for as many users transmitting through the channel simultaneously.

Definition 1 (Stability conditions). *Two unit signals $\phi \neq \varphi$ are called stably cross-correlated if $|m_{\varphi,\phi}(v)| \ll 1$ for every $v \in V$. A unit signal φ is called stably auto-correlated if $|A_\varphi(v)| \ll 1$, for $v \neq 0$. A system $\mathfrak{S}$ of signals is called a stable system if every signal $\varphi \in \mathfrak{S}$ is stably auto-correlated and any two different signals $\phi, \varphi \in \mathfrak{S}$ are stably cross-correlated.*

Formally what we require for CDMA is a stable system $\mathfrak{S}$. Let us explain why this corresponds to a reasonable solution to our problem. At a certain time t the antenna receives a message

$$u = \sum_{i \in J} u_i,$$

which is transmitted from a subset of users $J \subset I$. Each message u_i, $i \in J$, is of the form $u_i = b_i e^{\frac{2\pi i}{p} w_i t} \varphi_i(t + \tau_i) = b_i \pi(h_i) \varphi_i$, where $h_i \in H$. In order to extract the bit b_i we compute the matrix coefficient

$$m_{\varphi_i, u} = b_i R_{h_i} A_{\varphi_i} + \#(J - \{i\}) o(1),$$

where R_{h_i} is the operator of right translation $R_{h_i} A_{\varphi_i}(h) = A_{\varphi_i}(hh_i)$.

If the cardinality of the set J is not too big then by evaluating $m_{\varphi_i, u}$ at $h = h_i^{-1}$ we can reconstruct the bit b_i. It follows from (1) and (2) that the oscillator system $\mathfrak{S}_O$ can support order of p^3 users, enabling reliable reconstruction when order of $\sqrt{p}$ users are transmitting simultaneously.

A   Algorithm

We describe an explicit algorithm that generates the oscillator system $\mathfrak{S}_O^s$ associated with the collection of split tori in $Sp = SL_2(\mathbb{F}_p)$.

A.1   Tori

Consider the standard diagonal torus

$$A = \left\{ \begin{pmatrix} a & 0 \\ 0 & a^{-1} \end{pmatrix}; \ a \in \mathbb{F}_p^\times \right\}.$$

Every split torus in Sp is conjugated to the torus A, which means that the collection $\mathcal{T}$ of all split tori in Sp can be written as

$$\mathcal{T} = \{ gAg^{-1}; \ g \in Sp \}.$$

A.2   Parametrization

A direct calculation reveals that every torus in $\mathcal{T}$ can be written as gAg^{-1} for an element g of the form

$$g = \begin{pmatrix} 1 + bc & b \\ c & 1 \end{pmatrix}, \ b, c \in \mathbb{F}_p. \tag{3}$$

Unless $c = 0$, this presentation is not unique: In the case $c \neq 0$, an element $\tilde{g}$ represents the same torus as g if and only if it is of the form

$$\tilde{g} = \begin{pmatrix} 1 + bc & b \\ c & 1 \end{pmatrix} \begin{pmatrix} 0 & c^{-1} \\ -c & 0 \end{pmatrix}.$$

Let us choose a set of elements of the form (3) representing each torus in $\mathcal{T}$ exactly once and denote this set of representative elements by R.

A.3   Generators

The group A is a cyclic group and we can find a generator g_A for A. This task is simple from the computational perspective, since the group A is finite, consisting of $p - 1$ elements.

Now we make the following two observations. First observation is that the oscillator basis $\mathcal{B}_A$ is the basis of eigenfunctions of the operator $\rho(g_A)$.

The second observation is that, other bases in the oscillator system $\mathfrak{S}_O^s$ can be obtained from $\mathcal{B}_A$ by applying elements form the sets R. More specifically, for a torus T of the form $T = gAg^{-1}$, $g \in R$, we have

$$\mathcal{B}_{gAg^{-1}} = \{ \rho(g)\varphi; \ \varphi \in \mathcal{B}_A \}.$$

Concluding, we described the oscillator system

$$\mathfrak{S}_O^s = \{ \mathcal{B}_{gAg^{-1}}; g \in R \}.$$

A.4   Formulas

We are left to explain how to write explicit formulas (matrices) for the operators involved in the construction of $\mathfrak{S}_O^s$.

First, we recall that the group Sp admits a Bruhat decomposition $Sp = B \cup BwB$, where B is the Borel subgroup and w denotes the Weyl element

$$ w = \begin{pmatrix} 0 & 1 \\ -1 & 0 \end{pmatrix}. $$

Furthermore, the Borel subgroup B can be written as a product $B = AU = UA$, where A is the standard diagonal torus and U is the standard unipotent group

$$ U = \left\{ \begin{pmatrix} 1 & u \\ 0 & 1 \end{pmatrix} : u \in \mathbb{F}_p \right\}. $$

Therefore, we can write the Bruhat decomposition also as $Sp = UA \cup UAwU$.

Second, we give an explicit description of operators in the Weil representation, associated with different types of elements in Sp. The operators are specified up to a unitary scalar, which is enough for our needs.

- The standard torus A acts by (normalized) scaling: An element $a = \begin{pmatrix} a & 0 \\ 0 & a^{-1} \end{pmatrix}$, acts by

$$ S_a\,[f]\,(t) = \sigma\,(a)\,f\left(a^{-1}t\right), $$

  where $\sigma : \mathbb{F}_p^\times \to \{\pm 1\}$ is the Legendre character, $\sigma(a) = a^{\frac{p-1}{2}} \,(\mathrm{mod}\ p)$.
- The standard unipotent group U acts by quadratic characters (chirps): An element $u = \begin{pmatrix} 1 & u \\ 0 & 1 \end{pmatrix}$, acts by

$$ M_u\,[f]\,(t) = \psi(\frac{u}{2}t^2)f\,(t), $$

  where $\psi : \mathbb{F}_p \to \mathbb{C}^\times$ is the character $\psi(t) = e^{\frac{2\pi i}{p}t}$.
- The Weyl element w acts by discrete Fourier transform

$$ F\,[f]\,(w) = \frac{1}{\sqrt{p}} \sum_{t \in \mathbb{F}_p} \psi\,(wt)\,f\,(t). $$

Hence, we conclude that every operator $\rho\,(g)$, where $g \in Sp$, can be written either in the form $\rho\,(g) = M_u \circ S_a$ or in the form $\rho\,(g) = M_{u_2} \circ S_a \circ F \circ M_{u_1}$.

Example 1. For $g \in R$, with $c \neq 0$, the Bruhat decomposition of g is given explicitly by

$$ g = \begin{pmatrix} 1 & \frac{1+bc}{c} \\ 0 & 1 \end{pmatrix} \begin{pmatrix} \frac{-1}{c} & 0 \\ 0 & -c \end{pmatrix} \begin{pmatrix} 0 & 1 \\ -1 & 0 \end{pmatrix} \begin{pmatrix} 1 & \frac{1}{c} \\ 0 & 1 \end{pmatrix}, $$

and

$$ \rho\,(g) = M_{\frac{1+bc}{c}} \circ S_{\frac{-1}{c}} \circ F \circ M_{\frac{1}{c}}. $$

For $g \in R$, with $c = 0$, we have

$$g = \begin{pmatrix} 1 & b \\ 0 & 1 \end{pmatrix},$$

and

$$\rho(g) = M_b.$$

A.5 Pseudocode

Below, is given a pseudo-code description of the construction of the *oscillator system*.

1. Choose a prime p.
2. Compute generator g_A for the standard torus A.
3. Diagonalize $\rho(g_A)$ and obtain the basis $\mathcal{B}_A$.
4. For every $g \in R$:
5. Compute the operator $\rho(g)$ as follows:
   (a) Calculate the Bruhat decomposition of g, namely, write g in the form $g = u_2 \cdot a \cdot w \cdot u_1$ or $g = u \cdot a$.
   (b) Calculate the operator $\rho(g)$, namely, take $\rho(g) = M_{u_2} \circ S_a \circ F \circ M_{u_1}$ or $\rho(g) = M_u \circ S_a$.
6. Compute the vectors $\rho(g)\varphi$, for every $\varphi \in \mathcal{B}_A$ and obtain the basis $\mathcal{B}_{gAg^{-1}}$.

Remark 3 (Running time). It is easy to verify that the time complexity of the algorithm presented above is $O(p^4 \log p)$. This is, in fact, an optimal time complexity, since already to specify p^3 vectors, each of length p, requires p^4 operations.

Remark about field extensions. All the results in this survey were stated for the basic finite field $\mathbb{F}_p$ for the reason of making the terminology more accessible. However, they are valid for any field extension of the form $\mathbb{F}_q$ with $q = p^n$. Complete proofs appear in [6].

Acknowledgement. The authors would like to thank J. Bernstein for his interest and guidance in the mathematical aspects of this work. We are grateful to S. Golomb and G. Gong for their interest in this project. We appreciate the many talks we had with A. Sahai. We thank B. Sturmfels for encouraging us to proceed in this line of research. We would like to thank V. Anantharam, A. Grünbaum for interesting conversations. Finally, the second author is indebted to B. Porat for so many discussions where each tried to understand the cryptic terminology of the other.

References

1. Golomb, S.W., Gong, G.: Signal design for good correlation. For wireless communication, cryptography, and radar. Cambridge University Press, Cambridge (2005)
2. Howard, S.D., Calderbank, A.R., Moran, W.: The finite Heisenberg-Weyl groups in radar and communications. EURASIP J. Appl. Signal Process (2006)

3. Viterbi, A.J.: CDMA: Principles of Spread Spectrum Communication. Addison-Wesley Wireless Communications (1995)
4. Paterson, K.G., Tarokh, V.: On the existence and construction of good codes with low peak-to-average power ratios. IEEE Trans. Inform. Theory 46 (2000)
5. Howe, R.: Nice error bases, mutually unbiased bases, induced representations, the Heisenberg group and finite geometries. Indag. Math (N.S.) 16(3-4), 553–583 (2005)
6. Gurevich, S., Hadani, R., Sochen, N.: The finite harmonic oscillator and its applications to sequences, communication and radar. IEEE Trans. on Inform. Theory (accepted March 2008) (to appear)
7. Weil, A.: Sur certains groupes d'operateurs unitaires. Acta Math. 111, 143–211 (1964)
8. Chang, R.W.: Synthesis of Band-Limited Orthogonal Signals for Multichannel Data Transmission. Bell System Technical Journal 45 (1966)
9. Woodward, P.M.: Probability and Information theory, with Applications to Radar. Pergamon Press, New York (1953)

Projective de Bruijn Sequences[*]

Yuki Ohtsuka[1], Makoto Matsumoto[1], and Mariko Hagita[2]

[1] Dept. of Math., Hiroshima University, Hiroshima 739-8526, Japan
m-mat@math.sci.hiroshima-u.ac.jp
http://www.math.sci.hiroshima-u.ac.jp/~m-mat/eindex.html
[2] Dept. of Info., Ochanomizu University, Tokyo 112-8610, Japan

Abstract. Let $\mathbb{F}_q$ denote the q-element field. A q-ary de Bruijn sequence of degree m is a cyclic sequence of elements of $\mathbb{F}_q$, such that every element in $\mathbb{F}_q{}^m$ appears exactly once as a consecutive m-tuple in the cyclic sequence. We consider its projective analogue; namely, a cyclic sequence such that every point in the projective space $(\mathbb{F}_q{}^{m+1} - \{0\})/(\mathbb{F}_q{}^{\times})$ appears exactly once as a consecutive $(m+1)$-tuple. We have an explicit formula $(q!)^{\frac{q^m-1}{q-1}} q^{-m}$ for the number of distinct such sequences.

1 Introduction

Let $\mathbb{F}_q$ be the q-element field. Consider a cyclic sequence $C = (x_0, x_1, x_2, \ldots, x_p = x_0, x_{p+1} = x_1, \ldots)$ of period p over $\mathbb{F}_q$. For an integer k, the k-shadow of C is the multiset

$$C[k] := \{(x_i, x_{i+1}, \ldots, x_{i+k-1}) \mid i = 0, 1, 2, \ldots, p-1\}$$

of consecutive k-tuples in C for a full period. A cyclic sequence C is said to be a q-ary de Bruijn sequence of degree m, if $C[m]$ has no multiplicity and $C[m] = \mathbb{F}_q{}^m$. It follows that $p = q^m$. Such a sequence has interesting applications, for example in the theory of shift-registers, see [2]. The number of the de Bruijn sequences is known ([1] for the binary case, and [5] for the general case).

Theorem 1. *There exist* $(q!)^{q^{m-1}} q^{-m}$ *distinct q-ary de Bruijn sequences of degree m.*

Through a study on the error-correcting sequence [3], we needed to introduce a projectivised version of de Bruijn sequences. As usual, the $(m-1)$-dimensional projective space is defined as the set of non-zero vectors quotiented by the action of scalar multiplication:

$$P(q, m-1) := \{(x_0, x_1, \ldots, x_{m-1}) \in \mathbb{F}_q{}^m \setminus \{0\}\}/ \sim .$$

The class of $(x_0, \ldots, x_{m-1})$ is denoted by $[x_0 : \cdots : x_{m-1}]$. Thus, $[x_0 : \cdots : x_{m-1}] = [y_0 : \cdots : y_{m-1}]$ if and only if there is a non-zero $\lambda \in \mathbb{F}_q{}^{\times}$ such that $(x_0, \ldots, x_{m-1}) = \lambda(y_0, \ldots, y_{m-1})$.

[*] This work is supported in part by JSPS Grant-in-Aid #16204002, #18654021, #18740044, #19204002 and JSPS Core-to-Core Program No.18005.

S.W. Golomb et al. (Eds.): SETA 2008, LNCS 5203, pp. 167–174, 2008.

A projective de Bruijn sequence of degree m is a cyclic sequence of period p $x_0, x_1, \ldots, x_{p-1}, x_p = x_0, \ldots$ such that the multiset of the consecutive m-tuples

$$\{[x_i : \cdots : x_{i+m-1}] \mid 0 \le i \le p-1\}$$

has no multiplicity and equals to the point set of $P(q, m-1)$. Accordingly, $p = (q^m - 1)/(q - 1)$ holds.

To state a relation to coding theory, recall that a q-ary $[(q^\ell - 1)/(q-1), (q^\ell - 1)/(q-1) - \ell, 3]$ Hamming code is a linear code characterized by its parity check matrix whose columns are all points in $P(q, \ell - 1)$ (e.g. [6]).

Put $k := (q^\ell - 1)/(q - 1) - \ell$ and $n := (q^\ell - 1)/(q - 1)$. Consider the following linear recurrence

$$a_{j+k} = - \sum_{0 \le i \le k-1} c_i a_{i+j} \quad (j = 0, 1, \ldots, n-1) \tag{1}$$

with constant coefficients $c_0, c_1, \ldots c_{k-1} \in \mathbb{F}_q$.

The following easy theorem is proved in [3, Theorem 10].

Theorem 2. *The set of solutions of (1) is the q-ary $[n, k, 3]$ Hamming code if and only if the sequence $(0, 0, \ldots, 0, c_0, c_1, \ldots, c_{k-1}, 1)$ of length (and period) n is a projective de Bruijn sequence of degree ℓ.*

In [3, Remark 13], we raised the problem of determining the number of projective de Bruijn sequences. In this paper, we settle this problem:

Theorem 3. *Let m be a positive integer. The number of q-ary projective de Bruijn sequences of degree $m + 1$ is*

$$\frac{(q!)^{\frac{q^m - 1}{q-1}}}{q^m}.$$

Here, we identify such a sequence with its nonzero scalar multiples.

In the rest of this paper, we prove this theorem. The outline of the proof is: (1) define projective de Bruijn graphs $PD(q, m)$, whose Eulerian circuits are equivalent to projective de Bruijn sequences of degree $m + 1$, (2) to show that the line digraph $L(PD(q, m-1))$ is almost isomorphic to $PD(q, m)$, (3) use the formula for the number of Eulerian circuits for line digraphs [4].

2   Line Digraph

Definition 1. *Let D be a digraph, with vertex set $V(D)$ and the arrow set $A(D)$. Its line digraph $L(D)$ has $V(L(D)) := A(D)$, and two vertices $v, w \in V(L(D))$ has an arrow from v to w if and only if v is adjacent to w (with ordering considered) as arrows in D.*

Let $E(D)$ denote the number of Eulerian circuits of D. The following lemma is a direct consequence of [4, Theorem 2, Lemmas 4,5].

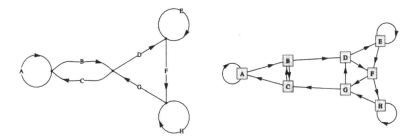

Fig. 1. Digraph (the left picture) and its line digraph (the right picture)

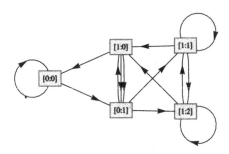

Fig. 2. The graph $PD(q, m)$ with $q = 3$, $m = 2$

Lemma 1. *Let D be a k-regular digraph with n vertices. Then, we have*

$$E(L(D)) = \frac{(k!)^{n(k-1)}}{k} E(D).$$

3   Projective de Bruijn Graph

Let q be a prime power, $\mathbb{F}_q$ the q-element field, and m a positive integer.

Definition 2. *We define the q-ary projective de Bruijn graph of degree m, denoted by $PD(q, m)$, as follows. Let $P(q, m-1)_O$ denotes the set of points in the m-dimensional projective space supplemented with a formal element $O_m :=$ $[0 : 0 : \cdots : 0]$ (an m-tuple of 0s). (1) $V(PD(q,m)) := P(q, m-1)_O$, (2) $A(PD(q,m)) := P(q, m)_O$, and (3) an arc $[x_0 : \cdots : x_m] \in A(PD(q,m))$ has its origin at $[x_0 : \cdots : x_{m-1}] \in V(PD(q,m))$ and its target at $[x_1 : \cdots : x_m] \in V(PD(q,m))$.*

There are three special vertices in $PD(q, m)$: $A_m = [1 : 0 : \cdots : 0]_m$, $B_m =$ $[0 : \cdots : 0 : 1]_m$, and $O_m := [0 : \cdots : 0]$ (and special arrows A_{m+1}, B_{m+1}, and O_{m+1}).

Remark 1. It is easy to check that the digraph $PD(q, m)$ has a loop at O_m, and $(q-1)$ multiple arrows from A_m to B_m. In addition, the out-degree and in-degree

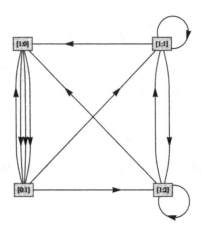

Fig. 3. Digraph $PD'(q,m)$ for $q = 3$, $m = 2$

of O_m are both two, and the other vertices have out-degree and in-degree equal to q.

Theorem 4. *There is a natural bijection from the set of Eulerian circuits $\mathcal{E}(PD(q,m))$ to the set of projective de Bruijn sequences of degree $m + 1$.*

Proof. An Eulerian circuit C gives a cyclic sequence of arrows, namely, a cyclic sequence of elements of $P(q,m)_O$. Since the indegree and the outdegree of O_m are two, C has the subsequence $A_{m+1}, O_{m+1}, B_{m+1}$. Take the representative $(0,\ldots,0,1)$ for B_{m+1}, and define $(x_0,\ldots,x_m) := (0,\ldots,0,1)$. We define $x_{m+1} \in \mathbb{F}_q$ so that $[x_1 : \cdots : x_{m+1}]$ is the next arrow to B_{m+1} in the sequence C. This is unique, since $(x_1,\ldots,x_m)$ is nonzero. Inductively we define $x_{m+2},\ldots$, according to the sequence of arrows in C. Then, this process will end just before reaching to O_{m+1} at A_{m+1}, where we have $[x_{p-1} : x_p : \cdots : x_{p+m-2}] = A_{m+1}$ for some p. Since the length of C is $(q^m - 1)/(q - 1) + 1$, we have $p = (q^m - 1)/(q - 1)$, and the sequence $x_0,\ldots,x_{p-1}$ with period p is a projective de Bruijn sequence, since by the construction $C[m]$ consists of $P(q,m)$. The converse construction of an Eulerian circuit from a projective de Bruijn sequence is equally easy. □

Definition 3. *Note that an Eulerian circuit of $PD(q,m)$ has the subsequence of vertices A_m, O_m, O_m, B_m, connected by arrows $A_{m+1}, O_{m+1}, B_{m+1}$. Let $PD'(q,m)$ be the digraph obtained from $PD(q,m)$ by removing the arrows $A_{m+1}, O_{m+1}, B_{m+1}$ and the vertex O_m, and adding an arrow AB_{m+1} from A_m to B_m.*

The following can be easily shown.

Lemma 2. *$PD'(q,m)$ is a q-regular digraph, and $E(PD(q,m)) = E(PD'(q,m))$ holds.*

Definition 4. *Suppose $m \geq 2$. We remove all the arrows from A_m to B_m in $PD'(q,m)$, and then identify A_m and B_m (the identified vertex is denoted by*

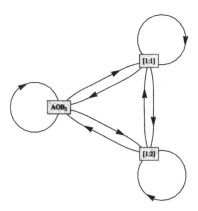

Fig. 4. Digraph $PD''(q,m)$ for $q = 3$, $m = 2$

AOB_m). *The obtained digraph is denoted by $PD''(q,m)$. Thus, the incoming arrows to AOB_m in $PD''(q,m)$ is the incoming arrows to A_m in $PD'(q,m)$, and outgoing arrows from AOB_m are those from B_m in $PD'(q,m)$. This digraph has no multiple arrows.*

Lemma 3. *Suppose $m \geq 2$. There is a natural $q!$ to 1 mapping*

$$\mathcal{E}(PD'(q,m)) \to \mathcal{E}(PD''(q,m)).$$

As a corollary, we have

$$E(PD'(q,m)) = q!E(PD''(q,m)).$$

Proof. Let C' be an Eulerian circuit of $PD'(q,m)$. Let C'' be a cyclic sequence of arrows of $PD'(q,m)$ obtained by simply removing the q arrows from A_m to B_m from C'. Since A_m and B_m are contracted in $PD''(q,m)$, C'' is a circuit, and easily seen to be an Eulerian circuit. Conversely, if we fix C'', then C' is obtained by adding an arrow from A_m to B_m at each place in C'' where C'' goes over AOB_m. Thus, this mapping is $q!$ to 1. □

Note that the above does not hold for $m = 1$, since $E(PD'(q,1)) = (q-1)!$.

Lemma 4

$$L(PD'(q,m)) = PD''(q,m+1).$$

Proof. Note that these graphs have no multiple arrows. As for the vertex sets,

$$V(L(PD'(q,m))) := A(PD'(q,m))$$

is obtained from $A(PD(q,m)) = P(q,m)$ by removing $A_{m+1}, O_{m+1}, B_{m+1}$ and adding an element named AB_{m+1}. By identifying AB_{m+1} with $AOB_{m+1} \in$

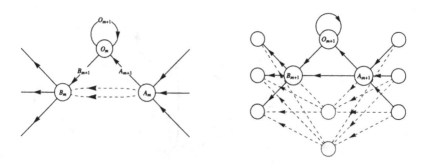

Fig. 5. Three arrows $O_{m+1}, A_{m+1}, B_{m+1}$ in $PD(q, m)$ (the left picture) and its line digraph $L(PD(q, m))$ (the right picture)

$V(PD''(q, m+1))$, we have identification $V(L(PD'(q, m))) = V(PD''(q, m+1))$. As for the arrow sets, there is a bijection

$$A(L(PD'(q, m))) \rightarrow A(PD''(q, m + 1))$$

defined as follows. The left hand side is a pair of adjacent arrows of $PD'(q, m)$, say P_{m+1}, Q_{m+1}. There are four cases.

- Case 1: $P_{m+1} \neq AB_{m+1}$ and $Q_{m+1} \neq AB_{m+1}$. Then, P_{m+1}, Q_{m+1} are in $P(q, m) \setminus \{A_{m+1}, O_{m+1}, B_{m+1}\}$, and are adjacent vertices in $PD(q, m + 1)$, hence $P_{m+1}Q_{m+1}$ is an arrow in $PD''(q, m + 1)$.
- Case 2: $P_{m+1} = AB_{m+1}$ and $Q_{m+1} \neq AB_{m+1}$. In this case, B_{m+1} is adjacent to Q_{m+1} in $A(PD(q, m))$. (See Definition 4). Then, an arrow in $PD''(q, m+1)$ that connects the vertex AOB_{m+1} to Q_{m+1} is assigned.
- Case 3: $P_{m+1} \neq AB_{m+1}$ and $Q_{m+1} = AB_{m+1}$. This case is similar (symmetric) to Case 2.
- Case 4: $P_{m+1} = AB_{m+1}$ and $Q_{m+1} = AB_{m+1}$. This does not happen since AB_{m+1} is not a loop.

It is straightforward to check that this mapping is a bijection: Case 1 is easy, and for the other cases by the case division, which is left to the reader. See Figures 5 and 6 for the structure for Cases 2 and 3.   □

Remark 2. A more conceptual proof is as follows. Let $D(q, m)$ be the q-ary de Bruijn graph of degree m [5]. By definition, we have $L(D(q, m)) = D(q, m + 1)$. The action of $\mathbb{F}_q^\times$ is free except for the neighborhood of A_m, O_m, B_m. Thus, by taking the quotient, we know that $L(PD(q, m))$ is isomorphic to $PD(q, m+1)$, except for the neighborhoods of these three vertices. Then, looking the non-regularity around the fixed points, we are lead to the above lemma.

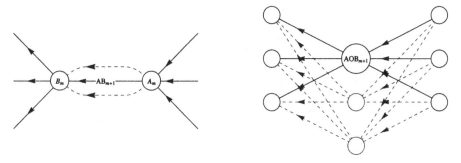

Fig. 6. Corresponding parts of $PD'(q,m)$ (the left picture) and its line digraph $L(PD'(q,m))$ (the right picture) in Fig 5

4   Proof of Main Theorem

By Theorem 4, our Main Theorem 3 is reduced to the following

Theorem 5

$$E(PD(q,m)) = \frac{(q!)^{\frac{q^m-1}{q-1}}}{q^m}.$$

Proof. By Lemmas 2, 3 and 4, we have for $m \geq 2$

$$E(PD(q,m)) = E(PD'(q,m)) = q!E(PD''(q,m)) = q!E(L(PD'(q,m-1))).$$

Then, by Lemma 1, we have

$$E(L(PD'(q,m-1))) = \frac{(k!)^{n(k-1)}}{k}E(PD'(q,m-1))$$

for $n = \frac{q^{m-1}-1}{q-1}$ and $k = q$. Thus, we have an induction formula

$$E(PD'(q,m)) = q! \cdot \frac{(q!)^{q^{m-1}-1}}{q}E(PD'(q,m-1)) = \frac{(q!)^{q^{m-1}}}{q}E(PD'(q,m-1)).$$

Since $E(PD(q,1)) = \frac{q!}{q}$, we have

$$E(PD(q,m)) = \frac{(q!)^{\frac{q^m-1}{q-1}}}{q^m}.$$

$\square$

References

1. de Bruijn, N.G.: A Combinatorial Problem. Koninklijke Nederlandse Akademie v. Wetenschappen 49, 758–764 (1946)
2. Golomb, S.W.: Shift register sequences. Holden-Day, Inc., San Francisco (1967)

3. Hagita, M., Matsumoto, M., Natsu, F., Ohtsuka, Y.: Error Correcting Sequence and Projective De Bruijn Graph. Graphs and Combinatorics 24(3), 185–192 (2008)
4. Huaxiao, Z., Fuji, Z., Qiongxiang, H.: On the number of spanning trees and Eulerian tours in iterated line digraphs. Discrete Appl. Math. 73(1), 59–67 (1997)
5. Li, X.L., Fuji, Z.: On the numbers of spanning trees and Eulerian tours in generalized de Bruijn graphs. Discrete Math. 94(3), 189–197 (1991)
6. MacWilliams, F.J., Sloane, N.J.A.: The theory of error-correcting codes. North-Holland, Amsterdam (1977)

Multiplicative Character Sums of Recurring Sequences with Rédei Functions

Domingo Gomez[1] and Arne Winterhof[2]

[1] Faculty of Sciences, University of Cantabria, E-39071 Santander, Spain
domingo.gomez@unican.es
[2] Johann Radon Institute for Computational and Applied Mathematics (RICAM)
Austrian Academy of Sciences, Altenbergerstr. 69, 4040 Linz, Austria
arne.winterhof@oeaw.ac.at

Abstract. We prove a new bound for multiplicative character sums of nonlinear recurring sequences with Rédei functions over a finite field of prime order. This result is motivated by earlier results on nonlinear recurring sequences and their application to the distribution of powers and primitive elements. The new bound is much stronger than the bound known for general nonlinear recurring sequences.

1 Introduction

Let p be a prime, $\mathbb{F}_p$ the finite field of p elements, and $F(X)$ a rational function over $\mathbb{F}_p$. Let (u_n) be the sequence of elements of $\mathbb{F}_p$ obtained by the recurrence relation

$$u_{n+1} = F(u_n), \quad n \geq 0, \tag{1}$$

with some initial value $u_0 \in \mathbb{F}_p$. Obviously, this sequence eventually becomes periodic with least period $T \leq p$, but we may restrict ourselves to the case where (u_n) is purely periodic since otherwise we can consider a shift of the sequence.

Let χ be a nontrivial multiplicative character of $\mathbb{F}_p$. We consider character sums

$$S_\chi = \sum_{n=0}^{T-1} \chi(u_n).$$

Bounds on S_χ can be applied to obtain results on the distribution of powers and primitive roots modulo p in the sequence (u_n) in a standard way, see e.g. [8].

In the special case of *linear recurring sequences* these sums are well-studied, see [3,4,11].

When $F(X) \in \mathbb{F}_p[X]$ is a polynomial of degree at least 2, in [8] under some natural restrictions on $F(X)$ the general but rather weak upper bound

$$S_\chi = O\left(T^{1/2} p^{1/2} (\log p)^{-1/2}\right)$$

and an analog for sums over parts of the period were given which are nontrivial whenever $T > p(\log p)^{-1}$. Here the implied constant depends only on the

S.W. Golomb et al. (Eds.): SETA 2008, LNCS 5203, pp. 175–181, 2008.
© Springer-Verlag Berlin Heidelberg 2008

degree of $F(X)$. Note that each mapping of $\mathbb{F}_p$ can be represented by a polynomial. However, a representation as rational mapping can provide much stronger results.

For example, in the special case

$$F(X) = aX^{p-2} + b, \quad a \in \mathbb{F}_p^*, \ b \in \mathbb{F}_p,$$

the much stronger result

$$S_\chi = O\left(T^{1/2}p^{1/4}\right)$$

was obtained in [7], which is nontrivial whenever $T > p^{1/2}$. Since $x^{p-2} = x^{-1}$ for all $x \in \mathbb{F}_p^*$, the sequence (u_n) can up to at most one sequence element be defined with the rational function $F(X) = aX^{-1} + b$.

In [2] we investigated another special class of nonlinear recurring sequences (1) constructed via *Dickson polynomials* $F(X) = D_e(X) \in \mathbb{F}_p[X]$ defined by the recurrence relation

$$D_e(X) = XD_{e-1}(X) - D_{e-2}(X), \quad e = 2, 3, \ldots,$$

with initial values

$$D_0(X) = 2, \quad D_1(X) = X.$$

Under the condition $\gcd(e, p^2 - 1) = 1$, which characterizes the Dickson permutation polynomials, see [5, Theorem 3.2], we obtained nontrivial bounds provided that $T > p^{1/2+\varepsilon}$ and if $T = p^{1+o(1)}$ we obtained

$$S_\chi = O\left(p^{7/8+o(1)}\right).$$

This article deals with the special class of nonlinear recurring sequences (1) constructed via *Rédei functions* defined in the sequel, which have some similar properties as Dickson polynomials.

Suppose that

$$r(X) = X^2 - \alpha X - \beta \in \mathbb{F}_p[X]$$

is an irreducible quadratic polynomial with the two different roots ξ and $\zeta = \xi^p$ in $\mathbb{F}_{p^2}$. Then any polynomial $b(X) \in \mathbb{F}_{p^2}[X]$ can uniquely be written in the form $b(X) = g(X) + h(X)\xi$ with $g(X), h(X) \in \mathbb{F}_p[X]$. For a positive integer e, we consider the elements

$$(X + \xi)^e = g_e(X) + h_e(X)\xi. \tag{2}$$

Note that $g_e(X)$ and $h_e(X)$ do not depend on the choice of the root ξ of $r(X)$. Evidently, e is the degree of the polynomial $g_e(X)$, and $h_e(X)$ has degree at most $e - 1$, where equality holds if and only if $\gcd(e, p) = 1$, see [5, p. 22] or [10]. The *Rédei function* $R_e(X)$ of degree e is then given by

$$R_e(X) = \frac{g_e(X)}{h_e(X)}.$$

The following facts can be found in [9]. The Rédei function $R_e(X)$ is a permutation of $\mathbb{F}_p$ if and only if $\gcd(e, p+1) = 1$, the set of these permutations is a group with respect to the composition which is isomorphic to the group of units of $\mathbb{Z}_{p+1}$. In particular for indices m, n with $\gcd(m, p+1) = \gcd(n, p+1) = 1$ we have

$$R_m(R_n(u)) = R_{mn}(u) = R_n(R_m(u)) \quad \text{for all } u \in \mathbb{F}_p. \tag{3}$$

For further background on Rédei functions we refer to [5,9,10].

We consider generators (u_n) defined by

$$u_{n+1} = R_e(u_n), \quad n \geq 0, \quad \gcd(e, p+1) = 1,$$

with a Rédei permutation $R_e(X)$ and some initial element $u_0 \in \mathbb{F}_p$. The sequences (u_n) are purely periodic and the period length T divides $\varphi(p+1)$, where φ denotes Euler's totient function. For details we refer to [9, Lemma 3.5].

Although we follow the same general method of bounding character sums as in [2], the details of the crucial step that a certain auxiliary polynomial, see (6) below, is not, up to a multiplicative constant, an sth power of a polynomial, where s denotes the order of χ, is more intricate.

2   Preliminaries

For an integer $t > 1$ we denote by $\mathbb{Z}_t$ the residue ring modulo t and always assume that it is represented by the set $\{0, 1, \ldots, t-1\}$. As usual, we denote by $\mathcal{U}_t$ the set of invertible elements of $\mathbb{Z}_t$.

We recall Lemma 2 from [1].

Lemma 1. *For any set $\mathcal{K} \subseteq \mathcal{U}_t$ of cardinality $\#\mathcal{K} = K$, any fixed $1 \geq \delta > 0$ and any integer $h \geq t^\delta$ there exists an integer $r \in \mathcal{U}_t$ such that the congruence*

$$rk \equiv y \pmod{t}, \quad k \in \mathcal{K}, \ 0 \leq y \leq h-1,$$

has

$$L_r(h) \gg \frac{Kh}{t}$$

solutions (k, y).

We also need the *Weil bound* for character sums which we present in the following form, see e.g. [6, Theorem 5.41].

Lemma 2. *Let χ be a multiplicative character of $\mathbb{F}_p$ of order $s > 1$ and let $F(X) \in \mathbb{F}_p[X]$ be a polynomial of positive degree that is not, up to a multiplicative constant, an sth power of a polynomial. Let d be the number of distinct roots of $F(X)$ in its splitting field over $\mathbb{F}_p$. Then we have*

$$\left| \sum_{x \in \mathbb{F}_p} \chi(F(x)) \right| \leq (d-1)p^{1/2}.$$

3   Main Result

Let t be the smallest positive integer for which $R_e(u_0) = R_f(u_0)$ whenever $e \equiv f$ (mod t). Note that $t|p+1$ and T is the multiplicative order of e modulo t.

Theorem 1. *For every fixed integer $\nu \geq 1$ and nontrivial multiplicative character χ of $\mathbb{F}_p$ we have*

$$S_\chi = O\left(T^{1 - \frac{2\nu+1}{2\nu(\nu+1)}} t^{\frac{1}{2(\nu+1)}} p^{\frac{\nu+2}{4\nu(\nu+1)}}\right),$$

where the implied constant depends only on ν.

Proof. We put

$$h = \left\lceil t^{\frac{\nu}{\nu+1}} T^{\frac{-\nu}{\nu+1}} p^{\frac{1}{2(\nu+1)}} \right\rceil.$$

Because $t \geq T$, for this choice of h we obtain $h \geq p^{1/2(\nu+1)}$, thus Lemma 1 applies.

Because the sequence (u_n) is purely periodic, for any $k \in \mathbb{Z}_t$, we have:

$$S_\chi = \sum_{n=0}^{T-1} \chi(R_{e^{n}+k}(u_0)). \tag{4}$$

Let $\mathcal{K}$ be the subgroup of $\mathcal{U}_t$ generated by e. Thus $\#\mathcal{K} = T$. We select r as in Lemma 1 and let $\mathcal{L}$ be the subset of $\mathcal{K}$ which satisfies the corresponding congruence. We denote $L = \#\mathcal{L}$. In particular, $L \gg hT/t$.

By (4) we have

$$LS_\chi = \sum_{n=0}^{T-1} \sum_{k \in \mathcal{L}} \chi\left(R_{e^{n}+k}(u_0)\right).$$

Applying the Hölder inequality, we derive

$$L^{2\nu}|S_\chi|^{2\nu} \leq T^{2\nu-1} \sum_{n=0}^{T-1} \left| \sum_{k \in \mathcal{L}} \chi\left(R_{e^{n}+k}(u_0)\right) \right|^{2\nu}. \tag{5}$$

Let $1 \leq r' \leq t-1$, be defined by the congruence $rr' \equiv 1$ (mod t). By (3) we obtain

$$R_{e^{n}+k}(u_0) \equiv R_{e^{n}+krr'}(u_0) \equiv R_{re^{k}}(R_{r'e^{n}}(u_0)) \pmod{p}.$$

Obviously, the values of $r'e^n$, $n = 0, \ldots, T-1$, are pairwise distinct modulo t. Thus, from the definition of t, we see that the values of $R_{r'e^n}(u_0)$ are pairwise distinct. Therefore, from (5) we derive

$$L^{2\nu}|S_\chi|^{2\nu} \leq T^{2\nu-1} \sum_{u \in \mathbb{F}_p} \left| \sum_{k \in \mathcal{L}} \chi\left(R_{re^{k}}(u)\right) \right|^{2\nu}.$$

Denoting $\mathcal{F} = \{re^k \mid k \in \mathcal{L}\}$ we deduct

$$L^{2\nu}|S_\chi|^{2\nu} \leq T^{2\nu-1} \sum_{u \in \mathbb{F}_p} \left| \sum_{f \in \mathcal{F}} \chi\left(R_f(u)\right) \right|^{2\nu}$$

$$\leq T^{2\nu-1} \sum_{f_1,\ldots,f_{2\nu} \in \mathcal{F}} \sum_{u \in \mathbb{F}_p} \chi\left(\prod_{j=1}^{\nu} \left(R_{f_j}(u) \left(R_{f_{\nu+j}}(u) \right)^{p-2} \right) \right).$$

For the case that no integer in the set $(f_1, \ldots, f_{\nu+1}, \ldots, f_{2\nu})$ appears only once we use the trivial bound p for the inner sum. Since in this case there are at most ν different values in $(f_1, \ldots, f_{\nu+1}, \ldots, f_{2\nu})$ there are at most L^ν such cases, which gives the total contribution

$$O(L^\nu p).$$

Otherwise, to apply Lemma 2 we have to show that the polynomial

$$\Psi_{f_1,\ldots,f_{2\nu}}(X) = \prod_{j=1}^{\nu} \left(g_{f_j}(X) h_{\nu+j}(X) (h_j(X) g_{f_{\nu+j}}(X))^{p-2} \right) \tag{6}$$

is not, up to a multiplicative constant, an sth-power of a polynomial, where $s > 1$ denotes the order of χ. We cancel all elements which appear in both sets $\{f_1, \ldots, f_\nu\}$ and $\{f_{\nu+1}, \ldots, f_{2\nu}\}$. Let $f > 1$ be the largest number in $\{f_1, \ldots, f_{2\nu}\}$ which is not eliminated. We may assume $f < p$. We show that $g_f(X)$ has a zero which is neither a zero of any $h_{f'}(X)$ with $f' \leq f$ nor a zero of any $g_{f'}(X)$ with $f' < f$.

With (2) we obtain

$$(X + \xi)^k - (X + \zeta)^k = (\xi - \zeta)h_k(X).$$

Hence, $h_k(x_0) = 0$ if and only if

$$\left(\frac{x_0 + \xi}{x_0 + \zeta} \right)^k = 1, \tag{7}$$

i.e., $(x_0 + \xi)/(x_0 + \zeta)$ is a k-th root of unity. Similarly using

$$\zeta(X + \xi)^k - \xi(X + \zeta)^k = (\zeta - \xi)g_k(X)$$

we get $g_k(x_0) = 0$ if and only if

$$\left(\frac{x_0 + \xi}{x_0 + \zeta} \right)^k = \frac{\xi}{\zeta} = \zeta^{p-1}.$$

Let $\alpha > 1$ be the order of ζ^{p-1}, i.e., $\alpha | p+1$, put $l = \alpha f$ and let ρ be a suitable primitive lth root of unity in an appropriate extension field of $\mathbb{F}_p$, which exists since $f < p$ and thus $\gcd(l, p) = 1$, such that

$$z_0 = \frac{\xi - \rho\zeta}{\rho - 1}$$

is a root of $g_f(X)$. Obviously we have $g_{f'}(z_0) \neq 0$ for $f' < f$. It can easily be seen with (7) that $h_{f'}(z_0) \neq 0$ for $f' \leq f$ since otherwise we had $\rho^{f'} = 1$.

Taking into account that the number of distinct roots of $\Psi_{f_1,\ldots,f_{2\nu}}(X)$ is less than $2\nu h$ and it cannot be an sth-power, Lemma 2 applies. We obtain that the total contribution from such terms is $O(L^{2\nu} h p^{1/2})$. Hence

$$L^{2\nu} |S_\chi|^{2\nu} = O\left(T^{2\nu-1}\left(L^\nu p + L^{2\nu} h p^{1/2}\right)\right).$$

So this leads us to the bound

$$|S_\chi|^{2\nu} = O\left(T^{2\nu-1}\left(L^{-\nu} p + h p^{1/2}\right)\right).$$

Recalling that $L \gg hT/t$, we derive

$$|S_\chi|^{2\nu} = O\left(T^{2\nu-1}\left(t^\nu T^{-\nu} h^{-\nu} p + h p^{1/2}\right)\right).$$

Substituting the selected value of h, which balances both terms in the above estimate, we finish the proof. $\square$

Remark. As for the bound for Dickson polynomials we mention the following simplifications. Assuming that $T = t^{1+o(1)}$, the bound of Theorem 1 takes the form

$$S_\chi = O\left(T^{1-1/2\nu+o(1)} p^{(\nu+2)/4\nu(\nu+1)}\right).$$

Therefore for any $\delta > 0$, choosing a sufficiently large ν we obtain a nontrivial bound provided $T \geq p^{1/2+\delta}$. On the other hand, if $t \geq T = p^{1+o(1)}$, then taking $\nu = 1$ we obtain

$$S_\chi = O\left(p^{7/8+o(1)}\right).$$

Acknowledgment

Parts of this paper were written during a pleasant stay of D.G. to Linz. He wishes to thank the Radon Institute for its hospitality. A.W. was supported in part by the Austrian Science Fund (FWF) Grant P19004-N18.

References

1. Friedlander, J.B., Hansen, J., Shparlinski, I.E.: On character sums with exponential functions. Mathematika 47, 75–85 (2000)
2. Gomez, D., Winterhof, A.: Character sums for sequences of iterations of Dickson polynomials. In: Mullen, G., et al. (eds.) Finite Fields and Applications Fq8, Contemporary Mathematics (to appear)
3. Korobov, N.M.: The distribution of non-residues and of primitive roots in recurrence series. Dokl. Akad. Nauk SSSR 88, 603–606 (1953) (in Russian)

4. Korobov, N.M.: Exponential sums and their applications. Mathematics and its Applications (Soviet Series), vol. 80. Kluwer Academic Publishers Group, Dordrecht (1992)
5. Lidl, R., Mullen, G.L., Turnwald, G.: Dickson Polynomials. Pitman Monographs and Surveys in Pure and Applied Math. Longman, London (1993)
6. Lidl, R., Niederreiter, H.: Finite fields. Cambridge University Press, Cambridge (1997)
7. Niederreiter, H., Shparlinski, I.E.: On the distribution of power residues and primitive elements in some nonlinear recurring sequences. Bull. London Math. Soc. 35, 522–528 (2003)
8. Niederreiter, H., Winterhof, A.: Multiplicative character sums for nonlinear recurring sequences. Acta Arith. 111, 299–305 (2004)
9. Nöbauer, R.: Rédei-Permutationen endlicher Körper. In: Czermak, J., et al. (eds.) Contributions to General Algebra 5, Hölder-Pichler-Tempsky, Vienna, pp. 235–246 (1987)
10. Rédei, L.: Über eindeutig umkehrbare Polynome in endlichen Körpern. Acta Sci. Math. 11, 85–92 (1946)
11. Shparlinskiĭ, I. E.: Distribution of nonresidues and primitive roots in recurrent sequences. Matem. Zametki 24, 603–613 (1978) (in Russian)

On the Connection between Kloosterman Sums and Elliptic Curves

Petr Lisoněk*

Department of Mathematics, Simon Fraser University
Burnaby, BC, Canada V5A 1S6
plisonek@sfu.ca

Abstract. We explore the known connection of Kloosterman sums on fields of characteristic 2 and 3 with the number of points on certain elliptic curves over these fields. We use this connection to prove results on the divisibility of Kloosterman sums, and to compute numerical examples of zeros of Kloosterman sums on binary and ternary fields of large orders. We also show that this connection easily yields some formulas due to Carlitz that were recently used to prove certain non-existence results on Kloosterman zeros in subfields.

Keywords: Kloosterman sum, elliptic curve, finite field.

1 Introduction

Kloosterman sums have recently enjoyed much attention. Some of this interest is due to their applications in cryptography and coding theory; see for example [4] and [14].

In the present paper we exploit the connection between Kloosterman sums on $\mathbb{F}_{p^m}$ and the number of $\mathbb{F}_{p^m}$-rational points on certain elliptic curves, where $p \in \{2, 3\}$. In the binary case this association was given in the well known paper by Lachaud and Wolfmann [9]; in the ternary case this is a recent result due to Moisio [13].

We use this connection to reprove the known characterization of binary Kloosterman sums divisible by 8, and we give a characterization of binary Kloosterman sums divisible by 16. We also quote our recent result on ternary Kloosterman sums modulo 4, which was proved elsewhere using similar methods.

There is considerable interest in non-zero elements of $\mathbb{F}_{p^m}$ at which the Kloosterman sum attains the value 0. The second objective of the paper is to provide a practical computational method for finding numerical examples of such elements. We also show that some older results due to Carlitz, which recently appered in a proof that such elements can not belong to certain subfields of $\mathbb{F}_{p^m}$, follow very easily from the connection with elliptic curves.

---

* Research partially supported by the Natural Sciences and Engineering Research Council of Canada (NSERC).

1.1 Kloosterman Sums

For a prime p and a positive integer m, let $q = p^m$ and let $\mathbb{F}_q$ denote the finite field of order q. Let $\mathrm{Tr} : \mathbb{F}_q \to \mathbb{F}_p$ be the absolute trace. The *Kloosterman sum* on $\mathbb{F}_q$ is the mapping $\mathcal{K} : \mathbb{F}_q \to \mathbb{R}$ defined by

$$\mathcal{K}(a) := 1 + \sum_{x \in \mathbb{F}_q^*} \omega^{\mathrm{Tr}(x^{-1}+ax)}, \tag{1}$$

where $\omega = e^{2\pi i/p}$ is a primitive p-th root of unity. (In some references the same mapping is defined by $\mathcal{K}(a) := \sum_{x \in \mathbb{F}_q} \omega^{\mathrm{Tr}(x^{-1}+ax)}$ with the proviso that $\mathrm{Tr}(0^{-1}) = 0$.)

Since $\omega + \omega^{-1} = -1$ for $p = 3$, it is easy to see that $\mathcal{K}(a)$ is an integer for $p \in \{2,3\}$. For integers s,t let $s|t$ denote that s divides t, and let $s \nmid t$ denote that s does not divide t.

Applying the Frobenius automorphism $x \mapsto x^p$ to (1) and using the properties of the trace map yields:

Lemma 1. *For all $a \in \mathbb{F}_{p^m}$ we have $\mathcal{K}(a) = \mathcal{K}(a^p)$.*

1.2 Elliptic Curves

Throughout the paper we will use standard definitions and results on elliptic curves over finite fields and on the Abelian groups associated with them. We recommend [12] as an accessible reference for these topics. Throughout this paper for an elliptic curve E defined over $\mathbb{F}_{p^m}$ we denote by $\#E$ the number of $\mathbb{F}_{p^m}$-rational points on E. Recall that the *order* of a point P on an elliptic curve is the smallest r such that $rP = \mathcal{O}$ and $sP \neq \mathcal{O}$ for $0 < s < r$, where $\mathcal{O}$ is the neutral element of the group of the curve (the point at infinity).

Theorem 1. [10] *Let $a \in \mathbb{F}_{2^m}^*$ and let $\mathcal{E}_2(a)$ be the elliptic curve over $\mathbb{F}_{2^m}$ defined by*

$$\mathcal{E}_2(a): \quad y^2 + xy = x^3 + a.$$

Then $\#\mathcal{E}_2(a) = 2^m + \mathcal{K}(a)$.

Theorem 2. [13] *Let $a \in \mathbb{F}_{3^m}^*$ and let $\mathcal{E}_3(a)$ be the elliptic curve over $\mathbb{F}_{3^m}$ defined by*

$$\mathcal{E}_3(a): \quad y^2 = x^3 + x^2 - a.$$

Then $\#\mathcal{E}_3(a) = 3^m + \mathcal{K}(a)$.

Let us now present our first result, which we be used in the next section.

Theorem 3. *Let $p \in \{2,3\}$, let $a \in \mathbb{F}_{p^m}^*$, and let $0 \leq k \leq m$. Then $p^k|\mathcal{K}(a)$ if and only if there exists a point of order p^k on $\mathcal{E}_p(a)$, where the curves $\mathcal{E}_2(a)$ and $\mathcal{E}_3(a)$ are defined in Theorems 1 and 2 above.*

Proof. By Theorems 1 and 2 we have that $p^k|\mathcal{K}(a)$ if and only if $p^k|\#\mathcal{E}_p(a)$. Recall that for any $a \in \mathbb{F}_{p^m}^*$ the Abelian group structure of $\mathcal{E}_p(a)$ is $\mathcal{E}_p(a) \simeq \mathbb{Z}_{n_1} \times \mathbb{Z}_{n_2}$, with $n_1|n_2$ and $n_1|p^m - 1$, and we have $\#\mathcal{E}_p(a) = n_1 n_2$. Suppose that $p^k|\#\mathcal{E}_p(a)$. Since $p \nmid n_1$, it follows that $p^k|n_2$ and $\mathcal{E}_p(a)$ contains a subgroup G isomorphic to $\mathbb{Z}_{p^k}$. A generator of G is a point of order p^k in $\mathcal{E}_p(a)$. Conversely, if $\mathcal{E}_p(a)$ contains a point of order p^k, then $p^k|\#\mathcal{E}_p(a)$ by the Lagrange Theorem.   □

2   Divisibility of Kloosterman Sums

We say that $a \in \mathbb{F}_{p^m}$ is a *zero of the Kloosterman sum*, or a *Kloosterman zero* for short, if $\mathcal{K}(a) = 0$. It is easy to see that $\mathcal{K}(0) = 0$, and we will often implicitly exclude the case $a = 0$ and deal only with $a \in \mathbb{F}_{p^m}^*$. There is a lot of interest in finding zeros of Kloosterman sums, see for example [3] for an application to construction of bent functions.

If we can find an integer s such that $s \nmid K(a)$, then $\mathcal{K}(a) \neq 0$. This gives one of motivations for studying divisibility properties of Kloosterman sums. The following two results are well known and easy to prove.

Lemma 2. *For each* $a \in \mathbb{F}_{2^m}$, $\mathcal{K}(a)$ *is divisible by 4.*

Lemma 3. *For each* $a \in \mathbb{F}_{3^m}$, $\mathcal{K}(a)$ *is divisible by 3.*

The following theorem was first proved in [7]. There are two proofs given in [7], and one proof is given in [4]. We give a new proof as an illustration of the methods applied in this article.

Theorem 4. [7,4] *Let* $m \geq 3$. *For any* $a \in \mathbb{F}_{2^m}$, $\mathcal{K}(a)$ *is divisible by 8 if and only if* $\mathrm{Tr}(a) = 0$.

Proof. The result holds for $a = 0$, so let us suppose that $a \in \mathbb{F}_{2^m}^*$. Since $m \geq 3$, by Theorem 3 we have that $8|\mathcal{K}(a)$ if and only if $\mathcal{E}_2(a)$ contains a point of order 8. By Lemma 1 we can replace $\mathcal{E}_2(a)$ with $\mathcal{E}_2(a^{2^l})$ for any positive integer l; let us for simplicity take the curve $\mathcal{E}_2(a^8) : y^2 + xy = x^3 + a^8$. By Lemma 7.4 in [12], an $x_0 \in \mathbb{F}_{2^m}$ is the x-coordinate of a point of order 8 on $\mathcal{E}_2(a^8)$ if and only if $X = x_0$ is a root of the polynomial $(X + a^2)^2 + aX$. This happens exactly if $\mathrm{Tr}(1 \cdot a^4/a^2) = \mathrm{Tr}(a) = 0$.   □

Lemma 7.4 of [12] used in the previous proof belongs to the theory of *division polynomials* for elliptic curves. A detailed treatment for characteristic 2 can be found in Chapter 7 of [12]. The following theorem is new.

Theorem 5. *Let* $m \geq 4$. *For any* $a \in \mathbb{F}_{2^m}$, $\mathcal{K}(a)$ *is divisible by 16 if and only if* $\mathrm{Tr}(a) = 0$ *and* $\mathrm{Tr}(y) = 0$ *where* $y^2 + ay + a^3 = 0$.

Proof. The result holds for $a = 0$, so let us suppose that $a \in \mathbb{F}_{2^m}^*$. Since $m \geq 4$, by Theorem 3 we have that $16|\mathcal{K}(a)$ if and only if $\mathcal{E}_2(a)$ contains a point of order 16. As in the previous proof, by Lemma 1 we can replace $\mathcal{E}_2(a)$ with

$\mathcal{E}_2(a^{16}) : y^2 + xy = x^3 + a^{16}$. By Lemma 7.4 in [12], an $x_0 \in \mathbb{F}_{2^m}$ is the x-coordinate of a point of order 16 on $\mathcal{E}_2(a^{16})$ if and only if $X = x_0$ is a root of the polynomial $((a^4 + X)^2 + a^2 X)^2 + aX(a^4 + X)^2$. This simplifies to the equation

$$X^4 + aX^3 + a^4 X^2 + a^9 X + a^{16} = 0. \tag{2}$$

Using the known theory of solvability of quartic polynomials in characteristic 2, see for example [10], some straightforward calculations show that (2) has a solution in $\mathbb{F}_{2^m}$ if and only if $\mathrm{Tr}(a) = 0$ and $\mathrm{Tr}(y_0) = 0$ where y_0 is any of the solutions of the quadratic equation $y^2 + ay + a^3 = 0$. □

Let us just remark that in the ternary case, one can prove using very similar ideas that for any $a \in \mathbb{F}_{3^m}$, $\mathcal{K}(a)$ is divisible by 9 if and only if $\mathrm{Tr}(a) = 0$. The necessary background on division polynomials in characteristic 3 can be found for example in the book [5].

One proof technique that has been used several times in the literature is to associate the value of the Kloosterman sum with the number of solutions to a certain system of equations; then a combinatorial argument (typically exploiting some symmetry present in the system of equations) is used to show that the number of solutions has to be divisible by some constant. This then yields a divisibility result for the associated Kloosterman sum and/or some other kind of exponential sum. For illustration let us quote our recent result obtained in this way, which appears in a different article.

Theorem 6. [6] *Let $a \in \mathbb{F}_{3^m}$. Then exactly one of the following cases occurs:*

1. *$a = 0$ or a is a square, $\mathrm{Tr}(\sqrt{a}) \neq 0$ and $\mathcal{K}(a) \equiv 0 \pmod{2}$.*
2. *$a = t^2 - t^3$ for some $t \in \mathbb{F}_{3^m} \setminus \{0, 1\}$, at least one of t and $1 - t$ is a square and $\mathcal{K}(a) \equiv 2m - 1 \pmod{4}$.*
3. *$a = t^2 - t^3$ for some $t \in \mathbb{F}_{3^m} \setminus \{0, 1\}$, both t and $1 - t$ are non-squares and $\mathcal{K}(a) \equiv 2m + 1 \pmod{4}$.*

3   Computation of Zeros of Kloosterman Sums

As we have already mentioned, there is a significant interest in finding $a \in \mathbb{F}_{p^m}^*$ for which $\mathcal{K}(a) = 0$. In the binary case, the existence of such $a \in \mathbb{F}_{2^m}^*$ for each $m > 1$ was the famous Dillon conjecture (related to the construction of bent functions) that was settled affirmatively by Lachaud and Wolfmann in [9]. More recently, Charpin and Gong [3] related the zeros of Kloosterman sums with hyperbent functions. Numerical examples found in the literature appear to be limited to relatively small field orders. Table 1 in [3] gives a complete classification of Kloosterman zeros in $\mathbb{F}_{2^m}$ for $m \leq 14$.

It is sometimes overlooked that the existence of very fast algorithms for counting points on elliptic curves [11], such as variants of the Schoof-Elkies-Atkin algorithm, allows one to compute binary and ternary Kloosterman sums on fields of very large orders by Theorems 1 and 2. For example, using the point counting procedure implemented in the computer algebra system Magma 2.14 running on Intel Xeon CPU at 3.0 GHz, for a random $a \in \mathbb{F}_{2^{800}}$ it takes only about 3 seconds to compute $\mathcal{K}(a)$.

Moreover, if one is only interested in proving that $\mathcal{K}(a) = 0$ for some given $a \in \mathbb{F}_{p^m}$ ($p \in \{2,3\}$), then one does not need to use the complicated algorithms for fast point counting. In this case the group of the elliptic curve must be isomorphic to $\mathbb{Z}_{p^m}$, and one can simply attempt to guess a generator P for the group. As there are $(p-1)p^{m-1}$ generators, there is a good chance of success. If P indeed happens to be a generator, then this can be easily verified by computing $p^i P$ for $i = 1, 2, \ldots, m$ via iterated multiplication by p. In the binary case this process can be further simplified using Lemma 7.4 of [12].

Using the algorithm described in the previous paragraph and starting with random elements $a \in \mathbb{F}_{p^m}$, we have found one Kloosterman zero for each field $\mathbb{F}_{2^m}$ where $m \leq 64$ and for each field $\mathbb{F}_{3^m}$ where $m \leq 34$. The computation took only a few days of CPU time. As was noted earlier, these results are easily verified for example using the system Magma [1].

Example 1. Let $\mathbb{F}_{2^{64}}$ be constructed as $\mathbb{F}_2[X]/(p(X))$ where $p(X) = 1 + X + X^3 + X^4 + X^{64}$. For $a = 1 + X^6 + X^9 + X^{12} + X^{16} + X^{17} + X^{20} + X^{22} + X^{24} + X^{27} + X^{30} + X^{31} + X^{32} + X^{36} + X^{37} + X^{40} + X^{41} + X^{43} + X^{45} + X^{47} + X^{48} + X^{50} + X^{51} + X^{53} + X^{56} + X^{59} + X^{60} \in \mathbb{F}_{2^{64}}$ we have $\mathcal{K}(a) = 0$.

4   Existence of Zeros in Subfields

Very recently, Charpin and Gong [3] proved that a Kloosterman zero in $\mathbb{F}_{2^m}$ can *not* belong to a certain set that is defined in terms of certain *subfields* of $\mathbb{F}_{2^m}$. For details please see the statement of Theorem 6 in [3]. Their proof is based on formulas by Carlitz [2] that express the value of a Kloosterman sum over an extension field in terms of the Kloosterman sums on subfields. For clarity let us now use the symbol $\mathcal{K}$ with a subscript to denote that the Kloosterman sum $\mathcal{K}_q(a)$ is computed over the field $\mathbb{F}_q$.

It is interesting to observe that some of the Carlitz formulas, which are proved by complicated calculations in [2], can be obtained very easily using the associated elliptic curves given in Theorems 1 and 2 above. Indeed, if $a \in \mathbb{F}_{p^m}$, then the Hasse-Weil Theorem (see, for example, [8, Theorem 2.26]) applied to the curves of Theorems 1 and 2 immediately yields a second order linear recurrence for $K_{p^{ms}}(a)$, where s is a positive integer, in terms of $K_{p^m}(a)$, where $K_q(a) := \mathcal{K}_q(a) - 1$. This is exactly the recurrence (5.7) of [2]. It would be interesting to see whether this approach can yield some further non-existence results beyond those quoted in the previous paragraph.

References

1. Bosma, W., Cannon, J., Playoust, C.: The Magma Algebra System. I. The User Language. J. Symbolic Comput. 24, 235–265 (1997)
2. Carlitz, C.: Kloosterman Sums and Finite Field Extensions. Acta Arithmetica XVI, 179–193 (1969)

3. Charpin, P., Gong, G.: Hyperbent Functions, Kloosterman Sums and Dickson Polynomials. IEEE Trans. Inform. Theory (to appear)
4. Charpin, P., Helleseth, T., Zinoviev, V.: Propagation Characteristics of $x \mapsto x^{-1}$ and Kloosterman Sums. Finite Fields Appl. 13, 366–381 (2007)
5. Enge, A.: Elliptic Curves and Their Applications to Cryptography: An Introduction. Kluwer Academic Publishers, Boston (1999)
6. Garaschuk, K., Lisoněk, P.: On Ternary Kloosterman Sums modulo 12. Finite Fields Appl. (to appear)
7. Helleseth, T., Zinoviev, V.: On Z_4-linear Goethals Codes and Kloosterman Sums. Des. Codes Cryptogr. 17, 269–288 (1999)
8. Hirschfeld, J.W.P.: Projective Geometries over Finite Fields, 2nd edn. The Clarendon Press, Oxford University Press, New York (1998)
9. Lachaud, G., Wolfmann, J.: The Weights of the Orthogonals of the Extended Quadratic Binary Goppa Codes. IEEE Trans. Inform. Theory 36, 686–692 (1990)
10. Leonard, P.A., Williams, K.S.: Quartics over GF(2^n). Proc. Amer. Math. Soc. 36, 347–350 (1972)
11. Lercier, R., Lubicz, D., Vercauteren, F.: Point Counting on Elliptic and Hyperelliptic Curves. In: Cohen, H., Frey, G. (eds.) Handbook of Elliptic and Hyperelliptic Curve Cryptography. Chapman & Hall/CRC, Boca Raton (2006)
12. Menezes, A.: Elliptic Curve Public Key Cryptosystems. Kluwer Academic Publishers, Boston (1993)
13. Moisio, M.: Kloosterman Sums, Elliptic Curves, and Irreducible Polynomials with Prescribed Trace and Norm. Acta Arithmetica (to appear)
14. Moisio, M., Ranto, K.: Kloosterman Sum Identities and Low-weight Codewords in a Cyclic Code with Two Zeros. Finite Fields Appl. 13, 922–935 (2007)

A Class of Optimal Frequency Hopping Sequences Based upon the Theory of Power Residues[*]

Daiyuan Peng[1], Tu Peng[2],
Xiaohu Tang[1], and Xianhua Niu[1]

[1] Key Laboratory of Information Coding and Transmission, Southwest
Jiaotong University, Chengdu, Sichuan 610031, P.R. of China
dypeng@swjtu.edu.cn, xhutang@ieee.org,
cow_flower1212@yahoo.com.cn
[2] Department of Computer Science, University of Texas at Dallas,
Ricardson, TX 75080-3021 USA
pengtutu@gmail.com

Abstract. The average Hamming correlation is an important performance indicator of the frequency hopping sequences. In this paper, the theoretical bound on the average Hamming correlation for frequency hopping sequences is established. Besides, a new family of frequency hopping sequences is proposed and investigated. The construction of new frequency hopping sequences is based upon the theory of power residues, and the new frequency hopping sequences are called the power residue frequency hopping sequences. It is shown that new frequency hopping sequences are optimal with respect to both average Hamming correlation family and the optimal maximum Hamming correlation.

Keywords: average Hamming correlation, Hamming correlation, frequency hopping sequence, theoretical bound.

1 Introduction

The design of frequency hopping (FH) sequences is suitable for frequency hopping or time hopping (TH) CDMA systems, multi-user radar and sonar systems which remain of great interest in recent years [1,2]. In FH CDMA systems, the address assignment must be achieved in such a way that there is no ambiguity about the sender and the information it transmits, and the received signal must interfere as negligible as possible to the reception of other users' signals. The mutual interference in FH CDMA is mainly controlled by the Hamming cross-correlation of the hopping sequences. The need for finding FH sequences which

[*] This work was supported by the National Science Foundation of China (NSFC, 60572142), and the Foundation for the Author of National Excellent Doctoral Dissertation of PR China (FANEDD) under Grants 200341, and the Application Fundamental Research Project of Sichuan Province (2006J13-112).

S.W. Golomb et al. (Eds.): SETA 2008, LNCS 5203, pp. 188–196, 2008.
© Springer-Verlag Berlin Heidelberg 2008

have simultaneously good Hamming autocorrelation functions, small Hamming crosscorrelation functions and large family size is therefore well motivated.

Frequency hopping sequence design normally involves the following parameters: the size q of the frequency slot set F, the sequence length L, the family size M, the maximum Hamming autocorrelation sidelobe H_a, the maximum Hamming crosscorrelation H_c, the average Hamming autocorrelation A_a, and the average Hamming crosscorrelation A_c. In applications, it is generally desired that the family of FH sequences has the following properties [1,3,4].

1). The maximum Hamming autocorrelation H_a should be as small as possible;

2). The maximum Hamming crosscorrelation H_c should be as small as possible;

3). The average Hamming autocorrelation A_a should be as small as possible;

4). The average Hamming crosscorrelation A_c should be as small as possible;

5). The family size M for given H_a, H_c, A_a, A_c, q and L should be as large as possible.

However, these parameters q, L, M, H_a, H_c, A_a and A_c are not independent, and are bounded by certain theoretical limits. In order to evaluate the theoretical performance of the FH sequences, it is important to find the tight theoretical limits which set bounded relation among these parameters. Early in 1974, Lempel and Greenberger [4] established some bounds on the periodic Hamming correlation of FH sequences for $M = 1$ or 2. In 2004, Peng and Fan [5] obtained the following theoretical limit which sets bounded relation among the parameters q, L, M, H_a and H_c.

$$(L - 1)qH_a + (M - 1)LqH_c \geq (LM - q)L. \tag{1}$$

If H_a and H_c are a pair of the minimum integer solutions of inequality (1), then the corresponding FH sequence set S is called optimal maximum Hamming correlation family.

As far as the authors aware, the theoretical limits haven't been obtained yet which set bounded relation among the parameters q, L, M, A_a and A_c.

There are a number of hopping sequences derived from the polynomials over finite fields, such as linear hopping sequences [6], general linear hopping sequences [7], quadratic hopping sequences [8], extended quadratic hopping sequences [9], cubic sequences [10], cubic + linear congruence sequences [11], extended cubic hopping sequences [12], general cubic hopping sequences[3], polynomial hopping sequences [12], etc.

In this paper, we will pay particular attention to the bounds on the average Hamming correlation for FH sequences, and the construction and the characteristics on the Hamming correlation for the power residue FH sequences.

The rest of the paper is organized as follows. Section 2 gives the bounds on the average Hamming correlation for FH sequences; Section 3 presents the construction of the family of power residue FH sequences and establishes the relevant properties; Section 4 gives several illustrative examples; then the final section concludes with a brief summary.

2   The Average Hamming Correlation for Frequency Hopping Sequences

We first introduce the necessary notations. Let p be a prime number, $GF(p)$ be a finite field with elements $\{0,1,\ldots,p-1\}$. Let $f(x)$ be a polynomial over the finite field $GF(p)$, we will use $N(f) = N(f(x)=0)$ to denote the number of solutions of the equation $f(x)=0$ in $GF(p)$.

Let $F=\{f_1, f_2,\ldots,f_q\}$ be a frequency slot set with size $|F|=q$, and S be a set of M frequency hopping sequences of length L. For any two frequency slots f_i, $f_j \in F$, let

$$h(f_i, f_j) = \begin{cases} 1, \text{ if } f_i = f_j, \\ 0, \text{ otherwise.} \end{cases}$$

For any two FH sequences $x = (x_0, x_1,\ldots, x_{L-1}), y = (y_0, y_1,\ldots,y_{L-1}) \in S$, and any integer $\tau \geq 0$, the periodic Hamming correlation function $H(x,y;\tau)$ of x and y at time delay τ is defined as follows:

$$H(x,y;\tau) = \sum_{i=0}^{L-1} h(x_i,y_{i+\tau}), (\tau = 0,1,\ldots,L-1)$$

where the subscript addition $i+\tau$ is performed modulo L. Moreover, the $H(x,x;\tau)$ is called the periodic Hamming autocorrelation function when $x = y$ and the periodic Hamming crosscorrelation function when $x \neq y$. For any given FH sequence set S, the maximum periodic Hamming autocorrelation sidelobe $H_a(S)$ and the maximum periodic Hamming crosscorrelation $H_c(S)$ are defined by

$$H_a(S) = \max\{H(x,x;\tau)|x \in S, \tau = 1,2,\ldots,L-1\}$$

$$H_c(S) = \max\{H(x,y;\tau)|x,y \in S, x \neq y, \tau = 0,1,\ldots,L-1\}$$

An important performance indicator of the hopping sequences is the average Hamming correlation defined as follows.

Definition 1 ([13]). *Let S be a set of M hopping sequences of length L over a given frequency slot set F with size q, we call*

$$S_a(S) = \sum_{x \in S,\ 1 \leq \tau \leq L-1} H(x,x;\tau),$$

$$S_c(S) = \frac{1}{2} \sum_{x,y \in S, x \neq y, 0 \leq \tau \leq L-1} H(x,y;\tau)$$

as the overall number of Hamming autocorrelation (auto-hits) and Hamming crosscorrelation (cross-hits) of S respectively, and call

$$A_a(S) = \frac{S_a(S)}{M(L-1)},$$

$$A_c(S) = \frac{2S_c(S)}{LM(M-1)}$$

as the average Hamming autocorrelation (average of auto-hits) and the average Hamming crosscorrelation (average of cross-hits) of S respectively.

Obviously, the average Hamming correlation (average of hits) indicates the average error (or interference) performance of the hopping systems. However, generally speaking, it is very difficult to derive the average Hamming correlation for a given hopping sequence. For simplicity, we will denote $S_a = S_a(S)$, $S_c = S_c(S)$, $A_a = A_a(S)$ and $A_c = A_c(S)$.

We now give the theoretical limits which set a bounded relation among the parameters q, L, M, A_a and A_c.

Theorem 1. *Let S be a set of M hopping sequences of length L over a given frequency slot set F with size q, A_a and A_c be the average Hamming autocorrelation and the average Hamming crosscorrelation of S respectively, then*

$$\frac{A_a}{L(M-1)} + \frac{A_c}{(L-1)} \geq \frac{LM-q}{q(L-1)(M-1)}. \tag{2}$$

Proof. We first have

$$\sum_{x,y \in S, \tau} H(x,y;\tau) = \sum_{x \in S} H(x,x;0) + \sum_{x \in S, \tau \neq 0} H(x,y;\tau) + \sum_{x \neq y \in S, \tau} H(x,y;\tau)$$
$$= LM + S_a + 2S_c.$$

In [5], a lower bound L^2M^2/q on $\sum_{x,y \in S, 0 \leq \tau \leq L-1} H(x,y;\tau)$ was given. Therefore,

$$LM + S_a + 2S_c \geq \frac{L^2M^2}{q},$$

and

$$\frac{1}{(L-1)(M-1)} + \frac{S_a}{LM(L-1)(M-1)} + \frac{2S_c}{LM(L-1)(M-1)} \geq \frac{LM}{q(L-1)(M-1)},$$

$$\frac{1}{(L-1)(M-1)} + \frac{A_a}{L(M-1)} + \frac{A_c}{(L-1)} \geq \frac{LM}{q(L-1)(M-1)},$$

and (2) follows immediately. This completes the proof.   □

If the parameters q, L, M, A_a, and A_c of the FH sequence set S satisfy inequality (2) with equality, then it is said that sequence set S is an optimal average Hamming correlation family.

3   Construction and Analysis of the Power Residue Frequency Hopping Sequences

This section gives the definition of the new family of FH sequences and then establishes the relevant properties.

Definition 2. *Let p be a prime number, k be any positive integer such that $1 \leq k \leq p$-2. A kth power residue FH sequence set $C(p, k)$ is defined as follows:*

$$\begin{cases} C(p,k) = \{c^{(i)} | 1 \leq i \leq p-1\}, \\ c^{(i)} = (c_0^{(i)}, c_1^{(i)}, ..., c_{p-1}^{(i)}), \\ c_n^{(i)} = i \cdot n^k \bmod p (0 \leq n \leq p-1). \end{cases}$$

As for their Hamming correlations, we have the following results.

Theorem 2. *Let p be a prime number, k be any positive integer such that $1 \leq k \leq p$-2, then the kth power residue FH sequence set $C(p, k)$ has the following characteristics.*

1) $M = p-1$, $L = q = p$.
2) The Hamming autocorrelation functions of $C(p, k)$ are given by

$$H(e^{(i)}, e^{(i)}; \tau) = \begin{cases} p, \tau = 0 \\ \gcd(p-1, k) - 1, 1 \leq \tau \leq p-1 \end{cases} \tag{3}$$

The average Hamming autocorrelation of $C(p, k)$ is

$$A_a = \gcd(p-1, k) - 1. \tag{4}$$

3) The Hamming crosscorrelation functions of $C(p, k)$ are given by

$$\begin{aligned} &H(e^{(i)}, e^{(j)}; \tau) \\ &= \begin{cases} 1, \tau = 0 \\ 0, \tau > 0 \text{ and } i \cdot j^{-1} \text{ is a kth power nonresidue modulo } p \\ \gcd(p-1, k), \tau > 0 \text{ and } i \cdot j^{-1} \text{ is a kth power residue modulo } p \end{cases} \end{aligned} \tag{5}$$

where $1 \leq i, j \leq p$-1, $i \neq j$.
The average Hamming crosscorrelation of $C(p, k)$ is

$$A_c = \frac{1}{p(p-2)} \left[p^2 - p - 1 - (p-1)\gcd(p-1, k) \right]. \tag{6}$$

4) The $C(p, k)$ is an optimal average Hamming correlation family.

Proof. The proof of 1) is straightforward. We prove now 2). For any FH sequence $e^{(i)} \in C(p, k)$, its Hamming autocorrelation functions can be written as

$$\begin{aligned} H(e^{(i)}, e^{(i)}; \tau) &= \sum_{n=0}^{p-1} h(e_n^{(i)}, e_{n+\tau}^{(i)}) = \sum_{n=0}^{p-1} h(e_{n+\tau}^{(i)} - e_n^{(i)}, 0) \\ &= N(i(n+\tau)^k - in^k = 0) = N((n+\tau)^k - n^k = 0). \end{aligned}$$

For any $1 \leq \tau \leq p-1$, we have from number theory [14]

$$H(e^{(i)}, e^{(i)}; \tau) = N((1 + \tau \cdot n^{-1})^k = 1) = \gcd(p-1, k) - 1,$$

hence (3) follows. Then (4) follows immediately from (3).

We prove now part 3). For any two kth power residue FH sequences $e^{(i)}, e^{(j)} \in C(p, k)$, their Hamming crosscorrelation functions can be written as

$$H(e^{(i)}, e^{(j)}; \tau) = \sum_{n=0}^{p-1} h(c_n^{(i)}, e_{n+\tau}^{(j)}) = \sum_{n=0}^{p-1} h(e_{n+\tau}^{(j)} - e_n^{(i)}, 0)$$
$$= N(j(n+\tau)^k - i \cdot n^k = 0) = N(j(n+\tau)^k = i \cdot n^k).$$

For $\tau = 0$, since $i \neq j$, we obtain

$$H(e^{(i)}, e^{(j)}; 0) = N(j \cdot n^k = i \cdot n^k) = 1.$$

For any $1 \leq \tau \leq p-1$, we have from number theory

$$H(e^{(i)}, e^{(j)}; \tau) = N((1 + \tau \cdot n^{-1})^k = i \cdot j^{-1})$$
$$= \begin{cases} \gcd(p-1, k), i \cdot j^{-1} \text{ is a } k\text{th power residue modulo } p \\ 0, i \cdot j^{-1} \text{ is a } k\text{th power nonresidue modulo } p \end{cases}$$

Hence (5) follows.

For any fixed $1 \leq j \leq p-1$, it is noted that $i \cdot j^{-1}$ runs independently over all elements in $\{1, 2, \ldots, p-1\}$ as i takes independently all elements in $\{1, 2, \ldots, p-1\}$. Therefore, $i \cdot j^{-1}$ takes kth power residues modulo p $(p-1)/\gcd(p-1, k)$ times, kth power nonresidues modulo $p p-1-(p-1)/\gcd(p-1,k)$ times from number theory. Therefore

$$S_c = \frac{1}{2} \sum_{1 \leq i,j \leq p-1, 0 \leq \tau \leq p-1, i \neq j} H(e^{(i)}, e^{(j)}; \tau)$$
$$= \frac{(p-1)(p-2)}{2} + \frac{1}{2} \sum_{1 \leq i,j \leq p-1, 1 \leq \tau \leq p-1, i \neq j} H(e^{(i)}, e^{(j)}; \tau)$$
$$= \frac{(p-1)(p-2)}{2} + \frac{1}{2} \sum_{1 \leq j \leq p-1, 1 \leq \tau \leq p-1} \left[\frac{p-1}{\gcd(p-1,k)} - 1 \right] \cdot \gcd(p-1, k)$$
$$= \frac{p-1}{2} \left[p^2 - p - 1 - (p-1)\gcd(p-1, k) \right].$$

The average Hamming crosscorrelation of $C(p, k)$ is

$$A_c = \frac{2S_c}{ML(M-1)}$$
$$= \frac{2}{(p-1)p(p-2)} \cdot \frac{p-1}{2} \left[p^2 - p - 1 - (p-1)\gcd(p-1, k) \right]$$
$$= \frac{1}{p(p-2)} \left[p^2 - p - 1 - (p-1)\gcd(p-1, k) \right].$$

Thus (6) is true. By applying 1), (4) and (5) to (2), it follows that

$$\frac{A_a}{L(M-1)} + \frac{A_c}{(L-1)}$$
$$= \frac{1}{p(p-2)} [\gcd(p-1, k) - 1] + \frac{1}{(p-1)} \frac{1}{p(p-2)} \left[p^2 - p - 1 - (p-1)\gcd(p-1, k) \right]$$
$$= \frac{1}{(p-1)} \geq \frac{LM - q}{q(L-1)(M-1)} = \frac{p(p-1) - p}{p(p-1)(p-2)} = \frac{1}{(p-1)}.$$

Thus, the FH sequence set $C(p, k)$ is an optimal average Hamming correlation set. This completes the proof. $\square$

Corollary 1. *Let p be a prime number, k be any positive integer such that $1 \leq k \leq p\text{-}2$ and $gcd(p\text{-}1,k)=1$, then the kth power residue FH sequence set $C(p, k)$ has the following characteristics.*

1) The Hamming autocorrelation functions of $C(p, k)$ are given by

$$H(e^{(i)}, e^{(i)}; \tau) = \begin{cases} p, \tau = 0 \\ 0, 1 \leq \tau \leq p-1 \end{cases}$$

The average Hamming autocorrelation of $C(p, k)$ is

$$A_a = 0.$$

2) The Hamming crosscorrelation functions of $C(p, k)$ are given by

$$H(e^{(i)}, e^{(j)}; \tau) = 1. \tag{7}$$

The average Hamming crosscorrelation of $C(p, k)$ is

$$A_c = 1. \tag{8}$$

3) $C(p, k)$ is an optimal maximum Hamming correlation family.

Proof. The proof of 1) is straightforward.

We prove now 2). For any fixed $1 \leq j \leq p-1$, it is noted that $i \cdot j^{-1}$ runs independently over all elements in $\{1,2,\dots,p\text{-}1\}$ as i takes independently all elements in $\{1,2,\dots,p\text{-}1\}$. Since $gcd(p\text{-}1,k)=1$, $i \cdot j^{-1}$ takes kth power residues modulo p altogether $(p-1)/gcd(p-1, k) = p-1$ times from number theory [14]. That is, for all i and j with $1 \leq i, j \leq p-1$, $i \cdot j^{-1}$ are kth power residues modulo p. Hence, (7) follows immediately from (5), and (8) follows immediately from (7).

It is easy to see that the FH sequence set $C(p, k)$ has the following parameters: $q = p$, $L = p$, $M = p-1$, $H_a=0$ and $H_c=1$. By applying these parameters to (1), it follows that

$$(L-1)qH_a + (M-1)LqH_c = (p-2)p^2 \geq (LM-q)L = (p-2)p^2,$$

hence, the FH sequence set $C(p, k)$ is an optimal maximum Hamming correlation family. This completes the proof. □

4   Illustrative Examples

For $p=7$, $k=3$, one can design 3rd power residue FH sequence set $C(7, 3)$ as shown below:

$C(7, 3)=\{e^{(1)}=(0,1,1,6,1,6,6)$, $e^{(2)}=(0,2,2,5,2,5,5)$, $e^{(3)}=(0,3,3,4,3,4,4)$, $e^{(4)}=(0,4,4,3,4,3,3)$, $e^{(5)}=(0,5,5,2,5,2,2)$, $e^{(6)}=(0,6,6,1,6,1,1)$ $\}$. The periodic Hamming correlations of $C(7, 3)$ are given by

$$H(e^{(i)}, e^{(j)}; \tau) = \begin{cases} (7,2,2,2,2,2,2), i = j, \\ (1,3,3,3,3,3,3), (i,j) = (1,6), (2,5), (3,4), \\ (1,0,0,0,0,0,0), \text{otherwise}. \end{cases}$$

The 3rd power residue FH sequence set $C(7, 3)$ is an optimal average Hamming correlation family by Theorem 2. It is easy to see that the FH sequence set $C(7, 3)$ has the following parameters: $q = 7$, $L=7$, $M=6$, $H_a=2$ and $H_c=3$. By applying these parameters to (1), it follows that

$$(L - 1)qH_a + (M - 1)LqH_c = 624 > (LM - q)L = 245,$$

hence, the FH sequence set $C(7, 3)$ is not an optimal maximum Hamming correlation family.

For $p = 7$, $k=4$, one can design 4th power residue FH sequence set $C(7, 4)$ as shown below:

$C(7, 4)=\{e^{(1)}=(0,1,2,4,4,2,1)$, $e^{(2)}=(0,2,4,1,1,4,2)$, $e^{(3)}=(0,3,6,5,5,6,3)$,
$e^{(4)}=(0,4,1,2,2,1,4)$, $e^{(5)}=(0,5,3,6,6,3,5)$, $e^{(6)}=(0,6,5,3,3,5,6)$ $\}$. The periodic Hamming correlations of $C(7, 4)$ are given by

$$H(e^{(i)}, e^{(j)}; \tau) = \begin{cases} (7,1,1,1,1,1,1), i=j, \\ (1,2,2,2,2,2,2), (i,j) = (1,2), (1,4), (2,4), (3,5), (3,6), (5,6), \\ (1,0,0,0,0,0,0), \text{otherwise}. \end{cases}$$

The 4th power residue FH sequence set $C(7, 4)$ is an optimal average Hamming correlation family by Theorem 2. It is easy to see that the FH sequence set $C(7, 4)$ has the following parameters: $q=7$, $L=7$, $M=6$, $H_a=1$ and $H_c=2$. By applying these parameters to (1), it follows that

$$(L - 1)qH_a + (M - 1)LqH_c = 532 > (LM - q)L = 245,$$

whence, the FH sequence set $C(7, 4)$ is not an optimal maximum Hamming correlation family.

For $p = 7$, $k=5$, one can design 5th power residue FH sequence set $C(7, 5)$ as shown below:

$C(7, 5)=\{e^{(1)}=(0,1,4,5,2,3,6)$, $e^{(2)}=(0,2,1,3,4,6,5)$, $e^{(3)}=(0,3,5,1,6,2,4)$,
$e^{(4)}=(0,4,2,6,1,5,3)$, $e^{(5)}=(0,5,6,4,3,1,2)$, $e^{(6)}=(0,6,3,2,5,4,1)$ $\}$. The periodic Hamming correlations of $C(7, 5)$ are given by

$$H(e^{(i)}, e^{(j)}; \tau) = \begin{cases} (7,0,0,0,0,0,0), i = j, \\ (1,1,1,1,1,1,1), \text{otherwise}. \end{cases}$$

Since $\gcd(5,7)=1$, the 5th power residue FH sequence set $C(7, 5)$ is an optimal average Hamming correlation family by Theorem 2, and is an optimal maximum Hamming correlation family by Corollary 1.

5   Concluding Remarks

In this paper, the lower bound on the average Hamming correlation for FH sequences with respect to the size p of the frequency slot set, the sequence length L, the family size M, the average Hamming autocorrelation A_a and the average Hamming crosscorrelation A_c, is established, and a class of power residue

FH sequences is proposed. It is shown that the new FH sequences set is an optimal average Hamming correlation family. Moreover, it is shown that the power residue FH sequence set $C(p, k)$ is an optimal maximum Hamming correlation family and a one-coincidence FH sequence family under the condition that $\gcd(p-1, k)=1$.

References

1. Fan, P.Z., Darnell, M.: Sequence Design for Communications Applications. Research Studies Press, John Wiley & Sons Ltd., London (1996)
2. Win, M.Z., Scholtz, R.A.: Ultra-Wide Bandwidth time-hopping spread-spectrum impulse radio for wireless multiple-access communications. IEEE Trans. Commun. 58(4), 679–691 (2002)
3. Peng, D.Y., Peng, T., Fan, P.Z.: Generalized Class of Cubic Frequency-Hopping Sequences with Large Family Size. IEE Proceedings on Communications 152(6), 897–902 (2005)
4. Lempel, A., Greenberger, H.: Families sequence with optimal hamming correlation properties. IEEE Trans. Inform. Theory IT-20, 90–94 (1974)
5. Peng, D.Y., Fan, P.Z.: Lower bounds on the Hamming auto- and cross- correlations of frequency-hopping sequences. IEEE Trans. Inform. Theory 50(9), 2149–2154 (2004)
6. Titlebaum, E.L.: Time-frequency hop signals part I: coding based upon the theory of linear congruences. IEEE Trans. Aerosp. Electron. Syst. 17(4), 490–493 (1981)
7. Iacobucci, M.S., Benedetto, M.G.D.: Multiple access design for impulse radio communication systems. In: IEEE International Conference on Communications, April 2002, pp. 817–820 (2002)
8. Titlebaum, E.L.: Time-frequency hop signals part II: coding based upon quadratic congruences. IEEE Trans. Aerosp. Electron. Syst. 17(4), 494–499 (1981)
9. Bellegarda, J.R., Titlebaum, E.L.: Time-frequency hop codes based upon extended quadratic congruences. IEEE Trans. Aerosp. Electron. Syst. 24(6), 726–742 (1988)
10. Maric, S.V.: Frequency hop multiple access codes based upon the theory of cubic congruences. IEEE Trans. Aerosp. Electron. Syst. 26(6), 1035–1039 (1990)
11. Jovancevic, A.V., Titlebaum, E.L.: A new receiving technique for frequency hopping CDMA systems: analysis and application. In: IEEE 47th Vehicular Technology Conference, May 1997, vol. 3, pp. 2187–2190 (1997)
12. Fan, P.Z., Lee, M.H., Peng, D.Y.: New Family of Hopping Sequences for Time/Frequency Hopping CDMA Systems. IEEE Trans. on Wireless Communications 4(6), 2836–2842 (2005)
13. Sarwate, D.V.: Reed-Solomon codes and the design of sequences for spread-spectrum multiple-access communications. In: Wicker, S.B., Bhargava, V.K. (eds.) Reed-Solomon Codes and Their Applications, IEEE Press, Piscataway (1994)
14. Niven, I.S., Zuckerman, H.: An Introduction to the Theory of Numbers. John Wiley Sons, Chichester (1980)

Sequences, DFT and Resistance against Fast Algebraic Attacks

(Invited Paper)

Guang Gong

Department of Electrical and Computer Engineering
University of Waterloo
Waterloo, Ontario N2L 3G1, Canada
ggong@calliope.uwaterloo.ca

Abstract. The discrete Fourier transform (DFT) of a boolean function yields a trace representation or equivalently, a polynomial representation, of the boolean function, which is identical to the DFT of the sequence associated with the boolean function. Using this tool, we investigate characterizations of boolean functions for which the fast algebraic attack is applicable. In order to apply the fast algebraic attack, the question that needs to be answered is that: for a given boolean function f in n variables and a pair of positive integers (d, e), when there exists a function g with degree at most d such that $h = fg \neq 0$ where h's degree is at most e. We give a sufficient and necessary condition for the existence of those multipliers of f. An algorithm for finding those multipliers is given in terms of a polynomial basis of 2^n dimensional space over $\mathbb{F}_2$ which is established by an arbitrary m-sequence of period $2^n - 1$ together with all its decimations and certain shifts. We then provide analysis for degenerated cases and introduce a new concept of resistance against the fast algebraic attack in terms of the DFT of sequences or boolean functions. Some functions which made the fast algebraic attack inefficient are identified.

Keywords: Discrete Fourier transform, fast algebraic attack, stream ciphers, LFSR, m-sequences, polynomials, bases, trace representations.

1 Introduction

Linear feedback shift register (LFSR) sequences are widely used as basic functional blocks in key stream generators in stream cipher models due to their fast implementation in hardware as well as in software in some cases. In LFSR based stream ciphers, there are mainly two types of operations which operate on the LFSRs: (a) outputs of one LFSR or multiple LFSRs are transformed by nonlinear functions with or without memory states (including those by using mixed finite field operations and integer modular operations); (b) change the clock of the LFSRs to an irregular clock or by deleting some output bits of the LFSRs. In theory, examples include filtering sequence generators (i.e., operate a boolean

S.W. Golomb et al. (Eds.): SETA 2008, LNCS 5203, pp. 197–218, 2008.
© Springer-Verlag Berlin Heidelberg 2008

function on one LFSR) and combinatorial sequences (i.e., operating a boolean function on multiple LFSRs), the clock control sequences and shrinking generators [12, 23]. In practice, such key steam generators include E0 [4], and most of the submissions in EStream Project [11]. In this type of key stream generators, the initial states the LFSRs are served as a cryptographic key in each communication session. The goal of an attack is to recover the key, i.e., initial states of the LFSRs, from some known bits of the key stream, then generates the rest of bits of the key stream used in that particular session. There are many attacks proposed in the literation for the LFSR based key stream generators. In this paper, we consider those attacks related to solve a system of linear equations, i.e., algebraic attacks and fast algebraic attacks.

1.1   Linearizations, Algebraic Attacks, Fast Algebraic Attacks, and Selective DFT Attacks

Algebraic attacks and fast algebraic attacks have been shown as an important cryptanalysis method for symmetric-key cryptographical systems in recent work by Shamir, Patrin, Courtois and Klimov [24], Courtois and Meier [8], and Courtois [7, 9]. Especially, it significantly improved efficiency of attacks on stream cipher systems in which key streams are generated by linear feedback shift register based systems. Those attacks usually contain three steps: (a) pre-computation, (b) substitution for establishing a system of linear equations over $\mathbb{F}_2$ or $\mathbb{F}_{2^n}$ from known key stream bits, and (c) solving the system.

A. Equations with Unknown Keys. In a stream cipher model, a ciphertext is a bit stream $c_0, c_1, \ldots$, obtained by exclusive-or a message bit stream $m_0, m_1, \ldots$, with the key stream $s_0, s_1, \ldots$, i.e., $c_i = m_i + s_i, i = 0, 1, \ldots$, in $\mathbb{F}_2$. One of strong attacks comes from *known plaintext attacks*, i.e., if a certain plaintext is known, then some bits of $\{s_t\}$ can be recovered. If the key can be recovered from those known bits of $\{s_t\}$, then the rest of bits of the key stream, i.e., all bits of $\{s_t\}$, can be reconstructed. Assume that $K = (k_0, \ldots, k_{n-1})$ be an initial state of an LFSR or a concatenation of the initial states of several LFSRs. In general, the key stream $\{s_t\}$ can be written as

$$s_t = f_t(k_0, k_1, \ldots, k_{n-1}), \quad t = 0, 1, \ldots, m - 1 \tag{1}$$

or

$$W_t(s_0, \ldots, s_{t-1}, k_0, \ldots, k_{n-1}) = 0, \quad t = 0, 1, \ldots, M - 1 \tag{2}$$

where f_t (or W_t) is a function given by f (or W), a boolean function in n variables, and L^t, where L is the (left) shift operator of a binary stream, and $L^t \mathbf{a}$ is a resulting sequence shifted by t bits from $\mathbf{a} = a_0, a_1, \ldots$. In the paper, we focus on the case (1).

B. Linearization and Algebraic Attacks. The system of the equations (1) can be linearized when each monomial in $k_{i_1} \ldots k_{i_s}$ is treated as a variable. The number of unknowns in (1) is varied, but it is dominated by the degree of f.

For filtering function generators, i.e., apply f on m tap positions of an LFSR of degree n, the number of unknowns in (1) is $T_{deg(f)}$ where $deg(f)$ is the degree of f and T_j is defined as

$$T_j = \sum_{i=0}^{j} \binom{n}{i}. \tag{3}$$

The algebraic attack is to multiply f by a function g with a degree lower than that of f such that the product fg is zero, see Courtois and Meier [8]. In other words, using g with $deg(g) < deg(f)$ such that $fg = 0$, we have the system of the linear equations as follows

$$s_t g_t(K) = 0, t = 0, 1, \ldots. \tag{4}$$

Thus, the number of the unknowns of (4) is now dominated by the degree of g instead of f. For example, in the filtering generators, this is equal to $T_{deg(g)}$ which is smaller than $T_{deg(f)}$. Compared with the system of the linear equations directly from linearization, the algebraic attack can reduce the complexity for solving a system of linear equations by decreasing the number of unknowns in the system of linear equations as well as reducing the number of the required known bits of the key stream.

C. Algebraic Immunity. From this result, study for algebraic immunity of boolean functions is in fashion for resistance against this attack, see [6, 10, 15, 17, 18, 19], just list a few. Let $\mathcal{B}_n$ be the set consisting of all boolean functions in m variables. The algebraic immunity of f is defined as the smallest degree $deg(g)$ such that $fg = 0$ or $(1 + f)g = 0$, denoted by $AI(f)$, i.e.,

$$AI(f) = \min_{g \in Ann(f)} deg(g), \text{ where } Ann(f) = \{g \in \mathcal{B}_n \mid fg = 0 \text{ or } g(f+1) = 0\}, \tag{5}$$

see Meier, Pasalic and Carlet in [17].

D. Fast Algebraic Attacks. In 2003, Courtois [9] also proposed the fast algebraic attack (FAA) on stream ciphers to accelerate the algebraic attack by introducing linear relations among the key stream bits. The consideration is whether we can find some g such that $h = fg$ where $deg(g) < AI(f)$. If so, one could further reduce the number of the unknowns in the linear equations. The reason is that from $h = fg$, we get $f(g + h) = 0$. So that $g + h \in Ann(f)$. Thus $deg(g + h)$ may be greater than $AI(f)$. FAA consists of two steps: (a) find a boolean function g with $d = deg(g) < deg(f)$ such that the product $h = fg \neq 0$ with $e = deg(h) > 0$; (b) compute $q(x)$ which is a linear recursive relation of the resulting sequence by replacing f in the same key stream generator, and apply $q(x) = \sum_i c_i x^i$ to $s_t g_t(K) = h_t(K)$ which results in

$$\sum_{i=0}^{r} c_i s_{i+t} g_{i+t}(K) = \sum_{i=0}^{r} c_i h_{i+t}(K) \tag{6}$$

where $\sum_{i=0}^{r} c_i h_{i+t}(K) = 0, t = 0, 1, \ldots$ in the original paper [9], and we have $\sum_{i=0}^{r} c_i h_{i+t}(K) \neq 0$ with the same unknowns in the left-hand side of (6) in the variant introduced by Armknecht and Ars [3]. The efficiency of the pre-computation and substitution in the fast algebraic attack is improved by Hawkes and Rose in [16] for the case of filtering sequence generators and Armknecht in [2] for combinatorial generators with or without memory. According to the above analysis, the number of unknowns in (6) is smaller than the number of unknowns in the algebraic attack. But in FAA, the known bits of the key stream should be consecutive. However this condition is not needed in the algebraic attack.

E. Selective DFT Attacks. More recently, Rønjom and Helleseth [20] proposed a method to recover an initial state in a filtering sequence generator by reducing the number of unknowns in the system of linear equations to the minimum, which is the degree of the LFSR by an increased complexity in the both pre-computation and substitution, especially, in the substitution step. Shortly after that, the work [21, 22] improved the efficiency of the pre-computation and substitution in terms of forming a system of linear equations over $\mathbb{F}_{2^n}$ with n unknowns instead of linear equations over $\mathbb{F}_2$ where n is the degree of the LFSR, and reduced the number of the required consecutive bits of the filtering sequence to the linear span of the sequence. Recently, these authors proposed a new attack by multiplying s_t by a sequence, say $\mathbf{b} = \{b_t\}$, with the linear span less than the linear span of the sequence $s_0, s_1, \ldots$, i.e., to recover the key from the relation $u_t = s_t b_t, t = 0, 1, \ldots$, which is referred to as the selective discrete Fourier transform (DFT) attack. The select DFT attack results in either a more efficient attack than FAA or it can work for the case that the number of the known consecutive bits of the key stream is too small to apply FAA by requiring the exact linear span of the key stream sequence $\{s_t\}$ and their respective DFT spectra of $\{b_t\}$ and $\{u_t\}$, which are not needed in FAA.

1.2   When Is the Fast Algebraic Attack Applicable?

In FAA, $h = fg$ where $deg(g)$ determines the number of the unknowns in the system of the linear equations in (6), and $deg(h)$ governs the number of the required consecutive bits of the sequence $\{s_t\}$ for establishing the system (6). These two parameters provides a trade-off between the number of the unknowns and the number of the required consecutive bits. Thus, in order to apply FAA in a way which is more efficient than the algebraic attack, one should investigate the following question: For a given boolean function f in n variables, and a pair of two positive integers (d, e), does there exist some boolean function g with $deg(g) \leq d$ such that $h = fg$ with $deg(h) \leq e$. The following assertion is given by Courtois and Meier for answering this question.

> **Assertion A. (Theorem 6.0.1 in [8], Theorem 7.2.1 in [9])** With the above f, d and e, there exists a boolean function g in n variables such that $h = fg$ with $deg(g) \leq d$ and $deg(h) \leq e$ when $d = \lceil \frac{n}{2} \rceil$ and $e = \lceil \frac{n+1}{2} \rceil$ in [8] where $\lceil x \rceil$ denotes the smallest integer that is greater or equal to x, or when $d + e \geq n$ in [9].

This sufficient condition is popularly recited in the literation for the study of algebraic immunity of boolean functions (for example [17]). In this paper, using an approach of the discrete Fourier transform (DFT) of boolean functions, we investigate the characterizations of the existence of those multipliers. For a given boolean function f in n variables and two positive integers d and e, we observed that the sufficient condition $d+e \geq n$, shown in Assertion A cannot guarantee the existence of a function g with $deg(g) \leq d$ such that $fg \neq 0$ with $deg(fg) \leq e$. Pursuing this approach, we find a sufficient and necessary condition for the existence of such a multiplier g. From this sufficient and necessary condition, we obtain an algorithm to construct these multipliers. The other fascinating result from this characterization is that there exist degenerated cases for which FAA lost its advantages. We then introduce the concept of *resistance against* FAA.

This paper is organized as follows. In Section 2, we introduce the (discrete) Fourier transform of boolean functions and sequences, their polynomial representations, and bases of a 2^n dimensional linear space in terms of m-sequences together with their decimations and shifts. In Section 3, we first present a sufficient condition for existence of multiplier g with $d = deg(g)$ such that $h = fg \neq 0$ with $deg(h) \leq e$ for a given function f and a pair of positive integers (d, e), then a sufficient and necessary condition, and thirdly, an algorithm for constructing such multipliers. A characterization of those multipliers through the Reed-Muller code is also provided. In Section 4, we discuss the degenerated cases and introduce the concept of resistance against FAA. Section 5 is devoted to conclusions and some discussions on trade-offs between polynomial representations and boolean representations.

2   Discrete Fourier Transform (DFT) of Boolean Functions and Sequences and Their Polynomial Bases

We use the following notation throughout the paper.

- $\mathbb{F}_q = GF(q)$, a finite field with q elements, and $\mathbb{F}_2^m = \{(x_0, x_1, \ldots, x_{m-1}) | x_i \in \mathbb{F}_2\}$ where m is a positive integer.
- A *boolean function* f in n variables is a function from $\mathbb{F}_2^n$ to $\mathbb{F}_2$. The algebraic normal form of f is given by

$$f(x_0, \ldots, x_{n-1}) = \sum a_{i_1, \ldots, i_t} x_{i_1} \cdots x_{i_t}, a_{i_1, \ldots, i_t} \in \mathbb{F}_2$$

  where the sum runs through all the t-subset $\{i_1, \ldots, i_t\} \subset \{0, 1, \ldots, n-1\}$. The *degree* of the boolean function f is the largest t for which $a_{i_1, \ldots, i_t} \neq 0$. $\mathcal{B}_n$ denotes the set of all boolean functions in n variables.
- If $S = \{e_0, \ldots, e_{r-1}\}$ is a linearly independent set of a linear space over $\mathbb{F}_2$, then $\langle S \rangle$ is a subspace generated by S.
- Let $S \subset \mathcal{B}_n$, then S_0 is the set consisting of functions in S with zero constant term.
- Let S and T be two subsets of $\mathbb{F}_2^m$, then $S + T = \{s + t | s \in S, t \in T\}$ and $ST = \{st | s \in S, t \in T\}$ where the addition $s+t$ and the multiplication st are

the term-by-term addition and multiplication, respectively. The elements 0 or 1 is the all zero or one vector in $\mathbb{F}_2^m$ depending on the context. In particular, if $s = 1 \in \mathbb{F}_2^m$, $1 + T$ is the complement of T, denoted as $\overline{T}$.

In this paper, the notation $w(s)$ represents the Hamming weight of s, i.e., the number of nonzeros in s, where s could be a positive integer represented in a binary form, or a k-dimensional binary vector, or a boolean function in n variables represented as a binary vector $(f(\mathbf{x}_0), f(\mathbf{x}_1), \ldots, f(\mathbf{x}_{2^n-1}))$ where $\mathbf{x}_i, i = 0, \ldots, 2^n - 1$ constitutes all elements in $\mathbb{F}_2^n$.

Note that $\mathbb{F}_2^n$ is isomorphic to the finite field $\mathbb{F}_{2^n}$, regarded as a linear space over $\mathbb{F}_2$ of dimension n. Any boolean function can be represented by a polynomial function from $\mathbb{F}_{2^n}$ to $\mathbb{F}_2$. For a polynomial function from $\mathbb{F}_{2^n}$ to $\mathbb{F}_2$, say $f(x) = \sum_{i=0}^{2^n-1} d_i x^i, d_i \in \mathbb{F}_{2^n}$, the *algebraic degree* of f is given by $\max_{i:d_i \neq 0} w(i)$. In this paper, the degree of a function f from $\mathbb{F}_{2^n}$ to $\mathbb{F}_2$ always means the degree of its boolean form or equivalently, the algebraic degree of its polynomial form, denoted by $deg(f)$.

2.1   Linear Feedback Shift Register (LFSR) Sequences

Let $v(x) = x^n + c_{n-1}x^{n-1} + \cdots + c_1 x + 1$ be a polynomial over $\mathbb{F}_2$. A sequence $\mathbf{a} = \{a_t\}$ is an *LFSR sequence* of degree n if it satisfies the following recursive relation

$$a_{n+t} = \sum_{i=0}^{n-1} c_i a_{t+i}, t = 0, 1, \ldots, \qquad (7)$$

$(a_0, \ldots, a_{n-1})$ is an *initial state* of the LFSR, $v(x)$ is called a *characteristic polynomial of* $\mathbf{a}$, the *reciprocal* of $t(x)$ is referred to as a *feedback polynomial of* $\mathbf{a}$, and we also say that $\mathbf{a}$ is *generated* by $v(x)$. The *minimal polynomial* of $\mathbf{a}$ is the characteristic polynomial with the smallest degree. The sequence $\mathbf{a}$ is an *m-sequence* if $v(x)$ is primitive over $\mathbb{F}_2$ (Golomb, 1954 [13]). For example, $\mathbf{a} = 1001011$ is an m-sequence of period 7 where $t(x) = x^3 + x + 1$.

Let L denote the (left) shift operator of a sequence $\mathbf{a}$, i.e., $L\mathbf{a} = a_1, a_2, \ldots$, $L^i \mathbf{a} = a_i, a_{i+1}, \ldots$. A k-*decimation* of $\mathbf{a}$ is the sequence $\mathbf{a}^{(k)} = \{a_{kt}\}_{t \geq 0}$, where the indices are reduced modulo N where N is the (least) period of $\mathbf{a}$.

2.2   DFT and Trace Representations of Boolean Functions, Polynomial Functions, and Sequences

A. DFT of Boolean Functions. The content of this subsection can be found in Chapter 6 in [14]. Using the Lagrange interpolation, we may define the (discrete) Fourier transform of boolean functions through their polynomial forms. Let f be a boolean function in n variables in a polynomial form. The *(discrete) Fourier Transform (DFT)* of f is defined as

$$F_k = \sum_{x \in \mathbb{F}_{2^n}^*} [f(x) + f(0)] x^{-k}, \ k = 0, 1, \ldots, 2^n - 1. \qquad (8)$$

The inverse DFT of f is given as follows:

$$f(x) = \sum_{k=0}^{2^n-1} F_k x^k, \quad x \in \mathbb{F}_{2^n}^*. \tag{9}$$

Fact 1 *(Lemma 6.3, [14])* $F_k \in \mathbb{F}_{2^n}$, and $F_{2^i k} = F_k^{2^i}, i = 0, 1, \ldots, n-1$.

B. Trace Representation. In the following, we show the trace representation of boolean functions in terms of their DFT. For doing so, we need the concept of cyclotomic cosets. A *(cyclotomic) coset* C_s modulo $2^n - 1$ is defined by

$$C_s = \{s, s \cdot 2, \ldots, s \cdot 2^{n_s-1}\},$$

where n_s is the smallest positive integer such that $s \equiv s2^{n_s} \pmod{2^n - 1}$. The subscript s is chosen as the smallest integer in C_s, and s is called the *coset leader* of C_s. For example, for $n = 4$, the cyclotomic cosets modulo 15 are:

$$C_0 = \{0\}, C_1 = \{1, 2, 4, 8\}, C_3 = \{3, 6, 12, 9\}, C_5 = \{5, 10\}, C_7 = \{7, 14, 13, 11\}$$

where $\{0, 1, 3, 5, 7\}$ are coset leaders modulo 15.

We then group monomial terms according to Fact 1, which results in a sum of trace monomial terms (see Theorem 6.1 [14]).

Proposition 1. (TRACE REPRESENTATION OF FUNCTIONS OF BOOLEAN FUNCTIONS) *Any non-zero function $f(x)$ from $\mathbb{F}_{2^n}$ to $\mathbb{F}_2$ can be represented as*

$$f(x) = \sum_{k \in \Gamma(n)} Tr_1^{n_k}(F_k x^k) + F_{2^n-1} x^{2^n-1}, F_k \in \mathbb{F}_{2^{n_k}}, F_{2^n-1} \in \mathbb{F}_2, x \in \mathbb{F}_{2^n} \tag{10}$$

where $\Gamma(n)$ is the set consisting of all coset leaders modulo $2^n - 1$, $n_k \mid n$ and n_k is the size of the coset C_k, and $Tr_1^{n_k}(x)$ the trace function from $\mathbb{F}_{2^{n_k}}$ to $\mathbb{F}_2$.

Recall that $w(k)$ denotes the Hamming weight of a positive integer k. If f is a boolean function with degree $deg(f) = r < n$, from Proposition 1, the trace representation of f is given by

$$f(x) = \sum_{w(k) \leq r} Tr_1^{n_k}(F_k x^k), \tag{11}$$

where $F_k \neq 0$ for at least one $k \in \Gamma(n)$ such that $w(k) = r$.

C. Relation to DFT of Sequences. Next we investigate the relationship between the DFTs of a boolean function and its associated binary sequence. Let α be a primitive element in $\mathbb{F}_{2^n}$. We assume that $f(0) = 0$ (if $f(0) \neq 0$, we replace f by $g = f - f(0)$ where $g(0) = 0$). We associate f with a binary sequence $a_f = \{a_t\}$ whose elements are given by

$$a_t = f(\alpha^t), t = 0, 1, \ldots, 2^n - 2. \tag{12}$$

Then the period of $\{a_t\}$, say N, is a factor of $2^n - 1$. The discrete Fourier transform of a_f is defined by

$$A_k = \sum_{t=0}^{2^n-2} a_t \alpha^{-tk}, 0 \le k < 2^n - 2 \tag{13}$$

and the inverse DFT is

$$a_t = \sum_{k=0}^{2^n-2} A_k \alpha^{kt} = \sum_{k \in \Gamma(n)} Tr_1^{n_k}(A_k \alpha^{kt}), 0 \le t < 2^n - 2. \tag{14}$$

By selecting α as a root of the primitive polynomial which defines $\mathbb{F}_{2^n}$ for computing the DFT of f in (8), then we have

$$A_k = F_k, 0 \le k < 2^n - 2.$$

In the following, we keep the notation $\{A_k\}$ for both f and a_f, since they are equal (with respect to α) for $1 \le k < 2^n - 1$ except for $A_0 = F_0 + F_{2^n-1}$. However, this difference does not effect any discussions proceeded in this paper.

Note that the spectral sequence $\{A_k\}$ has period $N \,|\, 2^n - 1$. Therefore, any function from $\mathbb{F}_{2^n}$ to $\mathbb{F}_2$, or equivalently a boolean function in n variables, corresponds a binary sequence with period $N \,|\, 2^n - 1$. Thus a boolean function and its associate sequence, given by (12), are related by their identical DFT spectra which results in the same polynomial representation in (10). This leads to another linearized method for finding multipliers g of f such that $fg \ne 0$, which will be shown in the next section.

D. DFT Convolution and Product Sequences

- Let **a** and **b** be two sequences of period N with their respective DFTs $A = \{A_k\}$ and $B = \{B_k\}$.
- For the term-by-term product $\mathbf{c} = \mathbf{a} \cdot \mathbf{b}$ where $c_t = a_t b_t$, $0 \le t < N$, let the DFT of $\{c_t\}$ be $C = \{C_k\}$. Then C is a convolution of A and B, denoted as $C = A * B$ where

$$C_k = \sum_{i+j=k(\bmod N)} A_i B_j, 0 \le k < N. \tag{15}$$

From the above treatment, we will not distinguish the set $\mathcal{F}_n$, the set consisting of all functions from $\mathbb{F}_{2^n}$ to $\mathbb{F}_2$ and the set $\mathcal{B}_n$, the set of all boolean functions in n variables. For the theory of sequences and their trace representations, the reader is referred to [14].

3   Polynomial Bases of $\mathcal{F}_n$ in Terms of LFSR Sequences

In this section, we introduce polynomial bases of $\mathcal{F}_n$ after we give a review for its boolean bases, and show a method for computing those bases in terms of applying shift and decimation operators to an m-sequence. We conclude this section by pointing out a relationship between the Reed-Muller code and the sequences with index bounded spectral sequences.

Note that $\mathcal{F}_n$ can be considered as a linear space of dimension 2^n over $\mathbb{F}_2$ when each function in $\mathcal{F}_n$ is represented by a binary vector of dimension q. For $f \in \mathcal{F}_n$, there are two common ways to represent f as a binary vector of dimension q ($q = 2^n$). One is the so-called the boolean bases, reviewed as follow.

Let f be of a boolean representation. We list the elements in $\mathbb{F}_2^n$ in the same order as the truth table of f. Thus,

$$f(x_0, x_1, \ldots, x_{n-1}) = (f(\mathbf{t}_0), f(\mathbf{t}_1), \ldots, f(\mathbf{t}_{q-1})),$$

where $\mathbf{t}_i = (t_{i,0}, t_{i,1}, \ldots, t_{i,n-1})$, $t_{ij} \in \mathbb{F}_2$ with $i = t_{i,0} + t_{i,1}2 + \cdots + t_{i,n-1}2^{n-1}, 0 \le i < q$.

For $\mathbf{x} = (x_0, \ldots, x_{n-1})$ and $\mathbf{c} = (c_0, \ldots, c_{n-1})$ in $\mathbb{F}_2^n$, we denote

$$\mathbf{x}^{\mathbf{c}} = x_0^{c_0} x_1^{c_1} \cdots x_{n-1}^{c_{n-1}}.$$

Then, the basis Δ of $\mathcal{F}_n$, regarded as a linear space over $\mathbb{F}_2$, consists of all monomial terms:

$$\Delta = \{\mathbf{x}^{\mathbf{c}} \mid \mathbf{c} \in \mathbb{F}_2^n\}.$$

This basis is referred to as a *boolean basis* of $\mathcal{F}_n$.

3.1   Polynomial Bases

Let f be of the polynomial form. We use the cyclic multiplicative group of $\mathbb{F}_{2^n}$, i.e.,

$$f(x) = (f(0), f(1), f(\alpha), \ldots, f(\alpha^{2^n-2})) = (f(0), a_0, a_1, \ldots, a_{2^n-2}) \qquad (16)$$

where $a_t = f(\alpha^t)$ is defined by (14). Let

$$\Pi_k = \{Tr_1^{n_k}(\beta_k(\alpha^i x)^k) \mid, i = 0, 1, \ldots, n_k - 1\}, \beta_k \in \mathbb{F}_{2^{n_k}} \qquad (17)$$

where n_k is the size of the coset containing k (how to select β_k will be given in the next subsection). Note that $\{\alpha^{ik} \mid i = 0, 1, \ldots, n_k - 1\}$ is a basis of $\mathbb{F}_{2^{n_k}}$ over $\mathbb{F}_2$, so is $\{c\alpha^{ik} \mid i = 0, 1, \ldots, n_k - 1\}$ for any nonzero $c \in \mathbb{F}_{2^{n_k}}$. From Proposition 1, any function in $\mathcal{F}_n$ can be represented as a sum of the trace monomial terms. For each trace monomial term $Tr_1^{n_k}(A_k x^k)$, since $A_k \in \mathbb{F}_{2^{n_k}}$, we have $A_k = \sum_{i=0}^{n_k-1} c_i \beta_k \alpha^{ik}, c_i \in \mathbb{F}_2$. Using the linear property of the trace function, we have

$$Tr_1^{n_k}(A_k x^k) = \sum_{i=0}^{n_k-1} Tr_1^{n_k}(c_i \beta_k(\alpha^i x)^k).$$

Thus, we have showed the following result.

Proposition 2. *Any trace monomial term can be represented as a linear combination of functions in Π_k for some k which is a coset leader modulo $2^n - 1$, and the following set is a basis of $\mathcal{F}_n$:*

$$\Pi = \bigcup_{k \in \Gamma(n)} \Pi_k. \tag{18}$$

This is referred to as a polynomial basis of $\mathcal{F}_n$.

Remark 1. Let $\mathbf{a}_k = \{a_{kt}\}_{t \geq 0}$. Then $\mathbf{a}_k$ is an LFSR sequence, generated by f_{α_k}, the minimal polynomial of α^k. Let $\{A_{k,j}\}$ be the DFT of $\mathbf{a}_k$. Then

$$A_{k,j} = \begin{cases} A_k & \text{if } j = k \\ 0 & \text{otherwise.} \end{cases}$$

In other words, the trace representation of $\mathbf{a}$ can be considered as a direct sum of the LFSR sequences with irreducible minimal polynomials for which the DFT sequences of any two of them are orthogonal.

3.2   Efficient Computation of the Polynomial Bases of $\mathcal{F}_n$

In the following, we assume that $\mathbf{a} = \{a_t\}$ is generated by $v(x)$ which is primitive over $\mathbb{F}_2$ of degree n, i.e., $\mathbf{a}$ is an m-sequence of degree n. Then the period $\mathbf{a}$ is $N = 2^n - 1$. Let α be a root of $v(x)$ in $\mathbb{F}_{2^n}$. Then we have $a_t = Tr_1^n(\beta\alpha^t)$ where $\beta \in \mathbb{F}_{2^n}^*$. If $\mathbf{a}^{(k)} = \mathbf{0}$, the k-decimation of $\mathbf{a}$, then we may choose r as the smallest integer such that the k-decimation of $L^i\mathbf{a}$ is a zero sequence for $i = 0, 1, \ldots, r - 1$, but the k-decimation of $L^r\mathbf{a}$ is not a zero sequence. Let β_k be the element corresponding to this shift (i.e., β_k in Π_k, defined in (17)). For simplicity, we still denote such a decimation as $\mathbf{a}^{(k)}$. Therefore, the elements of $\mathbf{a}^{(k)}$ are given by $a_{kt} = Tr_1^{n_k}(\beta_k\alpha^{kt}), t = 0, 1, \ldots$, where n_k is the size of the coset C_k. We denote

$$P_k = \begin{bmatrix} 0, & \mathbf{a}^{(k)} \\ 0, & L\mathbf{a}^{(k)} \\ \vdots & \\ 0, & L^{n_k-1}\mathbf{a}^{(k)} \end{bmatrix} \tag{19}$$

where $L^i\mathbf{a}^{(k)}$'s are regarded as binary vectors of dimension $2^n - 1$, and each row corresponds to a function in Π_k. Thus, for each k, a coset leader modulo $2^n - 1$, we only need to compute $\mathbf{a}^{(k)}$, the rest of the rows in P_k can be obtained by the shift operator which has no cost. Furthermore, $|\Gamma(n)|$, the number of the coset leaders modulo $2^n - 1$, is equal to the number of the irreducible polynomials over $\mathbb{F}_2$ of degrees dividing n. Thus, for computing the polynomial basis of $\mathcal{F}_n$, one only need to compute $|\Gamma(n)|$ decimation sequences from $\mathbf{a}$, approximately, $2^n/n$ decimation sequences from $\mathbf{a}$, in order to get the polynomial basis of $\mathcal{F}_n$, compared it with the boolean basis, where we have to compute 2^n vectors of dimension 2^n since there is no computational saving for these evaluations.

There is a relation between the Reed-Muller code and the row vectors of P_k, which will be introduced in the next section.

4   Characterizations of Boolean Functions with Low Degree Multipliers

Assume that a boolean function f employed in a cipher system is not of low degree.

Definition 1. *Given f a boolean function in n variables, i.e., $f \in \mathcal{F}_n$, a pair of positive integers $d < deg(f)$ and e, if there exists some function g with $deg(g) \le d$ such that the product $h = fg \ne 0$ or $h = (f+1)g \ne 0$ with $deg(h) \le e$, then g is said to be a low degree multiplier of f.*

In this section, we will show how to characterize those multipliers. We first present a sufficient condition for the existence of low degree multipliers. Following this approach, we establish a sufficient and necessary condition for the existence of low degree multipliers, which also yields an algorithm to construct them. It follows that Assertion A in Section 1 is not true from the results we derived in this section. Let

$$S_d = \{ \text{ row vectors of } P_k \,|\, k \in \Gamma(n), w(k) \le d\} \tag{20}$$

where P_k is defined by (19). Then S_d can be considered as either the set consisting of all the functions in the polynomial basis with $w(k) \le r$ or the set consisting of all boolean monomial terms of degrees less than or equal to d, i.e., we can rewrite S_d as follows.

$$S_d = \bigcup_{w(k) \le d} \Pi_k$$

where Π_k is defined in (17). Then we have $|S_d| = T_d$, defined by (3) in Section 1. Notice that any function in $\mathcal{F}_n$ of degree d is a linear combination of functions in S_d over $\mathbb{F}_2$. From Section 3.2, the sequence version of S_d, still denoted by S_d, is given by

$$S_d = \{L^i \mathbf{a}^{(k)} \,|\, 0 \le i < n_k, w(k) \le d\} \tag{21}$$

where n_k is the size of the coset C_k or the degree of the minimal polynomial of $\mathbf{a}^{(k)}$. For a given $f \in \mathcal{F}_n$, we denote

$$fS_d = \{f(x) \cdot g(x)|g \in S_d\}. \tag{22}$$

From (21), the sequence version of (22) is given by

$$\mathbf{u}S_d = \{\mathbf{u} \cdot \mathbf{b} \,|\, \mathbf{b} \in S_d\} \tag{23}$$

where $\mathbf{u}$ is the sequence associated with $f(x)$, defined by (16) in Section 2, and the multiplication of two sequences is the term-by-term multiplication. (Note. We keep the notation S_d for both functions and sequences, and the exact meaning depends on the context.)

Theorem 1. *With the above notation, for a given* $f \in \mathcal{F}_n$ *and two positive integers* d *and* e *with* $1 \leq d, e < n$, *if* fS_d *contains at least*

$$2^n + 1 - |S_e|$$

linear independent functions over $\mathbb{F}_2$, *then there exists a function* $g \in \mathcal{F}_n$ *with degree at most* d *such that* $h = fg \neq 0$ *with* $\deg(h) \leq e$.

Proof. Note that $e < n$ implies that $|S_e| < |S_n| = 2^n$, thus $2^n + 1 - |S_e| > 1$. Let $r = |fS_d|$ and $s = |S_e|$, then $r > 1$. Since $r + s > 2^n$ and the dimension of $\mathcal{F}_n$ is 2^n, the vectors in $fS_d \cup S_e$ are linearly dependent over $\mathbb{F}_2$. We list the elements in fS_d as $fg_1, \ldots, fg_r$, and the elements in S_e as $h_1, \ldots, h_s$. Then there exist $c_i \in \mathbb{F}_2, i = 1, \ldots, r + s$, which are not all zeros, such that

$$\sum_{i=1}^{r} c_i f g_i + \sum_{i=1}^{s} c_{r+i} h_i = 0. \tag{24}$$

Note that S_e is linearly independent over $\mathbb{F}_2$. Thus, there are some j where $1 \leq j \leq r$ and $i > r$ such that $c_j \neq 0$ and $c_i \neq 0$. We write $g = \sum_{i=1}^{r} c_i g_i$ and $h = \sum_{i=1}^{s} c_{r+i} h_i \neq 0$. Hence $fg = h$ where $\deg(g) \leq d$ and $\deg(h) \leq e$. □

Remark 2. The condition in Theorem 1 implies that $d + e \geq n$. Otherwise $|S_d| + |S_e| \leq 2^n$, therefore $|fS_d| + |S_e| \leq 2^n$, which is a contradiction. However, the converse is not true.

Note that it is possible that $|fS_d| < |S_d|$ and the elements in fS_d are linearly dependent over $\mathbb{F}_2$. In this case, possibly, there is no function g with $\deg(g) \leq d$ such that $fg \neq 0$ and $\deg(fg) \leq e$.

Example 1. Let $f(x) = Tr_1^3(\alpha^5 x + \alpha^6 x^3)$ be a function from $\mathbb{F}_{2^3}$ to $\mathbb{F}_2$ where α is a primitive element in $\mathbb{F}_{2^3}$ with $\alpha^3 + \alpha + 1 = 0$. Let $d = 2$ and $e = 1$, then $d + e = n$. The set S_2 contains the following seven functions

constant function c = 11111111 $Tr_1^3(x) = 01001011$

$Tr_1^3(\alpha x) = 00010111$ $Tr_1^3(\alpha^2 x) = 00101110$

$Tr_1^3(x^3) = 01110100$ $Tr_1^3((\alpha x)^3) = 01101001$

$Tr_1^3((\alpha^2 x)^3) = 01010011$

On the other hand, we have $f(x) = 00100001$. Thus the elements of fS_2 are

$$fTr_1^3(x) = fTr_1^3(\alpha x) = fTr_1^3((\alpha^2 x)^3) = 00000001$$
$$fTr_1^3(\alpha^2 x) = fTr_1^3(x^3) = 00100000$$
$$fc = fTr_1^3((\alpha x)^3) = 00100001$$

Thus $|fS_2| = 3$. However, $00100001 = 00000001 + 00100000$. Thus the elements of fS_2 are linearly dependent. The maximal linear independent set in fS_2 is given by $D = \{00000001, 00100001\}$. It is easy to see that $D \cup S_1$ is a linearly independent set. So there is no function g with degree 2 such that $h = fg \neq 0$ with $deg(h) = 1$.

This shows that the condition $d + e \geq n$ in Assertion A in Section 1 cannot guarantee that the existence of some function $g \neq 0$ with $deg(g) \leq d$ such that $h = fg \neq 0$ with $deg(h) \leq e$, since there exists a degenerated case for which $|fS_d| < |S_d|$.

Note that the boolean form of f is given by $f(x_0, x_1, x_2) = x_0x_1 + x_0x_2 + x_1x_2 + x_1$. We may use the boolean form of f to show the above result using a similar technique as above.

From the proof of Theorem 1, it is not necessary to require $r + s > 2^n$. In the following, we characterize this condition.

Theorem 2. (EXISTENCE OF LOW DEGREE MULTIPLIERS, *Function Version*) *With the above notation, for a given $f \in \mathcal{F}_n$ and $1 \leq d, e < n$, let D_d be a maximal linearly independent set of fS_d. Then there exists a function $g \in \mathcal{F}_n$ with degree at most d such that $h = fg \neq 0$ with $deg(h) \leq e$ if and only if $D_d \cup S_e$ is linearly dependent over $\mathbb{F}_2$.*

Proof. A proof for the sufficient condition is similar as we did for Theorem 1. We denote $|D_d| = t$ and $|S_e| = s$. Since $D_d \cup S_e$ is linearly dependent over $\mathbb{F}_2$, there exist $c_i \in \mathbb{F}_2, i = 1, \ldots, t + s$ such that

$$\sum_{i=1}^{r} c_i f g_i + \sum_{i=1}^{s} c_{t+i} h_i = 0$$

where $fg_i \in D_d$ and $h_i \in S_e$. Since both D_d and S_e are linear independent over $\mathbb{F}_2$, there exist some i with $1 \leq i \leq t$ and j with $1 \leq j \leq s$ such that $c_i \neq 0$ and $c_{t+j} \neq 0$, respectively. Thus $g = \sum_{i=1}^{t} c_i g_i \neq 0$ with $deg(g) \leq d$, and $fg = h = \sum_{j=1}^{s} c_{t+j} h_j \neq 0$ with $deg(h) \leq e$, which establishes the assertion.

Conversely, assume that there exists some $g \in \mathcal{F}_n$ with $deg(g) \leq d$ such that $fg \neq 0$ and $deg(fg) \leq e$. Since any function with degree less than or equal to d is a linear combination of functions in S_d, we may write $g = \sum_{i=1}^{l} d_i g_i$ where $d_i \in \mathbb{F}_2$ and $l = |S_d|$. For simplicity, we may assume that $g_i, i = 1, \ldots, t$ are functions in S_d such that $\{fg_i \mid i = 1, \ldots, t\}$ is a maximal linearly independent set of fS_d. Thus we have

$$fg = f \cdot \sum_{i=1}^{l} d_i g_i = \sum_{i=1}^{l} d_i (fg_i) = \sum_{i=1}^{t} e_i (fg_i) \neq 0, e_i \in \mathbb{F}_2.$$

Consequently, there exists some i with $1 \leq i \leq t$ such that $e_i \neq 0$. On the other hand, since $h = fg$ with degree $deg(h) \leq e$, then $h \in S_e$. Therefore, h is a linear combination of the functions in S_e, say $h = \sum_{i=1}^{s} k_i h_i \neq 0, k_i \in \mathbb{F}_2$. This shows that there is some $k_i \neq 0$ with $1 \leq i \leq s$. Therefore, we have

$$h = fg \iff \sum_{i=1}^{t} e_i f g_i + \sum_{i=1}^{e} k_i h_i = 0$$

where e_i's and k_i's are not all zero. Thus, $D_d \cup S_e$ is linear dependent over $\mathbb{F}_2$, which completes the proof.   □

From the sequence version of fS_d and the DFT of the sequences in S_j, we can express the result of Theorem 2 in a sequence version. Let $\{X_k\}$ be the DFT of $\{x_t\}$ in the following corollary.

Corollary 1. *(EXISTENCE OF LOW DEGREE MULTIPLIERS, Sequence Version)*
For a given binary sequence **u** *with period* $N \mid 2^n - 1$, *and a pair of integers* (d, e)
which satisfy that $1 \leq d, e < n$, *let* D_d *be a maximal linearly independent set*
of $\mathbf{u}S_d$. *Then there exists a sequence* **b** *with* $B_k = 0$ *for all* k *in the range of*
$w(k) > d$ *such that* $\mathbf{c} = \mathbf{ub} \neq 0$ *with* $C_k = 0$ *for all* k *in the range of* $w(k) > e$
if and only if $D_d \cup S_e$ *is linearly dependent over* $\mathbb{F}_2$.

Another characterization of low degree multipliers of f is through the Reed-Muller code. The Reed-Muller code of length 2^n and order r, $1 \leq r \leq n$, denoted as $R(r, n)$, is the set consisting of the binary sequences with trace representation $f(x)$ given by (11), i.e.,

$$R(r,n) = \{(f(0), f(1), f(\alpha), \ldots, f(\alpha^{2^n-2})) \mid f(x) = \sum_{w(k) \leq r} Tr_1^{n_k}(A_k x^k)\}. \quad (25)$$

Therefore,

$$R(r, n) = <S_r > \quad (26)$$

where S_r is defined in (20). Using (26), together with Theorem 2, the following assertion is immediate.

Corollary 2. *For a given* $f \in \mathcal{F}_n$, *and two integers* d *and* e, $1 \leq d, e < n$, *there*
exists a function g *with* $deg(g) \leq d$ *such that* $h = fg \neq 0$ *with* $deg(h) \leq e$ *if and*
only if the intersection of $fR(d, n)$ *and* $R(e, n)$ *contains functions which are not*
constant, i.e.,

$$\{0, 1\} \subset fR(d, n) \cap R(e, n) \quad \text{and} \quad |fR(d, n) \cap R(e, n)| > 2$$

where 0 *is understood as the all zeros vector or zero, which depends on the*
context, a similar notation for 1.

Note. For the cryptographic properties of Reed-Muller codes, the reader is referred to [5].

Note that there is an extremely degenerated case, i.e., $fS_d = \{f, 0\}$. From the proof of Theorem 2, in this case, for all nonconstant $g \in S_d$, $fg = 0$. Thus, we have established the following corollary for the case $fg = 0$.

Corollary 3. *For a given* $f \in \mathcal{F}_n$ *and* $0 < d < n$,

(a) *there exists some* $g \neq 0$ *with* $deg(g) \leq d$ *such that* $fg = 0$ *if and only if*
$|D| < |S_d|$, *and*

(b) the assertion that $fg = 0$ for all nonconstant $g \in \langle S_d \rangle_0$ is true if and only if $fS_d = \{f, 0\}$ where $\langle S_d \rangle_0$ is the subset of $\langle S_d \rangle$ which consists of the functions with the zero constant term.

Some questions for resistance against the fast algebraic attack (FAA) arise from the degenerated cases in Corollary 3. We will analysis those cases in depth and introduce some new concepts related to those new phenomena in the next section.

In the following, using Theorem 2, we provide an algorithm for determining whether there exists a function g with $deg(g) \leq d$ such that $fg \neq 0$ with $deg(fg) \leq e$ for a given f and two integers d and e with $1 \leq d, e < n$. If there exists such a multiplier g, the algorithm outputs g and fg. Otherwise, it outputs $g = 0$ and $fg = 0$. For simplicity, in the algorithm, if it is needed, a subset of $\mathcal{F}_n$ is also regarded as a matrix in which each row is a function in the set, represented as a 2^n-dimensional binary vector.

Algorithm 1. AN ALGORITHM FOR FINDING A LOW DEGREE MULTIPLIER

Input: $f \in \mathcal{F}_n$, a function from $\mathbb{F}_{2^n}$ to $\mathbb{F}_2$;

   $1 \leq d, e < n$; and

   $t(x) = x^n + \sum_{i=0}^{n-1} t_i x^i, t_i \in \mathbb{F}_2$, a primitive polynomial over $\mathbb{F}_2$ of degree n.

Output: $g \in \mathcal{F}_n$ with $deg(g) \leq d$ and $h = fg$ with $h \neq 0$ and $deg(h) \leq e$ if there exist such g and h. Otherwise, outputs $g = 0$ and $h = 0$.

PROCEDURE

1. *Randomly select an initial state $(a_0, a_1, \ldots, a_{n-1}), a_i \in \mathbb{F}_2$, and compute*

$$a_{n+i} = \sum_{j=0}^{n-1} t_j a_{j+i}, i = 0, 1, \ldots, 2^n - 1 - n.$$

   (Note $\mathbf{a} = (a_0, \ldots, a_{2^n-2})$ is a binary m-sequence of degree n.)
2. *Compute k, each coset leader modulo $2^n - 1$, and n_k, the size of C_k. Set $I = \{(k, n_k) | k \in \Gamma n\}$. (Recall that $\Gamma(n)$ is the set consisting of all coset leaders modulo $2^n - 1$.)*
3. *Set $m = max\{d, e\}$. Establish S_m:*
   $P_0 = (1, 1, \ldots, 1)$;
   for $0 \neq k$ in $\Gamma(n)$ with $w(k) \leq m$ do
   Compute $\mathbf{a}^{(k)} = (a_0, a_k, \ldots, a_{k(2^n-2)})$, a k-decimation of $\mathbf{a}$, then apply the shift operator to the decimated sequence.
   Loading P_k, defined by (19), as an $n_k \times 2^n$ sub-matrix of S_m for all k with $0 \leq w(k) \leq m$.
4. *Using the Gauss elimination, find the rank of fS_d, represented as an $|fS_d| \times 2^n$ matrix, and a maximal linearly independent set of fS_d, say D_d.*

5. *Apply the Gauss elimination to the following matrix whose entries are given by*

$$\begin{bmatrix} D_d \\ S_e \end{bmatrix}.$$

If the rank of the above matrix is equal to $t + s$ where $|D_d| = t$ and $|S_e| = s$, set $g = 0$ and $h = 0$, then go to Step 6. Otherwise, find $c_i, i = 1, \ldots, t$ such that

$$\sum_{i=1}^{t} c_i f g_i + \sum_{i=1}^{s} c_{t+i} h_i = 0.$$

Set $g = \sum_{i=1}^{t} c_i g_i$ and $h = \sum_{i=1}^{s} c_{t+i} h_i$.
6. *Return g and h.*

Remark 3. The algorithm can also be applied to a boolean representation of f. However, establishing S_m using the polynomial basis is more efficient than establishing S_m using the boolean function basis:

$$1, x_1, \ldots, x_n, x_1 x_2, \ldots, \mathbf{x^c}, \ldots, x_{i_1} x_{i_2} \ldots x_{i_m}, \text{ for all } \mathbf{c} \in \mathbb{F}_2^n \text{ with } w(\mathbf{c}) \leq m,$$

due to Step 3 which needs only the shift operation for the group of $n_k - 1$ rows of S_m for each $k \in \Gamma(n)$ with $w(k) \leq m$.

Remark 4. Algorithm 1 can be applied to find g such that $fg = 0$, i.e., the annihilators of f, when Step 5 is replaced by Step 5': If $|D| = |S_d|$, then return $g = 0$ which represents no annihilators with degree at most d. Otherwise, finding c_i such that $\sum_{i=1}^{t} c_i f g_i = 0$, set $g = \sum_{i=1}^{t} c_i g_i$, and return g. When Algorithm 1 is restricted to compute annihilators of f, the computational complexity of this algorithm is approximately determined by the cost for applying the Gaussian elimination algorithm to a $T_d \times 2^n$ matrix.

Remark 5. Any algorithm for finding annihilators in the boolean basis can be extended to compute low degree multipliers in the polynomial basis, for example, Algorithms 1 and 2 in [17]. We omit the details for the sake of space.

However, the algorithm that we proposed here is easier to characterize the degenerated cases and analysis of the resistance against FAA, which will be given in the next section. Another aspect is that it is easier to apply to those functions which directly are given in their polynomial forms.

5   Degenerated Cases and Resistance against FAA

In order to launch an efficient FAA attack, one needs to find a multiplier g with $deg(g) \leq d$ such that $h = fg \neq 0$ with $deg(h) \leq e$ for some pair (d, e) which is in favor of FAA.

Definition 2. *For a given function f and an integer pair (d, r), assume that there exists some function g with $deg(g) \leq d$ such that $h = fg \neq 0$ with $deg(h) \leq e$. Then we say that the pair (d, r) is an enable pair of f.*

5.1 Degenerated Cases

Case 1. $f S_d = \{f, 0\}$: Weak Functions.
From Corollary 3-(b), for given $f \in \mathcal{F}_n$ and d, a positive integer, there is an extremely degenerated case, i.e.,

$$fg = 0, \quad \text{for all } g \in< S_d >_0 \tag{27}$$

where $\langle S_d \rangle_0$ is defined as the subspace of $\langle S_d \rangle$ with zero constant term, i.e., $g(0) = 0$. Let M be a $T_d \times 2^n$ matrix for which the row vectors are given by P_k, $w(k) \le d$ (see (19) for the definition of P_k). We write $f = (f(0), x_0, \ldots, x_{2^n-2})$ where $x_i = f(\alpha^i)$. Then f satisfies the condition (27) if and only if

$$Mf = 0.$$

This is equivalent to that f is in the null space of M, denoted as V. Thus, any function in V satisfies (27). By the linear algebra, the dimension of V is given by $2^n - |S_d|$. On the other hand, the set consisting of all functions with degree $\le d$ is given by $\langle S_d \rangle = \langle S_d \rangle_0 \cup \langle S_d \rangle_1$, where $\langle S_d \rangle_1$ is the complement of $\langle S_d \rangle_0$. Therefore, for any $g_1 \in \langle S_d \rangle_0$, we have $f(g_1 + 1) = fg_1 + f = f$ since $fg_1 = 0$. Thus we have $g = g_1 + 1$ with $deg(g) = deg(g_1) \le d$ and $h = f$ with degree $deg(h) = deg(f)$. However, those functions have the lowest algebraic immunity, which is 1. Note that in this case we have $\langle S_d \rangle_0 = Ann(f)$, the set consisting of all functions g such that $fg = 0$ (see (5) in Section 1). Then any enable pair is given by $(i, deg(f))$ where $1 \le i \le d$. Those functions are very poor in terms of the resistance against both the algebraic attack and FAA. We summarize the above discussion in the following proposition.

Proposition 3. *Let V be the null space of M in $\mathbb{F}_2^{2^n}$.*

1. *For each $f \in V$, $fg = 0 \;\forall g \in< S_d >_0$.*
2. *All enable pairs of f are in the form of $(i, deg(f))$ where $1 \le i \le d$.*
3. *The algebraic immunity of f is equal to 1, and $Ann(f) =< S_d >_0$.*

Case 2. $|D_d| < |S_d|$ where D_d is the maximal independent subset of $f S_d$. From Theorem 2 and Corollary 3, if $|D_d| < |S_d|$, then we also have some enable pairs in the form of $(i, deg(f))$ with $1 \le i \le d$ for those $g \in \langle S_d \rangle$ such that $g = g_1 + 1$ where $fg_1 = 0, g_1 \in \langle S_d \rangle_0$. If $g \in \langle S_d \rangle_0$ such that $h = fg \neq 0$, there are two cases as follows.

(a) If $D_d \cup S_e$ are linearly dependent, then $deg(h) \le e$.
(b) If $D_d \cup S_e$ are linearly independent, then $deg(h) > e$.

Furthermore, any enable pair of f belongs to one of the following patterns:

$$(i, deg(f)) \text{ where } 1 \le i \le d \text{ or } (i, j) \text{ where } 1 \le i \le d; 1 \le j < n.$$

Case 3. $|D_d| = |S_d|$. In this case, for each $g \in \langle S_d \rangle$, $fg \neq 0$. It also has two phenomena as stated in (a) and (b) in Case 2.

From the above analysis for three cases of D_d, we formerly define a nondegenerated case for f. For a given a positive integer d, we say that f is *nondegenerated* with respect to d if $|D_d| = |S_d|$. In other words, f is nondegerated with respect to d if and only if all the elements in fS_d are distinct and linearly independent. We then obtain the condition which can remedy Assertion A, i.e., if f is nondegenerated with respect to d, then it is true, as stated below.

Proposition 4. *For given $f \in \mathcal{F}_n$, and a pair of positive integers (d, e), if f is nondegenerated with respect to d, then there exists a boolean function $g \in \mathcal{F}_n$ with $deg(g) \leq d$ such that $h = fg \neq 0$ with $deg(h) \leq e$ when $d + e \geq n$.*

Proof. Since D_d is a linearly independent set, then $|D_d| = |S_d| = T_d$ (recall T_j defined in (3)). If $T = D_d \cup S_e$ is independent, then $|T| > 2^n$ because $d + e \geq n$. Since T is a subset of $\mathbb{F}_2^{2^n}$, it has maximum 2^n linearly independent elements, which is a contradiction. So, the elements in T are linearly dependent. Thus, the assertion is true.   □

From the proof of Proposition 4, we know that the condition $d + e \geq n$ guarantees the existence of a multiplier g with $deg(g) \leq e$ such that $h = fg \neq 0$ with $deg(h) \leq e$ only if f is nondegenerated with respect to d.

5.2   Resistance against FAA

Recall that $AI(f)$ denotes the algebraic immunity of f. Note that for any g such that $fg = h \neq 0$, then $g + h \in Ann(f)$ implies $deg(g + h) \geq AI(f)$. If we assume that $deg(g) \leq d < AI(f)$, then $deg(h) \geq AI(f)$. So $AI(f) \leq deg(h) \leq e$ (otherwise, we swap g with h).

Definition 3. *For a given $f \in \mathcal{F}_n$, and a pair of positive integers (d, e), f is said to be (d, e)-resistance against FAA if and only if for all g with $deg(g) \leq d$ and $h = fg \neq 0$ with $deg(h) \geq e$. If the assertion is true for every $d : 1 \leq d < n$, then f is said to be e-resistance against FAA.*

From Theorem 2, we have the following condition to determine whether a given function is (d, e)-resistance against FAA.

Theorem 3. *Let $f \in \mathcal{F}_n$, (d, e) be an enable pair of f where $e > deg(f)$, and D_d be a maximal linear independent set of fS_d. Then f is (d, e)-resistance against FAA if and only if $|D_d| = |S_d|$ and $D_d \cup S_r$ is linearly independent over $\mathbb{F}_2$ for all r with $1 \leq r < e$ and $D_d \cup S_e$ is linearly dependent over $\mathbb{F}_2$.*

Proof. If $|D_d| = |S_d|$, i.e., all the elements in fS_d are linearly independent, then $h = fg \neq 0$ for $\forall g \in \langle S_d \rangle$. From the second condition in Theorem 3, we know that there are some g such that $h = fg$ with $deg(h) = e$. Fo the rest of g, $deg(h) > e$. Thus f is (d, e)-resistance against FAA.

Conversely, if there exists some r such that $D_d \cup S_r$, $1 \leq r < e$ is linear dependent over $\mathbb{F}_2$, according to Theorem 2, there is some $g \in \langle S_d \rangle$ such that $h = fg$ with $deg(h) \leq r < e$, which contradicts with $deg(h) > e$ for any g.

If $|D_d| < |S_d|$, according to Case 2 in Section 4.1, there is $g_1 \in \langle S_d \rangle$ such that $fg_1 = 0$. Let $g = g_1 + 1$ then $h = fg = f$ with $deg(h) = deg(f)$ which contradicts with the definition that $deg(h) \geq e > deg(f)$. $\qquad\square$

From Theorem 3, we know that Algorithm 1 can be used in two ways. One is to verify whether a given function is (d, e)-resistance against FAA. In other words, for a given function of degree r, we may run Algorithm 1 for all pairs of (d, e) where $1 \leq d < r$ and $1 \leq e < n$. If the outputs are zero for each pair of (d, e), then f is (d, e)-resistance to FAA. Note that we only need to load S_m in Algorithm 1 once. The complexity for verifying whether a given function is (d, e)-resistance is approximately equal to $n(r - 1)$ times the complexity of the Gauss reduction involved in Algorithm 1.

Example 2. Let $\mathbb{F}_{2^4}$ be defined by a primitive polynomial $t(x) = x^4 + x + 1$ and α a root of $t(x)$. Let $f(x) = Tr(\alpha x^3)$ where $Tr(x) = x + x^2 + x^4 + x^8$ is the trace function from $\mathbb{F}_{2^4}$ to $\mathbb{F}_2$. (Note that $f(x)$ is a bent function.) Then, $f(x)$ is 2-resistance against FAA.

Proof. From the DFT, any function $g(x) \in \mathcal{F}_n$ with $g(0) = 0$ can be written as

$$g(x) = Tr(bx) + Tr(cx^3) + Tr_1^2(dx^5) + Tr(ex^7) + wx^{15}, b, c, e \in \mathbb{F}_{2^4}, d \in \mathbb{F}_{2^2}, w \in \mathbb{F}_2$$

where $Tr_1^2(x) = x + x^2$ is the trace function from $\mathbb{F}_{2^2}$ to $\mathbb{F}_2$. Multiplying f by each monomial trace term in g, we have

$$Tr(bx)Tr(\alpha x^3) = Tr(b^4\alpha^4 x) + Tr_1^2((b^2\alpha + b^8\alpha^4)x^5) + Tr((b\alpha^2 + b^4\alpha)x^7)$$
$$Tr(cx^3)Tr(\alpha x^3) = Tr((c^2\alpha^4 + c^4\alpha^2 + c^8\alpha^8)x^3) + Tr(c\alpha^4)$$
$$Tr_1^2(dx^5)Tr(\alpha x^3) = Tr(d^2\alpha^2 x + d^2\alpha^4 x^7)$$
$$Tr(ex^7)Tr(\alpha x^3) = Tr((e^4\alpha^4 + e\alpha^8)x) + Tr_1^2((e^2\alpha^2 + e^8\alpha^8)x^5) + Tr(e^4\alpha^8 x^7).$$

In the following, we consider $w = 0$. The case $w = 1$ is similar. From the above identities, we have the expansion of $f(x)g(x)$ as follows

$$f(x)g(x) = Tr(Ax + Bx^3 + Dx^7) + Tr_1^2(Cx^5) + E$$

where

$$A = b^4\alpha^4 + d^2\alpha^2 + e^4\alpha + e\alpha^8$$
$$B = c^2\alpha^4 + c^4\alpha^2 + c^8\alpha^8$$
$$D = b\alpha^2 + b^4\alpha + d^2\alpha^4 + e^4\alpha^8$$
$$C = b^2\alpha + b^8\alpha^4 + e^2\alpha^2 + e^8\alpha^8$$
$$E = Tr(c\alpha^4).$$

Considering that $fg \neq 0$, then $deg(fg) = 1$ if and only if

$$B = C = D = 0.$$

It can be verified that the system of those equations has no solutions for any choices of b, c, d and e. Thus $deg(fg) \geq 2$. Therefore, f is 2-resistance to FAA. Alternatively, we can directly apply Algorithm 1 to f for obtaining this result. (Note that the boolean form of $f(x)$ is given by $x_2 + x_0 x_1 + x_0 x_2 + x_0 x_3 + x_1 x_2 + x_1 x_3 + x_2 x_3$. However, in order to verify whether f is 2-resistance to FAA, using the polynomial form of f is much easier than using the boolean form.)   □

Another application of an enable pair (d, r) of f is to find proper values for (d, e) which are in favor of FAA. For example, let $f(x)$ be a hyper-bent function with two trace terms from $\mathbb{F}_{2^8}$ to $\mathbb{F}_2$ (for the definition of hyper-bent functions, see [25]). Then $deg(f) = 4$. We have verified that there are many hyper-bent functions with the algebraic immunity 3, i.e., $AI(f) = 3$. However we could have an enable pair $(d, e) = (2, 3)$ for many such functions. In other words, there exists some g with degree $d = 2$ such that $h = fg \neq 0$ with degree $e = 3$. In this case, the required known bits from a key stream is similar for both the algebraic attack and FAA. But the number of unknowns in FAA is much smaller than in the algebraic attack (of course for FAA case, the required key stream bits should be consecutive). For some results about the algebraic immunity of hyper-bent functions, see an earlier version of this work [15].

Remark 6. In general, it is not easy to determine whether a given function f is (d, e) resistance against FAA or it is an enable pair of f, because the DFT of h is the convolution of the DFTs of f and g (see the definition of the convolution at Section 2). Even it is not easy for some particular functions, for example, hyper-bent functions. This could be an interesting question for further research.

6   Conclusions and Discussions

Courtois and Meier showed the effective algebraic attacks or the fast algebraic attack on several concrete stream cipher systems proposed in the literature. In order to enable the fast algebraic attack, it needs to find a function g with degree at most d such that $h = fg \neq 0$ with degree at most e. In terms of the DFT of sequences and boolean functions, we have showed the characterizations of existence of those multipliers for given function f and a pair of integer (d, e). Those results revealed that the sufficient condition, given by Courtois and Meier in [8] and [9], cannot guarantee the existence of such multipliers. An algorithm for constructing such multipliers is given in terms of the polynomial basis. We have provided a thorough analysis for degenerated cases of boolean functions and how it effects the fast algebraic attack. A new concept of resistance against the fast algebraic attack is introduced, and a set of functions in a degenerated case is identified. Those degenerated functions are weak for resistance against both the algebraic attack and the fast algebraic attack.

All results obtained in this paper are due to use the (discrete) Fourier transform (DFT), which gives a polynomial or trace representation of a boolean function in terms of technics of analysis of pseudo-random sequences. We would like to point out some trade-offs between the polynomial representation of a function from $\mathbb{F}_{2^n}$ to $\mathbb{F}_2$ and the boolean representation of the function. In this paper, we use the polynomial representation, which have advantages in analysis of cryptographical properties of functions. Another advantage is to prevent finding a low degree multiplier probabilistically. In general, for a boolean representation, if the number of monomial terms with high degrees are small, then one can easily remove these terms for obtaining a low degree function for which the probability that these two functions are equal is close to 1. However, if a function is in a polynomial form, the degree of the function is governed by the Hamming weights of exponents in monomial trace terms. Removing one or more monomial trace terms from the expression may result in a large change of the distance between the function and the resulting function, since it is equivalent to calculate the distance of two codewords of a cyclic code. So, the probability that these two functions are equal is not close to 1, possibly, close to $1/2$. However, boolean forms are easily used to analyze the correlation immunity/resiliency and propagation property, and they can also be efficiently implemented at hardware level.

Acknowledgement

The work is supported by NSERC Discovery Grant. The author wishes to thank the referee for his/her valuable comments.

References

1. Armknecht, F., Krause, M.: Algebraic attacks on combiners with memory. In: Boneh, D. (ed.) CRYPTO 2003. LNCS, vol. 2729, pp. 162–175. Springer, Heidelberg (2003)
2. Armknecht, F.: Improving fast algebraic attacks. In: Roy, B., Meier, W. (eds.) FSE 2004. LNCS, vol. 3017, pp. 65–82. Springer, Heidelberg (2004)
3. Armknecht, F., Ars, G.: Introducing a new variant of fast algebraic attacks and minimizing their successive data complexity. In: Dawson, E., Vaudenay, S. (eds.) Mycrypt 2005. LNCS, vol. 3715, pp. 16–32. Springer, Heidelberg (2005)
4. Bluetooth CIG, Specification of the Bluetooth system, Version 1.1 (February 22, 2001), www.bluetooth.com
5. Canteaut, A., Carlet, C., Charpin, P., Fontaine, C.: On cryptographic properties of the cosets of $R(1, m)$. IEEE Trans. on Inform. Theory 47(4), 1491–1513 (2001)
6. Carlet, C.: On the higher order nonlinearities of algebraic immune functions. In: Dwork, C. (ed.) CRYPTO 2006. LNCS, vol. 4117, pp. 584–601. Springer, Heidelberg (2006)
7. Courtois, N.: Higher order correlation attacks, XL algorithm and cryptanalysis of Toyocrypt. In: Lee, P.J., Lim, C.H. (eds.) ICISC 2002. LNCS, vol. 2587, pp. 549–564. Springer, Heidelberg (2003)
8. Courtois, N., Meier, W.: Algebraic attacks on stream ciphers with linear feedback. In: Biham, E. (ed.) EUROCRYPT 2003. LNCS, vol. 2656, pp. 345–359. Springer, Heidelberg (2003)

9. Courtois, N.: Fast algebraic attacks on stream ciphers with linear feedback. In: Boneh, D. (ed.) CRYPTO 2003. LNCS, vol. 2729, pp. 176–194. Springer, Heidelberg (2003)
10. Dalai, D.K., Gupta, K.C., Maitra, S.: Cryptographically Significant Boolean functions: Construction and Analysis in terms of Algebraic Immunity. In: Gilbert, H., Handschuh, H. (eds.) FSE 2005. LNCS, vol. 3557, pp. 98–111. Springer, Heidelberg (2005)
11. eSTREAM - The ECRYPT Stream Cipher Project, http://www.ecrypt.eu.org/stream/
12. Gollmann, D.: Pseudo random properties of cascade connections of clock controlled shift registers. In: Advances in Cryptology-Eurocrypt 1984. LNCS, pp. 93–98. Springer, Heidelberg (1984)
13. Golomb, S.W.: Shift Register Sequences. Holden-Day, Inc., San Francisco 1967, revised edition. Aegean Park Press, Laguna Hills, CA (1982)
14. Golomb, S.W., Gong, G.: Signal Design with Good Correlation: for Wireless Communications, Cryptography and Radar Applications. Cambridge University Press, Cambridge (2005)
15. Gong, G.: On existence and Invariant of algebraic attacks, Technical Report of University of Waterloo, CORR 2004-17 (2004)
16. Hawkes, P., Rose, G.G.: Rewriting variables: the complexity of fast algebraic attacks on stream ciphers. In: Franklin, M. (ed.) CRYPTO 2004. LNCS, vol. 3152, pp. 390–406. Springer, Heidelberg (2004)
17. Meier, W., Pasalic, E., Carlet, C.: Algebraic attacks and decomposition of boolean functions. In: Cachin, C., Camenisch, J.L. (eds.) EUROCRYPT 2004. LNCS, vol. 3027, pp. 474–491. Springer, Heidelberg (2004)
18. Nawaz, Y., Gong, G., Gupta, K.C.: Upper bounds on algebraic immunity of boolean power functions. In: Robshaw, M. (ed.) FSE 2006. LNCS, vol. 4047, pp. 375–389. Springer, Heidelberg (2006)
19. Nawaz, Y., Gupta, K.C., Gong, G.: Efficient techniques to find algebraic immunity of s-boxes based on power mappings. In: Proceedings of International Workshop on Coding and Cryptography, Versailles, France, April 16-20, 2007, pp. 237–246 (2007)
20. Rønjom, S., Helleseth, T.: A New Attack on the Filter Generator. IEEE Transactions on Information Theory 53(5), 1752–1758 (2007)
21. Rønjom, S., Gong, G., Helleseth, T.: On attacks on filtering generators using linear subspace structures. In: Golomb, S.W., Gong, G., Helleseth, T., Song, H.-Y. (eds.) SSC 2007. LNCS, vol. 4893, pp. 204–217. Springer, Heidelberg (2007)
22. Rønjom, S., Gong, G., Helleseth, T.: A survey of recent attacks on the filter generator. In: The Proceedings of AAECC 2007, pp. 7–17 (2007)
23. Rueppel, R.A.: Analysis and Design of Stream Ciphers. Springer, Heidelberg (1986)
24. Shamir, A., Patarin, J., Courtois, N., Klimov, A.: Efficient algorithms for solving overdefined systems of multivariate polynomial equations. In: Preneel, B. (ed.) EUROCRYPT 2000. LNCS, vol. 1807, pp. 392–407. Springer, Heidelberg (2000)
25. Youssef, A.M., Gong, G.: Hyper-Bent Functions. In: Pfitzmann, B. (ed.) EUROCRYPT 2001. LNCS, vol. 2045, pp. 406–419. Springer, Heidelberg (2001)

Expected π-Adic Security Measures of Sequences

(Extended Abstract)

Andrew Klapper*

Department of Computer Science, University of Kentucky,
Lexington, KY 40506-0046, USA
http://www.cs.uky.edu/~klapper/

Abstract. Associated with a class of AFSRs based on a ring R and $\pi \in R$, there is a security measure, the π-adic complexity of a sequence. To understand the normal behavior of π-adic complexity we can find the average π-adic complexity, averaged over all sequences of a given period. This has been done previously for linear and p-adic complexity. In this paper we show that when $\pi^2 = 2$, the average π-adic complexity of period n sequences is $n - O(\log(n))$.

Keywords: feedback with carry shift register, algebraic feedback shift register, pseudo-randomness, sequence, security measure.

1 Introduction

A variety of measures, such as linear span and 2-adic span, have been proposed for deciding the cryptographic security of stream ciphers. Recently there has been interest in understanding the average behavior of such measures. Several variations have been studied, both for linear and 2-adic span [10,11,12,13]. In this paper we extend this work to more general π-adic span, where π is an element of certain algebraic rings.

Many of these measures arise as follows: We are given a class $\mathcal{F}$ of sequence generators. There is a notion of *size* of a generator with a particular initial state. This integer should be approximately the number of elements of the output alphabet needed to represent all states in the execution of the generator starting at that initial state using some standard encoding of states.

The $\mathcal{F}$-*span* of a (finite or infinite) sequence is the smallest size of a generator in the given class that outputs the sequence. Thus the $\mathcal{F}$-span is an integer.

---

* Parts of this work were carried out while the author visited the Fields Inst. at the University of Toronto; The Inst. for Advanced Study; and Princeton Universiy. This material is based upon work supported by the National Science Foundation under Grant No. CCF-0514660. Any opinions, findings, and conclusions or recommendations expressed in this material are those of the author and do not necessarily reflect the views of the National Science Foundation.

S.W. Golomb et al. (Eds.): SETA 2008, LNCS 5203, pp. 219–229, 2008.

It is infinite if there is no such generator. Often there is another closely related measure that is defined algebraically and is much easier to work with. We typically call this related measure the $\mathcal{F}$-complexity. It may be based on a function defined on some related algebraic structure and satisfy properties such as $f(ab) = f(a) + f(b)$ or $f(a + b) \leq \max(f(a), f(b)) + c$ for some constant c. In the case when $\mathcal{F}$ is the set of *linear feedback shift registers* (LFSRs) the $\mathcal{F}$-span is called the *linear span*. The size of an LFSR with connection polynomial $q(x)$ is simply the number of cells it contains. If the generating function of the output sequence is $u(x)/q(x)$, then the *linear complexity* is $\max(\deg(q), \deg(u) + 1)$ and equals the linear span. In the case of *feedback with carry shift registers* (FCSRs) with output alphabet $\{0, 1, \cdots, N - 1\}$, $N \in \mathbf{Z}$ [5], the $\mathcal{F}$-span is called the N-*adic span*. The integer values of the carry are encoded by their base N expansions. Thus the size of an FCSR with connection integer q is the number of cells in the basic register plus the ceiling of the log base N of the maximum absolute value of the carry over its infinite execution. If the output sequence has associated N-adic number $\alpha = u/q$, then the related N-*adic complexity* is $\log_N(\max(|u|, |q|))$ and differs from the N-adic span by $O(\log(\text{the } N\text{-adic span}))$ [5].

More generally we may consider *algebraic feedback shift registers* (AFSRs) with respect to some particular algebraic ring R and parameter π. The $\mathcal{F}$-span associated with the class $\mathcal{F}$ of AFSRs based on R and π is called the π-*adic span*. However, we know of no algebraic definition of a "π-adic complexity" that makes sense for all (over even many) classes of AFSRs. AFSRs include both LFSRs ($R = F[x]$ with F a field, $\pi = x$) and FCSRs ($R = \mathbf{Z}$, $\pi \geq 2$) [6].

In more detail, let R be a commutative ring and $\pi \in R$. Assume that $R/(q)$ is finite for every nonzero $q \in R$. We consider sequences over the ring $R/(\pi)$, whose cardinality we denote by p. We also let $S \subseteq R$ be a complete set of representatives for $R/(\pi)$. We identify sequences over $R/(\pi)$ with sequences over S. An AFSR based on R, π, and S is a stated device D with output. It is determined by constants $q_0, \cdots, q_r \in R$ with the image of q_0 in $R/(\pi)$ invertible. Its state is an $(r+1)$-tuple $\sigma = (a_0, a_1, \cdots, a_{r-1}; m)$ where $a_i \in S$, $i = 0, \cdots, r-1$, and $m \in R$. From this state the AFSR outputs a_0 and changes state to $(a_1, \cdots, a_r; m')$ where $a_r \in S$ and

$$q_0 a_r + \pi m' = m + \sum_{i=1}^{r} q_i a_{r-i}.$$

By repeating the state change operation, the AFSR generates an infinite sequence, denoted by $D(\sigma)$. We refer to $D(\sigma)$ as the sequence generated by D with initial state σ. In this paper we are concerned with average behavior of the $\mathcal{F}$-span where $\mathcal{F}$ is the class of AFSRs based on R, π, S. The $\mathcal{F}$-span is commonly referred to as the π-span. See [6] for more details on AFSRs and their analysis by π-adic numbers.

The element

$$q^{(D)} = q = \sum_{i=1}^{r} q_i \pi^i - q_0 \tag{1}$$

is known as the *connection element* of the AFSR. If we start with q, we denote by D_q the AFSR with connection element q^1. We may also refer to it as a connection element for any sequence **a** generated by D_q. Any given sequence has many connection elements. The connection element plays a critical role in the analysis of AFSR sequences. We can associate a π-*adic integer*

$$\alpha = \sum_{i=0}^{\infty} a_i \pi^i$$

to the sequence $\mathbf{a} = a_0, a_1, \cdots, a_i \in S$. The set of all π-adic integers is a ring in a natural way. But note that this is not simply a power series ring — the sum or product of two elements of S may not be in S, so expressing sums and products of π-adic integers as π-adic integers in general involves complicated carries to higher degree terms, perhaps even carries to infinitely many terms. Again, we refer to [6] for more details on π-adic integers.

If q is as in equation (1), then any rational element $u/q \in R[1/q]$, $u \in R$ can also be expressed as such a π-adic integer[2] and it can be shown that **a** is the output from an AFSR with connection element q if and only if there exists $u \in R$ so that $u/q = \alpha$. In fact u can be expressed in terms of the initial state $(a_0, \cdots, a_{r-1}; m)$ by

$$\alpha = \frac{\sum_{n=0}^{r-1} \sum_{i=1}^{n} q_i a_{n-i} \pi^n - q_0 \sum_{n=0}^{r-1} a_n \pi^n - m\pi^r}{q} = \frac{u}{q}. \tag{2}$$

More discussion of AFSRs, π-adic numbers, and π-adic span can be found in a paper by Xu and Klapper [14]. We emphasize that the π-adic span of an AFSR depends both on the structure of the AFSR (that is, on the connection element q) and on a particular initial state.

The π-adic span of a sequence is important as a security measure if there is an *AFSR synthesis algorithm* for R, π, and S. This is an algorithm which takes a prefix of a sequence as input and outputs a minimal AFSR over R, π, and S that outputs the prefix. If the prefix is sufficiently large (typically a constant times the π-adic span of the sequence), the AFSR in fact generates the entire sequence. Thus if such an algorithm is known (register synthesis algorithms are only known for certain types of AFSRs [14]), any sequence used as a key in a stream cipher should have large π-adic span. It is important, then, to understand the average behavior of π-adic span, as has been done for linear span and 2-adic span [10,11,12,13].

The π-adic span of an infinite sequence is not always defined. As is the case with linear feedback shift registers and feedback with carry shift registers, some sequences are not generated by any AFSR over R, π, and S at all. For example,

---

[1] We are being a little sloppy here since a given q may have several representations in the form in equation (1), but this causes no problems.

[2] Actually, for this statement to be true we need a *separability condition* on R, that $\cap_{i=1}^{\infty}(\pi)^i = (0)$. This condition always holds when $R = \mathbb{Z}[\pi]$ and π is integral over $\mathbb{Q}$.

the sequences generated by LFSRs are exactly the eventually periodic sequences. Thus they make up a countable subset of the uncountable set of all sequences over S. The same is true of the sequences generated by FCSRs. However, some AFSRs generate sequences that are not eventually periodic. This implies that the memory is unbounded over an infinite execution of the AFSR, so the π-adic span is infinite. However, it is known that if R is an integral domain whose field of fractions is a number field, and for every embedding of R in the complex numbers, the complex norm of π is greater than 1, then the memory of every AFSR is bounded over every infinite execution [6]. When this happens, the sequences for which π-adic span is defined are exactly the eventually periodic sequences. This is the case for all rings studied in this paper.

Our goal in this paper is to find the average π-adic span of sequences with fixed period n. A sequence has period n if and only if it has $\pi^n - 1$ as a connection element. Several of the proofs are omitted for lack of space.

From here on we assume that $R = \mathbb{Z}[\pi]$, with $\pi^d = p$, $p > 0$ and $x^d - p$ irreducible over $\mathbb{Z}$. We take $S = \{0, 1, \cdots, p - 1\}$. Our goal is to find the expected π-adic complexity of sequences over S with fixed period n. We have $R = \{\sum_{i=0}^{d-1} b_i \pi^i : b_i \in \mathbb{Z}\}$ and $|R/(\pi)| = p$. AFSRs over such a ring are called d-FCSRs, and many of their properties have been studied [2,3,4,7,14]. Note that there are register synthesis algorithms for AFSRs over these rings and the output from such AFSRs is eventually periodic.

2   Size Measure and Algebraic Preliminaries

We define the size function

$$\lambda \left(\sum_{i=0}^{d-1} b_i \pi^i \right) = \max(d \log_p |b_i| + i).$$

This is the same size function that was used to establish the existence of a rational approximation algorithm for d-FCSRs [14].

Theorem 1. Let $c_1 = d \log_p(2)$, $c_2 = d \log_p(d)$, and $c_3 = d$. Then

S1: For all $a, b \in R$, we have $\lambda(a \pm b) \leq \max(\lambda(a), \lambda(b)) + c_1$;

S2: For all $a, b \in R$, we have $\lambda(ab) \leq \lambda(a) + \lambda(b) + c_2$;

S3: For all $a \in R$, we have $\lambda(\pi a) = 1 + \lambda(a)$;

S4: For all $a_0, \cdots, a_{r-1} \in S$ with $a_{r-1} \neq 0$, we have $\left| \lambda \left(\sum_{i=0}^{r-1} a_i \pi^i \right) - r \right| \leq c_3$;

S5: For any real number x, there are finitely many elements $a \in R$ with $\lambda(a) \leq x$;

S6: For any $a_1, \cdots, a_r \in R$, $\lambda(a_1 \pm \cdots \pm a_r) \leq \max\{\lambda(a_i), i = 1, \cdots, r\} + c_1 \lceil \log_2(r) \rceil$.

We make use of the algebraic norm function $N : R \rightarrow \mathbb{Z}$, defined by

$$N(\sum_{i=0}^{d-1} b_i \pi^i) = \prod_{j=0}^{d-1} \left(\sum_{i=0}^{d-1} b_i \zeta^{ij} \pi^i \right),$$

where ζ is a complex primitive dth root of 1. An element in R is a unit if and only if its norm is 1 or -1. Moreover, the norm function is multiplicative: $N(ab) = N(a)N(b)$ for all $a, b \in R$.

Lemma 1. *We have*

1. $\forall z \in R : |R/(z)| = |N(z)|$.
2. $\forall z \in R : \log_p |N(z)| \leq d\log_p(d) + \lambda(z)$.
3. *If $d = 2$, then $\forall z \in R : \log_p |N(z)| \leq \lambda(z)$.*
4. *If $d = 2$ and R is the full ring of integers in its fraction field, then there is a constant $c_p \in \mathbb{R}^+$ so that $\forall z \in R$ there is a unit $u \in R : \lambda(uz) \leq 2\log_p(|N(z)| + c_p) - \log_p(4)$.*
5. *If $d = p = 2$, then $\forall z \in R$ there is a unit $u \in R$ so that $\lambda(uz) \leq \log_2 |N(z)| + 1$.*

It follows that if $d = 2$ and $\lambda(z)$ is minimal among all associates of z, then $\log_p |N(z)| \leq \lambda(z) \leq 2\log_p(|N(z)| + c_p) - \log_p(4) \leq 2\log_p |N(z)| + e_p$ for some constant e_p. Moreover, if $p = 2$ and $\lambda(z)$ is minimal among all associates of z, then $\lambda(z) \leq \log_2 |N(z)| + 1 \leq \lambda(z) + 1$.

The ring R is Noetherian. Thus every element $q \in R$ can be written as a product $z = u \prod z_i^{e_i}$, where u is a unit, the z_i are irreducible, each e_i is positive, and z_i and z_j are not associates[3] if $i \neq j$ (this representation may not be unique). We may assume that $\lambda(z_i)$ is minimal among all associates of z_i. In particular we can write

$$\pi^n - 1 = u \prod_{i=1}^{t} z_i^{e_i}, \tag{3}$$

where u is a unit and each z_i is irreducible.

Suppose further that R is a unique factorization domain or UFD. The representation of q as a product of irreducibles is then unique up to permutation of the z_i and replacing a z_i by an associate (and so also changing the unit u). In this case there is a connection element for any sequence that is minimal in the sense that it divides all other connection elements for the sequence. Any two minimal connection elements are associates.

Lemma 2. *Suppose $R = \mathbb{Z}[\pi]$ is a UFD and $z \in R$ is not a unit. Let u/z and v/z be rational elements whose π-adic expansions are periodic. If z is the minimal connection element for either of these sequences, then u and v are not congruent modulo z.*

It follows that the set U_z of elements $u \in R$ such that u/z has a strictly periodic π-adic expansion and such that z is the minimal connection element is a subset of a complete set of representatives V_z modulo z. Let $u \in R$. Then u/z has an eventually periodic π-adic expansion, so there is an element $a \in R$ so that $y/z = a + u/z$ has a strictly periodic π-adic expansion. Then $y = u + az$, so we may assume that if $u \in V_z$, then u/z has a strictly periodic π-adic expansion.

---

[3] Elements a and b are *associates* if $a = vb$ for some unit v.

The question of when a ring $R = \mathbf{Z}[\pi]$ is a UFD is a delicate one, not fully understood. First, it is generally necessary to use the integral closure of R rather than R. The quadratic fields $\mathbf{Z}[\sqrt{p}]$ whose integral closures are UFDs are known for negative p, but they are not yet all known for positive p. The negative p for which the integral closure of $\mathbf{Z}[\sqrt{p}]$ is a UFD are

$$\{-163, -67, -43, -19, -11, -7, -3, -2, -1\}.$$

The known positive p for which the integral closure of $\mathbf{Z}[\sqrt{p}]$ is a UFD are

$$\{2,3,5,6,7,11,13,14,17,19,21,22,23,29,33,37,41,53,57,61,69,73,77,89,93,97\}.$$

3   Expected π-Adic Complexity

The function λ is extended to states of an AFSR D by $\lambda(a_0, \cdots, a_{r-1}; m) = r + \lambda(m)$. We extend λ to AFSRs by letting $\lambda(D, \sigma)$ denote the maximum of $\lambda(\tau)$ over all states τ that occur in the infinite execution of the AFSR starting with initial state σ. We extend λ to sequences by letting $\lambda(\mathbf{a})$, the π-adic span of $\mathbf{a}$, be the minimum of $\lambda(D, \sigma)$ over all (D, σ) with $D(\sigma) = \mathbf{a}$. Also, let Γ_n denote the set of pairs (D, σ) such that D is an AFSR, σ is a state of D, and $D(\sigma)$ has period n.

Now suppose that $\mathbf{a}$ is a sequence over S. Suppose that

$$\alpha = \sum_{i=0}^{\infty} a_i \pi^i = \frac{u}{q} \quad \text{and} \quad q = \sum_{i=1}^{r} q_i \pi^i - q_0,$$

where $q_i \in S$ for $i = 0, \cdots, r$, $D_q(\sigma) = \mathbf{a}$ for some initial state σ, and $\lambda(\mathbf{a}) = \lambda(D_q, \sigma)$.

Lemma 3. *If $\mathbf{a}$ is strictly periodic, then $\lambda(\mathbf{a}) \in \lambda(q) + O(\log(\lambda(q))) = r + O(\log(r))$.*

Thus the size of the carry m has little effect on the π-adic complexity.

Let Γ_n^* denote the set of pairs $(D, \sigma) \in \Gamma_n$ for which $\lambda(D, \sigma)$ is minimal among all pairs (C, τ) with $C(\tau) = D(\sigma)$. Suppose $(D, \sigma) \in \Gamma_n^*$ and $q^D = \sum_{i=1}^{r} q_i \pi^i - q_0$. Then

$$\lambda(D(\sigma)) \geq r \geq \lambda\left(\sum_{i=1}^{r} q_i \pi^i\right) + c_3$$

by S4. Furthermore,

$$\lambda(q^D) = \lambda\left(\sum_{i=1}^{r} q_i \pi^i - q_0\right) \leq \max\left(\lambda\left(\sum_{i=1}^{r} q_i \pi^i\right), \lambda(q_0)\right) + c_1$$

$$\leq \lambda\left(\sum_{i=1}^{r} q_i \pi^i\right) + \max(\lambda(s) : s \in S) + c_1.$$

It follows that $\lambda(D(\sigma)) \geq \lambda(q^D) - c$ for some constant c.

Let E_n^λ be the expected π-adic complexity of sequences with period n. That is,

$$E_n^\lambda = \frac{1}{p^n} \sum_{\mathbf{a}} \lambda(\mathbf{a}) = \frac{1}{p^n} \sum_{(D,\sigma) \in \Gamma_n^*} \lambda(D(\sigma))$$

$$\geq \frac{1}{p^n} \sum_{(D,\sigma) \in \Gamma_n^*} (\lambda(q^D) - c) = \frac{1}{p^n} \left(\sum_{(D,\sigma) \in \Gamma_n^*} \lambda(q^D) \right) - c, \qquad (4)$$

where the first sum is over all sequences over S with period n.

There are two notions of minimality that we use for connection elements of a sequence $\mathbf{a}$. A connection element $q \in R$ is λ-minimal if it minimizes $\lambda(q)$ over all connection elements of $\mathbf{a}$. It is R-minimal if it divides every other connection element of $\mathbf{a}$. The two notions are not equivalent, since we might have $\lambda(zq) < \lambda(q)$. However, when $d = p = 2$, if q is λ-minimal, then by Lemma 1 we have $\lambda(zq) \geq \log_2(|N(zq)|) = \log_2(|N(q)|) + \log_2(|N(z)|) \geq \log_2(|N(q)|) \geq \lambda(q) - 1$.

Let $\mu(\mathbf{a})$ denote the minimum $\lambda(q^C)$ over all C that can output $\mathbf{a}$. Even if $(D,\sigma) \in \Gamma_n^*$ and $D(\sigma) = \mathbf{a}$, we might not have $\mu(\mathbf{a}) = \lambda(q^D)$. However, $\lambda(q^D) \geq \mu(\mathbf{a})$. That is, for all D we have $\lambda(q^D) \geq \mu(D(\sigma))$. Let $\Delta_n(q)$ denote the number of period n sequences with q as a connection element, and let $\Delta_n^*(q)$ denote the number of period n sequences whose λ-minimal connection element is an associate of q. Note that

$$\sum_{q|^* \pi^n - 1} \Delta_n^*(q) = p^n, \qquad (5)$$

where $q|^* \pi^n - 1$ means we choose one q in each set of associates among the divisors of $\pi^n - 1$. Gathering terms in equation (4) together, we have

$$E_n^\lambda \geq \frac{1}{p^n} \sum_{(D,\sigma) \in \Gamma_n^*} \mu(D(\sigma)) - c \geq \frac{1}{p^n} \sum_{q|\pi^n - 1} \lambda(q) \Delta_n^*(q) - c.$$

4   When $p = d = 2$

In this section we assume that $p = d = 2$. Thus R is a PID and UFD, so that the representation in equation (3) is unique (in the usual sense). It follows that, up to multiplying by a unit, every divisor of $\pi^n - 1$ is of the form $w = \prod_{i=1}^{t} z_i^{m_i}$ with $0 \leq m_i \leq e_i$. Every period n sequence $\mathbf{a}$ has some such w as R-minimal connection element. If q is the λ-minimal connection element of $\mathbf{a}$, then $\lambda(w) \geq \lambda(q) - 1$.

Let $I = \{(m_1, \cdots, m_t) : 0 \le m_i \le e_i\} \subseteq \mathbb{Z}^t$. By part (3) of Lemma 1 we have

$$E_n^\lambda \ge \frac{1}{2^n} \sum_{q|^*\pi^n - 1} (\lambda(q) - 1)\Delta_n^*(q) - c$$

$$\ge \frac{1}{2^n} \sum_{q|^*\pi^n - 1} \log_2 |N(q)|\Delta_n^*(q) - c - 1 \tag{6}$$

$$= \frac{1}{2^n} \sum_{\substack{(m_1,\cdots,m_k) \in I \\ m_i \text{not all } 0}} \log_2 \left| N\left(\prod_{i=1}^t z_i^{m_i}\right) \right| \Delta_n^*\left(\prod_{i=1}^t z_i^{m_i}\right) - c - 1.$$

Lemma 4. *For any divisor* $z \in R$ *of* $\pi^n - 1$, $\Delta_n^*(z) = |(R/(z))^*|$.

Proof: By Lemma 2 the map f that takes the sequence associated with u/z to u is an injection from U_z into $(R/(z))^*$. On the other hand, if $u \in R$ is a unit modulo z, then by the argument following Lemma 2, there is a $y \in R$ so that $y \equiv u \mod z$ and y/z has a strictly periodic π-adic expansion. Since y is relatively prime to z, we have $y \in U_z$, so that f maps U_z onto $(R/(z))^*$. □

Lemma 5. *If* $z_1, \cdots, z_t \in R$ *are not units and are pairwise relatively prime, then*

$$\Delta_n^*\left(\prod_{i=1}^t z_i\right) = \prod_{i=1}^t \Delta_n^*(z_i).$$

Proof Sketch: By induction it suffices to prove the result when $t = 2$. Let $z = z_1 z_2$. We define $\Gamma(u, v)$ to be the element of U_z that is congruent to $uz_2 + vz_1$ modulo z. It can be shown that Γ induces a one to one, onto function from $U_{z_1} \times U_{z_2}$ to U_z. It follows that $\Delta_n^*(z_1 z_2) = \Delta_n^*(z_1)\Delta^*(z_2)$. □

Lemma 6. *If* $z \in R$ *is irreducible and* $t \ge 1$, *then*

$$\Delta_n^*(z^t) = |N(z)|^{t-1}(|N(z)| - 1).$$

Therefore

$$\sum_{i=0}^t \Delta_n^*(z^i) = |N(z)|^t + 3.$$

Proof: By Lemma 4, it suffices to show that $|(R/(z))^*| = |N(z)|^{t-1}(|N(z)| - 1)$. Let $V_{t-1} \subseteq R$ be a complete set of representatives for $R/(z^{t-1})$. No two elements of V_{t-1} are congruent modulo z^{t-1}, so no two elements of zV_{t-1} are congruent modulo z^t. Let $V_t \subseteq R$ be a complete set of representatives for $R/(z^t)$ containing zV_t. If $zx \in V_t - zV_{t-1}$, then x is not congruent to any element of V_{t-1} modulo z^{t-1}. Thus zV_{t-1} contains all the elements of V_t that are multiples of z. Since z is irreducible, the elements of V_t that are not relatively prime to z^t are exactly those elements of V that are multiples of z. Thus by part (1) of Lemma 1, $|(R/(z))^*| = |V_t| - |V_{t-1}| = |N(z)|^t - |N(z)|^{t-1} = |N(z)|^{t-1}(|N(z)| - 1)$.

The second claim follows from the fact that $\Delta_n^*(1) = 4$ (see the proof of Lemma 2). $\square$

Lemma 7. *We have*

$$\lim_{n \to \infty} \frac{|N(\pi^n - 1)|}{2^n} = 1.$$

Thus for every $\epsilon > 0$, if n is sufficiently large $\log_2 |N(\pi^n - 1) - n| \leq \epsilon$.

Proof: Fix $y \in \{0, 1, \cdots, d-1\}$ and consider just the n of the form $n = xd + y$, $x \geq 0$. Then

$$\frac{|N(\pi^n - 1)|}{2^n} = \frac{|\prod_{i=0}^{d-1}(p^x \zeta^{y_i} \pi^y - 1)|}{2^n} = \left|\prod_{i=0}^{d-1}(1 - p^{-x}\zeta^{-y_i}\pi^{-y})\right|.$$

This has limit 1 as x tends to infinity since p^{-x} tends to zero. The last statement is immediate. $\square$

Suppose $\pi^n - 1$ is irreducible, so in the sum in equation (6), q equals $\pi^n - 1$ or 1. The term with $q = 1$ is zero, so by Lemma 7, for any $\epsilon > 0$, if n is sufficiently large, the sum is at least $n - \epsilon$. Thus $E_n^\lambda \geq n - O(1)$.

Theorem 2. *In general we have $E_n^\lambda \in n - O(\log(n))$.*

Proof: We have

$$E_n^\lambda = \frac{1}{2^n} \sum_{\substack{(m_1, \cdots, m_k) \in I \\ m_i \text{ not all } 0}} \log_2 \left|N\left(\prod_{i=1}^k z_i^{m_i}\right)\right| \Delta_n^*\left(\prod_{i=1}^k z_i^{m_i}\right) - c - 1. \quad (7)$$

Let

$$\Delta_{0,n}^*(x) = \begin{cases} 1 & \text{if } x \text{ is a unit} \\ \Delta_n^*(x) & \text{otherwise.} \end{cases}$$

Thus

$$\sum_{i=0}^t \Delta_{0,n}^*(z^i) = |N(z)|^t.$$

Then note that the right hand side of equation (7) is unchanged if we replace Δ_n^* by $\Delta_{0,n}^*$. Thus (after a long calculation that we omit due to lack of space)

$$E_n^\lambda \geq \frac{1}{2^n} \sum_{(m_1, \cdots, m_k) \in I} \sum_{i=1}^k m_i \log_2(|N(z_i)|) \prod_{i=1}^k \Delta_{0,n}^*(z_i^{m_i}) - O(1)$$

$$\geq \frac{|N(\pi^n - 1)|}{2^n} \log_p(|N(\pi^n - 1)|) - \frac{|N(\pi^n - 1)|}{2^n} \sum_{\ell=1}^k \frac{\log_p(|N(z_\ell)|)}{|N(z_\ell)| - 1} - O(1).$$

To bound the summation in the last line, it suffices to bound

$$A \stackrel{def}{=} \sum_{\ell=1}^{k} \frac{\log_2(|N(z_\ell)|)}{|N(z_\ell)|}$$

in terms of $|N(\pi^n - 1)| \geq |N(w)|$, where $w = \prod_i z_i$. The function $\log_2(x)/x$ is decreasing for integers $x \geq 3$. Thus we may assume that $\{z_\ell\}$ consists of the k irreducible and pairwise relatively prime elements of R with smallest norms.

Let j be a positive integer and consider the irreducible elements with norms between 2^j and 2^{j-1}. By Landau's prime ideal theorem there are asymptotically

$$\frac{2^j}{\ln(2^j)} = \frac{2^j}{j \ln(2)}$$

such elements [8]. Each contributes at most $(j-1)/2^{j-1}$ to A, so the total contribution to A for each j is at most $2/\ln(2)$. It also follows that the largest 2^j we must consider is asymptotically $\Theta(k \log(k))$. It follows that $A \in O(\log(k))$.

Now consider $|N(w)|$. The contribution to $|N(w)|$ from the irreducible elements with norms between 2^m and 2^{m+1} is asymptotically at least

$$(2^m)^{2^m/(m \ln(2))} = 2^{2^m/\ln(2)}.$$

The largest m we need is at most $\log(k \ln(k)) < 2 \log(k)$. Thus

$$|N(w)| \in \Omega \left(\prod_{m=1}^{\log(k)} 2^{2^m/\ln(2)} \right) = \Omega(2^{(\sum_{m=1}^{\log(k)} 2^m/\ln(2))}) = \Omega(2^{2k/\ln(2)}).$$

Therefore $A \in O(\log \log |N(w)|) \subseteq O(\log \log |N(\pi^n - 1)|) = O(\log(n))$.   □

5   Conclusions

We have expanded our understanding of average behavior of generator-based security measures for sequences to the case of AFSRs based on rings of the form $\mathbf{Z}[p^{1/d}]$. We have only fully analyzed the case when $p = d = 2$. This analysis used specialized algebraic properties of this ring that do not hold in general. It is not clear what can be said for more general p and d. Even more problematic is the general case of AFSRs. In the most general setting there may not even be an algebraic notion of $\mathcal{F}$-complexity that is closely tied to $\mathcal{F}$-span.

In the case $p = d = 2$, we have seen that E_n^λ is asymptotically n. This says that random long sequences are resistant to register synthesis attacks based on these AFSRs. It would be very interesting to find a ring R for which E_n^λ is asymptotically smaller than n, so that random sequences do not resist attack.

Acknowledgement

Thanks to Hugh Williams for suggesting the proof of part 3 of Lemma 1.

References

1. Borevich, Z.I., Shafarevich, I.R.: Number Theory. Academic Press, New York (1966)
2. Goresky, M., Klapper, A.: Feedback registers based on ramified extensions of the 2-adic numbers. In: Advances in Cryptology — Eurocrypt 1994. LNCS, vol. 718, pp. 215–222. Springer, Heidelberg (1995)
3. Goresky, M., Klapper, A.: Fibonacci and Galois Mode Implementation of Feedback with Carry Shift Registers. IEEE Trans. Info. Thy. 48, 2826–2836 (2002)
4. Goresky, M., Klapper, A.: Periodicity and Correlations of d-FCSR Sequences. Designs, Codes, and Crypt. 33, 123–148 (2004)
5. Klapper, A., Goresky, M.: Feedback Shift Registers, Combiners with Memory, and 2-Adic Span. J. Crypt. 10, 111–147 (1997)
6. Klapper, A., Xu, J.: Algebraic feedback shift registers. Theor. Comp. Sci. 226, 61–93 (1999)
7. Klapper, A.: Distributional properties of d-FCSR sequences. J. Complexity 20, 305–317 (2004)
8. Landau, E.: Ueber die zu einem algebraischen Zahlkörper gehörige Zetafunction und die Ausdehnung der Tschebyschefschen Primzahlentheorie auf das Problem der Vertheilung der Primideale. J. für die reine und angewandte Math. 125, 64–188 (1902)
9. Lidl, R., Niederreiter, H.: Finite Fields, 2nd edn. Cambridge University Press, Cambridge (1997)
10. Meidl, W., Niederreiter, H.: Counting functions and expected values for the k-error linear complexity. Finite Fields Appl. 8, 142–154 (2002)
11. Meidl, W., Niederreiter, H.: On the expected value of linear complexity and the k-error linear complexity of periodic sequences. IEEE Trans. Info. Thy. 48, 2817–2825 (2002)
12. Meidl, W., Niederreiter, H.: The expected value of joint linear complexity of multisequences. J. Complexity 19, 61–72 (2003)
13. Niederreiter, H.: A combinatorial approach to probabilistic results on the linear complexity profile of random sequences. J. Crypt. 2, 105–112 (1990)
14. Xu, J., Klapper, A.: Register synthesis for algebraic feedback shift registers based on non-primes. Designs, Codes, and Crypt. 31, 225–227 (2004)

Distance-Avoiding Sequences for Extremely Low-Bandwidth Authentication

Michael J. Collins and Scott Mitchell

Sandia National Laboratories[*]
Albuquerque, NM 87185, USA
mjcolli@sandia.gov

Abstract. We develop a scheme for providing strong cryptographic authentication on a stream of messages which consumes very little bandwidth (as little as one bit per message) and is robust in the presence of dropped messages. Such a scheme should be useful for extremely low-power, low-bandwidth wireless sensor networks and "smart dust" applications. The tradeoffs among security, memory, bandwidth, and tolerance for missing messages give rise to several new optimization problems. We report on experimental results and derive bounds on the performance of the scheme.

1   Introduction and Previous Work

We consider the following scenario: we wish to send a stream of many short messages $m_1, m_2, m_3, \cdots$ on a channel with very limited bandwidth, and we need to provide strong cryptographic authentication for this data. Because bandwidth is so limited, we assume that we must use almost all transmitted bits for delivering payload data: say we can append no more than r bits of authentication to each message, where r is too small to provide adequate security. Such a situation might arise for power-scavenging or energy harvesting systems, since communication is generally energy-intensive relative to computation.

Suppose we have decided that qr authentication bits are needed for security; a simple solution would be to send q consecutive messages $m_1, m_2, \cdots m_q$, followed by a message authentication tag t of length qr for the concatenated message $(m_1|m_2| \cdots m_q)$ (repeating this process for the next block of q messages and so on). This achieves the desired data rate, but it is unsatisfactory for several reasons. In an extremely low-power environment (such as a wireless network of very small sensors), we expect that many messages will be dropped or corrupted, making it impossible for the receiver to verify the correctness of t. Also, we are transmitting no data at all during the relatively long time needed to transmit the tag. We seek a more robust solution which will tolerate some missing messages

---

[*] Sandia is a multiprogram laboratory operated by Sandia Corporation, a Lockheed Martin Company, for the United States Department of Energy's National Nuclear Security Administration under contract DE-AC04-94AL85000.

S.W. Golomb et al. (Eds.): SETA 2008, LNCS 5203, pp. 230–238, 2008.

(without the additional cost of applying an erasure code to already-redundant data), and which does not interrupt the flow of payload data.

Perrig et al. [1,2,3] have considered the problem of efficient authentication for lossy data streams, but our work considers a somewhat different set of issues. Protocols such as μTESLA [2] have low overhead compared to earlier authentication methods for such streams, but "low overhead" in their context means tens of bytes; our work considers a scenario in which we may add only a few bits to each message. In order to attain such extreme bandwidth efficiency, it is necessary to constrain the problem somewhat. The μTESLA protocol specifically addresses the problem of *broadcast* in sensor networks, but our work is only applicable to point-to-point communication, because we assume that the sender and receiver share a symmetric key.

2   Subset Authentication

Our basic approach is to append a short authentication tag a_i to each message m_i; each a_i is an r-bit authentication tag for some appropriately-chosen subset S_i of the previous messages. Let $\mathcal{A}_K$ be a message authentication code (MAC) with key K that produces an r-bit output. Thus if $S_i = \{j_1^i < j_2^i < \cdots j_k^i\}$ we have[1]

$$a_i = \mathcal{A}_K(i|m_{j_1^i}|\cdots|m_{j_k^i}) \tag{1}$$

and we transmit $m_1, a_1, m_2, a_2, \cdots$. If each message is contained in q sets, then each message is used in the computation of q different tags, and we will eventually accumulate the required qr bits of authentication for each message. If $\mathcal{A}_K$ is a pseudorandom function, an adversary cannot cause an invalid m_i to be accepted without guessing qr random bits. In practice, $\mathcal{A}_K$ could be implemented by truncating the output of a full-length authentication code such as HMAC with SHA-1 [4]. The design of an equally secure but less computationally intensive MAC which inherently produces a short output would pose an interesting and challenging problem.

It is essential to include the sequence number i in the computation of a_i, so that an attacker cannot replay the same data with different authentication bits and attack each r bit tag separately. We are here making a non-standard security assumption; an attacker only has access to a *stateful* verification oracle. Once this oracle has answered one query for a message with sequence number i, it will not answer queries (or will always answer "no") for any message with sequence number $\leq i$. This is a plausible assumption in many cases, especially for data with a short lifetime, which is likely to be the case in our intended applications. Some non-standard assumption of this type is necessary to achieve our very strong efficiency requirements.

Given that bandwidth is very constrained, we would seek to avoid explicitly transmitting the entire sequence number with each message. In particular, if

---

[1] It is convenient to ignore the distinction between a message and its index, writing $j \in S_i$ instead of $m_j \in S_i$.

messages are guaranteed to be received in order (i.e. if there is a direct radio link from sender to receiver), and if we assume that no more than 2^k consecutive messages will ever be lost, then it is only necessary to transmit the k low-order bits of i.

Of course it is not enough to simply require that each message appears q times. We are assuming a very low-power network with no acknowledgement or retransmission protocol, no error-correction mechanism, and occasional loss of connectivity. Thus we must expect that some messages will be lost, and if m_j is lost, all tags a_i such that $j \in S_i$ will be useless. Therefore every message must be contained in more than q sets, to provide robustness against the expected missing messages. The question then becomes, what conditions must we impose on the sets S_i, and what is the optimal way to achieve those conditions?

We first consider the following requirement (more general requirements are considered in Sect. 4): if any one message is lost, this must not prevent full authentication of any other message. This means that for any pair of messages m_i, m_j, we must have at least q sets which contain m_i but not m_j. Thus if m_j is lost, we still have enough good tags to authenticate m_i with the desired degree of security.

This "set-cover" approach requires the sender to remember many old messages. If a node can remember at most v old messages, then we must have $S_i \subset [i - v, i]$ for all i. Memory is presumably quite limited since we are dealing with very low-power nodes. Note that v is also the maximum delay before a message finally achieves full authentication, which is another reason to limit v.

Thus we have the following problem: Given memory bound v, find sets S_i that maximize q where

- For each $i \in \mathbb{N}$, $S_i \subset [i - v, i]$.
- For each $i \neq j$ there are at least q sets S with $i \in S$, $j \notin S$.

Given a collection of sets S, define the *strength* of the collection as

Definition 1
$$\mu(S) = \min_{i,j} \#\{t | i \in S_t, j \notin S_t\}$$

(here $\#A$ denotes the size of a set A). We have defined S as an infinite collection; such a collection would of course be specified either by rotating through a finite collection of given sets, or more generally by specifying a way to generate S_i as a function of i. To get the process started, we can implicitly have dummy messages $m_{-v}, \cdots m_{-1}, m_0 = 0$.

3   Sliding-Window Construction

We first consider the special case in which each set S_i is defined by a "sliding window"; we select a set of distances $\delta = \{\delta_1 < \cdots \delta_k \leq v\}$ and let each $S_i = \{i - \delta_1, \cdots i - \delta_k\}$. Without loss of generality we can assume $\delta_1 = 0$ and $\delta_k = v$.

It will be convenient to identify the vector of distances d with a binary sequence b of length v which is zero except on δ. Then

$$S_i = \{i - d : b_d = 1\}.$$

We may also treat b as an infinite sequence with $b_j = 0$ for j outside of the interval $[0, v - 1]$. We say that difference d is "realized (at j)" if $b_j = 1, b_{j+d} = 0$ and call the ordered pair $(j, j + d)$ a "realization of d". We define

Definition 2

$$\mu_b(d) = \#\{i | b_i = 1, b_{i+d} = 0\}, \tag{2}$$

so $\mu_b(d)$ is the number of times d is realized in b (we may drop the subscript b when the context is clear). We then define the strength of the vector b as

$$\mu(b) = \min_d \mu_b(d), \tag{3}$$

consistent with the definition given above for $\mu(S)$. Then b is a *t-distance avoiding sequence* if $\mu(b) \geq t$.

We can assume with no loss of generality that $b_0 = b_{v-1} = 1$. Changing b_0 from zero to one does not destroy any realizations of any d; changing b_{v-1} from zero to one creates one new realization of d for every d, while destroying one realization of each d with $b_{v-d-1} = 1$. Thus $\mu(b)$ might increase and cannot decrease.

Note that we do not need to consider differences with absolute value greater than v; for such differences we clearly have $\mu_d = \sum_i b_i$, which is a trivial upper bound on all μ_d. In fact we can limit our attention to positive differences:

Lemma 1. *For all d, $\mu_d = \mu_{-d}$.*

Proof. $\mu_d - \mu_{-d} = \sum_i (b_i - b_{i+d}) = 0$ ☐

We can bound the maximum strength of a sequence for a given memory size v as follows:

Theorem 1. *For all b of length v,*

$$\mu(b) \leq \frac{v + 2}{3}. \tag{4}$$

Proof. We in fact prove the stronger result that

$$\min(\mu_1, \mu_2) \leq \frac{v + 2}{3}. \tag{5}$$

Let R_ℓ^s be the number of runs of $s \in \{0, 1\}$ of length ℓ. With no loss of generality we may assume that $\ell \leq 2$; in a long run of ones or zeros, the third value can be changed without decreasing μ_1 or μ_2. Then we have

$$v = R_1^0 + R_1^1 + 2R_2^0 + 2R_2^1. \tag{6}$$

Runs of zeros and ones alternate, and we can assume with no loss of generality that the sequence starts and ends with 1, so we also have

$$R_1^1 + R_2^1 = 1 + R_1^0 + R_2^0, \tag{7}$$

and combining these we obtain

$$v = 2(R_1^1 + R_2^1) + R_2^0 + R_2^1 - 1. \tag{8}$$

Now $\mu_1 = R_1^1 + R_2^1$ since this is the number of runs of ones. Furthermore, the distance 2 will fail to be realized at $b_i = 1$ if and only if this is immediately followed by a zero-run of length one; thus from (7) we have

$$\mu_2 = R_1^1 + 2R_2^1 - R_1^0 = 1 + R_2^0 + R_2^1, \tag{9}$$

therefore

$$v = 2\mu_1 + \mu_2 - 2 \tag{10}$$

and the theorem follows.   □

In fact, the same bound applies to any collection of sets, without the sliding-window assumption:

Theorem 2. *For any collection of sets S with memory bound v,*

$$\mu(S) \le \frac{v+2}{3}. \tag{11}$$

Proof. Let b^i be the binary sequence corresponding to the set S_i, i.e. $b_t^i = 1$ if and only if $i - t \in S_i$. Consider v consecutive sets $S_i, \cdots S_{i+v-1}$. These are the only sets which can contain i; thus for any distance d, at least $\mu(S)$ of these sets contain i but not $i+d$. Thus the sequences $b^i \cdots b^{i+v-1}$ together contain at least $\mu(S)$ realizations of d, where in sequence b^{i+t} we only count a realization at bit position t.

Similarly, the sequences $b^{i+1} \cdots b^{i+v}$ contain at least $\mu(S)$ *different* realizations of d and so, for any L, the $v + L - 1$ sequences $b^i \cdots b^{i+v+L-2}$ contain qL different realizations of d. Thus as L approaches infinity, the average value of $\mu_{b^j}(d)$ for $i \le j \le i + v + L - 2$ approaches (at least) $\mu(S)$. In particular this holds for $d = 1, 2$. Now from the proof of Theorem 1, we know that

$$2\mu_{b^j}(1) + \mu_{b^j}(2) \le v + 2$$

for each sequence b^j, thus the same must be true of the averages, i.e.

$$3\mu(S) \le v+2.   □$$

It is not known whether the strength of an arbitrary collection of sets can exceed the maximum strength achievable by a sliding window. The proof of Theorem 2 shows that if this is the case, we must have a collection of sliding windows in which the average value of each μ_d exceeds the maximum strength of any single sliding window.

We also have the following relationship among different distances:

Theorem 3. *For all d, d'*

$$\mu_d + \mu_{d'} \geq \mu_{d+d'}. \tag{12}$$

In particular,

$$2\mu_d \geq \mu_{2d}.$$

Proof. If $d \neq d'$ define a mapping from realizations of $d + d'$ to realizations of d and d' as follows: for each $b_i > b_{i+d+d'}$, map $(i, i+d+d')$ to $(i, i+d)$ if $b_{i+d} = 0$, else map to $(i+d, i+d+d')$. Clearly this map is injective.

If $d = d'$ then similarly every realization of $2d$ can be mapped to exactly one realization of d, and no more than two realizations of $2d$ can map to the same point. □

3.1   Lower Bounds for the Sliding Window Construction

To obtain lower bounds, we recall the following well-known [5] concept:

Definition 3. *A (v, k, λ)-cyclic difference set is a subset $D \subset \mathbb{Z}_v$ such that $\#D = k$ and, for each $d \in \mathbb{Z}_v \backslash \{0\}$, there are exactly λ pairs $a, b \in D$ with $a - b = d$.*

If such a set exists we must have $\lambda(v - 1) = k(k - 1)$. In particular, if $k = \frac{v-1}{2}$, then for each $d \neq 0$ there are exactly $\frac{k-1}{2}$ pairs $x, y \in D$ with $x - y = d \mod v$; thus there are exactly $\frac{k+1}{2}$ pairs $x \in D, y \notin D$ with $x - y = d \mod v$. So if b is a binary sequence of length v with $b_i = 1$ precisely when $i \in D$, we have

$$\mu(b) \geq \frac{k+1}{2} = \frac{v+1}{4}. \tag{13}$$

We can have strict inequality in (13), because we may have $y = x - d < 0$ when the indices are not taken modulo v, giving a realization of d even though $y \in D$.

A *translate* of D is a set $D + t = \{a + t | a \in D\}$. Clearly any translate of a difference set is another difference set with the same parameters, but different translates can give different values of $\mu(b)$. However, we have $\mu(b) \leq 1 + \frac{v+1}{4}$ because $\mu_1 \leq 1 + \frac{v+1}{4}$; the only way to get a non-cyclic realization of $d = 1$ is at $b_0 = 1$ when $b_{v-1} = 1$. Experimentation suggests the following

Conjecture 1. for every difference set of size v, there exists a translate with $\mu(b) = 1 + \frac{v+1}{4}$.

Going in the other direction, a difference set always gives us a sequence b attaining $\mu(b) = q$ with length strictly less than $4q - 1$; taking a translate with $b_{v-t} = b_{v-t+1} = \cdots = b_{v-1} = 0$, let us truncate b to length $v - t = 4q - 1 - t$.

Cyclic difference sets do not take advantage of the edge effects inherent in this problem, and they do not necessarily provide optimal solutions. It appears to be possible to do somewhat better than $\frac{v+1}{4}$ for all v (see Sect. 3.3), although it also seems that the maximum $\mu(b)/v$ approaches $\frac{1}{4}$ as v approaches infinity.

3.2   Optimal Sequences for Small Memory Bounds

For small values of v, optimal sequences can be found by exhaustive search; results are summarized in Table 1. Only "critical" values are shown, i.e. v at which the maximum $\mu(b)$ changes. These results show that the bound of Theorem 1 can be attained for small v. For all lengths except $v = 35$, the table gives the lexicographically smallest vector attaining $\max \mu(b)$. Exhaustive search was not completed for $v = 35$, but $\mu(b) = 11$ is still known to be optimal; if we had b of length 35 attaining $\mu(b) = 12$, we could remove one bit to obtain $\mu(b) = 11$ at length 34, which has been ruled out by exhaustive search.

Note that most of these optimal values cannot be attained by the difference-set constructions of Sect. 3.1. For example, a difference set of size 31 would give $\mu(b) \leq 9$. Furthermore, a sequence attaining $\mu(b) = 10$ with $v = 31$ could not be obtained by truncating a block of consecutive zeros from a larger difference set; the larger difference set would have $\frac{v+1}{4} = 10$ thus $v = 39$, but no cyclic difference set of that size exists.

As a secondary objective, we could seek to minimize the Hamming weight of b: this weight is the number of messages that must be combined to compute each authentication tag, so reducing this weight may reduce the amount of work needed to compute a_i. For all v in this table (except $v = 21$ and possibly $v = 35$) exhaustive search confirms that there are no optimal vectors with weight less than $\frac{v}{2}$.

3.3   Iterative Improvement of Windows

Starting with a random binary vector b^0, we can attempt to maximize q by iterative local improvement. At each step, we change one bit of the current solution b^i. If we can attain $\mu(b^{i+1}) > \mu(b^i)$ by flipping a single bit, we do this (note that a single bit change cannot increase the strength of the vector by more than 1, since it cannot change any μ_d by more than 1). If such immediate improvement is not possible, we consider the set of distances d which are tight, i.e. which have $\mu_d = \mu(b^i)$. The local optimization criteria is to reduce the size

Table 1. Optimal windows for small v

v	$\max \mu(b)$	an optimal vector
4	2	1 1 0 1
7	3	1 1 0 0 1 0 1
10	4	1 1 0 1 0 1 0 0 1 1
14	5	1 1 1 0 0 1 0 1 0 1 1 0 0 1
17	6	1 1 1 0 0 1 0 1 1 0 0 1 1 0 1 0 1
21	7	1 1 1 0 0 0 1 0 1 0 1 1 0 1 0 0 1 1 0 0 1
24	8	1 1 1 0 0 0 1 0 1 1 0 1 0 0 1 1 0 0 1 1 0 1 0 1
27	9	1 1 1 0 0 1 0 1 0 1 0 1 1 0 0 1 0 1 1 0 0 0 1 1 0 1 1
31	10	1 1 1 1 0 0 0 1 1 0 1 0 1 0 0 1 1 0 0 1 1 0 1 0 0 1 1 0 1 0 1
35	11	1 1 0 1 0 1 0 0 1 0 0 1 1 1 1 0 0 1 1 0 0 0 1 0 1 1 0 0 1 0 1 0 1 1 1

Table 2. Best known $\mu(b)$ for large v

v	max known $\mu(b)$	Min weight attaining max $\mu(b)$
40	12	19
60	18	30
100	28	48
200	55	100
300	81	150

of this set as much as possible, subject to the condition that strength does not decrease (i.e. that there is no d for which μ_d decreases to $\mu(b^i) - 1$). If local improvement is impossible, we flip two bits at random.

In order to implement this search, note that it is not necessary to recompute q from scratch for every vector at Hamming distance 1 from b^i. Instead, for each bit position i and for each tight distance d, we can compute in constant time the effect on μ_d of flipping bit i. Table 2 gives the strengths of the best vectors found in this manner.

4   More General Independence Conditions

More generally we may consider conditions of the following form: for parameters (r, r'), require that loss of any r messages does not prevent authentication of more than r' remaining messages. The problem considered above is the special case $r = 1, r' = 0$. In the general case we have the following: for any set A with $\#A = r$, there can be no more than r' indices $i \notin A$ such that

$$\#\{j | i \in S_j, A \cap S_j = \emptyset\} < q.$$

This is a difficult condition to deal with in general, so we still consider some special cases, and still consider only the sliding-window approach. If we have $r = 1$ but $r' > 0$, then we no are no longer maximizing the minimum value of μ_d; instead we seek to maximize the $(r' + 1)^{\text{th}}$ smallest value. The r' smallest values correspond to the r' messages for which we are allowed to loose full authentication.

If we have $r > 1$ and $r' = 0$, then we require that the loss of any set of r messages does not prevent authentication of any other message. For this case we define $\mu_{(d_1, d_2, \cdots d_r)}$ as the number of indices j where $b_j = 1, b_{j+d_1} = \cdots = b_{j+d_r} = 0$, and maximize $q_r(b) = \min \mu_d$ over all vectors d, where we may assume $i < j$ implies $d_i < d_j$ since order does not matter. Note that the d_i may be negative. Trivialy we have

$$q_r(b) \leq \frac{v + r}{r + 1} \tag{14}$$

since every realization of $d = (1, 2, \cdots r)$ (except at $b_{v-1} = 1$) consists of a 1 followed by r zeros, and these cannot overlap. Table 3 gives the best known values of q_2 for various memory bounds v; in general these can be attained while simultaneously coming close to the best known $q = q_1$.

Table 3. Best known $q_2(b)$ for some v

v	max known $q_2(b)$	best $q_1(b)$ for this q_2
10	2	3
20	4	6
30	5	9
40	7	12
60	10	16
100	16	26

Acknowledgment

The authors would like to thank Carl Diegert and Roberto Tamassia for helpful discussions, and Austin McDaniel for implementing iterative search.

References

1. Perrig, A., Canetti, R., Tygar, D., Song, D.: The TESLA broadcast authentication protocol. Crypto. Bytes 5(2), 2–13 (2002)
2. Perrig, A., Canetti, R., Tygar, J.D., Song, D.X.: Efficient authentication and signing of multicast streams over lossy channels. In: IEEE Symposium on Security and Privacy, pp. 56–73 (2000)
3. Luk, M., Perrig, A., Whillock, B.: Seven cardinal properties of sensor network broadcast authentication. In: SASN 2006: Proceedings of the fourth ACM workshop on Security of ad hoc and sensor networks, pp. 147–156. ACM, New York (2006)
4. Bellare, M., Canetti, R., Krawczyk, H.: Message authentication using hash functions: the HMAC construction. CryptoBytes 2(1), 12–15 (1996)
5. Hall Jr., M.: Combinatorial theory, 2nd edn. John Wiley & Sons, Inc., New York (1998)

On the Number of Linearly Independent Equations Generated by XL

Sondre Rønjom and Håvard Raddum

Department of Informatics, University of Bergen, N-5020 Bergen, Norway
sondrer@ii.uib.no, haavardr@ii.uib.no

Abstract. Solving multivariate polynomial equation systems has been the focus of much attention in cryptography in the last years. Since most ciphers can be represented as a system of such equations, the problem of breaking a cipher naturally reduces to the task of solving them. Several papers have appeared on a strategy known as *eXtended Linearization* (XL) with a view to assessing its complexity. However, its efficiency seems to have been overestimated and its behaviour has yet to be fully understood. Our aim in this paper is to fill in some of these gaps in our knowledge of XL. In particular, by examining how dependencies arise from multiplication by monomials, we give a formula from which the efficiency of XL can be deduced for multivariate polynomial equations over $\mathbb{F}_2$. This confirms rigorously a result arrived at by Yang and Chen by a completely different approach. The formula was verified empirically by investigating huge amounts of random equation systems with varying degree, number of variables and number of equations.

Keywords: XL, Gröbner bases, Stream Ciphers.

1 Introduction

The problem of solving multivariate polynomial equations is encountered in many different fields. This problem has in particular received a great deal of attention in cryptography. The problem of breaking a cipher is reformulated as a problem of solving a (very large) system of polynomial equations. Solving multivariate polynomial equations over $\mathbb{F}_2$ is known to be NP-hard.

XL was proposed in [10], as a new algorithm for solving multivariate polynomial equations. A parameter D is associated with XL, and the complexity of XL is exponential in D. In [10] and [11], the authors try to evaluate for which D XL works. For a system with m equations in n variables they use the estimation $D \approx \frac{n}{\sqrt{m}}$, which seems to work fine for special cases, but for general systems this approximation becomes very inaccurate. For the AES, we estimate that $D \approx \frac{n}{\sqrt{m}} \cdot 2.5$, and for Serpent it is more like $D \approx \frac{n}{\sqrt{m}} \cdot 4.9$, which makes a huge difference, as running time is exponential in D. So it is evident that the early estimations of D are inaccurate. This is also shown in [13] where a formula for estimating D is given. This formula applies to quadratic systems over large fields and is proved to be correct given one more assumption.

S.W. Golomb et al. (Eds.): SETA 2008, LNCS 5203, pp. 239–251, 2008.

The link between XL and computing Gröbner bases was established in [12]. Their computer simulations show that XL do not seem to follow the proposed bound $D \approx \frac{n}{\sqrt{m}}$, but behave much worse. They also compare Faugères F_4 Gröbner basis algorithm with XL, and show that F_4 computes a Gröbner basis in less time, using less space. This conclusion is not very surprising, as XL tries to compute a Gröbner basis, but spends huge amount of time computing dependent equations, such that the reduction step spends an equal amount of time computing zero. The difference from other Gröbner basis algorithms is that XL contains no strategies nor avoid linear dependencies. So XL can be said to be a worst kind of Gröbner basis algorithm.

In [5], Yang and Chen describe a formula for estimating D, but as they mention, their formula is neither trivial to use nor proved to be correct.

2   Preliminaries

Let $\lambda = \mathbb{F}_2[x_1, ..., x_n]/\{x_i^2 + x_i\}_{1 \leq i \leq n}$ denote the ring of Boolean functions in n variables $x_1, x_2, \ldots, x_n$ which satisfy the relation $x_i^2 + x_i = 0$ since $x_i \in \mathbb{F}_2$. Then an $f \in \lambda$ defines a function $\mathbb{F}_2^n \xrightarrow{f} \mathbb{F}_2$, on the n-dimensional vector space $\mathbb{F}_2^n$, with values in $\mathbb{F}_2$. We define a set of polynomials by

$$F = \{f_1(x_1, ..., x_n), ..., f_m(x_1, ..., x_n)\} \subseteq \lambda$$

and the associated system of equations

$$E = \{f_1(x_1, ..., x_n) = 0, ..., f_m(x_1, ..., x_n) = 0\}.$$

We associate a function over $\mathbb{F}_2$ with its coefficient vector, so to avoid confusion we introduce notation for this.

Definition 1. *Let $f \in \lambda$. We denote by $[f]$ the vector indexed by the 2^n monomials in λ, which contains a 1 at indices where the corresponding monomial appear in f and 0 otherwise, with respect to some monomial ordering. We call $[f]$ the coefficient vector of f. When writing $\mathfrak{m}[f]$ for a monomial $\mathfrak{m}$, we will mean $[\mathfrak{m}f]$. This is well-defined, so for a polynomial g, $g[f]$ (a sum of vectors) is equal to $[gf]$.*

2.1   The XL Algorithm

Let F be a system of m polynomials of degree D_e. We associate a number $D_\mathfrak{m} \in \mathbb{N}$, such that XL is constrained by $1 \leq \deg(\mathfrak{m}) \leq D_\mathfrak{m}$, where $\mathfrak{m}$ is a monomial. Further let $D = D_e + D_\mathfrak{m}$, and let $\lambda_{\leq d}$ denote the span of the monomials of degree less than or equal to d. XL will construct a new system U_D, formed by multiplying each polynomial in F with all monomials of degree less than or equal to $D_\mathfrak{m}$, that is, $U_D = \lambda_{\leq D_\mathfrak{m}} \otimes F$. Then U_D will contain polynomials satisfying $0 \leq \deg(f_i) \leq D$. The version of XL defined in [10] consider only quadratic systems, but we generalize our analysis and consider polynomial systems of any degree D_e. XL constructs a set of polynomials U_D, satisfying the following properties:

- $F \subseteq U_D \subseteq \lambda_{\leq D}$: F is contained in U_D which is a subset of all polynomials of degree less or equal to D.
- The total number of polynomials in U_D after an execution of XL is $|U_D| = m \cdot \sum_{i=0}^{D_m} \binom{n}{i}$, but these are not necessarily linearly independent.

The XL algorithm can now be described as follows (adapted from original paper):

Algorithm (The XL algorithm). For a positive integer D_m and a set of polynomial equations $F = \{f_1, \ldots, f_m\}$, execute the following steps:

1. *Multiply*: Generate all the products $m \cdot f_i \in U_D, 1 \leq \deg(m) \leq D_m, f_i \in F$
2. *Linearize*: Consider each monomial m with $deg(m) \leq D$ as a new variable and perform Gaussian elimination on the equations obtained in Step 1.
3. *Solve or Repeat*: If the system is not solved, increase D_m and redo step 1 and 2.

Remark 1. Setting $D_e = 2$ gives the algorithm from [10].

The idea behind the algorithm is that new linearly independent equations are in fact generated. But this comes at the expense of increasing the total degree D of the system U_D. In [12] it is shown that XL is basically the construction of the Macaulay matrix. Lazard (see [17] and [18]) describes the link between linear algebra and computing Gröbner bases and proves that Gaussian elimination on the Macaulay matrix for D large enough returns a Gröbner basis. Thus the problem presented above can be reformulated as:

Problem 1. For which D will the XL algorithm, after a row reduction of U_D with respect to a monomial ordering, return a matrix with full rank?

It should be noted that the fastest Gröbner basis algorithm, Faugeres F_5 algorithm, is based on the same type of computation as XL. However, F_5 differs from XL in that it avoids computing unnecessary linear dependencies. But our result applies also to the case of F_5, as both algorithms reach the same degree D before a linear basis can be computed.

2.2 Restrictions on F

Exact analysis becomes very hard, if not impossible, if we assume nothing about the polynomials in F. The best approach is to define a sufficiently general restriction on the system F in which the algorithm can be analyzed in fair terms. In this section we define some restrictions on F, which is almost always fulfilled for systems coming from cryptography.

First, we put the following restrictions on F:

- Total degree: $\deg(f_i) = D_e, \forall f_i \in F$.
- All $[f_i]$ are linearly independent.

We will also restrict F by defining the types of relations which occur between the polynomials.

Construct the coefficient matrix H_0 of the set U_D, and assume $D_m \geq D_e$. The rows of H_0 consist of all vectors $m[f_j]$, with $1 \leq j \leq m$ and $0 \leq \deg(m) \leq D_m$. Since we have formed products of f_j with all monomials of degree $0 \leq \deg(m) \leq \deg(f_i)$, some of the rows in H_0 will add together to $[f_j]f_i$. We can do the same starting with f_i and construct some coefficient vectors that add up to $[f_i]f_j$. Since $[f_i]f_j = [f_j]f_i$ we have identified a linear dependency among some rows of H_0.

During the execution of XL, as long as $D_m \geq D_e$, we will always create linear dependencies of the form

$$[f_j] \cdot f_i + [f_i] \cdot f_j = 0.$$

The relation

$$f_i \cdot [f_i] + [f_i] = 0$$

also occurs since we have that $x_i^2 + x_i = 0$. These relations are trivial in that they will always exist for the equation systems over λ.

3   Equation Systems from LFSR-Based Stream Ciphers

Equations coming from LFSR-based stream ciphers are examples of equations that are very interesting with respect to an XL-type algorithm. Assume we are given a set of equations $F = \{f_1 = 0, f_2 = 0, \ldots, f_m = 0\}$ coming from a regularly clocked filter generator. Let $g(x) \in \mathbb{F}_2[x]$ denote a primitive generator of $\mathbb{F}_{2^n}$ such that the binary sequence $s_t = \sum_{i=0}^{n-1} s_{t-n+i} c_i$ is a sequence with maximal period $2^n - 1$ (an m-sequence). Then let $f(x_1, \ldots, x_n)$ denote a Boolean filter function of degree d. Then the keystream-sequence $z = \{z_t\}_{t=0,1,\ldots}$ is generated by

$$z_t = f(s_t, s_{t+1}, \ldots, s_{t+n-1}),$$

which can be represented by the equations $F = f_t(s_0, \ldots, s_{n-1}) + z_t = 0_{t=0,1,\ldots}$, where $f(s_t, \ldots, s_{t+n-1}) = f_t(s_0, \ldots, s_{n-1})$, since s_t can be written in terms of the initial state bits $s_0, s_1, \ldots, s_{n-1}$. Let $W_d = \sum_{i=1}^{d} \binom{n}{i}$ denote the number of monomials of degree less or equal to d. Then the coefficient vector $[f]$ may be restricted to a length W binary vector. Notice that in the direct algebraic attack we need W equations in order to solve the system using a naive linear algebra attack. Following [RonHel], there exist a $W \times W$ linear matrix operator T, which is invariant of the filtering function $f(x_1, \ldots, x_n)$, but variant of the polynomial $g(x)$. Using their notation, we may instead write the sequence z_t by

$$\begin{aligned}
z_t &= f(s_t, s_{t+1}, \ldots, s_{t+n-1}) \\
&= s_t^*[f_0] \\
&= s_0^* T^t[f] \\
&= s_0^*[f_t],
\end{aligned}$$

where s_t^* is the expanded state vector $[s_t, \ldots, s_{t+n-1}, s_t s_{t+1}, \ldots, s_{t+n-1} \cdots s_{t+n-d}]$. Thus, the equations in F are simply generated by $F = \{T^t f\}_{0 \leq t \leq m}$ for some $m \leq W$. In [19], the exact rank of the equation system F is determined. If $l(z)$ denotes the degree of the minimal polynomial of the sequence $z = \{z_t\}_{t=0,1,\ldots}$, then we have that $\dim(F) = m$ where $0 \leq m \leq ls(z) \leq W$. There are basically two cases to consider:

- A. if $m = l(z) \leq W$, then the system can be solved using a linear subspace attack or variations (see for instance [20]).
- B. if $m < l(z) \leq W$, then the system of m linearly independent equations may be solved using an XL-type algorithm.

This just tells us that the systems coming from the filter generator satisfy the two first restrictions in our analysis; that they are linearly independent and have the same degree. But there is a possibility that the system may contain nonlinear dependencies to begin with. For instance, the sequence z may satisfy a nonlinear recursion $z_{t+r} = h(z_t, z_{t+1}, \ldots z_{t+r-1}), 0 \leq r \leq m$, meaning that $f_{t+r} = h(f_t, f_{t+1}, \ldots, f_{t+r-1}), 0 \leq r \leq m$. If such a relation exist, we do not need to use a method like XL, since we may generate as many equations as we need using the nonlinear recursion h. On the other hand, determining such relations on practical systems seems to be a very hard problem.

But for a D_m and D_e where $D = D_m + D_e$, then if there are no nonlinear relations between the m equations in F of degree $\lfloor \frac{D_m}{D_e} \rfloor$, our estimate of the number of linearly independent equations will be exact since the system will only contain relations that we construct ourselves. Thus if the smallest degree of a nonlinear relation between the equations in F is k, then our analysis will be exact when applying XL with $D_m = k \cdot D_e - 1$, but not for $D_m = k \cdot D_e$.

4 Preliminary Observations

Before we present an explanation of XLs behaviour, we give an intuitive presentation of how to count the dependencies which occur in the application of an algorithm such as XL.

We assume that the systems we study behave according to our assumptions in Section 2 and use the notation introduced in Section 2. We will look at the analysis of XL on quadratic systems presented in [11], and correct a mistake made there. This will help to see how to systematically count the number of linear dependencies created by XL.

In the following, we set $D_e = 2$ and $D = D_m + D_e$. In [11] they use R to count the total number of equations, and set S to be the number of dependencies and I the number of linearly independent equations generated by XL, such that

$$I = R - S.$$

The authors do computer simulations on quadratic polynomial equations for $3 \leq D \leq 6$ and identify some relations. For instance, for $D = 4$ they identify two dependencies:

1. $f_i[f_j] + f_j[f_i] = 0$,
2. $f_i[f_i] + [f_i] = 0$.

For $D = 5$ they identify additionally two types of dependencies:

3. $f_i[f_j]x_k + [f_i]f_jx_k = 0$
4. $f_i[f_i]x_k + [f_i]x_k = 0$

The authors conclude that these are the only existing dependencies for $D = 4$ and $D = 5$, and verify that their estimations are coherent with computer simulations. For $D = 4$ there are $\binom{m}{2}$ ways of constructing relations on the form 1 and m ways of constructing relations of the form 2. For $D = 5$ these numbers are multiplied with the $n+1$ monomials of degree ≤ 1 to form additional types of dependencies:

- Case $D = 4$: $I = R - \binom{m}{2} - m$
- Case $D = 5$: $I = R - (n+1)\binom{m}{2} - (n+1)m$.

For $D = 6$ they state that the number of linearly independent equations is

$$I = R - \left(\binom{n}{2} + \binom{n}{1} + 1\right) \cdot \left(\binom{m}{2} + m\right). \tag{1}$$

At this point we step in and show where their analysis becomes wrong. They conclude that the only relations will be multiples of $f_i[f_j] + f_j[f_i] = 0$. It seems reasonable to assume that they have drawn this conclusion based on Buchberger's two criterion, but this is not correct and will turn out fatal in further analysis for larger D. Their formulas are indeed correct for $3 \leq D \leq 5$, but for $D = 6$ they forget to count the *dependencies among the dependencies*. By a slight abuse of notation, which will be clarified in the next section, these dependencies may be expressed as follows (the number of such dependencies is indicated in the brackets to the right):

5. $f_i[f_j[f_k]] + [f_i[f_j]]f_k + [[f_i]f_k]f_j = 0$ $(\binom{m}{3})$
6. $[[f_i]f_j]f_j + [[f_i]f_j] = 0$ $(2 \cdot \binom{m}{2})$
7. $[[f_i]f_i]f_i + [[f_i]] = 0$ $(\binom{m}{1})$

This means we count $\binom{m+2}{3}$ of the dependent equations twice, so we need to balance this by calculating from an inclusion/exclusion point of view. Using the authors notation, the correct bound should have been:

$$I = R - \left(\binom{n}{2} + \binom{n}{1} + 1\right) \cdot \left(\binom{m}{2} + m\right) + \left(\binom{m}{3} + 2\binom{m}{2} + \binom{m}{1}\right) \tag{2}$$

Note that the formulas above for $3 \leq D \leq 6$ works only for quadratic equations. The dependencies behave with respect to the degree D_e of the initial system E. If for instance $D_e = 7$ there would be constructed no dependencies applying XL with $D_m \leq 6$. If we work with linear equations, XL will introduce new dependencies for each time we increase D_m.

5   The Number of Linearly Independent Equations in XL

In this section we will estimate the number of linearly independent equations one generates with the XL-method. As we will see, the linear dependencies we can identify is governed by products of polynomials from F, and the number of dependencies is calculated by counting the number of such products. We multiply each f_i with all monomials of degree $\leq D_m$ (including the monomial 1, which keeps the original polynomials) to form the set U_D.

Let H_0 be the matrix whose rows are the coefficient vectors of all polynomials in U_D. The columns of H_0 are indexed by all monomials of degree $\leq D_m + D_e$. To avoid confusion in the generalization that follows, the rows of H_0 are indexed by $\mathfrak{m} \cdot r(f_i)$, instead of $\mathfrak{m}[f_i]$. The entry $(\mathfrak{m} \cdot r(f_i), \mathfrak{m}')$ is 1 if $\mathfrak{m}'$ occurs as a term in $\mathfrak{m} \cdot f_i$ and 0 otherwise.

We will now recursively construct a sequence of binary matrices $H_i, i \geq 1$. The rows of H_i will be indexed by $\mathfrak{m} \cdot r(f_{i_1}^{e_1} f_{i_2}^{e_2} \cdots f_{i_s}^{e_s})$, where $\sum_{j=1}^{s} e_j = i+1$ and $\mathfrak{m}$ is a monomial with $\deg(\mathfrak{m}) \leq D_m - i \cdot D_e$. The columns of H_i will have the same indices as the rows of H_{i-1}. The degree of $\mathfrak{m}$ needs to be non-negative, so the final H_i we construct is for $i = \lfloor \frac{D_m}{D_e} \rfloor$. When writing $g \cdot r(\cdots)$ for a polynomial $g = \mathfrak{m}_1 + \ldots + \mathfrak{m}_k$, we will mean the sum $\mathfrak{m}_1 \cdot r(\cdots) + \ldots + \mathfrak{m}_k \cdot r(\cdots)$. The row $\mathfrak{m} \cdot r(f_{i_1}^{e_1} f_{i_2}^{e_2} \cdots f_{i_s}^{e_s})$ will contain a 1 in all columns that occur as terms in the following sum:

$$\mathfrak{m} \cdot \sum_{j=1}^{s}(f_{i_j} + (e_{i_j} - 1 \mod 2))r(f_{i_1}^{e_1} \cdots f_{i_j}^{e_j - 1} \cdots f_{i_s}^{e_s}). \tag{3}$$

The rest of the entries in row $\mathfrak{m} \cdot r(f_{i_1}^{e_1} f_{i_2}^{e_2} \cdots f_{i_s}^{e_s})$ will contain 0.

Let $\mathbf{v}$ be a binary vector indexed by the rows of H_i. If $\mathbf{v} \cdot H_i = \mathbf{0}$, $\mathbf{v}$ identifies a linear dependency between the rows of H_i. With the matrices $H_i, i = 0, \ldots, \lfloor \frac{D_m}{D_e} \rfloor$ defined, we are ready to prove the first result.

Theorem 1. $H_i \cdot H_{i-1} = [\mathbf{0}]$, the all-zero matrix for $i = 1, \ldots, \lfloor \frac{D_m}{D_e} \rfloor$. That is, each row of H_i identifies a linear dependency among the rows of H_{i-1}.

Proof. The row $\mathfrak{m} \cdot r(f_{i_1}^{e_1} f_{i_2}^{e_2} \cdots f_{i_s}^{e_s})$ in H_i contains a 1 in the columns indexed by the terms in the following sum

$$\mathfrak{m} \cdot \sum_{j=1}^{s}(f_{i_j} + (e_{i_j} - 1 \mod 2))r(f_{i_1}^{e_1} \cdots f_{i_j}^{e_j - 1} \cdots f_{i_s}^{e_s}). \tag{4}$$

If $e_j = 1$, it means that $f_{i_j}^{e_j - 1}$ vanishes from the expression $r(\cdots)$. Assume without loss of generality that $e_{t+1} = \ldots = e_s = 1$, and that $e_j > 1, j = 1, \ldots, t$. We need to show that (4) is the all-zero vector when the terms are regarded as rows in H_{i-1}. We substitute $r(f_{i_1}^{e_1} \cdots f_{i_j}^{e_j - 1} \cdots f_{i_s}^{e_s})$ with the expression given by (3) to find which columns in H_{i-1} that contain a 1 in the rows found in (4).

We then examine the parity of the number of 1's in these columns. After substituting, (4) can be written as

$$
m \cdot \sum_{j=1}^{s} (f_{i_j} + (e_j - 1 \mod 2)) \sum_{k=1, k \neq j}^{s} (f_{i_k} + (e_k - 1 \mod 2)) r(\cdots f_{i_j}^{e_j-1} \cdots f_{i_k}^{e_k-1} \cdots)
$$

$$
+ m \cdot \sum_{j=1}^{t} (f_{i_j} + (e_j - 1 \mod 2))(f_{i_j} + (e_j - 2 \mod 2)) r(\cdots f_{i_j}^{e_j-2} \cdots). \quad (5)
$$

Each term in (5) represents a 1 in the column in H_{i-1} corresponding to the same term. In the double sum where $j \neq k$, each $r(\cdots f_{i_j}^{e_j-1} \cdots f_{i_k}^{e_k-1} \cdots)$ will occur exactly twice, once when $j < k$ and once when $j > k$. Both times the polynomials to be multiplied with $r(\cdots f_{i_j}^{e_j-1} \cdots f_{i_k}^{e_k-1} \cdots)$ will be $m(f_{i_j} + (e_j - 1 \mod 2))(f_{i_k} + (e_k - 1 \mod 2))$, so the number of 1's in columns involving $r(\cdots f_{i_j}^{e_j-1} \cdots f_{i_k}^{e_k-1} \cdots)$ is even. For the remaining single sum, we note that $f_{i_j} f_{i_j} = f_{i_j}$ and that exactly one of $e_j - 1$ and $e_j - 2$ is 1 mod 2 and the other is 0 mod 2. Multiplying out the brackets in the single sum we get $m(f_{i_j} + f_{i_j}) = 0$ mod 2 in front of each $r(\cdots f_{i_j}^{e_j-2} \cdots)$, so the number of 1's in columns involving $r(\cdots f_{i_j}^{e_j-2} \cdots)$ is also even.   □

We are now ready to proceed to the main result, an estimation of the number of linearly independent equations generated by the XL-method.

Theorem 2. *Let I be the number of linearly independent equations generated by the XL-method on a system of m equations in n variables, where each equation has degree D_e and we multiply with all monomials of degree $\leq D_m$. If the only linear dependencies among the rows of H_{i-1} are the ones indicated by H_i, then*

$$
I = \sum_{i=0}^{\lfloor \frac{D_m}{D_e} \rfloor} (-1)^i \binom{m+i}{i+1} \sum_{j=0}^{D_m - i \cdot D_e} \binom{n}{j}.
$$

Proof. By the construction of the matrices H_i, we have $I = \text{rank}(H_0)$. Let the number of rows in H_i be b_i. If all the rows of H_i are linearly independent we will have $\text{rank}(H_{i-1}) = b_{i-1} - b_i$. However, there will in general be linear dependencies also between the rows of H_i, so the correct expression will be

$$
\text{rank}(H_{i-1}) = b_{i-1} - \text{rank}(H_i), i = 1, \ldots \lfloor \frac{D_m}{D_e} \rfloor. \quad (6)
$$

The matrix $H_{\lfloor \frac{D_m}{D_e} \rfloor}$ will have full rank $b_{\lfloor \frac{D_m}{D_e} \rfloor}$ since there is no H-matrix coming after it. By recursively using (6) for substituting the expressions for $\text{rank}(H_i)$ we have the following formula

$$
I = \text{rank}(H_0) = \sum_{i=0}^{\lfloor \frac{D_m}{D_e} \rfloor} (-1)^i b_i. \quad (7)
$$

To finish, we need to compute b_i, the number of rows in H_i. The rows of H_i are indexed by $\mathfrak{m} \cdot r(f_{i_1}^{e_1} f_{i_2}^{e_2} \cdots f_{i_s}^{e_s})$, where $\deg(\mathfrak{m}) \leq D_\mathfrak{m} - i \cdot D_e$ and $\sum_{j=1}^s e_j = i+1$. The number of choices for $\mathfrak{m}$ is $\sum_{j=0}^{D_\mathfrak{m} - i \cdot D_e} \binom{n}{j}$. The number of ways to make a product of $i+1$ equations by picking equations from the total set of m equations is the number of ways to throw $i+1$ balls into m bins. This number is $\binom{m+i}{i+1}$ by [9]. We then get

$$b_i = \binom{m+i}{i+1} \sum_{j=0}^{D_\mathfrak{m} - i \cdot D_e} \binom{n}{j},$$

and substituting this into (7) gives us the desired expression. $\square$

We restrict our analysis to systems which only contain trivial dependencies. This means that our formula is not correct for initial systems containing other types of dependencies. One example of non-trivial dependencies are systems containing nonlinear dependencies.

6   Linking Theorem 2 to Theorem/Conjecture from Related Work

We are aware of three papers by Yang and Chen (one also with Courtois) [5], [6] and [7] which, among other things, try to estimate the number of linearly independent equations generated by the XL-method. Their result uses the notation $[t^D]p(t)$, which represents the coefficient of the D'th-degree term in the polynomial (or series) $p(t)$. For an instance of the XL algorithm where the initial equations all have degree k, let T be the number of monomials generated, and let I be the number of linearly independent equations. For $\mathbb{F}_2$, their result from [5] is as follows.

Theorem 3

$$T - I \geq [t^D]\left(\frac{1}{1-t}\left(\frac{1-t^2}{1-t}\right)^n \left(\frac{1-t^k}{1-t^{2k}}\right)^m\right)$$

when $D < D_{reg}$, where D_{reg} is the degree of the first term with a negative coefficient in the series.

In [5] there is a proof of the theorem for $k = 2$, but this proof has been shown to be flawed. In [6] they write "As pointed out by C. Diem, the [5] proof is inaccurate.... In any event, since it is also confirmed by many simulations we will henceforth assume Theorem 3 holds in general..." In [7] they write "The [5] proof was faulty..."

   Theorem 3 seeks to give an upper bound on the number of linearly independent equations we can get from XL. With the extra assumption in Theorem 2, saying that the only linear dependencies occurring are the ones we can identify, Theorem 3 could be stated with equality. Below we will find the link between Theorems 2 and 3, showing that these two results indeed say the same things when assuming only trivial linear dependencies. However, we thing that it is easier to use Theorem 2 as one can plug in the numbers for a particular system and do the simple arithmetic to get the expected number of linearly independent equations. To find the same thing using Theorem 3, one needs to expand a complicated series to find the coefficient of a particular term. We start by computing the D'th degree coefficient in the series from Theorem 3

Proposition 1

$$[t^D]p(t) = [t^D]\left(\frac{1}{1-t}\left(\frac{1-t^2}{1-t}\right)^n\left(\frac{1-t^k}{1-t^{2k}}\right)^m\right)$$

$$= \sum_{i=0}^{\lfloor\frac{D}{k}\rfloor}(-1)^i\binom{m+i-1}{i}\sum_{j=0}^{D-ik}\binom{n}{j}.$$

Proof The first fraction in $p(t)$ can be written as $\frac{1}{1-t} = \sum_{l=0}^{\infty}t^l$. The second fraction can be expressed as $\left(\frac{1-t^2}{1-t}\right)^n = (1+t)^n = \sum_{j=0}^{n}\binom{n}{j}t^j$. The third fraction can be expressed as $\left(\frac{1-t^k}{1-t^{2k}}\right)^m = (1-(-t^k))^{-m}$. By [8], this is equal to $\sum_{i=0}^{\infty}(-1)^i\binom{m+i-1}{i}t^{ik}$. Since we are only interested in $[t^D]p(t)$ where $D \leq n$, we can cut away terms of degree higher than D in the three sums to get

$$[t^D]p(t) = [t^D]\left(\sum_{l=0}^{D}t^l\right)\left(\sum_{j=0}^{D}\binom{n}{j}t^j\right)\left(\sum_{i=0}^{\lfloor\frac{D}{k}\rfloor}(-1)^i\binom{m+i-1}{i}t^{ik}\right) =$$

$$[t^D]\left(\sum_{l=0}^{D}t^l\sum_{j=0}^{l}\binom{n}{j}\right)\left(\sum_{i=0}^{\lfloor\frac{D}{k}\rfloor}(-1)^i\binom{m+i-1}{i}t^{ik}\right).$$

We are looking for the coefficient of the D'th degree term, so we need only multiply the two sums together with the constraint $l = D - ik$. Taking away the sum over l, substituting $D - ik$ for l in the rest of the first sum and multiplying together we get the desired expression for the D'th degree coefficient.   □

The number D is the maximum degree of a monomial when running the XL algorithm. Relating this to our notation we have $D = D_e + D_m$ and $k = D_e$. We then get the following result.

Corollary 1. *Let $D = D_e + D_m$ and $k = D_e$, then*

$$T - I = [t^D]\left(\frac{1}{1-t}\left(\frac{1-t^2}{1-t}\right)^n\left(\frac{1-t^k}{1-t^{2k}}\right)^m\right)$$

is equivalent to

$$I = \sum_{i=0}^{\lfloor\frac{D_m}{D_e}\rfloor}(-1)^i\binom{m+i}{i+1}\sum_{j=0}^{D_m-ik}\binom{n}{j}.$$

Proof. Rearranging the first expression, using the proposition and substituting for D and k we get

$$I = T - \sum_{i=0}^{\lfloor \frac{D_m}{D_e} \rfloor + 1} (-1)^i \binom{m+i-1}{i} \sum_{j=0}^{D_m - (i-1)D_e} \binom{n}{j}.$$

The T can be expressed as $T = \sum_{j=0}^{D_e + D_m} \binom{n}{j}$, which is the term for $i = 0$ in the sum. Canceling these terms we are left with

$$I = - \sum_{i=1}^{\lfloor \frac{D_m}{D_e} \rfloor + 1} (-1)^i \binom{m+i-1}{i} \sum_{j=0}^{D_m - (i-1)D_e} \binom{n}{j}.$$

We can let the sum start at $i = 0$ by subtracting 1 from the upper limit for i and increasing each i in the rest of the expression by 1. Compensating for the minus sign in front of the sum we get

$$I = \sum_{i=0}^{\lfloor \frac{D_m}{D_e} \rfloor} (-1)^i \binom{m+i}{i+1} \sum_{j=0}^{D_m - iD_e} \binom{n}{j}. \qquad \square$$

The corollary shows that Theorems 3 and 2 are basically the same result. With the proof of Theorem 2 we have also proved Theorem 3 with equality when assuming that only trivial linear dependencies occur. Many computer experiments on random systems (with no restrictions on the dependencies), counting the number of linearly independent equations have been done, both by us and the authors of Theorem 3. In these experiments it has never occurred that Theorem 3 fails, so we believe that the result is correct and end this section with the following conjecture.

Conjecture 1. Let I be the number of linearly independent polynomials in n variables generated by step 1 of the XL algorithm, where all m initial equations have degree D_e and we multiply with all monomials up to degree D_m. Then

$$I \leq \sum_{i=0}^{\lfloor \frac{D_m}{D_e} \rfloor} (-1)^i \binom{m+i}{i+1} \sum_{j=0}^{D_m - ik} \binom{n}{j}.$$

7   Conclusions

The work in this paper comes from the field of cryptography, in particular algebraic attacks on symmetric key ciphers. The complexity of such attacks has been hard to estimate, but this paper shows the XL-algorithm generates a lot of linearly dependent equations and is not as efficient as initially hoped.

Using the formula in Theorem 2, we can predict the smallest D for which the XL algorithm will work over $\mathbb{F}_2$. The formula is easier to use than that of Yang and Chen, since no multiplication of polynomials or series is involved, but only simple arithmetic.

References

1. Courtois, N., Pieprzyk, J.: Cryptanalysis of Block Ciphers with Overdefined Systems of Equations. In: Zheng, Y. (ed.) ASIACRYPT 2002. LNCS, vol. 2501, pp. 267–287. Springer, Heidelberg (2002)

2. Courtois, N.: Higher Order Correlation Attacks,XL Algorithm and Cryptanalysis of Toyocrypt. In: Lee, P.J., Lim, C.H. (eds.) ICISC 2002. LNCS, vol. 2587, pp. 182–199. Springer, Heidelberg (2003)

3. Courtois, N.: Fast Algebraic Attacks on Stream Ciphers with Linear Feedback. In: Boneh, D. (ed.) CRYPTO 2003. LNCS, vol. 2729, pp. 176–194. Springer, Heidelberg (2003)

4. Courtois, N., Meier, W.: Algebraic Attacks on Stream Ciphers with Linear Feedback. In: Biham, E. (ed.) EUROCRYPT 2003. LNCS, vol. 2656, pp. 345–359. Springer, Heidelberg (2003)

5. Yang, B.-Y., Chen, J.-M.: Theoretical Analysis of XL over Small Fields. In: Wang, H., Pieprzyk, J., Varadharajan, V. (eds.) ACISP 2004. LNCS, vol. 3108, pp. 277–288. Springer, Heidelberg (2004)

6. Yang, B.-Y., Chen, J.-M., Courtois, N.: On Asymptotic Security Estimates in XL and Gröbner Bases-Related Algebraic Cryptanalysis. In: López, J., Qing, S., Okamoto, E. (eds.) ICICS 2004. LNCS, vol. 3269, pp. 401–413. Springer, Heidelberg (2004)

7. Yang, B.-Y., Chen, J.-M.: All in the XL Family: Theory and Practice. In: Park, C.-s., Chee, S. (eds.) ICISC 2004. LNCS, vol. 3506, pp. 67–86. Springer, Heidelberg (2005)

8. Anderson, I.: A First Course in Combinatorial Mathematics, Theorem 2.6, 2nd edn., p. 16. Oxford University Press, Oxford (1989)

9. van Lint, J.H., Wilson, R.M.: A Course in Combinatorics, Theorem 13.1, 2nd edn., p. 119. Cambridge University Press, Cambridge (2001)

10. Courtois, N., Klimov, A., Patarin, J., Shamir, A.: Efficient Algorithms for Solving Overdefined Systems of Multivariate Polynomial Equations. In: Preneel, B. (ed.) EUROCRYPT 2000. LNCS, vol. 1807, pp. 392–407. Springer, Heidelberg (2000)

11. Courtois, N., Patarin, J.: About the XL Algorithm over GF(2). In: Joye, M. (ed.) CT-RSA 2003. LNCS, vol. 2612, pp. 141–157. Springer, Heidelberg (2003)

12. Ars, G., Faugre, J.C., Imai, H., Kawazoe, M., Sugita, M.: Comparison Between XL and Gröbner Basis Algorithms. In: Lee, P.J. (ed.) ASIACRYPT 2004. LNCS, vol. 3329, pp. 338–353. Springer, Heidelberg (2004)

13. Moh, T.: On The Method of "XL" And Its Inefficiency to TTM, IACR eprint server (2001), http://eprint.iacr.org/2001/047

14. Kipnis, A., Shamir, A.: Cryptanalysis of the HFE Public Key Cryptosystem by Relinearization. In: Wiener, M. (ed.) CRYPTO 1999. LNCS, vol. 1666, pp. 19–30. Springer, Heidelberg (1999)

15. Macaulay, F.S.: On some formula in elimination. In: Proceedings of London Mathematical Society, pp. 3–27 (1902)

16. Bardet, M., Faugre, J.C., Salvy, B.: Complexity of Gröbner basis computation for Semi-regular Overdetermined sequences over F_2, with solutions in F_2 Rapport de recherche de l'INRIA, No. 5049 (2003)

17. Lazard, D.: Gröbner-Bases, Gaussian Elimination and Resolution of Systems of Algebraic Equations. In: van Hulzen, J.A. (ed.) EUROCAL 1983. LNCS, vol. 162, pp. 146–156. Springer, Heidelberg (1983)

18. Lazard, D.: Rèsolution des systémes d'équations algébriques. Theoretical Computer Science 15(1) (1981)
19. Rønjom, S., Helleseth, T.: The Linear Vector Space Spanned by the Nonlinear Filter Generator. In: Golomb, S.W., Gong, G., Helleseth, T., Song, H.-Y. (eds.) SSC 2007. LNCS, vol. 4893, pp. 169–183. Springer, Heidelberg (2007)
20. Rønjom, S., Gong, G., Helleseth, T.: On attacks on filtering generators using linear subspace structures. In: Golomb, S.W., Gong, G., Helleseth, T., Song, H.-Y. (eds.) SSC 2007. LNCS, vol. 4893, pp. 204–217. Springer, Heidelberg (2007)

2ⁿ-Periodic Binary Sequences with Fixed k-Error Linear Complexity for k = 2 or 3⋆

2^n-Periodic Binary Sequences with Fixed
k-Error Linear Complexity for $k = 2$ or $3^\star$

Ramakanth Kavuluru

Department of Computer Science, University of Kentucky,
Lexington, KY 40506, USA
ramakanth.kavuluru@uky.edu

Abstract. The linear complexity of sequences is an important measure to gauge the cryptographic strength of key streams used in stream ciphers. The instability of linear complexity caused by changing a few symbols of sequences can be measured using k-error linear complexity. In their SETA 2006 paper, Fu, Niederreiter, and Su [3] studied linear complexity and 1-error linear complexity of 2^n-periodic binary sequences to characterize such sequences with fixed 1-error linear complexity. In this paper we study the linear complexity and the k-error linear complexity of 2^n-periodic binary sequences in a more general setting using a combination of algebraic, combinatorial, and algorithmic methods. This approach allows us to characterize 2^n-periodic binary sequences with fixed 2-error or 3-error linear complexity L, when the Hamming weight of the binary representation of $2^n - L$ is $w_H(2^n - L) \neq 2$. Using this characterization we obtain the counting function for the number of 2^n-periodic binary sequences with fixed k-error linear complexity L for $k = 2$ and 3 when $w_H(2^n - L) \neq 2$.

Keywords: Periodic sequence, linear complexity, k-error linear complexity.

1   Introduction

The linear complexity of a sequence is the length of the shortest linear feedback shift register (LFSR) that can generate the sequence. The LFSR that generates a given sequence can be determined using the Berlekamp-Massey algorithm using only the first $2L$ elements of the sequence, where L is the linear complexity of the sequence. Hence for cryptographic purposes sequences with high linear complexity are essential as an adversary would then need large initial segments of the sequences to recover the LFSRs that generate them using the Berlekamp-Massey algorithm.

⋆ This material is based upon work supported by the National Science Foundation under Grant No. CCF-0514660. Any opinions, findings, and conclusions or recommendations expressed in this material are those of the author and do not necessarily reflect the views of the National Science Foundation.

S.W. Golomb et al. (Eds.): SETA 2008, LNCS 5203, pp. 252–265, 2008.

A system is insecure if all but a few symbols of the key stream can be extracted. Hence for a cryptographically strong sequence, the linear complexity should not decrease drastically if a few symbols are changed. If it did, an attacker could modify the known prefix of the key stream and try to decrypt the result using the Berlekamp-Massey algorithm. If the resulting sequence differed from the actual key stream by only a few symbols, the attacker could extract most of the message. This observation gives rise to k-error linear complexity of sequences introduced in [8] based on the earlier concepts of sphere complexity and weight complexity, see [2]. The k-error linear complexity of a periodic sequence is the smallest linear complexity achieved by making k or fewer changes per period. Besides having large linear complexity, cryptographically strong sequences should, thus, also have large k-error linear complexity at least for small k.

Let $\mathbf{S} = (s_0, s_1, \cdots, s_{T-1})^\infty$ be a periodic binary sequence with period T. We associate the polynomial $\mathbf{S}(x) = s_0 + s_1 x + \cdots + s_{T-1} x^{T-1}$ and the corresponding T-tuple $\mathbf{S}^{(T)} = (s_0, s_1, \cdots, s_{T-1})$ to $\mathbf{S}$. The relationship between the linear complexity, denoted $L(\mathbf{S})$, of $\mathbf{S}$ and the associated polynomial $\mathbf{S}(x)$ is given by

$$L(\mathbf{S}) = T - \deg(\gcd(x^T - 1, \mathbf{S}(x))), \tag{1}$$

see e.g. [1], Lemma 8.2.1. Let $w_H(\mathbf{S})$ denote the Hamming weight of the T-tuple $\mathbf{S}^{(T)}$. For $0 \le k \le T$, the k-error linear complexity of $\mathbf{S}$, denoted $L_k(\mathbf{S})$, is given by

$$L_k(\mathbf{S}) = \min_{\mathbf{E}} L(\mathbf{S} + \mathbf{E}),$$

where the minimum is over all T-periodic binary sequences $\mathbf{E}$ with $w_H(\mathbf{E}) \le k$. Since we consider only 2^n-periodic sequences, we use $T = 2^n$ and the observation

$$x^T - 1 = x^{2^n} - 1 = (x-1)^{2^n} \tag{2}$$

for the rest of the paper.

Let $merr(\mathbf{S})$ denote the minimum value k such that the k-error linear complexity of a 2^n-periodic sequence $\mathbf{S}$ is strictly less than its linear complexity. That is

$$merr(\mathbf{S}) = \min\{k : L_k(\mathbf{S}) < L(\mathbf{S})\}.$$

Kurosawa et al. [5] derived the formula for the exact value of $merr(\mathbf{S})$.

Lemma 1. *For any nonzero 2^n-periodic sequence $\mathbf{S}$, we have*

$$merr(\mathbf{S}) = 2^{w_H(2^n - L(\mathbf{S}))},$$

where $w_H(j)$, $0 \le j \le 2^n - 1$, denotes the Hamming weight of the binary representation of j.

The counting function of a sequence complexity measure gives the number of sequences with a given complexity measure value. Rueppel [7] determined the counting function of linear complexity for 2^n-periodic binary sequences. Using equations (1) and (2) it is straightforward to characterize the 2^n-periodic sequences with fixed linear complexity.

Lemma 2 ([3]). *Let $\mathcal{N}(L)$ and $\mathcal{A}(L)$ denote, respectively, the number of and the set of 2^n-periodic binary sequences with given linear complexity L, $0 \le L \le 2^n$. Then*

$$\mathcal{N}(0) = 1 \text{ and } \mathcal{N}(L) = 2^{L-1} \text{ for } 1 \le L \le 2^n.$$

Also, $\mathcal{A}(0) = \{(0, 0, \cdots)\}$ and $\mathcal{A}(L)$, where $1 \le L \le 2^n$, is equal to the set of 2^n-periodic binary sequences $\mathbf{S}$ with the corresponding polynomials

$$\mathbf{S}(x) = (1 - x)^{2^n - L} a(x),$$

where $a(x)$ is a binary polynomial with $\deg(a(x)) \le L - 1$ and $a(1) \ne 0$.

Recently, using algebraic and combinatorial methods Fu et al. [3] characterized 2^n-periodic binary sequences with fixed 1-error linear complexity. They derived some properties on the structure of the set $\mathcal{A}(L)$ that deal with two symbol changes in sequences in $\mathcal{A}(L)$ and used them to obtain the characterization. In this paper we extend and in some cases generalize those properties to handle four symbol changes and use them to obtain the characterization of 2^n-periodic binary sequences with fixed 2-error or 3-error linear complexity L when $w_H(2^n - L) \ne 2$. Using the characterization we also give the counting function for the number of 2^n-periodic binary sequences with fixed k-error linear complexity L for $k = 2$ and 3 when $w_H(2^n - L) \ne 2$.

2   Extended Properties of $\mathcal{A}(L)$

We recall that $\mathcal{A}(L)$ is the set of 2^n-periodic sequences with fixed linear complexity L, $0 \le L \le 2^n$. For any two 2^n-periodic sequences $\mathbf{S}_1$ and $\mathbf{S}_2$, let $d_H(\mathbf{S}_1, \mathbf{S}_2)$ denote the Hamming distance between the tuples $\mathbf{S}_1^{(2^n)}$ and $\mathbf{S}_2^{(2^n)}$. In this section we derive some properties of $\mathcal{A}(L)$ which extend those in Fu et al.'s paper [3] using the Games-Chan algorithm [4]. First we state two well known related results on 2^n-periodic binary sequences.

Lemma 3 ([3]). *For any 2^n-periodic sequence $\mathbf{S}$, $L(\mathbf{S}) = 2^n$ if and only if $w_H(\mathbf{S})$ is odd.*

Lemma 4 ([3]). *For any 2^n-periodic sequence $\mathbf{S}$, if $w_H(\mathbf{S})$ is even then $L_1(\mathbf{S}) = L(\mathbf{S})$. If $w_H(\mathbf{S})$ is odd, then $L_2(\mathbf{S}) = L_1(\mathbf{S}) < L(\mathbf{S}) = 2^n$.*

We give a generalization of [3, Theorem 1] using a more straightforward approach. The proof is given in the appendix.

Theorem 1. *For a given $r \in \{1, \cdots, n-1\}$, let $1 \le L < 2^{n-r}$. Then for any two distinct sequences $\mathbf{S}_1, \mathbf{S}_2 \in \mathcal{A}(L)$ we have*

$$d_H(\mathbf{S}_1, \mathbf{S}_2) = t \cdot 2^{r+1} \text{ for some } t \in \{1, 2, 3, \cdots, 2^{n-r-1}\},$$

which implies $d_H(\mathbf{S}_1, \mathbf{S}_2) \ge 2^{r+1}$.

By Lemma 1 the linear complexity of any 2^n-periodic sequence $\mathbf{S}$ with $2^{n-1} < L(\mathbf{S}) < 2^n$ and $merr(\mathbf{S}) = 2^{m+1}$, $m \in \{1, \cdots, n-2\}$, can be uniquely expressed as

$$L(\mathbf{S}) = 2^n - \sum_{i=1}^{m+1} 2^{n-r_i}, \qquad (3)$$

where $1 < r_1 < \cdots < r_{m+1} \leq n$. From equation (3), the linear complexity of any 2^n-periodic binary sequence $\mathbf{S}$ with $2^{n-1} < L(\mathbf{S}) < 2^n$ and $merr(\mathbf{S}) \geq 2^{m+1}$, $m \in \{1, \cdots, n-2\}$, can be bounded as

$$2^n - \left(\sum_{i=1}^{m-1} 2^{n-r_i} + 2^{n-r_m+1} \right) < L(\mathbf{S}) < 2^n - \sum_{i=1}^{m} 2^{n-r_i}, \qquad (4)$$

for some $r_i \in \{2, \cdots, n\}$, $i = 1 \cdots m$, satisfying $1 < r_1 < \cdots < r_m$. We note that the lower bound in inequality (4) can never be less than 2^{n-1} conforming to the original condition $2^{n-1} < L(\mathbf{S}) < 2^n$. Also, for any sequence $\mathbf{S}$ satisfying the inequality (4), we have $merr(\mathbf{S}) \geq 2^{m+1}$. We also note that the bounds in (4) are unique in the sense that the linear complexity of any 2^n-periodic sequence $\mathbf{S}$ with $merr(\mathbf{S}) \geq 2^{m+1}$ satisfies exactly one inequality of the particular form given in equation (4).

For a 2^n-periodic sequence $\mathbf{S}$ and two integers i and j with $0 \leq i, j \leq 2^n - 1$, denote by $\mathbf{S}_{i,j}$ the 2^n-periodic binary sequence with corresponding polynomial $\mathbf{S}_{i,j}(x) = \mathbf{S}(x) + x^i + x^j$. We use the following result by Fu et al. [3] for the main result of this section.

Lemma 5. *For any sequence $\mathbf{S} \in \mathcal{A}(L)$, where $2^n - 2^{n-r} < L < 2^n - 2^{n-r-1}$ for some $1 \leq r \leq n-2$, and for any integer $0 \leq i \leq 2^n - 1$, the number of sequences $\mathbf{S}_{i,j} \in \mathcal{A}(L)$, where $0 \leq j \leq 2^n - 1$ and $j \neq i$, is exactly $2^r - 1$ corresponding to all $j \in \{i \oplus t2^{n-r} : 1 \leq t \leq 2^r - 1\}$ where $\oplus$ is the operation of addition modulo 2^n.*

The rest of this section deals with extending Lemma 5 to the case when four symbols per period are changed. For a 2^n-periodic binary sequence $\mathbf{S}$ and four integers i, j, k, and l with $0 \leq i, j, k, l \leq 2^n - 1$, denote by $\mathbf{S}_{i,j,k,l}$ the 2^n-periodic binary sequence with the corresponding polynomial

$$\mathbf{S}_{i,j,k,l}(x) = \mathbf{S}(x) + x^i + x^j + x^k + x^l.$$

The Games-Chan algorithm [4] is a fast algorithm to compute the linear complexity of a 2^n-periodic binary sequence, which we use for the rest of this section.

For any $\mathbf{S} \in \mathcal{A}(L)$ with period $\mathbf{S}^{(2^n)} = (s_0, \cdots, s_{2^n-1})$, denote the left and right halves of $\mathbf{S}^{(2^n)}$ by

$$\mathbf{S}_L^{(2^{n-1})} = (s_0, \cdots, s_{2^{n-1}-1}) \quad \text{and} \quad \mathbf{S}_R^{(2^{n-1})} = (s_{2^{n-1}}, \cdots, s_{2^n-1}).$$

Let $\mathbf{S}_L$ and $\mathbf{S}_R$ denote the 2^{n-1} periodic sequences

$$\mathbf{S}_L = (s_0, \cdots, s_{2^{n-1}-1})^\infty \quad \text{and} \quad \mathbf{S}_R = (s_{2^{n-1}}, \cdots, s_{2^n-1})^\infty.$$

Games-Chan Algorithm ([4]). Let $\mathbf{S}$ be a 2^n-periodic binary sequence.

(i) If $\mathbf{S}_L^{(2^{n-1})} = \mathbf{S}_R^{(2^{n-1})}$, then $L(\mathbf{S}) = L(\mathbf{S}_L)$.

(ii) If $\mathbf{S}_L^{(2^{n-1})} \neq \mathbf{S}_R^{(2^{n-1})}$, then $L(\mathbf{S}) = 2^{n-1} + L(\mathbf{S}_L + \mathbf{S}_R)$.

(iii) Apply the above procedure recursively to the 2^{n-1}-periodic binary sequence $\mathbf{S}_L$ in (i), or the 2^{n-1}-periodic binary sequence $\mathbf{S}_L + \mathbf{S}_R$ in (ii).

We make some observations and establish notation we use for the rest of the section. We note that the procedure of the Games-Chan algorithm as stated here is executed a total of $n + 1$ times to compute the linear complexity of any $\mathbf{S} \in \mathcal{A}(L)$. In the ith step, $i = 0, \cdots, n$, the algorithm computes the linear complexity of a 2^{n-i}-periodic binary sequence. Let $\psi^i(\mathbf{S})$, $i = 0, \cdots, n$, denote the first period of the 2^{n-i}-periodic binary sequence considered in the ith step of the algorithm when run with input sequence $\mathbf{S}$. Let $\psi_L^i(\mathbf{S})$ and $\psi_R^i(\mathbf{S})$ denote, respectively, the left and right halves of $\psi^i(\mathbf{S})$. Let $m^i(\mathbf{S})$ denote the total value contributed to $L(\mathbf{S})$ in the algorithm during the execution from the 0-th step to the i-th step of the algorithm. For any two finite binary sequences of same length, $\mathbf{S}$ and $\mathbf{S}'$, let $d_H(\mathbf{S}, \mathbf{S}')$ denote the Hamming distance between $\mathbf{S}$ and $\mathbf{S}'$. We slightly abuse the notation because we also use $d_H(\mathbf{S}, \mathbf{S}')$ to denote the Hamming distance between the first periods of $\mathbf{S}, \mathbf{S}' \in A(L)$. The next lemma follows from the Games-Chan algorithm.

Lemma 6. *Let $\mathbf{S}$ be a 2^n-periodic binary sequence. For any t integers $r_1, \cdots, r_t$ such that $0 < r_1 < r_2 < \cdots < r_t \leq n$, we have*

$$L(\mathbf{S}) = 2^n - (2^{n-r_1} + 2^{n-r_2} + \cdots + 2^{n-r_t}) \tag{5}$$

if and only if

$$\psi_L^{u-1}(\mathbf{S}) = \psi_R^{u-1}(\mathbf{S}) \quad \text{exactly when} \quad u \in \{r_1, \cdots, r_t\}. \tag{6}$$

Theorem 2. *Let $\mathbf{S} \in \mathcal{A}(L)$ where*

$$2^n - (2^{n-r_1} + 2^{n-r_2}) < L < 2^n - (2^{n-r_1} + 2^{n-r_2-1}), \tag{7}$$

for some $r_1, r_2 \in \{2, \cdots, n-1\}$ satisfying $1 < r_1 \leq r_2$. Then, any sequence $\mathbf{S}_{i,j,k,l} \in \mathcal{A}(L)$, $0 \leq i, j, k, l \leq 2^n - 1$, if and only if the positions i, j, k, and l are in exactly one of the following two forms.

(i) *For any $i, j \in \{0, \cdots, 2^n - 1\}$, set*

$$k = (i + m_1 2^{n-r_1+1}) \bmod 2^n \quad and \quad l = (j + m_2 2^{n-r_1+1}) \bmod 2^n, \tag{8}$$

where $1 \leq m_1, m_2 \leq 2^{r_1-1} - 1$.

(ii) *For any $t \in \{0, \cdots, 2^n - 1\}$, set*

$$i = (t + b_1 2^{n-r_1+1}) \bmod 2^n,$$
$$j = (t + g 2^{n-r_2} + b_2 2^{n-r_1+1}) \bmod 2^n,$$
$$k = (t + 2^{n-r_1} + b_3 2^{n-r_1+1}) \bmod 2^n, \quad and \tag{9}$$
$$l = (t + g 2^{n-r_2} + 2^{n-r_1} + b_4 2^{n-r_1+1}) \bmod 2^n,$$

where $1 \leq g \leq 2^{r_2-r_1} - 1$, $0 \leq b_1, b_2, b_3, b_4 \leq 2^{r_1-1} - 1$.

Proof. Due to space constraints, we only prove the forward direction of the theorem. The other direction is straightforward and can be proved by reversing the argument used for the forward case.

Consider any sequence

$$\mathbf{S}_{i,j,k,l} \in \mathcal{A}(L), \quad \text{where} \quad i, j, k, l \in \{0, \cdots, 2^n - 1\}. \tag{10}$$

We assume i, j, k, and l are all different as two symbol changes are already covered by Lemma 5. From equation (7) we have

$$w_H(2^n - L) \geq 3 \quad \text{and} \quad L = 2^n - (2^{n-r_1} + 2^{n-r_2-1} + c), \tag{11}$$

for some $0 < c < 2^{n-r_2-1}$. From equations (5), (6), and (11), we have

$$\forall \, \mathbf{S} \in \mathcal{A}(L), \quad \psi_L^{r_1-1}(\mathbf{S}) = \psi_R^{r_1-1}(\mathbf{S}) \quad \text{and} \quad \psi_L^{r_2}(\mathbf{S}) = \psi_R^{r_2}(\mathbf{S}). \tag{12}$$

By $\mathcal{P}_{i,j,k,l}$ denote the set of all permutations of i, j, k, and l. We first assume

$$\exists (a, b, c, d) \in \mathcal{P}_{i,j,k,l} : \quad L(\mathbf{S}_{a,b}) = L(\mathbf{S}_{c,d}) = L(\mathbf{S}). \tag{13}$$

From equation (7) we have

$$2^n - 2^{n-r_1+1} < L < 2^n - 2^{n-r_1}. \tag{14}$$

From our assumption in equation (13), by Lemma 5 and equation (14), ignoring order, i, j, k, and l must be in the first form as in equation (8).

If equation (13) does not hold, then from equation (10) and Lemma 5 we have

$$\forall (a, b, c, d) \in \mathcal{P}_{i,j,k,l} : \quad L(\mathbf{S}_{a,b}) > L(\mathbf{S}) \quad \text{and} \quad L(\mathbf{S}_{c,d}) > L(\mathbf{S}). \tag{15}$$

Equation (15) implies

$$\forall a, b \in \{i, j, k, l\}, \quad a \not\equiv b \bmod 2^{n-r_1+1}. \tag{16}$$

Equation (16) implies that the integers $i \bmod 2^{n-r_1+1}$, $j \bmod 2^{n-r_1+1}$, $k \bmod 2^{n-r_1+1}$, and $l \bmod 2^{n-r_1+1}$ are all different. Also, by Lemma 6 and equation (11) the left and right halves are not equal during the first $r_1 - 2$ steps of the Games-Chan procedure for any $\mathbf{S} \in \mathcal{A}(L)$. Thus, by the procedure of the Games-Chan algorithm and equation (16) we get

$$d_H(\psi^{r_1-1}(\mathbf{S}), \psi^{r_1-1}(\mathbf{S}_{i,j,k,l})) = 4. \tag{17}$$

By equations (12) and (17), the four positions where the vectors $\psi^{r_1-1}(\mathbf{S})$, $\psi^{r_1-1}(\mathbf{S}_{i,j,k,l})$ differ are of the form

$$c_1, \quad c_2, \quad c_1 + 2^{n-r_1}, \quad \text{and} \quad c_2 + 2^{n-r_1}, \tag{18}$$

for some $0 \leq c_1 < c_2 \leq 2^{n-r_1} - 1$. From equations (12) and (17), we have $d_H(\psi_L^{r_1-1}(\mathbf{S}), \psi_L^{r_1-1}(\mathbf{S}_{i,j,k,l})) = 2$. This implies

$$d_H(\psi^{r_1}(\mathbf{S}), \psi^{r_1}(\mathbf{S}_{i,j,k,l})) = 2. \tag{19}$$

Now we treat $\psi^{r_1}(\mathbf{S})$ and $\psi^{r_1}(\mathbf{S}_{i,j,k,l})$ as the first periods of 2^{n-r_1}-periodic binary sequences $\mathbf{S}'$ and $\mathbf{S}'_{i,j,k,l}$, respectively, that differ at 2 positions. With this notation, from equations (18) and (19) we have $\mathbf{S}' = (\psi^{r_1}(\mathbf{S}))^\infty$, $\mathbf{S}'_{i,j,k,l} = (\psi^{r_1}(\mathbf{S}_{i,j,k,l}))^\infty$, and

$$S'_{i,j,k,l}(x) = S'(x) + x^{c_1} + x^{c_2}. \tag{20}$$

From the procedure of the Games-Chan algorithm, since the left and right halves are different in the first $r_1 - 2$ steps for both $\mathbf{S}$ and $\mathbf{S}_{i,j,k,l}$, we have

$$m^{r_1-1}(\mathbf{S}) = m^{r_1-1}(\mathbf{S}_{i,j,k,l}) = 2^{n-1} + \cdots + 2^{n-r_1+1} = 2^n - 2^{n-r_1+1}. \tag{21}$$

From Lemma 6 and equations (10), (20) and (21) we have

$$\mathbf{S}', \mathbf{S}'_{i,j,k,l} \in \mathcal{A}(L') \quad \text{where} \quad L' = L - (2^n - 2^{n-r_1+1}). \tag{22}$$

From equations (7) and (22) L' satisfies

$$2^{n-r_1} - 2^{n-r_2} < L' < 2^{n-r_1} - 2^{n-r_2-1}. \tag{23}$$

From Lemma 5 and equation (23), the positions c_1 and c_2 in equations (18) and (20) must be in the form

$$c_i = u + g_i 2^{n-r_2}, \quad 0 \le u \le 2^{n-r_2} - 1, \quad 0 \le g_1 < g_2 \le 2^{r_2-r_1} - 1. \tag{24}$$

From equations (18) and (24), the four positions, denoted f_1, f_2, f_3, and f_4, where $\psi^{r_1-1}(\mathbf{S})$ and $\psi^{r_1-1}(\mathbf{S}_{i,j,k,l})$ differ are of the form

$$f_1 = c_1, \quad f_2 = c_2, \quad f_3 = c_1 + 2^{n-r_1}, \quad \text{and} \quad f_4 = c_2 + 2^{n-r_1}, \tag{25}$$

where c_1 and c_2 are as in equation (24).

Using equation (25) and the procedure of Games-Chan algorithm, it can be shown that i, j, k, and l should be exclusively in the second form as given in equation (9). This completes the proof of the theorem. $\square$

Remark 1. The proof of Theorem 2 works even when $r_1 = 1$ and $r_1 < r_2$, that is when $2^{n-2} \le L < 2^{n-1}$. In this case we see that the four changes made to obtain a sequence of the same linear complexity as any given sequence are always in the second form as in equation (9). Also, from equation (4) we can see that there exist cases when $r_1 = r_2$ in Theorem 2 when expressing the linear complexity uniquely as in equation (7).

3   Characterization When $w_H(2^n - L) \ne 2$

In this section we characterize the 2^n-periodic binary sequences with fixed 2-error linear complexity when the linear complexity is not of the form $2^n - (2^i + 2^j)$, $1 \le i < j \le n - 1$, by using the results from the previous section.

For $0 \le L \le 2^n$ and $1 \le k \le 2^n$, denote by $\mathcal{A}_k(L)$ the set of 2^n-periodic binary sequences with given k-error linear complexity L and let $\mathcal{N}_k(L) = |\mathcal{A}_k(L)|$, the cardinality of $\mathcal{A}_k(L)$. Let $\mathbf{0}$ denote the zero sequence. For any $1 \leq t \leq 2^n$, let $\mathbf{E}_{i_1,\cdots,i_t}$, $0 \leq i_1 < \cdots < i_t \leq 2^n - 1$, denote the 2^n-periodic binary sequence of weight t with a 1 at positions with subscripts $i_1,\cdots,i_t$ in the first period and 0 elsewhere. We denote by $\mathcal{A}(L) + \mathbf{E}_{i_1,\cdots,i_t}$ the set $\{\mathbf{S} + \mathbf{E}_{i_1,\cdots,i_t} : \mathbf{S} \in \mathcal{A}(L)\}$. For the rest of the paper, for any set $\mathcal{R}$ of 2^n-periodic binary sequences, by $\mathcal{A}(L) + \mathcal{R}$ denote the set of sets $\{\mathcal{A}(L) + \mathbf{R} : \mathbf{R} \in \mathcal{R}\}$. We define sets $\mathbb{E}_t$, $t = 1,\cdots,2^n$, which are used for the rest of the paper. Let $\mathbb{E}_t = \{\mathbf{E}_{i_1,\cdots,i_t} : 0 \leq i_1 < i_2 < \cdots < i_t \leq 2^n - 1\}$. It is straightforward to see that

$$\mathcal{A}_2(0) = \mathbb{E}_1 \cup \mathbb{E}_2 \cup \{\mathbf{0}\} \quad \text{and} \quad \mathcal{N}_2(0) = \binom{2^n}{2} + 2^n + 1. \tag{26}$$

From Lemmas 3 and 4 we have

$$\mathcal{A}_2(2^n) = \emptyset \quad \text{and} \quad \mathcal{N}_2(2^n) = 0. \tag{27}$$

For any 2^n-periodic binary sequence $\mathbf{S}$, from [6, Proposition 1] we know that for $k \geq 2$, $L_k(\mathbf{S})$ is different from $2^n - 2^t$ for every integer t with $0 \leq t < n$. Hence we get

$$\mathcal{A}_2(L) = \emptyset \quad \text{and} \quad \mathcal{N}_2(L) = 0 \quad \text{for} \quad L = 2^n - 2^t, \quad 0 \leq t < n. \tag{28}$$

A characterization of 2^n-periodic binary sequences with fixed 2-error linear complexity L such that $w_H(2^n - L) = 0$ or 1 is given in equations (26)-(28). Next, we give the characterization when $w_H(2^n - L) \geq 3$. We start with a result that will be used for the rest of the section. The proof is provided in the appendix.

Theorem 3. *Let $\{i_1,\cdots,i_{t_1}\}$ and $\{j_1,\cdots,j_{t_2}\}$ denote two sets of subscripts where $0 \leq i_l, j_m \leq 2^n - 1$, $l = 1,\cdots,t_1$, $m = 1,\cdots,t_2$. Then*

$$(\mathcal{A}(L) + \mathbf{E}_{i_1,\cdots,i_{t_1}}) \cap (\mathcal{A}(L) + \mathbf{E}_{j_1,\cdots,j_{t_2}}) = \emptyset$$

or

$$\mathcal{A}(L) + \mathbf{E}_{i_1,\cdots,i_{t_1}} = \mathcal{A}(L) + \mathbf{E}_{j_1,\cdots,j_{t_2}}.$$

We need the following generalization of [3, Theorem 4] to obtain the basic characterization of $\mathcal{A}_2(L)$.

Lemma 7. *Let $\mathbf{S}$ be a 2^n-periodic binary sequence. Consider any two positive integers u, v such that $0 < v \leq u$ and $u+v < merr(\mathbf{S})$. Then for any 2^n-periodic binary sequence $\mathbf{E}$ such that $w_H(\mathbf{E}) = v$ we have*

$$L_u(\mathbf{S} + \mathbf{E}) = L(\mathbf{S}).$$

The basic characterization can be obtained by using the definition of k-error linear complexity and Lemma 7.

Theorem 4. *If $w_H(2^n - L) \geq 3$, then*

$$\mathcal{A}_2(L) = \mathcal{A}(L) \bigcup \left(\bigcup_{\mathbf{E}_i \in \mathbb{E}_1} (\mathcal{A}(L) + \mathbf{E}_i) \right) \bigcup \left(\bigcup_{\mathbf{E}_{i,j} \in \mathbb{E}_2} (\mathcal{A}(L) + \mathbf{E}_{i,j}) \right). \tag{29}$$

For any $1 \leq L < 2^{n-1}$, from Theorem 1 we know that for any two sequences $\mathbf{S}$, $\mathbf{S}' \in \mathcal{A}(L)$, $d_H(\mathbf{S}, \mathbf{S}') \geq 4$. Hence we have

$$\mathcal{A}(L) \cap (\mathcal{A}(L) + \mathbf{E}_t) = \emptyset, \tag{30}$$

$$\mathcal{A}(L) \cap (\mathcal{A}(L) + \mathbf{E}_{i,j}) = \emptyset, \quad \text{and} \tag{31}$$

$$(\mathcal{A}(L) + \mathbf{E}_t) \cap (\mathcal{A}(L) + \mathbf{E}_{i,j}) = \emptyset, \tag{32}$$

for all $\mathbf{E}_t \in \mathbb{E}_1$ and $\mathbf{E}_{i,j} \in \mathbb{E}_2$.

We enumerate the disjoint sets in equation (29) and obtain the counting function when $1 \leq L < 2^{n-2}$ using the fact that $d_H(\mathbf{S}, \mathbf{S}') \geq 8$ from Theorem 1 for any two sequences $\mathbf{S}$, $\mathbf{S}' \in \mathcal{A}(L)$.

Theorem 5. *If $w_H(2^n - L) \geq 3$ and $1 \leq L < 2^{n-2}$, then the sets $\mathcal{A}(L)$, $\mathcal{A}(L) + \mathbf{E}_i$, $\mathbf{E}_i \in \mathbb{E}_1$, and $\mathcal{A}(L) + \mathbf{E}_{i,j}$, $\mathbf{E}_{i,j} \in \mathbb{E}_2$, are disjoint and*

$$\mathcal{N}_2(L) = \left(\binom{2^n}{2} + 2^n + 1 \right) 2^{L-1}.$$

We enumerate the disjoint sets in equation (29) and give the counting function when $2^{n-1} \leq L < 2^n$.

Theorem 6. *Let $w_H(2^n - L) \geq 3$ where*

$$2^n - (2^{n-r_1} + 2^{n-r_2}) < L < 2^n - (2^{n-r_1} + 2^{n-r_2-1}), \tag{33}$$

for some r_1, r_2 satisfying $1 < r_1 \leq r_2 \leq n - 1$. Define the sets

$$\mathbb{D}_1(L) = \{\mathbf{E}_i : 0 \leq i \leq 2^{n-r_1+1} - 1\} \quad \text{and}$$

$$\mathbb{D}_2(L) = \{\mathbf{E}_{i,j} : 0 \leq i < j \leq 2^{n-r_1+1} - 1\}.$$

For $u = 0, \cdots, 2^{n-r_2} - 1$ define the sets

$$\mathcal{D}_u^1(L) = \bigcup_{1 \leq t \leq 2^{r_2-r_1}-1} \{\mathbf{E}_{i,i+2^{n-r_1}} : i = u + t2^{n-r_2}\} \tag{34}$$

and

$$\mathcal{D}_u^2(L) = \bigcup_{\substack{t_2=1 \\ 0 \leq t_1 < t_2}}^{2^{r_2-r_1}-1} \{\mathbf{E}_{i,j}, \mathbf{E}_{i,j+2^{n-r_1}} : i = u + t_1 2^{n-r_2}, j = u + t_2 2^{n-r_2}\}. \tag{35}$$

Consider the set $\mathcal{D}(L)$ formed from sets in equations (6), (34), and (35) by

$$\mathcal{D}(L) = \mathbb{D}_2(L) - \bigcup_{u=0}^{2^{n-r_2}-1} (\mathcal{D}_u^1(L) \cup \mathcal{D}_u^2(L)). \tag{36}$$

Then the sets $\mathcal{A}(L)$, $\mathcal{A}(L) + \mathbf{E}_i$, $\mathbf{E}_i \in \mathbb{D}_1(L)$, and $\mathcal{A}(L) + \mathbf{E}_{i,j}$, $\mathbf{E}_{i,j} \in \mathcal{D}(L)$, are disjoint. Furthermore,

$$\mathcal{N}_2(L) = \left(\binom{2^{n-r_1+1}}{2} - 2^{n-r_2}(2^{2r_2-2r_1} - 1) + 2^{n-r_1+1} + 1 \right) 2^{L-1}. \tag{37}$$

Proof. First we enumerate the disjoint sets in $\mathcal{A}(L) + \mathbb{E}_1$. By equation (33) we have

$$2^n - 2^{n-r_1+1} < L < 2^n - 2^{n-r_1}. \tag{38}$$

Using Theorem 3 and Lemma 5, from equation (38) we have

$$(\mathcal{A}(L) + \mathbf{E}_u) \cap (\mathcal{A}(L) + \mathbf{E}_v) = \emptyset, \quad 0 \le u < v \le 2^{n-r_1+1} - 1, \tag{39}$$

and for $u = 0, \cdots, 2^{n-r_1+1} - 1$,

$$\mathcal{A}(L) + \mathbf{E}_u = \mathcal{A}(L) + \mathbf{E}_{u+t2^{n-r_1+1}}, \quad t = 0, \cdots, 2^{r_1-1} - 1. \tag{40}$$

Thus, from equation (39) there are 2^{n-r_1+1} disjoint sets $\mathcal{A}(L) + \mathbf{E}_i$, $\mathbf{E}_i \in \mathbb{D}_1(L)$, in $\mathcal{A}(L)+\mathbb{E}_1$. To obtain the disjoint sets in $\mathcal{A}(L)+\mathbb{E}_2$, we only have to enumerate the disjoint sets in $\mathcal{A}(L) + \mathbb{D}_2(L)$ because from equation (40) we have $\mathcal{A}(L) + \mathbf{E}_{i,j,i+u2^{n-r_1+1},j+v2^{n-r_1+1}} = \mathcal{A}(L)$, for $0 \le i < j \le 2^{n-r_1+1} - 1, 0 \le u, v \le 2^{r_1-1} - 1$.

From Theorem 3, we know that $\mathcal{A}(L) + \mathbf{E}_{i,j} = \mathcal{A}(L) + \mathbf{E}_{k,l}$ if and only if there exists a sequence $\mathbf{S} \in \mathcal{A}(L)$ such that the new sequence $\mathbf{S} + \mathbf{E}_{i,j,k,l} \in \mathcal{A}(L)$. Hence we observe that doubly counted sets in $\mathcal{A}(L)+\mathbb{D}_2(L)$ arise if only if there exist integers i, j, k, and l, $0 \le i, j, k, l \le 2^{n-r_1+1} - 1$, that are in the second form of Theorem 2. Exactly all such i, j, k, and l are derived as part of the proof of Theorem 2 and are enumerated in equation (25). Here we repeat the enumeration for clarity. Integers i, j, k, and l, $0 \le i, j, k, l \le 2^{n-r_1+1} - 1$, that are in the second form of Theorem 2 are

$$i = u + g_1 2^{n-r_2}, \quad j = u + g_2 2^{n-r_2}, \quad k = i + 2^{n-r_1}, \quad l = j + 2^{n-r_1}, \tag{41}$$

where

$$0 \le u \le 2^{n-r_2} - 1 \quad \text{and} \quad 0 \le g_1 < g_2 \le 2^{r_2-r_1} - 1. \tag{42}$$

From equations (41) and (42) there are exactly $2^{n-r_2}\binom{2^{r_2-r_1}}{2}$ distinct pairs i, j and hence distinct sets $\{i, j, k, l\}$ such that $0 \le i, j, k, l \le 2^{n-r_1+1} - 1$ and $\mathcal{A}(L) + \mathbf{E}_{i,j,k,l} = \mathcal{A}(L)$.

Hence for all settings of i and j in equation (41) we have the set equalities

$$\mathcal{A}(L) + \mathbf{E}_{i,j} = \mathcal{A}(L) + \mathbf{E}_{i+2^{n-r_1},j+2^{n-r_1}} \tag{43}$$

and

$$\mathcal{A}(L) + \mathbf{E}_{i,j+2^{n-r_1}} = \mathcal{A}(L) + \mathbf{E}_{i+2^{n-r_1},j}. \tag{44}$$

Also, for each $u = 0, \cdots, 2^{n-r_2} - 1$, we have $2^{r_2-r_1} - 1$ set equalities

$$\mathcal{A}(L) + \mathbf{E}_{u,u+2^{n-r_1}} = \mathcal{A}(L) + \mathbf{E}_{i,i+2^{n-r_1}}, \quad \text{where} \quad i = u + t2^{n-r_2} \tag{45}$$

for $1 \le t \le 2^{r_2-r_1} - 1$.

By the settings of i and j in equation (41), the set equalities in equations (43) and (44) result in $2^{n-r_2}(2 \cdot \binom{2^{r_2-r_1}}{2}))$ doubly counted sets in $\mathcal{A}(L) + \mathbb{D}_2(L)$ enumerated as $\mathcal{A}(L) + \mathcal{D}_u^2(L)$, $u = 0, \cdots, 2^{n-r_2} - 1$. Similarly, the set equalities in equation (45) result in $2^{n-r_2}(2^{r_2-r_1}-1)$ doubly counted sets in $\mathcal{A}(L)+\mathbb{D}_2(L)$ enumerated as $\mathcal{D}_u^1(L)$, $u = 0, \cdots, 2^{n-r_2}-1$. For each $u = 0, \cdots, 2^{n-r_2}-1$ we have

$$|\mathcal{D}_u^1(L)| = 2^{r_2-r_1}-1 \quad \text{and} \quad |\mathcal{D}_u^2(L)| = 2\binom{2^{r_2-r_1}}{2}. \tag{46}$$

Note that any L such that $2^{n-1} \leq L < 2^n$ and $w_H(2^n - L) \geq 3$, satisfies equations (30) and (32). From Lemma 5 and equation (38) we have

$$\mathcal{A}(L) \cap (\mathcal{A}(L) + \mathbf{E}_{i,j}) = \emptyset, \quad \mathbf{E}_{i,j} \in \mathbb{D}_2(L). \tag{47}$$

Thus, from equations (29), (30)-(32), (41)-(45), and (47), the sets $\mathcal{A}(L)$, $\mathcal{A}(L) + \mathbf{E}_i$, $\mathbf{E}_i \in \mathbb{D}_1(L)$, and $\mathcal{A}(L) + \mathbf{E}_{i,j}$, $\mathbf{E}_{i,j} \in \mathcal{D}(L)$, are disjoint. The number of disjoint sets in $\mathcal{A}(L) + \mathbb{E}_2$ is equal to $|\mathcal{D}(L)|$. From equations (36) and (46) we have

$$|\mathcal{D}(L)| = |\mathbb{D}_2(L)| - \sum_{u=0}^{2^{n-r_2}-1} (|\mathcal{D}_u^1(L)| + |\mathcal{D}_u^2(L)|)$$

$$= \binom{2^{n-r_1+1}}{2} - 2^{n-r_2}\left(2^{r_2-r_1} - 1 + 2\binom{2^{r_2-r_1}}{2}\right). \tag{48}$$

From Lemma 2 we have $|\mathcal{A}(L)| = 2^{L-1}$, $1 \leq L \leq 2^n$. Hence the counting function in equation (37) follows from equations (29),(30)-(32), (39),(47), and (48). This completes the proof of the theorem. □

Using Remark 1 with $r_1 = 1$ and $r_1 < r_2$ in the statement of Theorem 6, we get the characterization when $2^{n-2} \leq L < 2^{n-1}$.

We also characterized 2^n-periodic binary sequences with fixed 3-error linear complexity L when $w_H(2^n - L) \neq 2$. Using the characterization we obtained the corresponding counting function. Due to space constraints we state our results here without proofs and also use some of the notation established in the statement of Theorem 6. The approach used is the same as that used for the 2-error case and the proofs use Lemma 5, Theorem 2, Lemma 7, and some intermediate findings of Theorem 6. It is straightforward to see that

$$\mathcal{A}_3(0) = \mathbb{E}_1 \cup \mathbb{E}_2 \cup \mathbb{E}_3 \cup \{0\} \quad \text{and} \quad \mathcal{N}_3(0) = \binom{2^n}{3} + \binom{2^n}{2} + 2^n + 1,$$

$$\mathcal{A}_3(2^n) = \emptyset \quad \text{and} \quad \mathcal{N}_3(2^n) = 0, \quad \text{and}$$

$$\mathcal{A}_3(L) = \emptyset \quad \text{and} \quad \mathcal{N}_3(L) = 0 \quad \text{for} \quad L = 2^n - 2^t, \quad 0 \leq t < n.$$

Theorem 7. *Let $L < 2^n$ be a positive integer such that $w_H(2^n - L) \geq 3$. Then*

$$\mathcal{A}_3(L) = \mathcal{A}_2(L) \bigcup \left(\bigcup_{\mathbf{E}_{i,j,k} \in \mathbb{E}_3} (\mathcal{A}(L) + \mathbf{E}_{i,j,k})\right).$$

Furthermore,

(i) If $1 \leq L < 2^{n-2}$, then the sets $\mathcal{A}(L)$, $\mathcal{A}(L) + \mathbf{E}_i$, $\mathbf{E}_i \in \mathbb{E}_1$, $\mathcal{A}(L) + \mathbf{E}_{i,j}$, $\mathbf{E}_{i,j} \in \mathbb{E}_2$, and $\mathcal{A}(L) + \mathbf{E}_{i,j,k}$, $\mathbf{E}_{i,j,k} \in \mathbb{E}_3$ are disjoint and

$$N_3(L) = \left(\binom{2^n}{3} + \binom{2^n}{2} + 2^n + 1 \right) 2^{L-1}.$$

(ii) Let L be such that

$$2^n - (2^{n-r_1} + 2^{n-r_2}) < L < 2^n - (2^{n-r_1} + 2^{n-r_2-1}),$$

for some r_1, r_2 satisfying $1 < r_1 \leq r_2 \leq n-1$. Let $\mathcal{D}(L)$ be as in equation (36). Define the sets $\mathbb{D}_3(L)$, $\mathcal{D}_u^3(L)$, $u = 0, \cdots, 2^{n-r_2}$, and $\mathcal{E}(L)$ by

$$\mathbb{D}_3(L) = \{ \mathbf{E}_{i,j,k} : 0 \leq i < j < k \leq 2^{n-r_1+1} - 1 \},$$

$$\mathcal{D}_u^3(L) = \bigcup_{0 \leq g_1 < g_2 < 2^{r_2-r_1}} \{ \mathbf{E}_{i_1,j_1,i_2}, \mathbf{E}_{i_1,j_1,j_2}, \mathbf{E}_{i_1,i_2,j_2}, \mathbf{E}_{i_2,j_1,j_2}$$

$$: i_t = u + g_t 2^{n-r_2}, j_t = i_t + 2^{n-r_1}, t = 1, 2 \},$$

and

$$\mathcal{E}(L) = \mathbb{D}_3(L) - \bigcup_{u=0}^{2^{n-r_2}-1} \mathcal{D}_u^3(L).$$

Then the sets $\mathcal{A}(L)$, $\mathcal{A}(L) + \mathbf{E}_i$, $\mathbf{E}_i \in \mathbb{D}_1(L)$, $\mathcal{A}(L) + \mathbf{E}_{i,j}$, $\mathbf{E}_{i,j} \in \mathcal{D}(L)$, and $\mathcal{A}(L) + \mathbf{E}_{i,j,k}$, $\mathbf{E}_{i,j,k} \in \mathcal{E}(L)$ are all disjoint and

$$N_3(L) = N_2(L) + \left(\binom{2^{n-r_1+1}}{3} - 4 \cdot 2^{n-r_2} \binom{2^{r_2-r_1}}{2} \right) 2^{L-1}.$$

Using Remark 1 with $r_1 = 1$ and $r_1 < r_2$ in the statement of Theorem 7, we get the characterization when $2^{n-2} \leq L < 2^{n-1}$.

4   Conclusion

In this paper, we characterize 2^n-periodic binary sequences with fixed 2-error or 3-error linear complexity L when $w_H(2^n - L) \neq 2$. First, we derive some properties of 2^n-periodic binary sequences with fixed linear complexity. We use the Games-Chan algorithm to find the exact form of four symbol changes that can be made in a 2^n-periodic sequence so that the resulting sequence has the same linear complexity as the original sequence. We use these properties to obtain the characterizations and the corresponding counting functions.

We believe that our approach in Theorem 2 can be used to generalize the results to the case when the number of changes made is any power of 2. We emphasize that we can also obtain the characterization in the case when $w_H(2^n - L) = 2$ by further analysis of the Games-Chan algorithm when the linear complexity is of the form $L = 2^n - (2^i + 2^j)$, $0 \leq i < j \leq n - 1$. Statistical properties like expected value and variance can also be considered for $L_k(\mathbf{S})$ for several small k. Extension to p^n-periodic sequences over $\mathbb{F}_p$ can also be considered.

Acknowledgements

Thanks to Dr. Andrew Klapper for several helpful suggestions during the preparation of this paper. Thanks to Dr. Zongming Fei for providing office space and resources while researching for this paper. Thanks to anonymous reviewers for their careful proofreading.

References

1. Cusick, T.W., Ding, C., Renvall, A.: Stream Ciphers and Number Theory. North-Holland, Amsterdam (1998)
2. Ding, C., Xiao, G., Shan, W.: The Stability Theory of Stream Ciphers. Springer, Heidelberg (1991)
3. Fu, F.-W., Niederreiter, H., Su, M.: The characterization of 2^n-periodic binary sequences with fixed 1-error linear complexity. In: Gong, G., Helleseth, T., Song, H.-Y., Yang, K. (eds.) SETA 2006. LNCS, vol. 4086, pp. 88–103. Springer, Heidelberg (2006)
4. Games, R.A., Chan, A.H.: A fast algorithm for determining the complexity of a pseudo-random sequence with period 2^n. IEEE Trans. Inform. Theory 29(1), 144–146 (1983)
5. Kurosawa, K., Sato, F., Sakata, T., Kishimoto, W.: A relationship between linear complexity and k-error linear complexity. IEEE Trans. Inform. Theory 46(2), 694–698 (2000)
6. Meidl, W.: On the stability of 2^n-periodic binary sequences. IEEE Trans. Inform. Theory 51(3), 1151–1155 (2005)
7. Rueppel, R.A.: Analysis and Design of Stream Ciphers. Springer, Heidelberg (1986)
8. Stamp, M., Martin, C.F.: An algorithm for the k-error linear complexity of binary sequences with period 2^n. IEEE Trans. Inform. Theory 39(4), 1398–1401 (1993)

A   Proof of Theorem 1

For any sequence $\mathbf{S} \in \mathcal{A}(L)$, consider the corresponding polynomial $\mathbf{S}(x) = (1 - x)^{2^n - L} a(x)$, where $a(x) \in \mathbb{F}_2[x]$ such that $\deg(a(x)) \leq L - 1$ and $a(1) \neq 0$. Since $1 \leq L < 2^{n-r}$, we have $2^n - L > 2^n - 2^{n-r}$. The generating function for $\mathbf{S}$ is given by

$$
\begin{aligned}
\frac{\mathbf{S}(x)}{x^{2^n} - 1} &= \frac{(x - 1)^{(2^n - 2^{n-r}) + (2^{n-r} - L)} a(x)}{(x - 1)^{2^n}} \\
&= \frac{(x - 1)^{2^{n-r} - L} a(x)}{x^{2^{n-r}} - 1},
\end{aligned}
$$

which implies 2^{n-r} is a period of $\mathbf{S}$.

Corresponding to any sequence $\mathbf{S} \in \mathcal{A}(L)$, let $\mathbf{S}'$ denote the 2^{n-r}-periodic sequence $(s_0, s_1, \cdots, s_{2^{n-r}-1})^{\infty}$. Since $1 \leq L < 2^{n-r}$, from Lemma 3 we know that $w_H(\mathbf{S}'_1)$ and $w_H(\mathbf{S}'_2)$ are even. Hence the Hamming distance between $\mathbf{S}'_1$ and $\mathbf{S}'_2$ is even. That is

$$
d_H(\mathbf{S}'_1, \mathbf{S}'_2) = 2t \quad \text{for some} \quad t \in \{1, 2, 3, \cdots, 2^{n-r-1}\}.
$$

Since 2^{n-r} is a period of $\mathbf{S}_1$ and $\mathbf{S}_2$, we have $d_H(\mathbf{S}_1, \mathbf{S}_2) = 2^r \cdot d_H(\mathbf{S}'_1, \mathbf{S}'_2) = t \cdot 2^{r+1}$. This completes the proof of the theorem.   □

B   Proof of Theorem 3

We assume

$$0 < L \leq 2^n \tag{49}$$

since the result holds trivially for $L = 0$.

Suppose $(\mathcal{A}(L) + \mathbf{E}_{i_1, \cdots, i_{t_1}}) \cap (\mathcal{A}(L) + \mathbf{E}_{j_1, \cdots, j_{t_2}}) \neq \emptyset$. So there exists sequences $\mathbf{S}, \mathbf{S}' \in \mathcal{A}(L)$ such that $\mathbf{S} + \mathbf{E}_{i_1, \cdots, i_{t_1}} = \mathbf{S}' + \mathbf{E}_{j_1, \cdots, j_{t_2}}$. This implies that

$$\mathbf{S} + \mathbf{E}_{i_1, \cdots, i_{t_1}} + \mathbf{E}_{j_1, \cdots, j_{t_2}} = \mathbf{S}'. \tag{50}$$

Consider the corresponding polynomials of $\mathbf{S}$ and $\mathbf{S}'$ given by

$$\mathbf{S}(x) = (1 - x)^{2^n - L} a(x) \quad \text{and} \quad \mathbf{S}'(x) = (1 - x)^{2^n - L} a'(x), \tag{51}$$

where $a(1) = a'(1) = 1$. From equations (49) and (51) we have

$$\deg(\gcd((1 - x^{2^n}), \mathbf{S}(x) + \mathbf{S}'(x))) > 2^n - L. \tag{52}$$

From equations (50) and (52) we have

$$\deg(\gcd((1 - x^{2^n}), x^{i_1} + \cdots + x^{i_{t_1}} + x^{j_1} + \cdots + x^{j_{t_2}})) > 2^n - L. \tag{53}$$

To prove the theorem we first show that every sequence in $\mathcal{A}(L) + \mathbf{E}_{i_1, \cdots, i_{t_1}}$ is in $\mathcal{A}(L) + \mathbf{E}_{j_1, \cdots, j_{t_2}}$. Consider any $\mathbf{R} \in \mathcal{A}(L)$ with the corresponding polynomial

$$\mathbf{R}(x) = (1 - x)^{2^n - L} b(x), \quad \text{where} \quad b(1) = 1. \tag{54}$$

Then let $\mathbf{R}' = \mathbf{R} + \mathbf{E}_{i_1, \cdots, i_{t_1}} + \mathbf{E}_{j_1, \cdots, j_{t_2}}$ with the corresponding polynomial $\mathbf{R}'(x)$. By equations (53) and (54) we have

$$\begin{aligned} &\deg(\gcd((1 - x^{2^n}), \mathbf{R}'(x))) \\ &= \deg(\gcd((1 - x)^{2^n}, \mathbf{R}(x) + x^{i_1} + \cdots + x^{i_{t_1}} + x^{j_1} + \cdots + x^{j_{t_2}})) \\ &= 2^n - L. \end{aligned} \tag{55}$$

From equation (55) using the definition of linear complexity we have $\mathbf{R}' \in \mathcal{A}(L)$, which implies $\mathcal{A}(L) + \mathbf{E}_{i_1, \cdots, i_{t_1}} \subseteq \mathcal{A}(L) + \mathbf{E}_{j_1, \cdots, j_{t_2}}$. By symmetry $\mathcal{A}(L) + \mathbf{E}_{j_1, \cdots, j_{t_2}} \subseteq \mathcal{A}(L) + \mathbf{E}_{i_1, \cdots, i_{t_1}}$, which proves the theorem.   □

Generalized Joint Linear Complexity of Linear Recurring Multisequences

Wilfried Meidl[1] and Ferruh Özbudak[2]

[1] Faculty of Engineering and Natural Sciences,
Sabancı University, Tuzla, 34956, İstanbul, Turkey
wmeidl@sabanciuniv.edu
[2] Department of Mathematics and Institute of Applied Mathematics,
Middle East Technical University, İnönü Bulvarı, 06531, Ankara, Turkey
ozbudak@metu.edu.tr

Abstract. The joint linear complexity of multisequences is an important security measure for vectorized stream cipher systems. Extensive research has been carried out on the joint linear complexity of N-periodic multisequences using tools from Discrete Fourier transform. Each N-periodic multisequence can be identified with a single N-periodic sequence over an appropriate extension field. It has been demonstrated that the linear complexity of this sequence, the so called generalized joint linear complexity of the multisequence, may be considerably smaller than the joint linear complexity, which is not desirable for vectorized stream ciphers. Recently new methods have been developed and results of greater generality on the joint linear complexity of multisequences consisting of linear recurring sequences have been obtained. In this paper, using these new methods, we investigate the relations between the generalized joint linear complexity and the joint linear complexity of multisequences consisting of linear recurring sequences.

1 Introduction

A sequence $S = s_0, s_1, \ldots$ with terms in a finite field $\mathbb{F}_q$ with q elements (or over the finite field $\mathbb{F}_q$) is called a *linear recurring sequence* over $\mathbb{F}_q$ with *characteristic polynomial*

$$f(x) = \sum_{i=0}^{d} c_i x^i \in \mathbb{F}_q[x]$$

of degree d, if

$$\sum_{i=0}^{d} c_i s_{n+i} = 0 \quad \text{for } n = 0, 1, \ldots .$$

Without loss of generality we can always assume that f is monic, i.e. $c_d = 1$. In accordance with the notation in [4] we denote the set of sequences over $\mathbb{F}_q$ with characteristic polynomial f by $\mathcal{M}_q^{(1)}(f)$. Let S be a linear recurring sequence over

S.W. Golomb et al. (Eds.): SETA 2008, LNCS 5203, pp. 266–277, 2008.

$\mathbb{F}_q$, i.e. $S \in \mathcal{M}_q^{(1)}(f)$ for some $f \in \mathbb{F}_q[x]$, then the *minimal polynomial* of S is defined to be the (uniquely determined) monic polynomial $d \in \mathbb{F}_q[x]$ of smallest degree such that $S \in \mathcal{M}_q^{(1)}(d)$. We remark that then d is a divisor of f. The degree of d is called the *linear complexity* $L(S)$ of the sequence S. Alternatively the linear complexity of a recurring sequence over $\mathbb{F}_q$ can be described as the length L of the shortest linear recurring relation with coefficients in $\mathbb{F}_q$ the sequence satisfies.

The concept of linear complexity is crucial in the study of the security of stream ciphers [13,14,15]. A keystream used in a stream cipher must have a high linear complexity to resist an attack by the Berlekamp-Massey algorithm [7].

Motivated by the study of vectorized stream cipher systems (see [2,5]) we consider the set $\mathcal{M}_q^{(m)}(f)$ of m-fold *multisequences* over $\mathbb{F}_q$ with joint characteristic polynomial f, i.e. m parallel sequences over $\mathbb{F}_q$ each of them being in $\mathcal{M}_q^{(1)}(f)$. The *joint minimal polynomial* of an m-fold multisequence $\boldsymbol{S} = (\sigma_1, \sigma_2, \ldots, \sigma_m)$ is then defined to be the (uniquely determined) monic polynomial d of least degree which is a characteristic polynomial for all sequences σ_r, $1 \leq r \leq m$. The *joint linear complexity* $L_q^{(m)}(\boldsymbol{S})$ of $\boldsymbol{S}$ is then the degree of d.

Extensive research has been carried out on the average behaviour of the linear complexity of a random sequence S and a random m-fold multisequence $\boldsymbol{S}$ in $\mathcal{M}_q^{(1)}(f)$ and $\mathcal{M}_q^{(m)}(f)$, respectively, for the special case that $f = x^N - 1$. Then $\mathcal{M}_q^{(1)}(f)$ and $\mathcal{M}_q^{(m)}(f)$ are precisely the sets of N-periodic sequences and N-periodic m-fold multisequences over $\mathbb{F}_q$. For the case of single N-periodic sequences we can refer to [1,9,10], for the case of N-periodic multisequences we refer to [3,11]. For the N-periodic case discrete Fourier transform turned out to be a convenient research tool.

Recently Fu, Niederreiter and Özbudak [4] developed new methods which made it possible to obtain results of greater generality. In fact in [4] expected value and variance for a random multisequence $\boldsymbol{S} \in \mathcal{M}_q^{(m)}(f)$ are presented for an arbitrary characteristic polynomial f.

Let $\boldsymbol{S} = (\sigma_1, \sigma_2, \ldots, \sigma_m) \in \mathcal{M}_q^{(m)}(f)$ be an m-fold multisequence over $\mathbb{F}_q$, and for $r = 1, \ldots, m$ let $s_{r,i} \in \mathbb{F}_q$ denote the ith term of the rth sequence of $\boldsymbol{S}$, i.e. $\sigma_r = s_{r,0} s_{r,1} s_{r,2} \cdots$.

Since the $\mathbb{F}_q$-linear spaces $\mathbb{F}_q^m$ and $\mathbb{F}_{q^m}$ are isomorphic, the multisequence $\boldsymbol{S}$ can be identified with a single sequence $\mathcal{S}$ having its terms in the extension field $\mathbb{F}_{q^m}$, namely $\mathcal{S} = s_0, s_1, \ldots$ with

$$s_n = \xi_1 s_{1,n} + \cdots + \xi_m s_{m,n} \in \mathbb{F}_{q^m}, \quad n \geq 0, \tag{1}$$

where $\boldsymbol{\xi} = (\xi_1, \ldots, \xi_m)$ is an ordered basis of $\mathbb{F}_{q^m}$ over $\mathbb{F}_q$. It is clear that $\mathcal{S}$ depends on the m-fold multisequence $\boldsymbol{S} \in \mathcal{M}_q^{(m)}(f)$ and the ordered basis $\boldsymbol{\xi}$. Therefore we also denote $\mathcal{S}$ as $\mathcal{S}(\boldsymbol{S}, \boldsymbol{\xi})$.

Let $L_{q^m, \boldsymbol{\xi}}(\boldsymbol{S})$ be the linear complexity of the sequence $\mathcal{S} = \mathcal{S}(\boldsymbol{S}, \boldsymbol{\xi}) \in \mathcal{M}_{q^m}^{(1)}(f)$. In accordance with [8] we call $L_{q^m, \boldsymbol{\xi}}(\boldsymbol{S})$ the *generalized joint linear complexity of $\boldsymbol{S}$ (depending on $\boldsymbol{\xi}$)*. The generalized joint linear complexity $L_{q^m, \boldsymbol{\xi}}(\boldsymbol{S})$ may be considerably smaller than $L_q^{(m)}(\boldsymbol{S})$ which is clearly not desirable for vectorized stream ciphers.

In [8] joint linear complexity and generalized joint linear complexity have been compared for the case of N-periodic multisequences. In particular conditions on the period have been presented for which generalized joint linear complexity always equals joint linear complexity, and a tight lower bound for the generalized joint linear complexity of an N-periodic multisequence with a given joint linear complexity has been established. As investigation tool a generalized discrete Fourier transform has been utilized. However this method is only applicable for the case of periodic sequences. In this article we will use the new approach and the methods of [4] to obtain similar results as in [8] for the much more general case of multisequences in $\mathcal{M}_q^{(m)}(f)$ with arbitrary characteristic polynomial f.

2 Preliminaries

Let $\boldsymbol{S} = (\sigma_1, \sigma_2, \ldots, \sigma_m) \in \mathcal{M}_q^{(m)}(f)$ be an m-fold multisequence with characteristic polynomial f, and suppose that $\sigma_r = s_{r,0}s_{r,1}s_{r,2}\ldots$, $1 \leq r \leq m$. Then there exist unique polynomials $g_r \in \mathbb{F}_q[x]$ with $\deg(g_r) < \deg(f)$ and $g_r/f = s_{r,0} + s_{r,1}x + s_{r,2}x^2\ldots$, $1 \leq r \leq m$. By [12, Lemma 1] this describes a one-to-one correspondence between the set $\mathcal{M}_q^{(m)}(f)$ and the set of m-tuples of the form $\left(\frac{g_1}{f}, \frac{g_2}{f}, \ldots, \frac{g_m}{f}\right)$, $g_r \in \mathbb{F}_q[x]$ and $\deg(g_r) < \deg(f)$ for $1 \leq r \leq m$.

If $\boldsymbol{S} \in \mathcal{M}_q^{(m)}(f)$ corresponds to $(g_1/f, g_2/f, \ldots, g_m/f)$ then the joint minimal polynomial d of $\boldsymbol{S}$ is the unique polynomial in $\mathbb{F}_q[x]$ for which there exist $h_1, \ldots, h_m \in \mathbb{F}_q[x]$ with $g_r/f = h_r/d$ for $1 \leq r \leq m$ and $\gcd(h_1, \ldots, h_m, d) = 1$. The joint linear complexity of $\boldsymbol{S}$ is then given by $L_q^{(m)}(\boldsymbol{S}) = \deg(f) - \deg(\gcd(g_1, g_2, \ldots, g_m, f))$.

Let again $\boldsymbol{S} \in \mathcal{M}_q^{(m)}(f)$ correspond to $(g_1/f, g_2/f, \ldots, g_m/f)$, then it is easily seen that the single sequence $\mathcal{S} \in \mathcal{M}_{q^m}^{(1)}(f)$ defined as in (1) corresponds to the 1-tuple (G/f) with

$$G = g_1\xi_1 + g_2\xi_2 + \cdots + g_m\xi_m.$$

The minimal polynomial of $\mathcal{S}$ is then $D = f/\gcd(G, f) \in \mathbb{F}_{q^m}[x]$ and $L_{q^m, \boldsymbol{\xi}}(\mathcal{S}) = \deg(f) - \deg(\gcd(G, f))$, where the greatest common divisor is now calculated in $\mathbb{F}_{q^m}[x]$.

It is clear that divisibility of polynomials in $\mathbb{F}_q[x]$ and $\mathbb{F}_{q^m}[x]$ plays a crucial role. We will use the following two propositions on divisibility.

Proposition 1. *Let m be a positive integer and $r \in \mathbb{F}_q[x]$ be an irreducible polynomial. Let $u = \gcd(m, \deg(r))$. Then the canonical factorization of r into irreducibles over $\mathbb{F}_{q^m}$ is of the form*

$$r = r_1 r_2 \ldots r_u,$$

where $r_1, \ldots, r_u \in \mathbb{F}_{q^m}[x]$ are distinct irreducible polynomials with

$$\deg(r_1) = \cdots = \deg(r_u) = \frac{\deg(r)}{u}.$$

Proof. This is just a restatement of [6, Theorem 3.46]. We refer to [6] for a proof. □

Proposition 2. *Let m be a positive integer, let $\boldsymbol{\xi} = (\xi_1, \ldots, \xi_m)$ be an ordered basis of $\mathbb{F}_{q^m}$ over $\mathbb{F}_q$, and let $h_1, \ldots, h_m \in \mathbb{F}_q[x]$ be arbitrary polynomials. For $h \in \mathbb{F}_q[x]$, there exists $s \in \mathbb{F}_{q^m}[x]$ such that*

$$sh = \xi_1 h_1 + \cdots + \xi_m h_m$$

if and only if there exist $s_1, \ldots, s_m \in \mathbb{F}_q[x]$ such that

$$s_i h = h_i \text{ for } 1 \le i \le m.$$

Proof. For a polynomial $s \in \mathbb{F}_{q^m}[x]$ let $s_1, \ldots, s_m \in \mathbb{F}_q[x]$ be the uniquely determined polynomials in $\mathbb{F}_q[x]$ such that

$$s = \xi_1 s_1 + \cdots + x_m s_m.$$

Then

$$sh = \xi_1 s_1 h + \cdots + x_m s_m h$$

is the unique representation in the basis $\boldsymbol{\xi}$ of the polynomial sh and the claim immediately follows. □

Finally we recall an important definition from [4]. For a monic polynomial $f \in \mathbb{F}_q[x]$ and a positive integer m we let $\Phi_q^{(m)}(f)$ denote the number of m-fold multisequences over $\mathbb{F}_q$ with minimal joint polynomial f. Note that $\Phi_q^{(m)}(f)$ can be considered as a function on the set of monic polynomials in $\mathbb{F}_q[x]$. In [4, Section 2] several important properties of $\Phi_q^{(m)}(f)$ have been derived, which we will use in this paper. We refer to [4] for further details.

3   Generalized Joint Linear Complexity

In this section we obtain our main results and we give illustrative examples. The following three lemmas will be used in the proof of the next theorem.

Lemma 1. *For an integer $n \ge 2$, let $H_n(x)$ be the real valued function on $\mathbb{R}$ defined by*

$$H_n(x) = x^n - 1 - (x - 1)^n.$$

For a real number $x > 1$, we have $H_n(x) > 0$.

Proof. We prove by induction on n. The case $n = 2$ is trivial and hence we assume that $n \geq 3$ and the lemma holds for $n - 1$. For the derivative we have

$$\frac{dH_n}{dx} = nx^{n-1} - n(x-1)^{n-1} = n\left(H_{n-1}(x) + 1\right). \tag{2}$$

By the induction hypothesis we have that $H_{n-1}(x) > 0$, for $x > 1$. Therefore using (2) we complete the proof. □

Lemma 2. *Let $q \geq 2$ be a prime power. Let a and $n \geq 2$ be positive integers. Then*

$$1 - \frac{1}{q^{na}} > \left(1 - \frac{1}{q^a}\right)^n.$$

Proof. Let $H_n(x)$ be the real valued function on $\mathbb{R}$ defined in Lemma 1. Note that $q^a > 1$ and

$$H_n\left(q^a\right) = q^{na} - 1 - \left(q^a - 1\right)^n.$$

Therefore using Lemma 1 we obtain that

$$q^{na} - 1 > \left(q^a - 1\right)^n. \tag{3}$$

Dividing both sides of (3) by q^{na} we complete the proof. □

Lemma 3. *Let $r \in \mathbb{F}_q[x]$ be an irreducible polynomial. For positive integers m and e we have*

$$\Phi_q^{(m)}(r^e) = \Phi_{q^m}^{(1)}(r^e) \text{ if } \gcd(\deg(r), m) = 1, \quad \text{and}$$

$$\Phi_q^{(m)}(r^e) > \Phi_{q^m}^{(1)}(r^e) \text{ if } \gcd(\deg(r), m) > 1.$$

Proof. It follows from [4, Lemma 2.2, (iii)] that

$$\Phi_q^{(m)}(r^e) = q^{me\deg(r)}\left(1 - \frac{1}{q^{m\deg(r)}}\right). \tag{4}$$

If $\gcd(\deg(r), m) = 1$, then, by Proposition 1, r is irreducible over $\mathbb{F}_{q^m}$ as well and hence using [4, Lemma 2.2, (iii)] again we obtain that $\Phi_{q^m}^{(1)}(r^e) = \Phi_q^{(m)}(r^e)$.

Assume that $u := \gcd(\deg(r), m) > 1$. It follows from Proposition 1 that the canonical factorization of r into irreducibles over $\mathbb{F}_{q^m}$ is of the form

$$r = t_1 t_2 \ldots t_u,$$

and $\deg(t_1) = \cdots = \deg(t_u) = \deg(r)/u$. Using [4, Lemma 2.2, (iii)] we have

$$\Phi_{q^m}^{(1)}(r^e) = q^{me\deg(r)}\left(1 - \frac{1}{q^{m\deg(r)/u}}\right)^u. \tag{5}$$

Therefore using Lemma 2, (4) and (5) we complete the proof. □

The following theorem determines the exact conditions on m and $f \in \mathbb{F}_q[x]$ for which the joint linear complexity and the generalized joint linear complexity on $\mathcal{M}_q^{(m)}(f)$ are the same.

Theorem 1. *Let m be a positive integer, let $f \in \mathbb{F}_q[x]$ be a monic polynomial with $\deg(f) \geq 1$, let*

$$f = r_1^{e_1} r_2^{e_2} \ldots r_k^{e_k}$$

be the canonical factorization of f into irreducibles, and let $\boldsymbol{\xi} = (\xi_1, \ldots, \xi_m)$ be an ordered basis of $\mathbb{F}_{q^m}$ over $\mathbb{F}_q$. Then we have

$$L_q^{(m)}(\boldsymbol{S}) = L_{q^m, \boldsymbol{\xi}}(\boldsymbol{S}) \quad \text{for each } \boldsymbol{S} \in \mathcal{M}_q^{(m)}(f),$$

if and only if

$$\gcd(m, \deg(r_i)) = 1, \quad \text{for } i = 1, 2, \ldots, k. \tag{6}$$

Proof. We first assume that $\gcd(m, \deg(r_i)) = 1$ for $i = 1, 2, \ldots, k$. Let $\boldsymbol{S} = (\sigma_1, \sigma_2, \ldots, \sigma_m)$ be an arbitrary multisequence in $\mathcal{M}_q^{(m)}(f)$, and let $g_1, g_2, \ldots, g_m$ be the polynomials in $\mathbb{F}_q[x]$ such that $\boldsymbol{S}$ corresponds to the m-tuple $(g_1/f, g_2/f, \ldots, g_m/f)$ as described in Section 2. The joint minimal polynomial of $\boldsymbol{S}$ is then the (uniquely determined) monic polynomial $d \in \mathbb{F}_q[x]$ dividing f such that

$$h_i/d = g_i/f, \quad \text{for } i = 1, 2, \ldots, m, \quad \text{and} \quad \gcd(h_1, h_2, \ldots, h_m, d) = 1, \tag{7}$$

for certain polynomials $h_1, h_2, \ldots, h_m$ in $\mathbb{F}_q[x]$. The sequence $S = S(\boldsymbol{S}, \boldsymbol{\xi})$ defined as in Section 1 depending on $\boldsymbol{S}$ and $\boldsymbol{\xi}$ then corresponds to

$$\frac{G}{f} = \frac{\xi_1 g_1 + \xi_2 g_2 + \cdots + \xi_m h_m}{f} = \frac{\xi_1 h_1 + \xi_2 h_2 + \cdots + \xi_m h_m}{d}.$$

We have to show that d is also the minimal polynomial of $S \in \mathcal{M}_{q^m}^{(1)}(f)$, or equivalently that d and $\xi_1 h_1 + \xi_2 h_2 + \cdots + \xi_m h_m$ are relatively prime in $\mathbb{F}_{q^m}[x]$. From (6) and Proposition 1 the canonical factorizations of f are the same over both fields, $\mathbb{F}_q$ and $\mathbb{F}_{q^m}$. Consequently this also applies to the divisor d of f. If d and $\xi_1 h_1 + \xi_2 h_2 + \cdots + \xi_m h_m$ are not relatively prime in $\mathbb{F}_{q^m}[x]$ then there exists a common factor in $\mathbb{F}_q[x]$ which contradicts (7) by Proposition 2.

We show the converse with a simple counting argument. Let $\boldsymbol{S}_1$ and $\boldsymbol{S}_2$ be distinct multisequences in $\mathcal{M}_q^{(m)}(f)$ both having minimal polynomial f. If $L_q^{(m)}(\boldsymbol{S}) = L_{q^m, \boldsymbol{\xi}}(\boldsymbol{S})$ for all elements $\boldsymbol{S} \in \mathcal{M}_q^{(m)}(f)$, then the distinct sequences $S_1, S_2 \in \mathcal{M}_{q^m}^{(1)}(f)$ corresponding to $\boldsymbol{S}_1$ and $\boldsymbol{S}_2$, respectively, will also have f as their minimal polynomial. By [4, Theorem 4.1] the numbers $\Phi_q^{(m)}(f)$ and $\Phi_{q^m}^{(1)}(f)$ of elements in $\mathcal{M}_q^{(m)}(f)$ and $\mathcal{M}_{q^m}^{(1)}(f)$, respectively, with minimal polynomial f are given by

$$\Phi_q^{(m)}(f) = \prod_{i=1}^k \Phi_q^{(m)}(r_i^{e_i}) \quad \text{and} \quad \Phi_{q^m}^{(1)}(f) = \prod_{i=1}^k \Phi_{q^m}^{(1)}(r_i^{e_i}).$$

With Lemma 3 we see that $\Phi_{q^m}^{(1)}(f) < \Phi_q^{(m)}(f)$ if condition (6) does not hold, which completes the proof.   □

Remark 1. For each $S \in \mathcal{M}_q^{(m)}(f)$, we always have

$$L_{q^m,\xi}(S) \leq L_q^{(m)}(S).$$

In Theorem 2 below we also derive tight lower bounds on $L_{q^m,\xi}(S)$ (see also Proposition 3 below).

Remark 2. Theorem 1 implies that the choice of f as a product of powers of irreducible polynomials $r_1, r_2, \ldots, r_k$ such that $\deg(r_1) = \cdots = \deg(r_k)$ is a (large) prime guarantees that generalized joint linear complexity is not smaller than joint linear complexity for any multisequence $S \in \mathcal{M}_q^{(m)}(f)$ if $m < \deg(r_i)$.

The following theorem gives a lower bound for the generalized joint linear complexity of a multisequence $S \in \mathcal{M}_q^{(m)}(f)$ with given minimal polynomial d.

Theorem 2. *Let f be a monic polynomial in $\mathbb{F}_q[x]$ with canonical factorization into irreducible monic polynomials over $\mathbb{F}_q$ given by*

$$f = r_1^{e_1} r_2^{e_2} \cdots r_k^{e_k},$$

and let $S \in \mathcal{M}_q^{(m)}(f)$ be an m-fold multisequence over $\mathbb{F}_q$ with joint minimal polynomial

$$d = r_1^{a_1} r_2^{a_2} \cdots r_k^{a_k}, \quad 0 \leq a_i \leq e_i \text{ for } 1 \leq i \leq k.$$

The generalized joint linear complexity $L_{q^m,\xi}(S)$ of S is then lower bounded by

$$L_{q^m,\xi}(S) \geq \sum_{i=1}^{k} a_i \frac{\deg(r_i)}{\gcd(\deg(r_i), m)}.$$

Proof. As the multisequence $S \in \mathcal{M}_q^{(m)}(f)$ has joint minimal polynomial d, we can uniquely associate S with an m-tuple $\left(\frac{h_1}{d}, \frac{h_2}{d}, \ldots, \frac{h_m}{d}\right)$ with $h_t \in \mathbb{F}_q[x]$, $\deg(h_t) < \deg(d)$ for $1 \leq t \leq m$, and $\gcd(h_1, \ldots, h_m, d) = 1$. If $a_i > 0$ then r_i does not divide all of the polynomials $h_1, \ldots, h_m$. Hence by Proposition 2 the polynomial r_i does not divide the polynomial $H = h_1\xi_1 + h_2\xi_2 + \cdots + h_m\xi_m$ over the extension field $\mathbb{F}_{q^m}$. Therefore if $r_i = t_{i,1}t_{i,2}\cdots t_{i,u_i}$ is the canonical factorization of r_i over $\mathbb{F}_{q^m}$, where $u_i = \gcd(\deg(r_i), m)$ and $\deg(t_{i,j}) = \deg(r_i)/u_i$ by Proposition 1, at least for one j, $1 \leq j \leq u_i$, we have $t_{i,j} \nmid H$. Consequently $t_{i,j}^{a_i}$ and H are relatively prime in $\mathbb{F}_{q^m}[x]$ which yields the lower bound for $L_{q^m,\xi}(S)$.   □

The following proposition shows that the lower bound of Theorem 2 is tight.

Proposition 3. *Let f be a monic polynomial in $\mathbb{F}_q[x]$ with canonical factorization into irreducible monic polynomials over $\mathbb{F}_q$ given by*

$$f = r_1^{e_1} r_2^{e_2} \cdots r_k^{e_k}.$$

Let $a_1, a_2, \ldots, a_k$ be integers with $0 \le a_i \le e_i$ for $1 \le i \le k$. Let $m \ge 2$ be an integer and $\boldsymbol{\xi} = (\xi_1, \ldots, \xi_m)$ be an ordered basis of $\mathbb{F}_{q^m}$ over $\mathbb{F}_q$. There exists an m-fold multisequence $\boldsymbol{S} \in \mathcal{M}_q^{(m)}(f)$ over $\mathbb{F}_q$ such that its joint minimal polynomial d is

$$d = r_1^{a_1} r_2^{a_2} \ldots r_k^{a_k},$$

and its generalized joint linear complexity $L_{q^m, \boldsymbol{\xi}}(\boldsymbol{S})$ is

$$L_{q^m, \boldsymbol{\xi}}(\boldsymbol{S}) = \sum_{i=1}^{k} a_i \frac{\deg(r_i)}{\gcd(\deg(r_i), m)}.$$

Proof. By reordering $r_1, \ldots, r_k$ suitably, we can assume without loss of generality that there exists an integer l, $1 \le l \le k$, with $\gcd(m, \deg(r_i)) = u_i \ge 2$ for $1 \le i \le l$ and $\gcd(m, \deg(r_i)) = 1$ for $l+1 \le i \le k$. Indeed otherwise $\gcd(m, \deg(r_i)) = 1$ for $1 \le i \le k$ and hence the result is trivial by Theorem 1. Using Proposition 1 we obtain that the canonical factorizations of r_i, $1 \le i \le l$, into irreducibles over $\mathbb{F}_{q^m}$ are of the form

$$r_i = t_{i,1} t_{i,2} \ldots t_{i,u_i}.$$

Let S be the sequence in $\mathcal{M}_{q^m}^{(1)}(f)$ corresponding to the polynomial

$$G = \frac{f}{d} \prod_{i=1}^{l} (t_{i,2} \ldots, t_{i,u_i})^{a_i} \in \mathbb{F}_{q^m}[x]$$

and let $h_1, h_2, \ldots, h_m \in \mathbb{F}_q[x]$ be the uniquely determined polynomials in $\mathbb{F}_q[x]$ such that

$$\prod_{i=1}^{l} (t_{i,2} \ldots, t_{i,u_i})^{a_i} = \xi_1 h_1 + \xi_2 h_2 + \cdots + \xi_m h_m. \tag{8}$$

Let $\boldsymbol{S} = (\sigma_1, \ldots, \sigma_m) \in \mathcal{M}_q^{(m)}(f)$ be the m-fold multisequence such that the sequence σ_i corresponds to $g_i = h_i f / d \in \mathbb{F}_q[x]$ for $1 \le i \le m$. We observe that we have $S = \mathcal{S}(\boldsymbol{S}, \boldsymbol{\xi})$ and

$$L_{q^m, \boldsymbol{\xi}}(\boldsymbol{S}) = \sum_{i=1}^{l} a_i \deg(t_{i,1}) + \sum_{i=l+1}^{k} a_i \deg(r_i).$$

Moreover d is the joint minimal polynomial of $\boldsymbol{S}$. Indeed, otherwise using (8) we obtain that there exists $1 \le i \le k$ with

$$r_i \mid \prod_{i=1}^{l} (t_{i,2} \ldots, t_{i,u_i})^{a_i} \text{ in } \mathbb{F}_{q^m}[x].$$

This is a contradiction, which completes the proof. $\qquad\square$

In the following corollary we consider $\dfrac{L_q^{(m)}(S) - L_{q^m,\boldsymbol{\xi}}(S)}{L_q^{(m)}(S)}$, the difference of joint linear complexity and generalized joint linear complexity in relation to the value for the joint linear complexity. We give a uniform and tight upper bound which applies to arbitrary nonzero multisequences in $\mathcal{M}_q^{(m)}(f)$.

Corollary 1. *Let $m \geq 2$ be an integer and f be a monic polynomial in $\mathbb{F}_q[x]$ with canonical factorization into irreducible monic polynomials over $\mathbb{F}_q$ given by*

$$f = r_1^{e_1} r_2^{e_2} \cdots r_k^{e_k}$$

with

$$u_{\max} = \max\{\gcd(\deg(r_i), m) : 1 \leq i \leq k\}.$$

Then for an arbitrary nonzero multisequence $S \in \mathcal{M}_q^{(m)}(f)$ and an ordered basis $\boldsymbol{\xi}$ of $\mathbb{F}_{q^m}$ over $\mathbb{F}_q$ we have

$$\frac{L_q^{(m)}(S) - L_{q^m,\boldsymbol{\xi}}(S)}{L_q^{(m)}(S)} \leq 1 - \frac{1}{u_{\max}}. \tag{9}$$

Moreover the bound in (9) is tight.

Proof. For any nonzero m-fold multisequence $S \in \mathcal{M}_q^{(m)}(f)$, its joint minimal polynomial d is of the form

$$d = r_1^{a_1} r_2^{a_2} \cdots r_k^{a_k},$$

where $0 \leq a_i \leq e_i$ are integers and $(a_1, \ldots, a_k) \neq (0, \ldots, 0)$. Therefore, using Theorem 2, for its joint linear complexity $L_q^{(m)}(S)$ and its generalized joint linear complexity $L_{q^m,\boldsymbol{\xi}}(S)$ we obtain that

$$L_q^{(m)}(S) = \sum_{i=1}^{k} a_i \deg(r_i), \quad \text{and} \quad L_{q^m,\boldsymbol{\xi}}(S) \geq \sum_{i=1}^{k} a_i \frac{\deg(r_i)}{\gcd(\deg(r_i), m)}. \tag{10}$$

It follows from the definition of $u_{\max}$ that

$$\frac{1}{u_{\max}} a_i \deg(r_i) \leq a_i \frac{\deg(r_i)}{\gcd(\deg(r_i), m)} \tag{11}$$

for $1 \leq i \leq k$. Combining (10) and (11) we obtain (9). Moreover let $a_1, \ldots, a_k$ be integers such that

$$a_i = \begin{cases} 0 & \text{if } \gcd(\deg(r_i), m) \neq u_{\max}, \\ \neq 0 & \text{if } \gcd(\deg(r_i), m) = u_{\max}. \end{cases} \tag{12}$$

For integers $a_1, \ldots, a_k$ as in (12) we have equality in (11). Using Proposition 3 we obtain an m-fold multisequence $S_{u_{\max}} \in \mathcal{M}_q^{(m)}(f)$ such that we have equality for $L_{q^m,\boldsymbol{\xi}}(S)$ in (10), where the integers $a_1, \ldots, a_k$ are as in (12). Hence we conclude that the bound in (9) is attained by $S_{u_{\max}}$, which completes the proof. $\square$

Remark 3. If condition (6) is satisfied then (9) will be zero for all multisequences $S \in \mathcal{M}_q^{(m)}(f)$. As $\gcd(\deg(r_i), m)$ can at most be m the largest possible relative distance between joint linear complexity and generalized joint linear complexity of an m-fold multisequence is given by $(m-1)/m$.

We give two examples illustrating our results.

Example 1. Let N, m be positive integers and consider the N-periodic m-fold multisequences over $\mathbb{F}_q$. Equivalently, let $f = x^N - 1 \in \mathbb{F}_q[x]$ and we can consider the multisequences in $\mathcal{M}_q^{(m)}(f)$. Let p be the characteristic of the finite field $\mathbb{F}_q$ and $N = p^v n$ with $\gcd(n, p) = 1$. Then we have $x^N - 1 = (x^n - 1)^{p^v}$, and the canonical factorization of $x^n - 1$ in $\mathbb{F}_q[x]$ is given by

$$x^n - 1 = \prod_{i=1}^{k} r_i(x) \quad \text{with} \quad r_i(x) = \prod_{j \in C_i} (x - \alpha^j),$$

where $C_1, \ldots, C_k$ are the different cyclotomic cosets modulo n relative to powers of q and α is a primitive nth root of unity in some extension field of $\mathbb{F}_q$. Let S be an N-periodic m-fold multisequence over $\mathbb{F}_q$ with minimal polynomial $d = r_1^{\rho_1} r_2^{\rho_2} \cdots r_k^{\rho_k}$, where $0 \le \rho_i \le p^v$. Then using Theorem 2 we have

$$L(S) \ge \sum_{i=1}^{k} \rho_i \frac{l_i}{\gcd(l_i, m)}, \tag{13}$$

where l_i denotes the cardinality of the cyclotomic coset C_i. Equation (13) coincides with the corresponding result in [8, Theorem 2].

Example 2. Let $r_1, \ldots, r_k \in \mathbb{F}_q[x]$ be distinct irreducible polynomials and let $e_1, \ldots, e_k$ be positive integers. For a positive integer m, let

$$f = r_1^{e_1} r_2^{e_2} \ldots r_k^{e_k},$$

and consider the multisequences in $\mathcal{M}_q^{(m)}(f)$. It is not difficult to observe that there exists a multisequence $S \in \mathcal{M}_q^{(m)}(f)$ with joint linear complexity $L_q^{(m)}(S) = t$ if and only if t can be written as

$$t = i_1 \deg(r_1) + i_2 \deg(r_2) + \cdots + i_k \deg(r_k), \tag{14}$$

where $0 \le i_1 \le e_1, \ldots, 0 \le i_k \le e_k$ are integers. Let $\boldsymbol{\xi} = (\xi_1, \ldots, \xi_m)$ be an ordered basis of $\mathbb{F}_{q^m}$ over $\mathbb{F}_q$. Let $0 \le i_1 \le e_1, \ldots, 0 \le i_k \le e_k$ be chosen integers. Consider the nonempty subset $T(i_1, \ldots, i_k)$ of $\mathcal{M}_q^{(m)}(f)$ consisting of S such that $L_q^{(m)}(S) = t$, where t is as in (14). Using the methods of this paper we obtain that, among the multisequences in $T(i_1, \ldots, i_k)$, there exists a multisequence S with generalized joint linear complexity $L_{q^m, \boldsymbol{\xi}}(S) = \tilde{t}$ if and only if $\tilde{t}$ can be written as

$$\tilde{t} = i_1 j_1 \frac{\deg(r_1)}{\gcd(\deg(r_1), m)} + i_2 j_2 \frac{\deg(r_2)}{\gcd(\deg(r_2), m)} + \cdots + i_k j_k \frac{\deg(r_k)}{\gcd(\deg(r_k), m)},$$

where $1 \le j_1 \le \gcd(\deg(r_1), m), \ldots, 1 \le j_k \le \gcd(\deg(r_k), m)$ are integers.

Remark 4. The results above do not depend on the choice of the basis. However the generalized joint linear complexity actually depends on the basis. The following simple example illustrates this fact.

Example 3. Let $S = (\sigma_1, \sigma_2, \sigma_3)$ be the 7-periodic 3-fold multisequence over $\mathbb{F}_2$ given by

$$\sigma_1 = 1\,0\,0\,1\,0\,1\,1 \cdots$$
$$\sigma_2 = 0\,1\,0\,1\,1\,1\,0 \cdots$$
$$\sigma_3 = 0\,0\,1\,0\,1\,1\,1 \cdots .$$

Let $\alpha \in \mathbb{F}_8$ with $\alpha^3 + \alpha + 1 = 0$. Consider the ordered bases $\boldsymbol{\xi}_1 = (1, \alpha, \alpha^2)$ and $\boldsymbol{\xi}_2 = (\alpha, 1, \alpha^2 + 1)$ of $\mathbb{F}_8$ over $\mathbb{F}_2$. The 7-periodic sequences over $\mathbb{F}_8$ obtained from S using the bases $\boldsymbol{\xi}_1$ and $\boldsymbol{\xi}_2$ are

$$S_1 := S(S, \boldsymbol{\xi}_1) = 1, \alpha, \alpha^2, \alpha + 1, \alpha^2 + \alpha, \alpha^2 + \alpha + 1, \alpha^2 + 1, \cdots \quad \text{and}$$
$$S_2 := S(S, \boldsymbol{\xi}_2) = \alpha, 1, \alpha^2 + 1, \alpha + 1, \alpha^2, \alpha^2 + \alpha, \alpha^2 + \alpha + 1, \cdots .$$

For the terms of S_1 we have $s_{n+1} = \alpha s_n$, where $n \geq 0$, and hence $L_{8, \boldsymbol{\xi}_1}(S) = 1$.
The first three terms of S_2 are $s_0 = \alpha$, $s_1 = (\alpha^2 + 1)s_0$ and $s_2 = (\alpha^2 + 1)s_1$. However for the third term of S_2 we have $s_3 \neq (\alpha^2 + 1)s_2$ and hence $L_{8, \boldsymbol{\xi}_2}(S) > 1$.

Acknowledgments

We would like to thank Arne Winterhof for very useful suggestions. The second author was partially supported by TÜBİTAK under Grant No. TBAG-107T826.

References

1. Davies, D.W. (ed.): EUROCRYPT 1991. LNCS, vol. 547, pp. 168–175. Springer, Heidelberg (1991)
2. Dawson, E., Simpson, L.: Analysis and design issues for synchronous stream ciphers. In: Niederreiter, H. (ed.) Coding Theory and Cryptology, pp. 49–90. World Scientific, Singapore (2002)
3. Fu, F.W., Niederreiter, H., Su, M.: The expectation and variance of the joint linear complexity of random periodic multisequences. J. Complexity 21, 804–822 (2005)
4. Fu, F.W., Niederreiter, H., Özbudak, F.: Joint Linear Complexity of Multisequences Consisting of Linear Recurring Sequences, Cryptography and Communications - Discrete Structures, Boolean Functions and Sequences (to appear)
5. Hawkes, P., Rose, G.G.: Exploiting multiples of the connection polynomial in word-oriented stream ciphers. In: Okamoto, T. (ed.) ASIACRYPT 2000. LNCS, vol. 1976, pp. 303–316. Springer, Heidelberg (2000)
6. Lidl, R., Niederreiter, H.: Finite Fields. Cambridge University Press, Cambridge (1997)
7. Massey, J.L.: Shift-register synthesis and BCH decoding. IEEE Trans. Inform. Theory 15, 122–127 (1969)

8. Meidl, W.: Discrete Fourier transform, joint linear comoplexity and generalized joint linear complexity of multisequences. In: Helleseth, T., Sarwate, D., Song, H.-Y., Yang, K. (eds.) SETA 2004. LNCS, vol. 3486, pp. 101–112. Springer, Heidelberg (2005)
9. Meidl, W., Niederreiter, H.: Linear complexity, k-error linear complexity, and the discrete Fourier transform. J. Complexity 18, 87–103 (2002)
10. Meidl, W., Niederreiter, H.: On the expected value of the linear complexity and the k-error linear complexity of periodic sequences. IEEE Trans. Inform. Theory 48, 2817–2825 (2002)
11. Meidl, W., Niederreiter, H.: The expected value of the joint linear complexity of periodic multisequences. J. Complexity 19, 61–72 (2003)
12. Niederreiter, H.: Sequences with almost perfect linear complexity profile. In: Chaum, D., Price, W.L. (eds.) Advances in Cryptology-EUROCRYPT 1987. LNCS, vol. 304, pp. 37–51. Springer, Berlin (1988)
13. Niederreiter, H., Johansson, T., Maitra, S. (eds.): INDOCRYPT 2003. LNCS, vol. 2904, pp. 1–17. Springer, Berlin (2003)
14. Rueppel, R.A.: Analysis and Design of Stream Ciphers. Springer, Berlin (1986)
15. Rueppel, R.A.: Stream ciphers. In: Simmons, G.J. (ed.) Contemporary Cryptology: The Science of Information Integrity, pp. 65–134. IEEE Press, New York (1992)

A Lattice-Based Minimal Partial Realization Algorithm

Li-Ping Wang

Center for Advanced Study, Tsinghua University,
Beijing 100084, People's Republic of China
wanglp@mail.tsinghua.edu.cn

Abstract. In this paper we extend a minimal partial realization algorithm for vector sequences to matrix sequences by means of a lattice basis reduction algorithm in function fields. The different ways of transforming a given basis into a reduced one lead to different partial realization algorithms and so our technique provides a unified approach to the minimal partial realization problem.

1 Introduction

As one of the most fundamental problem in linear systems theory, the minimal partial realization problem has attracted considerable attention since the early 1960s, which brings a lot of minimal partial realization algorithms [1, 2, 3, 5, 9, 10, 12, 18]. In information theory the minimal partial realization problem is also called the linear feedback register synthesis problem, which plays an important role in the analysis and design of cryptosystems since especially in recent years there has been an increasing interest in the study of multivariable cryptosystems [4, 7, 11].

Let us recall the problem. Consider an infinite sequence T of $p \times m$ matrices $\{T_1, T_2, \cdots, \}$ over an arbitrary field $\mathbb{F}$, and so we regard as specifying the transfer function in the linear system via

$$T(z) = \sum_{i=1}^{\infty} T_i z^{-i}.$$

For a positive integer N, the aim is to find an $m \times m$ nonsingular polynomial matrix $M_N(z)$ and a polynomial matrix $Y_N(z)$ such that the first N terms of the Laurent expansion of $Y_N(z)M_N^{-1}(z)$ are equal to $\{T_1, T_2, \ldots, T_N\}$. Therefore $Y_N(z)M_N^{-1}(z)$ will be called an Nth (right) partial realization of $T(z)$ or T. If the degree of $\det(M_N(z))$ (also called the McMillan degree) is minimal, $Y_N(z)M_N^{-1}(z)$ is an Nth (right) minimal partial realization of $T(z)$ or T.

For the single-input-single-output (SISO) systems $p = m = 1$, as we all know, there is an effective algorithm (Berlekamp-Massey algorithm) [16]. For the systems $m = 1$ and $p > 1$, that is, vector sequences, there are several synthesis algorithms [6, 8, 19, 20].

S.W. Golomb et al. (Eds.): SETA 2008, LNCS 5203, pp. 278–289, 2008.
© Springer-Verlag Berlin Heidelberg 2008

In [19, 20] a minimal partial realization algorithm for vector sequences was proposed based on a lattice reduction algorithm in function fields. However, the generalization to the MIMO systems is not clear. Therefore in this paper we extend the algorithm to the matrix sequences. Furthermore, we characterize all the minimal partial realizations and give the sufficient and necessary condition for the unique issue. Our point of view is rather algebraic, however, the different ways of transforming a given basis into a reduced one lead to different partial realization algorithms and so our technique provides a unified approach to the minimal partial realization problem.

2   The Realization Algorithm

We shall restrict our attention to an arbitrary field $\mathbb{F}$, the polynomial ring $\mathbb{F}[z]$, the rational function field $\mathbb{F}(z)$, the field of formal Laurent series $K = \mathbb{F}((z^{-1}))$. There is a valuation v on K whereby for $\alpha = \sum_{j=j_0}^{\infty} a_j z^{-j} \in K$ we put $v(\alpha) = \max\{-j \in \mathbb{Z} : a_j \neq 0\}$ if $\alpha \neq 0$ and $v(\alpha) = -\infty$ if $\alpha = 0$. It can be seen as the generalization for the degree of a polynomial. For any two positive integers k and n, the *valuation* $v(A)$ of a $k \times n$ matrix $A = (\alpha_{ij})_{k \times n}$ over K is defined as $\max\{v(\alpha_{ij}) : 1 \leq i \leq k, 1 \leq j \leq n\}$ and define $Az^{-h} = (a_{ij}z^{-h})_{1 \leq i \leq k, 1 \leq j \leq n}$ where h is an arbitrary integer so that $T(z)$ can be seen as a $p \times m$ matrix over K. In this paper we mainly use the valuation of a column vector. In the sequel we often use the *projection* $\theta : K^n \to \mathbb{F}^n$ such that $\gamma = (\alpha_i)_{1 \leq i \leq n} \mapsto (a_{1,-v(\gamma)}, \ldots, a_{n,-v(\gamma)})^t$, where $\alpha_i = \sum_{j=j_0}^{\infty} a_{i,j} z^{-j}$, $1 \leq i \leq n$, and t denotes the transpose of a vector.

For any positive integers k and s, we denote the identity matrix of order k by $I_{k \times k}$, the $k \times s$ zero matrix by $\mathbf{0}_{k \times s}$ and the zero vector with k components by $\mathbf{0}_k$.

A nonzero polynomial column vector $\mathbf{c}(z)$ in $\mathbb{F}[z]^m$ can be written as $\mathbf{c}(z) = \sum_{i=0}^{d} \mathbf{c}_i z^i$ where $\mathbf{c}_0, \ldots, \mathbf{c}_d \in \mathbb{F}^m$ and $d = v(\mathbf{c}(z))$. Define m classes $[\beta_1], \ldots, [\beta_m]$ by $[\beta_i] = \{(\underbrace{0, \ldots, 0}_{i-1}, 1, b_1, \ldots, b_{m-i})^t \in \mathbb{F}^m : b_j \in \mathbb{F} \text{ for } 1 \leq j \leq m - i\}$, $i = 1, \ldots, m$.

Definition 1. *A nonzero polynomial column vector* $\mathbf{c}(z) = \sum_{i=0}^{d} \mathbf{c}_i z^i$ *is called an Nth right i-annihilating polynomial column vector of T if the kth discrepancy* $\delta_k(\mathbf{c}(z), T) = \mathbf{0}_p$ *for all* $1 \leq k \leq N$, *where*

$$\delta_k(\mathbf{c}(z), T) = T_k \mathbf{c}_d + T_{k-1} \mathbf{c}_{d-1} + \ldots + T_{k-d} \mathbf{c}_0 = \sum_{i=0}^{d} T_{k-d+i} \mathbf{c}_i,$$

and $\theta(\mathbf{c}(z)) \in [\beta_i]$. *The Nth right i-minimal polynomial column vector of T is the Nth right i-annihilating polynomial column vector with the least valuation.*

Similarly, we can define the *left i-annihilating polynomial row vector* and *left i-minimal polynomial row vector*. In this paper we only discuss the right case and it is easy to get the corresponding results for the left case.

For each i, $1 \leq i \leq m$, it is clear that there always exists an Nth right i-minimal polynomial column vector of T and we denote them by $M_{N,1}(z), \cdots,$ $M_{N,m}(z)$ respectively. Let $M_N(z) = (M_{N,1}(z) \cdots M_{N,m}(z))$ be a matrix, which is invertible since $M_{N,1}(z), \cdots, M_{N,m}(z)$ are linear independent over $\mathbb{F}[z]$, and we have

$$v(T(z)M_{N,i}(z) - \text{Pol}(T(z)M_{N,i}(z))) \leq -N - 1 + v(M_{N,i}(z)), \quad 1 \leq i \leq m, \quad (1)$$

where $\text{Pol}(T(z)M_{N,i}(z))$ is the polynomial part of $T(z)M_{N,i}(z)$.

We can write
$$M_N(z) = (M_N(z))_{hc}D(z) + R(z),$$

where

$$D(z) = Diag\{z^{v(M_{N,i}(z))}, i = 1, \ldots, m\}$$
$$(M_N(z))_{hc} = (\theta(M_{N,1}(z)) \quad \cdots \quad \theta(M_{N,m}(z))),$$

and $R(z)$ denotes the remaining terms with the valuations of its column vectors strictly less than those of $D(z)$. Then

$$\det(M_N(z)) = \det((M_N(z)_{hc}))z^{\sum_{i=1}^{m} v(M_{N,i}(z))} + \text{ terms of lower degree in z.}$$

Since $(M_N(z))_{hc}$ is nonsingular and so we obtain the following lemma.

Lemma 1. *We have*

$$\deg(\det(M_N(z))) = \sum_{i=1}^{m} v(M_{N,i}(z)).$$

Lemma 2. *We have* $v(T(z) - pol(T(z)M_N(z))M_N^{-1}(z)) \leq -N - 1.$

Proof. We write $M_N^{-1}(z) = D'(z)(M_N^{-1}(z))_{hr} + B(z)$ where

$$D'(z) = Diag\{z^{v(R_{N,i}(z))}, i = 1, \ldots, m\}$$
$$(M_N^{-1}(z))_{hr} = \begin{pmatrix} (\theta(R_{N,1}(z)))^t \\ \vdots \\ (\theta(M_{N,m}(z)))^t \end{pmatrix}$$

and $R_{N,i}(z)$ is the ith row of $M_N^{-1}(z)$ for $1 \leq i \leq m$ and $B(z)$ is the remaining terms with the valuations of its row vectors strictly less than those of $D'(z)$. Since $M_N(z)M_N^{-1}(z) = I_{m \times m}$ and $(M_N(z))_{hc}$ is nonsingular, we have

$$v(R_{N,i}(z)) = -v(M_{N,i}(z)). \quad (2)$$

By (1) and (2), the desired result follows. $\qquad\square$

Therefore $\mathrm{pol}(T(z)M_N(z))M_N^{-1}(z)$ is an Nth partial realization of T. Conversely, let $\mathrm{Pol}(T(z)M_N(z))M_N^{-1}(z)$ be an Nth partial realization of T. Since $M_N(z)$ is nonsingular, it is easy to multiply $M_N(z)$ by elementary matrices such that the projection of each column of the obtained matrix is in $[\beta_i]$ respectively. Therefore the ith column is an Nth right i-annihilating polynomial column vector of T for $1 \le i \le m$. With the above discussions we obtain the following theorem.

Theorem 1. *If $M_{N,i}(z)$, $1 \le i \le m$, is an Nth right i-minimal polynomial column vector of T and $M_N(z) = (M_{N,i}(z))_{1 \le i \le m}$, then $\mathrm{pol}(T(z)M_N(z))M_N^{-1}(z)$ is an Nth minimal partial realization of T.*

Therefore the minimal partial realization problem is reduced to finding m minimal polynomial column vectors with their projections in distinct classes.

Here we need to use the lattice theory.

In the sequel we always let $n = m+p$. A subset Λ of K^n is called an $\mathbb{F}[z]$-*lattice* if there exists a basis $\omega_1, \ldots, \omega_n$ of K^n such that

$$\Lambda = \sum_{i=1}^{n} \mathbb{F}[z]\,\omega_i = \left\{ \sum_{i=1}^{n} f_i\,\omega_i : f_i \in \mathbb{F}[z],\, i = 1, \ldots, n \right\}.$$

In this situation we say that $\omega_1, \ldots, \omega_n$ form a *basis* for Λ. A basis $\omega_1, \ldots, \omega_n$ is *reduced* if $\theta(\omega_1), \ldots, \theta(\omega_n)$ are linearly independent over $\mathbb{F}$. The *determinant* of the lattice is defined by $\det(\Lambda) = v(\det(\omega_1, \ldots, \omega_n))$ and is independent of the choice of the basis. In [14, 15] it was proved that

$$\sum_{i=1}^{n} v(\omega_i) = \det(\Lambda) \qquad (3)$$

if $\omega_1, \ldots, \omega_n$ are reduced for a lattice Λ. For any vector γ, let $\bar{\gamma}$ be the vector containing only the last m components of γ. The reduced basis is normal if $v(\omega_1) \le \ldots \le v(\omega_p)$, $\overline{\theta(\omega_i)} = 0_m$ for $1 \le i \le p$ and $\overline{\theta(\omega_{p+i})} \in [\beta_i]$ for $1 \le i \le m$.

Consider the matrix

$$\begin{pmatrix} -I_{p\times p} & T(z) \\ 0_{m\times p} & I_{m\times m}z^{-N-1} \end{pmatrix},$$

and denote its n columns by $-\varepsilon_1, \cdots, -\varepsilon_p, \alpha_{N,1}, \cdots, \alpha_{N,m}$, which span an $\mathbb{F}[z]$-lattice, simply denoted by Λ later.

By means of a lattice basis reduction algorithm [14, 17], we can transform the initial basis into a reduced one. Then it is easy to obtain a normal basis by performing some elementary transformations on the reduced basis. In the following we will show that the information we want about T must appear in a normal basis of Λ.

The mapping $\eta : \Lambda \to \mathbb{F}[z]^m$ is given by

$$\eta(-f_1(z)\varepsilon_1 - \cdots - f_p(z)\varepsilon_p + f_{p+1}(z)\alpha_{N,1} + \cdots + f_n(z)\alpha_{N,m}) = (f_{p+1}(z), \cdots, f_n(z))^t,$$

where $f_1(z), \ldots, f_n(z) \in \mathbb{F}[z]$. Conversely, any polynomial column vector $\mathbf{c}(z) \in \mathbb{F}[z]^m$ completely determines an associated element in Λ given by

$$\sigma(\mathbf{c}(z))|_\Lambda := \begin{pmatrix} T(z)\mathbf{c}(z) - \mathrm{Pol}(T(z)\mathbf{c}(z)) \\ \mathbf{c}(z)z^{-N-1} \end{pmatrix}.$$

For $1 \le i \le m$, let

$$S_i(\Lambda) = \{\gamma \in \Lambda : \overline{\theta(\gamma)} \in [\beta_i]\}$$

and

$$\Gamma_i(T) = \{\mathbf{c}(z) \in \mathbb{F}[z]^m : \mathbf{c}(z) \text{ is an } N\text{th right } i\text{-annihilating polynomial}$$
$$\text{column vector of } T\}.$$

Furthermore, we define a natural ordering, namely, both $S_i(\Lambda)$ and $\Gamma_i(T)$ are ordered by the valuation of an element in them.

Theorem 2. *The mapping η restricted on $S_i(\Lambda)$ is an order-preserving one-to-one correspondence from $S_i(\Lambda)$ to $\Gamma_i(T)$ for $1 \le i \le m$, and σ restricted on $\Gamma_i(T)$ is its inverse.*

Proof. Denote η restricted on $S_i(\Lambda)$ by $\eta|_{S_i(\Lambda)}$, σ restricted on $\Gamma_i(T)$ by $\sigma|_{\Gamma_i(T)}$, respectively. First we need to show that $\eta|_{S_i(\Lambda)}$ is well-defined. For any vector $\gamma \in S_i(\Lambda)$, we have

$$\gamma = \begin{pmatrix} T(z)\eta(\gamma) - \mathrm{Pol}(T(z)\eta(\gamma)) \\ \eta(\gamma)z^{-N-1} \end{pmatrix}.$$

Since $\gamma \in S_i(\Lambda)$, we have $v(T(z)\eta(\gamma) - \mathrm{Pol}(T(z)\eta(\gamma))) \le -N - 1 + v(\eta(\gamma))$ and so $\eta(\gamma) \in \Gamma_i(T)$. Similarly $\sigma|_{\Gamma_i(T)}$ is well-defined. It is easy to check that $\eta|_{S_i(\Lambda)}\sigma|_{\Gamma_i(T)} = 1_{\Gamma_i(T)}$ and $\sigma|_{\Gamma_i(T)}\eta|_{S_i(\Lambda)} = 1_{S_i(\Lambda)}$. For any two elements $\gamma_1, \gamma_2 \in S_i(\Lambda)$ with $v(\gamma_1) \le v(\gamma_2)$, we have $v(\eta(\gamma_1)) \le v(\eta(\gamma_2))$ for all $1 \le i \le m$. $\qquad\square$

Theorem 3. *Let $\omega_1, \cdots, \omega_{p+m}$ be a normal basis for Λ. Then $\eta(\omega_{p+i})$, $1 \le i \le m$, is an Nth right i-minimal polynomial column vector of T.*

Proof. By Theorem 2, it suffices to show that ω_{p+i} is a minimal element in $S_i(\Lambda)$ for $1 \le i \le m$. Suppose there exists a vector $\gamma \in S_i(\Lambda)$ such that $v(\gamma) < v(\omega_{p+i})$ for some i, $1 \le i \le m$. In view of $\gamma \in \Lambda$, we can write $\gamma = f_1(z)\omega_1 + \cdots + f_n(z)\omega_n$ with $f_j(z) \in \mathbb{F}[z]$ for $1 \le j \le n$. Since $\theta(\omega_1), \cdots, \theta(\omega_n)$ are linearly independent over $\mathbb{F}$ and by convention $\deg(0) = -\infty$, we have $v(\gamma) = \max\{v(\omega_j) + \deg(f_j(z)) : 1 \le j \le n\}$. Let $I(\gamma) = \{1 \le j \le n : v(\omega_j) + \deg(f_j(z)) = v(\gamma)\}$ and so $f_{p+i}(z) = 0$ because of $v(\gamma) < v(\omega_{p+i})$. Let $\mathrm{lc}(f(x))$ denote the leading coefficient of a polynomial $f(x)$. Thus it follows that

$$\theta(\gamma) = \sum_{j \in I(\gamma)} \mathrm{lc}(f_j(z)) \theta(\omega_j) \notin [\beta_i],$$

a contradiction with $\gamma \in S_i(\Lambda)$. $\qquad\square$

So far we can solve the minimal partial realization problem by Theorem 1, Theorem 2 and Theorem 3. We give our algorithm below, in which we give the initialization in step 1, transform the initial basis into a reduced one in step 2, and into a normal one in step 3. Finally we output the result in step 4.

Algorithm 2.1

Input: the first N terms of a matrix sequence $T = (T_1, T_2, \ldots)$.
Output: an Nth minimal realization of T.

1. Initialize $\omega_1 \leftarrow -\varepsilon_1, \cdots, \omega_p \leftarrow -\varepsilon_p, \omega_{p+1} \leftarrow \alpha_{N,1}, \cdots, \omega_n \leftarrow \alpha_{N,m}$.
2. **While** $\theta(\omega_1), \cdots, \theta(\omega_n)$ are linearly dependent over $\mathbb{F}$ **do**
    (Reduction step)There is a vector $(a_1, \cdots, a_n)$ such that $\sum_{i=1}^{n} a_i \theta(\omega_i) = 0_n$
    and find an integer h such that $v(\omega_h) = \max\{v(\omega_i) : 1 \le i \le m, a_i \ne 0\}$.
    Set $\xi \leftarrow \sum_{i=1}^{n} a_i z^{-v(\omega_i)+v(\omega_h)} \omega_i$ and $\omega_h \leftarrow \xi$.
    end-While
3. **For** $j = 1, \ldots, m$ **do**
    Find an integer k such that $v(\omega_k) = \min\{v(\omega_i) : 1 \le i \le n$ and the $(p+j)$th component of $\theta(\omega_i)$ is nonzero$\}$ and set $c_k \leftarrow$ the $(p+j)$-th component of $\theta(\omega_k)$.
    For $i = 1, \ldots, n$ **do**
        If $i \ne k$ and the $(p+j)$-th component c of $\theta(\omega_i)$ is not zero **then**
        set $\omega_i \leftarrow \omega_i - \frac{c}{c_k} z^{-v(\omega_k)+v(\omega_i)} \omega_k$ and $\omega_k \leftarrow \frac{1}{c_k} \omega_k$.
        end-If
    end-For
    end-For
    Arrange the ω_i in such a way such that $\overline{\theta(\omega_i)} = 0_m$, $i = 1, \ldots, p$, $v(\omega_1) \le \cdots$
    $\le v(\omega_p)$ and $\overline{\theta(\omega_{p+i})} \in [\beta_i]$ for $1 \le i \le m$.
4. Set $M_N(z) \leftarrow (\eta(\omega_{p+i}))_{1 \le i \le m}$, output $\mathrm{pol}(T(z)M_N(z))M_N^{-1}(z)$ and terminate the algorithm.

Remark 1. When $m = 1$ the above algorithm becomes the synthesis algorithm for vector sequences as in [20]. Similar to [20] we also show that the algorithm will terminate in the finite steps. We introduce a function $\Psi(\omega_1, \cdots, \omega_n) = -m(N + 1) - \sum_{i=1}^{n} v(\omega_i)$. Whenever a reduction step takes place, the value of $\Psi(\omega_1, \cdots, \omega_{m+1})$ strictly increases by the above discussions and when $\Psi(\omega_1, \cdots, \omega_{m+1}) = 0$ the corresponding basis becomes reduced [17]. Thus the number of reduction steps is at most $m(N + 1)$.

3   Parametrization of All Minimal Partial Realizations and the Uniqueness Issue

Given a normal basis $\omega_1, \ldots, \omega_n$ of the lattice Λ, put $\pi_i = v(\omega_i)$. Then the set $\{\pi_1, \ldots, \pi_n\}$ is completely determined by the lattice Λ and does not depend on the particular normal basis $\omega_1, \ldots, \omega_n$ [17]. It is easy to see that the lattice Λ is completely determined by T and N, and so the set can be as the invariance of the first N terms of the matrix sequence T. Since $\omega_{p+i} \in S_i(\Lambda)$, it is easy to get

$$v(\eta(\omega_{p+i})) = N + 1 + \pi_{p+i}, 1 \le i \le m. \tag{4}$$

So we have the following theorem.

Theorem 4. *If $M_N(z)$ is an Nth minimal partial realization of T then*

$$\deg(\det(M_N(z))) = -\sum_{j=1}^{p} \pi_j.$$

Proof. Let $\omega_1, \dots, \omega_n$ be a normal basis of Λ. By (3) we have

$$\sum_{j=1}^{n} v(\omega_j) = \det(\Lambda) = -m(N+1).$$

Thus the result follows from (4) and Lemma 1.   □

As we know, the realization pair $(\mathrm{Pol}(T(z)M_N(z)), M_N(z))$ is defined at best only up to right multiplication by an $m \times m$ unimodular matrix. In the following we first give all solutions of the Nth right i-minimal polynomial column vectors of T for $1 \le i \le m$.

Theorem 5. *Let $\omega_1, \cdots, \omega_n$ be a normal basis for the lattice Λ. Then all the Nth right i-minimal polynomial column vectors of T, $1 \le i \le m$, are obtained from*

$$\eta(\omega_{p+i} + \sum_{j=1, j \ne p+i}^{n} f_{i,j}(z)\,\omega_j),$$

where $f_{i,j}(z) \in \mathbb{F}[z]$ and $\deg(f_{i,j}(z)) \le \pi_{p+i} - \pi_j$ with $1 \le j \le n$ and $j \ne p+i$.

Proof. Assume $\mathbf{c}(z)$ is an Nth right i-minimal polynomial column vector of T for $1 \le i \le m$. By Theorem 2 we have $\eta^{-1}(\mathbf{c}(z)) = \gamma \in S_i(\Lambda)$ and

$$v(\gamma) = v(\omega_{p+i}). \tag{5}$$

So γ can be written as the form $\gamma = \sum_{j=1}^{n} f_{i,j}(z)\omega_i$. Since $\omega_1, \dots, \omega_n$ are reduced, $\theta(\omega_1), \dots, \theta(\omega_n)$ are linearly independent over $\mathbb{F}$. By (5), we have $f_{i,p+i}(z) = 1$ and $\deg(f_{i,j}(z)) + v(\omega_j) = v(\omega_{p+i})$ for all j, $1 \le j \le n$ and $j \ne p+i$. The result is easily obtained since η restricted on $S_i(\Lambda)$ is one-to-one correspondence.   □

In addition we can parameterize all minimal partial realizations as in [5, 13, 18], that is,

$$M_N(z) = (\eta(\omega_{p+i} + \sum_{j=1}^{p} f_{i,j}(z)\omega_j))_{1 \le i \le m},$$

where $f_{i,j}(z)$ is defined as above.

From Theorem 5 we can give the sufficient and necessary condition for the uniqueness of the minimal partial realizations [5, 13, 18].

Theorem 6. *The Nth minimal partial realization of T is unique if and only if $\pi_{p+i} < \pi_1$ for $i = 1, \dots, m$.*

4 A Special Realization Algorithm

In Algorithm 2.1, if $m = 1$ there is a unique vector $(a_1, \cdots, a_n)$ in reduction step such that $\sum_{j=1}^{n} a_j \theta(\omega_j) = \mathbf{0}_n$. However in general there are many choices about the vector $(a_1, \cdots, a_n)$ and so we give a special choice such that the total number of reduction steps is as small as possible.

First we introduce length parameters, that is, $N_i = v(\eta(\omega_i)) - v(\omega_i)$ for $1 \leq i \leq n$. Hence we have $\delta_j(\eta(\omega_i), T) = \mathbf{0}_p$ for $j = 1, \ldots, N_i - 1$, but $\delta_{N_i}(\eta(\omega_i), T) \neq \mathbf{0}_p$. In particular, $N_i = N + 1$ if and only if $\overline{\theta(\omega_i)} \neq \mathbf{0}_m$. According to these parameters, combine certain reduction steps into a small while-loop and let r denote the current number of such small while-loops. It will be showed that at each r, the recursively updated basis $\omega_1, \ldots, \omega_n$ satisfies the following conditions.

1. $\theta(\omega_1), \ldots, \theta(\omega_p)$ are linearly independent over $\mathbb{F}$ and $\overline{\theta(\omega_j)} = \mathbf{0}_m$ for $j = 1, \ldots, p$.
2. $\overline{\theta(\eta(\omega_{p+1}))}, \ldots, \overline{\theta(\eta(\omega_{p+m}))}$ are linearly independent over $\mathbb{F}$.
3. For each s, $p + 1 \leq s \leq n$, $N_s > N_j$ for $j = 1, \ldots, p$.

Put $r \leftarrow 0$ when the same initialization is given as Algorithm 2.1 and at this time the above conditions are trivially satisfied.

In the above situation one proceeds as follows. If $\overline{\theta(\omega_{p+i})} \neq \mathbf{0}_m$ for $i = 1, \ldots, m$, the basis becomes normal just by rearranging the order of the basis elements since the basis satisfies Condition 1, 2 and 3, and so let $M_N(z) = (\eta(\omega_{p+i}))_{1 \leq i \leq m}$ and $\mathrm{pol}(T(z)M_N(z))M_N^{-1}(z)$ is an Nth minimal partial realization of T. Therefore the algorithm terminates.

Otherwise, set $I = \{p + 1 \leq s \leq n : \overline{\theta(\omega_s)} = \mathbf{0}_m \text{ and } N_s \text{ is minimum}\}$.

Choose a number $s \in I$ and do the following reduction step. Since $\overline{\theta(\omega_s)} = \mathbf{0}_m$ and so there exists a unique vector $(a_1, \ldots, a_p)$ such that $\theta(\omega_s) = \sum_{j=1}^{p} a_j \theta(\omega_j)$. Let h be an integer such that $v(\omega_h) = \max\{v(\omega_j) : 1 \leq j \leq p, a_i \neq 0\}$. We need to consider two cases.

Case 1. Suppose $v(\omega_s) \geq v(\omega_h)$. In this case we let

$$\xi = \omega_s - \sum_{j=1}^{p} a_j z^{-v(\omega_j) + v(\omega_s)} \omega_j. \tag{6}$$

Clearly, $\eta(\xi) = \eta(\omega_s) - \sum_{j=1}^{p} a_j z^{-v(\omega_j) + v(\omega_s)} \eta(\omega_j)$. From (6) we have $v(\xi) < v(\omega_s)$. Then ω_s is replaced by ξ and the other ω_j are unchanged.

Case 2. Suppose $v(\omega_s) < v(\omega_h)$. Then we let

$$\xi = z^{-v(\omega_s) + v(\omega_h)} \omega_s - \sum_{j=1}^{p} a_j z^{-v(\omega_j) + v(\omega_h)} \omega_j. \tag{7}$$

Similarly, we have $\eta(\xi) = z^{-v(\omega_s) + v(\omega_h)} \eta(\omega_s) - \sum_{j=1}^{p} a_j z^{-v(\omega_j) + v(\omega_h)} \eta(\omega_j)$. From (7), we have $v(\xi) < v(\omega_h)$. Then we replace ω_h by ω_s, ω_s by ξ, and leave the other ω_j unchanged.

Set $I \leftarrow I/\{s\}$ and if I is not an empty set, we do the above reduction step. Therefore it is easy to check that the new updated basis returns to the situation satisfying Condition 1, 2 and 3 when I becomes an empty set. Set $r \leftarrow r+1$ and we proceed with the algorithm from there. We give the algorithm as follows.

Algorithm 4.1

Input: the first N terms of a matrix sequence $T = (T_1, T_2, \ldots)$.
Output: an Nth minimal partial realization of T.

1. Initialize $\omega_1 \leftarrow -\varepsilon_1, \cdots, \omega_p \leftarrow -\varepsilon_p, \omega_{p+1} \leftarrow \alpha_{N,1}, \cdots, \omega_n \leftarrow \alpha_{N,m}, r \leftarrow 0$.
2. **While** there is an element $\omega_s, p+1 \leq s \leq n$, with $\overline{\theta(\omega_s)} = \mathbf{0}_m$ **do**
   Set $I = \{p+1 \leq s \leq n : \theta(\omega_s) = \mathbf{0}_m$ and $N_s = v(\eta(\omega_s)) - v(\omega_s)$ is minimal$\}$.
   While $I \neq \emptyset$ **do**
   Choose an integer $s \in I$ and so there exists a vector $(a_1, \ldots, a_p)$ such that $\theta(\omega_s) = \sum_{j=1}^{p} a_j \theta(\omega_j)$ and find an integer h such that $v(\omega_h) = \max\{v(\omega_j) : 1 \leq j \leq p, a_j \neq 0\}$.
   If $v(\omega_s) \geq v(\omega_h)$ **then**
   Set $\xi \leftarrow \omega_s - \sum_{j=1}^{p} a_j z^{-v(\omega_j)+v(\omega_s)} \omega_j$
   else
   set $\xi \leftarrow z^{-v(\omega_s)+v(\omega_h)} \omega_s - \sum_{j=1}^{p} a_j z^{-v(\omega_j)+v(\omega_h)} \omega_j, \omega_h \leftarrow \omega_s$
   end-If
   set $\omega_s \leftarrow \xi$ and $I \leftarrow I/\{s\}$.
   end-While
   $r \leftarrow r+1$.
   end-While
3. Set $M_N(z) \leftarrow (\eta(\omega_{p+i}))_{1 \leq i \leq m}$, output $\mathrm{pol}(T(z)M_N(z))M_N^{-1}(z)$, and terminate the algorithm.

The above algorithm requires much memory and so it is not efficient in practice when N is very large. Similar to the method in [20], we deduce an efficient iterative algorithm from Algorithm 4.1. Here we only simply describe the idea. Note that for every basis element $\omega_i, 1 \leq i \leq n$, we only use its three parameters, that is, $v(\omega_i), \theta(\omega_i)$ and $\eta(\omega_i)$ and so we suffice to keep track of those values, which can be represented from the information of the given matrix sequence T, that is, $v(\omega_i) = -N_i + v(\eta(\omega_i)), \theta(\omega_i) = (\delta_{N_i}(\eta(\omega_i), T), \mathbf{0}_m)^t$. Thus we can derive the following algorithm from algorithm 4.1.

Algorithm 4.2

Input: the first N terms of a matrix sequence $T = (T_1, T_2, \ldots)$.
Output: an Nth minimal partial realization of T.

1. For $i = 1, \ldots, p$, set $\mathbf{c}_i^*(z) \leftarrow \mathbf{0}_m, \delta_i^* \leftarrow (\underbrace{0, \ldots, 0}_{i-1}, -1, 0, \ldots, 0) \in \mathbb{F}^p$, and

$v_i^* \leftarrow 0$. For $i = 1, \ldots, m$, set $M_{0,i}(z) = (\underbrace{0, \ldots, 0}_{i-1}, 1, 0, \ldots, 0) \in \mathbb{F}^m$.

$k \leftarrow 0$.

2. **For** $i = 1, \ldots, m$ **do**

   If $\delta_{k+1}(M_{k,i}(z), T) = \mathbf{0}_p$ then set $M_{k+1,i}(z) = M_{k,i}(z)$.

   else

   Set $\delta_i \leftarrow \delta_{k+1}(M_{k,i}(z), T)$, $v_i \leftarrow -k - 1 + v(M_{k,i}(z))$.

   Find a vector $(a_1, \cdots, a_p)$ such that $\delta_i = \sum_{j=1}^{p} a_j \delta_j^*$.

   Find an integer h such that $v_h^* = \max\{v_j^* : 1 \leq j \leq p, a_j \neq 0\}$.

   If $v_i \geq v_h^*$ then

   Set $M_{k+1,i}(z) = M_{k,i}(z) - \sum_{j=1}^{p} a_j\, z^{-v_j^* + v_i} \mathbf{c}_j^*(z)$.

   else

   Set $M_{k+1,i}(z) = z^{-v_i + v_h^*} M_{k,i}(z) - \sum_{j=1}^{p} a_j\, z^{-v_j^* + v_h^*} \mathbf{c}_j^*(z)$

   and $v_h^* \leftarrow v_i$, $\mathbf{c}_h^*(z) \leftarrow M_{k,i}(z)$, $\delta_h^* \leftarrow \delta_i$.

   end-If

   end-If

   end-For

3. **If** $k + 1 = N$ **then**

   set $M_N(z) \leftarrow (M_{N,i})_{1 \leq i \leq m}$, output $\mathrm{pol}(T(z)M_N(z))M_N^{-1}(z)$ and terminate the algorithm.

   else

   set $k \leftarrow k + 1$, go to 2.

   end-If

5 An Example

We use an example in [5]. Given a matrix sequence

$$T(z) = \begin{pmatrix} -1 & 1 \\ 0 & 0 \end{pmatrix} z^{-1} + \begin{pmatrix} 0 & 1 \\ 0 & 0 \end{pmatrix} z^{-2} + \begin{pmatrix} 1 & 1 \\ -1 & 1 \end{pmatrix} z^{-3} + \begin{pmatrix} 1 & 2 \\ 0 & 1 \end{pmatrix} z^{-4} + \begin{pmatrix} 1 & 4 \\ 1 & 1 \end{pmatrix} z^{-5} + \begin{pmatrix} 2 & 7 \\ 1 & 2 \end{pmatrix} z^{-6} + \cdots,$$

the goal is to find a 6th minimal partial realization of T.

We use Algorithm 4.1 to compute it.

$r = 0$. The initial basis is $\omega_1 \leftarrow (-1, 0, 0, 0)^t$, $\omega_2 \leftarrow (0, -1, 0, 0)^t$, $\omega_3 \leftarrow (-z^{-1} + z^{-3} + z^{-4} + z^{-5} + 2z^{-6}, -z^{-3} + z^{-5} + z^{-6}, z^{-7}, 0)^t$, and $\omega_4 \leftarrow (z^{-1} + z^{-2} + z^{-3} + 2z^{-4} + 4z^{-5} + 7z^{-6}, z^{-3} + z^{-4} + z^{-5} + 2z^{-6}, 0, z^{-7})^t$.

$r = 1$. $I = \{3, 4\}$.

Since $\theta(\omega_3) = \theta(\omega_1)$ and $v(\omega_3) < v(\omega_1)$, we have $\xi \leftarrow z\omega_3 - \omega_1 = (z^{-2} + z^{-3} + z^{-4} + 2z^{-5}, -z^{-2} + z^{-4} + z^{-5}, z^{-6}, 0)^t$, $\omega_1 \leftarrow \omega_3$, $\omega_3 \leftarrow \xi$.

Since $\theta(\omega_4) = -\theta(\omega_1)$ and $v(\omega_4) = v(\omega_1)$, we have $\omega_4 \leftarrow \omega_4 + \omega_1 = (z^{-2} + 2z^{-3} + 3z^{-4} + 5z^{-5} + 9z^{-6}, z^{-4} + 2z^{-5} + 3z^{-6}, z^{-7}, z^{-7})^t$, and the others are unchanged.

$r = 2$. $I = \{4\}$.

Since $\theta(\omega_4) = -\theta(\omega_1)$ and $v(\omega_4) \leq v(\omega_1)$, we have $\xi \leftarrow z\omega_4 + \omega_1 = (2z^{-2} + 4z^{-3} + 6z^{-4} + 10z^{-5} + 2z^{-6}, 2z^{-4} + 4z^{-5} + z^{-6}, z^{-6} + z^{-7}, z^{-6})^t$, $\omega_1 \leftarrow \omega_4$, $\omega_4 \leftarrow \xi$ and the others are not changed.

$r = 3.$ $I = \{3, 4\}.$

Since $\theta(\omega_3) = \theta(\omega_1) + \theta(\omega_2)$ and $v(\omega_3) \leq v(\omega_2)$, we have $\xi = -z^2\omega_3 + z^2\omega_1 + \omega_2 = (z^{-1} + 2z^{-2} + 3z^{-3} + 9z^{-4}, z^{-3} + 3z^{-4}, -z^{-4} + z^{-5}, z^{-5})^t$, $\omega_2 \leftarrow \omega_3$, $\omega_3 \leftarrow \xi$ and the others are unchanged.

Since $\theta(\omega_4) = 2\theta(\omega_1)$ and $v(\omega_4) = v(\omega_1)$, we have $\omega_4 \leftarrow \omega_4 - 2\omega_1 = (-16z^{-6}, -5z^{-6}, z^{-6} - z^{-7}, z^{-6} - 2z^{-7})^t$, and the others are not changed.

$r = 4.$ $I = \{3\}.$

Since $\theta(\omega_3) = \theta(\omega_1)$ and $v(\omega_3) < v(\omega_1)$, we have $\omega_3 \leftarrow \omega_3 - z\omega_1 = (4z^{-4} - 9z^{-5}, z^{-4} - 3z^{-5}, -z^{-4} + z^{-5} - z^{-6}, z^{-5} - z^{-6})^t$, $\omega_4 \leftarrow (-16z^{-6}, -5z^{-6}, z^{-6} - z^{-7}, z^{-6} - 2z^{-7})^t$, $\omega_2 \leftarrow (z^{-2} + z^{-3} + z^{-4} + 2z^{-5}, -z^{-2} + z^{-4} + z^{-5}, z^{-6}, 0)^t$ and $\omega_1 \leftarrow (z^{-2} + 2z^{-3} + 3z^{-4} + 5z^{-5} + 9z^{-6}, z^{-4} + 2z^{-5} + 3z^{-6}, z^{-7}, z^{-7})^t$.

At this time they become normal and so

$$M_6(z) = \begin{pmatrix} -z^3 + z^2 - z & z - 1 \\ z^2 - z & z - 2 \end{pmatrix}.$$

6   Conclusions

In this paper we extend a minimal partial realization algorithm for vector sequences to matrix sequences by means of the lattice basis reduction. The main idea of the algorithm lies in finding the reduction relation of the basis elements such that their valuations become smaller and smaller till they become reduced. Therefore different reduction ways lead to different realization algorithms. As we see, Algorithm 4.2 is similar to the minimal partial realization algorithms in [5, 18].

In Algorithm 4.1, if the matrix $(\theta(\omega_1) \ldots \theta(\omega_p))$ is kept lower-triangular, we have a special method to solve the equation $\theta(\omega_{p+i}) = \sum_{j=1}^{p} a_j \theta(\omega_j)$, that is, eliminating the nonzero jth component, $1 \leq j \leq p$, of $\theta(\omega_{p+i})$ with the corresponding $\theta(\omega_j)$ step by step. Similar to the method of deriving Algorithm 4.2 from Algorithm 4.1, we can deduce the algorithm in [13].

Therefore our algorithm provides a greater insight into the minimal partial realization problem and gives a unified algorithm for this problem.

Acknowledgment

The research is supported by the National Natural Science Foundation of China (Grant No. 60773141 and No. 60503010) and the National 863 Project of China (Grant No. 2006AA 01Z420). The author also would like to thank the reviewers for their helpful suggestions and comments.

References

1. Antoulas, A.C.: On recursiveness and related topics in linear systems. IEEE Trans. Automat. Control 31(12), 1121–1135 (1986)
2. Antoulas, A.C.: Recursive modeling of discrete-time time series. In: Van Dooren, P., Wyman, B. (eds.) Linear Algebra for Control Theory, IMA, vol. 62, pp. 1–20 (1994)
3. Bultheel, A., De Moor, B.: Rational approximation in linear systems and control. J. Comput. Appl. Math. 121, 355–378 (2000)
4. Dawson, E., Simpson, L.: Analysis and design issues for synchronous stream ciphers. In: Niederreiter, H. (ed.) Coding Theory and Cryptology, pp. 49–90. World Scientific, Singapore (2002)
5. Dickinson, B.W., Morf, M., Kailath, D.: A minimal realization algorithm for matrix sequences. IEEE Trans. Automat. Control 19(1), 31–38 (1974)
6. Ding, C.S.: Proof of Massey's conjectured algorithm, Advances in Cryptology. LNCS, vol. 330, pp. 345–349. Springer, Berlin (1988)
7. ECRYPT stream cipher project, http://www.ecrypt.eu.org/stream
8. Feng, G.L., Tzeng, K.K.: A generalization of the Berlekamp-Massey algorithm for multisequence shift-register synthesis with applications to decoding cyclic codes. IEEE Trans. Inform. Theory 37, 1274–1287 (1991)
9. Forney, G.D.: Minimal bases of rational vector spaces, with applications to multivariable linear systems. SIAM J. Control 13, 493–520 (1975)
10. Gragg, W.B., Lindquist, A.: On the partial realization problem. Linear Alg. Appl. 50, 277–319 (1983)
11. Hawkes, P., Rose, G.G.: Exploiting multiples of the connection polynomial in word-oriented stream ciphers. In: Okamoto, T. (ed.) ASIACRYPT 2000. LNCS, vol. 1976, pp. 303–316. Springer, Heidelberg (2000)
12. Kalman, R.E.: On minimal partial realizations of a linear input/output map. In: Kalman, R.E., DeClaris, N. (eds.) Aspects of network and System Theory, New York, pp. 385–407 (1971)
13. Kuijper, M.: An algorithm for constructing a minimal partail realization in the multivariable case. Syst. Contr. Lett. 31, 225–233 (1997)
14. Lenstra, A.K.: Factoring multivariate polynomials over finite fields. J. Comp. Sys. Sci. 30, 235–248 (1985)
15. Mahler, K.: An analogue to Minkowski's geometry of numbers in a field of series. Ann. of Math. 42, 488–522 (1941)
16. Massey, J.L.: Shift-register synthesis and BCH decoding. IEEE Trans. Inform. Theory 15(1), 122–127 (1969)
17. Schmidt, W.M.: Construction and estimation of bases in function fields. J. Number Theory 39, 181–224 (1991)
18. Van Barel, M., Bultheel, A.: A generalized minimal partial realization problem. Linear Alg. Appl. 254, 527–551 (1997)
19. Wang, L.-P., Zhu, Y.-F.: $F[x]$-lattice basis reduction algorithm and multisequence synthesis. Science in China (Series F) 44, 321–328 (2001)
20. Wang, L.-P., Zhu, Y.-F., Pei, D.-Y.: On the lattice basis reduction multisequence synthesis algorithm. IEEE Trans. Inform. Theory 50, 2905–2910 (2004)

A Fast Jump Ahead Algorithm for Linear Recurrences in a Polynomial Space*

Hiroshi Haramoto[1], Makoto Matsumoto[1], and Pierre L'Ecuyer[2]

[1] Dept. of Math., Hiroshima University, Hiroshima 739-8526 Japan
m-mat@math.sci.hiroshima-u.ac.jp
http://www.math.sci.hiroshima-u.ac.jp/~m-mat/eindex.html
[2] Département d'Informatique et de Recherche Opérationnelle,
Université de Montréal, Montréal, Canada
lecuyer@iro.umontreal.ca
http://www.iro.umontreal.ca/~lecuyer

Abstract. Linear recurring sequences with very large periods are widely used as the basic building block of pseudorandom number generators. In many simulation applications, multiple streams of random numbers are needed, and these multiple streams are normally provided by jumping ahead in the sequence to obtain starting points that are far apart. For maximal-period generators having a large state space, this jumping ahead can be costly in both time and memory usage. We propose a new jump ahead method for this kind of situation. It requires much less memory than the fastest algorithms proposed earlier, while being approximately as fast (or faster) for generators with a large state space such as the Mersenne twister.

1 Introduction

Pseudorandom number generators (PRNGs) are widely used in many scientific areas, such as simulation, statistics, and cryptography. Generating multiple disjoint streams of pseudorandom number sequences is important for the smooth implementation of variance-reduction techniques on a single processor (see [1,2,3] for illustrative examples), as well as in conjunction with parallel computing. Generators with multiple streams and substreams have been already adopted, or are in the process of being adopted, in leading edge simulation software tools such as Arena, Automod, MATLAB, SAS, Simul8, SSJ, Witness, and ns2, for example. The substreams are normally obtained by splitting the sequence of a large-period generator into long disjoint subsequences whose starting points are equidistant in the original sequence, say J steps apart. To obtain the initial state (or starting point) of a new substream, we must jump ahead by J steps in the original sequence from the initial state of the most recently created substream. In many

---

* This work was supported in part by JSPS Grant-In-Aid #16204002, #18654021, #19204002, JSPS Core-to-Core Program No.18005, NSERC-Canada, and a Canada Research Chair to the third author.

S.W. Golomb et al. (Eds.): SETA 2008, LNCS 5203, pp. 290–298, 2008.

cases, this must be done thousands of times (or more) in a simulation, so we need an efficient algorithm for this jump ahead. Unfortunately, for huge-period generators such as the Mersenne twister and the WELL [4,5], for example, efficient jump ahead is difficult. For this reason, most current implementations use a base generator whose state space is not so large (e.g., 200 bits or so), and this limits the period length.

When the PRNG is based on a linear recurrence, a simple way to jump ahead is to express the recurrence in matrix form, precompute and store the J-th power of the relevant matrix, and jump ahead via a simple matrix-vector multiplication. But for large-period generators, this method requires an excessive amount of memory and is much too slow. Haramoto et al. [6] introduced a reasonably fast jumping-ahead algorithm based on a sliding-window method. One drawback of this method, however, is that it requires the storage of a large precomputed table, at least in its fastest version.

The new method proposed in this paper is based on a representation of the linear recurrence in a space of formal series, where jumping ahead corresponds to multiplying the series by a polynomial. It requires much less memory than the previous one, and is competitive in terms of speed. Under a certain condition on the output function, the speed is actually $O(k^{\log_3 2}) \approx O(k^{1.59})$ for a k-bit state space, compared with $O(k^2)$ for the previous method. For the Mersenne twister with period length $2^{19937} - 1$, this condition is satisfied and the new method turns out to be faster, according to our experiments.

The remainder is organized as follows. In Section 2, we define the setting in which these jumping-ahead methods are applied, and we briefly summarize the previously proposed techniques. The new method is explained in Section 3. Its application to the Mersenne twister is discussed in Section 5. Section 6 reports timing experiments.

2   Setting and Summary of Existing Methods

Many practical generators used for simulation are based on linear recurrences, because important properties such as the period and the high-dimensional distribution can then be analyzed easily by linear algebra techniques. For notational simplicity, our description in this paper is in the setting of a linear recurrence in a field of characteristic 2, i.e., we assume that the base field is the two-element field $\mathbb{F}_2$. However, the proposed method is valid for any finite field.

We consider a PRNG with state space $S := \mathbb{F}_2^k$, for some integer $k > 0$, and (state) transition function $f : S \to S$, linear over $\mathbb{F}_2$. Thus, f can be identified with its representation matrix F, a binary matrix of size $k \times k$, and a state $s \in S$ is then a k-dimensional column vector. For a given initial state s_0, the state evolves according to the recurrence

$$s_{m+1} := f(s_m) = F s_m, \qquad m = 0, 1, 2, \ldots. \tag{1}$$

An output function $o : S \to O$ is specified, where O is the set of output symbols, and $o(s)$ is the output when we are in state s. Thus, the generator produces a sequence of elements $o(s_0), o(s_1), \ldots,$ of O. For example, in the stream cipher, o is called the filter, and high nonlinearity is required. In principle, jumping ahead should depend only on f, and not on o. However, the algorithm introduced in this paper assumes that o has a specific (linear) form and that one can easily reconstruct the state from a sequence of k successive output values.

Our purpose is to provide an efficient procedure Jump that computes $f^J(s)$ for arbitrary states s, for a given huge integer J. Typically, J is fixed and taken large enough to make sure that a stream will never use more than J numbers in a simulation.

A naive implementation of Jump is to precompute the matrix power $A := F^J$ (in $\mathbb{F}_2$) and store it (this requires k^2 bits of memory). Then, $f^J(s)$ is just the matrix-vector multiplication As. However, modern generators with huge state spaces, such as the Mersenne Twister [4] and the WELL [7], for which $k = 19937$ or more, merely storing A requires too much memory, and the matrix-vector multiplication is very time-consuming.

The alternative proposed in [6] works as follows. Let $\varphi_F(t)$ be the minimal polynomial of F. (The method also works if we use the characteristic polynomial $\det(tI - F)$ instead of $\varphi_F(t)$.) First, we precompute and store the coefficients of

$$g(t) := t^J \bmod \varphi_F(t) = \sum_{i=0}^{k-1} a_i t^i. \tag{2}$$

This requires only k bits of memory. Then, Jump can be implemented using Horner's method:

$$F^J s_0 = g(F)s_0 = F(\cdots F(F(F(a_{k-1}s_0) + a_{k-2}s_0) + a_{k-3}s_0) + \cdots) + a_0 s_0. \tag{3}$$

An important remark is that the matrix-vector multiplication Fs corresponds to advancing the generator's state by one step as in (1). This operation is usually very fast: good generators are designed so that it requires only a few machine instructions. Thus, when computing the right side of (3), the addition of vectors dominates the computing effort. In this procedure, assuming that $g(t)$ is precomputed, k applications of F and approximately $k/2$ vector additions are required for implementing Jump. Since these are k-bit vectors, this means an $O(k^2)$ computing time.

The speed can be improved by a standard method called the sliding window algorithm, which precomputes a table that contains $h(F)s_0$ for all polynomials $h(t)$ of degree less than or equal to some constant q, and uses this table to compute (3) $(q+1)$ digits at a time. This requires $2^q k$ bits of memory for the table (this can be significant when k is huge), but the number of (time-consuming) vector additions is decreased to roughly

$$2^q + \lceil k/(q+1) \rceil. \tag{4}$$

The integer q can be selected to optimize the speed, while paying attention to the memory consumption of $2^q k$ bits; see [6] for the details. The new method proposed in the next section does not require such a large table.

3   Jumping by Fast Polynomial Multiplication

A linear recurrence over $\mathbb{F}_2$ can be represented in different spaces and it is not difficult (at least in principle) to switch from one representation to the other [8,9]. The basic PRNG implementation usually represents the state as a k-bit vector and computes the matrix-vector product in (1) by just a few elementary operations. In other representations, used for example to verify maximal period conditions and to analyze the multidimensional uniformity of the output values, the state is represented as a polynomial or as a formal series [10,9]. Here, we will use a formal series representation of the state, switch to that representation to perform the jumping ahead, and then recover the state in the original representation. A key practical requirement is the availability of an efficient method to perform this last step.

For our purpose, we assume that the linear output function o returns a single bit; that is, we have $o : S \rightarrow \mathbb{F}_2$. Here, we may choose any o satisfying the injective condition stated below, for the purpose of jump computation. For the Mersenne twister, for example, the output at each step is a block of 32 bits and we can just pick up the most significant bit. We also assume that the mapping $S \rightarrow \mathbb{F}_2^k$ which maps the generator's state to the next k bits of output is one-to-one, so we can recover the state from k successive bits of output. This assumption is not restrictive: for example, if the period of this single-bit output is $2^k - 1$, which is usually the case in practice, then the assumption is satisfied by comparing the cardinality of the state space and the set of the k successive bits.

More specifically, let

$$G(s, t) = \sum_{i=1}^{\infty} o(s_{i-1})t^{-i},$$

which is the generating function of the output sequence when the initial state is $s_0 = s$. Note that $G(s_1, t) = t\, G(s_0, t) \pmod{\mathbb{F}_2[t]}$, so that

$$G(s_J, t) = t^J G(s_0, t) \pmod{\mathbb{F}_2[t]} = g(t)G(s_0, t) \pmod{\mathbb{F}_2[t]},$$

because $\varphi_F(t) \in \mathbb{F}_2[t]$. To recover the state s_J, we only need the coefficients of $t^{-1}, \ldots, t^{-k}$ in $G(s_J, t) = g(t)G(s_0, t)$, i.e., the truncation of $G(s_J, t)$ to its first k terms. This means that we can replace $G(s_0, t)$ by its truncation to its first $2k$ terms, or equivalently by the truncation of $t^{2k}G(s_0, t)$ to its first $2k$ terms, which gives the polynomial

$$h(s_0, t) = \sum_{i=0}^{2k-1} o(s_i)t^{2k-1-i}.$$

We can then compute the polynomial product $g(t)h(s_0, t)$, and observe that the coefficients of $t^{2k-1}, \ldots, t^k$ in this polynomial are exactly the output bits $o(s_J), \ldots, o(s_{J+k-1})$, from which we can recover the state s_J.

With the classical (standard) method, we need $O(k^2)$ bit operations just to multiply the polynomials $g(t)$ and $h(s_0, t)$, so we are doing no better than with the method of [6]. Polynomial multiplication can be done with only $O(k \log k)$ bit operations using fast Fourier transforms, but the hidden constants are larger and the corresponding algorithm turns out to be slower when implemented, for the values of k that we are interested in. A third approach, implemented in the NTL library [11], is Karatsuba's algorithm (see, e.g., [12]), which requires $O(k^{\log_2 3}) \approx O(k^{1.59})$ bit operations. This algorithm is faster than the classical method even for moderate values of k, and this is the one we adopt for this step of our method.

The last ingredient we need is a fast method to compute the inverse image of the mapping

$$o^{(k)} : s \mapsto (o(s), o(Fs), o(F^2 s), \ldots, o(F^{k-1} s)),$$

to be able to recover the state s_J from the coefficients of $g(t)h(s_0, t)$. For important classes of PRNGs such as the twisted GFSR and Mersenne twister, there is a simple algorithm to compute this inverse image in $O(k)$ time. Then, our entire procedure works in $O(k^{\log_2 3}) \approx O(k^{1.59})$ time.

The procedure is summarized in Algorithm 1.. It assumes that $g(t)$ has been precomputed in advance.

Algorithm 1. Jump ahead by polynomial multiplication

Input the state $s = s_0$;
Compute the polynomial $h(s_0, t)$ by advancing the generator for $2k$ steps;
Compute the product $g(t)h(s_0, t)$ and extract the coefficients $o(s_J), \ldots, o(s_{J+k-1})$;
Compute the state s_J from the bits $o(s_J), \ldots, o(s_{J+k-1})$;
Return s_J.

---

4   Illustrative Example by LFSR

The Linear Feedback Shift Register (LFSR) is a most classical and widely spread generator. Here, we use the term LFSR in the following limited sense (see [13]), although sometimes LFSR refers to a wider class of generators.

The state space is the row vector space $S := \mathbb{F}_2^k$, and the state transition function is defined by

$$(x_0, \ldots, x_{k-1}) \mapsto (x_1, x_2, \ldots, x_{k-1}, \sum_{i=0}^{k-1} a_i x_i),$$

where $a_0, \ldots, a_{k-1}$ are constants in $\mathbb{F}_2$. If we choose $o : (x_0, \ldots, x_{k-1}) \mapsto x_0$, then it directly follows that $o^{(k)} : S \to \mathbb{F}_2^k$ is the identity function. Thus, we can skip the computation of its inverse.

Proposition 1. *The computational complexity of PM-Jump for LFSR is the same with that for the polynomial multiplications of degree $2k$. As a result, jumping ahead can be done in $O(k^{1.59})$ time if we use Karatsuba's polynomial multiplication, and in $O(k \log k)$ time if we use a fast Fourier transform.*

Note that it is irrelevant to use $(x_i, \ldots, x_{i+k-1})$ as the i-th output k-bit integer of the pseudorandom number generator, since the consecutive outputs are overlapped. However, such an LFSR is used in stream cipher (pseudorandom bit generator), with suitably chosen nonlinear output function $o : S \to \mathbb{F}_2$, see for example [14].

5   Application to the Mersenne Twister

The Mersenne Twister (MT) generator can be described as follows [4]. Let w be the word size of the machine (e.g., $w = 32$). The row vector space $W := \mathbb{F}_2^w$ is identified with the set of words. For fixed integers $n > m$, MT generates a sequence $\mathbf{x}_0, \mathbf{x}_1, \ldots$ of elements of W by the following recurrence:

$$\mathbf{x}_{j+n} := \mathbf{x}_{j+m} \oplus (\mathbf{x}_j^{w-r} | \mathbf{x}_{j+1}^r) A, \qquad j = 0, 1, \ldots,$$

where $(\mathbf{x}_j^{w-r} | \mathbf{x}_{j+1}^r)$ denotes the concatenation of the upper $(w - r)$ bit $(\mathbf{x}_j^{w-r})$ of $\mathbf{x}_j$ and the lower r bit $(\mathbf{x}_{j+1}^r)$ of $\mathbf{x}_{j+1}$, and the $w \times w$ matrix A is defined indirectly as follows: For any w-dimensional row vector $\mathbf{x}$,

$$\mathbf{x}A = \begin{cases} \text{shiftright}(\mathbf{x}) & \text{if the LSB of } \mathbf{x} = 0, \\ \text{shiftright}(\mathbf{x}) \oplus \mathbf{a} & \text{if the LSB of } \mathbf{x} = 1, \end{cases}$$

where LSB means the least significant bit (i.e., the rightmost bit), and $\mathbf{a}$ is a suitably chosen constant. This generator has the state transition function

$$f(\mathbf{x}_0^{w-r}, \mathbf{x}_1, \ldots, \mathbf{x}_{n-1}) = (\mathbf{x}_1^{w-r}, \mathbf{x}_2, \ldots, \mathbf{x}_n),$$

where $\mathbf{x}_n$ is determined by the above recursion with $j = 0$, and the state space is $S = \mathbb{F}_2^{nw-r}$.

The most popular instance, named MT19937, uses the parameters $n = 624$, $w = 32$, $r = 31$. Its sequence has a maximal period equal to the Mersenne prime $2^{19937} - 1$. Because of its high speed and good distribution property, MT19937 is widely used as a standard PRNG. However, is had been lacking an efficient jumping-ahead method, and this was a motivation for the work of [6].

Proposition 2. *For the MT generator, if we choose the output function*

$$o : (\mathbf{x}_0^{w-r}, \mathbf{x}_1, \ldots, \mathbf{x}_{n-1}) \mapsto \text{the LSB of } \mathbf{x}_1,$$

then the inverse image by $o^{(k)}$ is computable with time complexity $O(k)$. As a result, jumping ahead can be done in $O(k^{1.59})$ time if we use Karatsuba's polynomial multiplication, and in $O(k \log k)$ time if we use a fast Fourier transform.

Proof. A tricky algorithm that does that is described in [4, Section 4.3, Proposition 4.2].

The twisted GFSR generator [15] is based on the same construction as MT, but with $r = 0$. Thus, the proposition also applies to it.

6   Timings

We made an experiment to compare the speeds of three jumping-ahead methods: the one that directly implements Horner's method (3) (Horner), the method of [6] with a sliding window with parameter q (SW), and our new method based on polynomial multiplication with Karatsuba's algorithm (PM). For the latter, we used the NTL implementation [11]. We applied these methods to MT generators with the Mersenne exponents $k = 19937, 21701, 23209, 44497, 86243, 110503, 132049$.

For each value of k, we repeated the following 1000 times, on two different computers: We generated a random polynomial $g(t)$ uniformly over the polynomials of degree less than k in $\mathbb{F}_2[t]$ and a random state s uniformly in $S = \mathbb{F}_2^k$, then we computed $f^J(s)$ by each of the three algorithms, and we measured the required CPU time. We then summed those CPU times over the 1000 replications. The results are given in Table 1 for the Intel Core Duo 32-bit processor and in Table 2 for the AMD Athlon 64 3800+ 64-bit processor. The total CPU times are in seconds. For the SW method, we selected the parameter q that gave the highest speed; this parameter is listed, together with the required amount of memory in Kbytes.

Table 1. Comparison of CPU time (in seconds) for 1000 jumps ahead for MT generators of various sizes, with the Horner, SW, and PM methods. This experiment was done on an Intel Core Duo (2.0 GHz) with 1 Gbytes of Memory

k	Horner	SW			PM
	CPU (sec)	q	memory (KB)	CPU (sec)	CPU (sec)
19937	17.7	7	312	5.1	6.7
21701	20.8	8	679	5.9	7.4
23209	23.7	8	726	6.5	7.9
44497	84.0	8	1391	20.4	19.8
86243	309.3	8	2696	71.7	52.3
110503	445.4	9	6908	113.6	71.1
132049	648.1	9	8254	158.3	89.8

Table 2. The same experiment as in Table 1, but on a 64-bit Athlon 64 3800+ processor with 1 Gbyte of memory

k	Horner	SW			PM
	CPU (sec)	q	memory (KB)	CPU (sec)	CPU (sec)
19937	8.6	7	312	2.7	2.2
21701	10.1	7	340	3.1	2.4
23209	11.4	7	363	3.5	2.6
44497	40.4	6	348	11.7	6.0
86243	148.3	6	674	44.8	14.5
110503	242.2	5	432	73.2	19.7
132049	345.0	5	516	106.2	24.7

We see that for the 32-bit computer, PM is faster than SW when $k \geq 44497$, whereas for the 64-bit machine, PM is faster for $k \geq 19937$. The results agree with the computational complexity approximations, which are $O(k^2)$ for SW and $O(k^{1.59})$ for PM.

7 Conclusion

The proposed jump ahead algorithm based on polynomial multiplication is advantageous over the sliding window method when the dimension k of state space is large enough, since the new PM method has time complexity $O(k^{1.59})$, compared with $O(k^2)$ for the sliding window method. Our empirical experiments confirm this and show that this large enough k corresponds roughly to the value of k used in the most popular implementation of MT. Much more importantly, the new PM method has space complexity of $O(k)$, which is much smaller than that of the sliding window method, namely $O(k2^q)$. The main limitation is that the new method requires the availability of an efficient algorithm to compute the inverse of $o^{(k)}$.

References

1. Law, A.M., Kelton, W.D.: Simulation Modeling and Analysis, 3rd edn. McGraw-Hill, New York (2000)
2. L'Ecuyer, P., Buist, E.: Simulation in Java with SSJ. In: Proceedings of the 2005 Winter Simulation Conference, pp. 611–620. IEEE Press, Los Alamitos (2005)
3. L'Ecuyer, P.: Pseudorandom number generators. In: Platen, E., Jaeckel, P. (eds.) Simulation Methods in Financial Engineering. Encyclopedia of Quantitative Finance. Wiley, Chichester (forthcoming, 2008)
4. Matsumoto, M., Nishimura, T.: Mersenne twister: A 623-dimensionally equidistributed uniform pseudo-random number generator. ACM Transactions on Modeling and Computer Simulation 8(1), 3–30 (1998)
5. Panneton, F., L'Ecuyer, P., Matsumoto, M.: Improved long-period generators based on linear recurrences modulo 2. ACM Transactions on Mathematical Software 32(1), 1–16 (2006)
6. Haramoto, H., Matsumoto, M., Nishimura, T., Panneton, F., L'Ecuyer, P.: Efficient jump ahead for $\mathbf{F}_2$-linear random number generators. INFORMS Journal on Computing (to appear, 2008)
7. Panneton, F., L'Ecuyer, P.: Infinite-dimensional highly-uniform point sets defined via linear recurrences in $\mathbf{F}_{2^w}$. In: Niederreiter, H., Talay, D. (eds.) Monte Carlo and Quasi-Monte Carlo Methods 2004, pp. 419–429. Springer, Berlin (2006)
8. L'Ecuyer, P.: Uniform random number generation. Annals of Operations Research 53, 77–120 (1994)
9. L'Ecuyer, P., Panneton, F.: $\mathbf{F}_2$-linear random number generators. In: Alexopoulos, C., Goldsman, D. (eds.) Advancing the Frontiers of Simulation: A Festschrift in Honor of George S. Fishman, Spinger, New York (to appear, 2007)
10. Couture, R., L'Ecuyer, P.: Lattice computations for random numbers. Mathematics of Computation 69(230), 757–765 (2000)

11. Shoup, V.: NTL: A Library for doing Number Theory. Courant Institute, New York University, New York (2005), http://shoup.net/ntl/
12. von zur Gathen, J., Gerhard, J.: Modern Computer Algebra. Cambridge University Press, Cambridge (2003)
13. Golomb, S.W.: Shift-Register Sequences. Holden-Day, San Francisco (1967)
14. Rueppel, R.A.: Analysis and Design of Stream Ciphers. Springer, Heidelberg (1986)
15. Matsumoto, M., Kurita, Y.: Twisted GFSR generators II. ACM Transactions on Modeling and Computer Simulation 4(3), 254–266 (1994)

Parallel Generation of ℓ-Sequences

Cédric Lauradoux[1] and Andrea Röck[2]

[1] Princeton University, Department of electrical engineering
Princeton, NJ 08544, USA
claurado@princeton.edu
[2] Team SECRET, INRIA Paris-Rocquencourt,
78153 Le Chesnay Cedex, France
andrea.roeck@inria.fr

Abstract. The generation of pseudo-random sequences at a high rate is an important issue in modern communication schemes. The representation of a sequence can be scaled by decimation to obtain parallelism and more precisely a sub-sequences generator. Sub-sequences generators and therefore decimation have been extensively used in the past for linear feedback shift registers (LFSRs). However, the case of automata with a non linear feedback is still in suspend. In this paper, we have studied how to transform of a feedback with carry shit register (FCSR) into a sub-sequences generator. We examine two solutions for this transformation, one based on the decimation properties of ℓ-sequences, *i.e.* FCSR sequences with maximal period, and the other one based on multiple steps implementation. We show that the solution based on the decimation properties leads to much more costly results than in the case of LFSRs. For the multiple steps implementation, we show how the propagation of carries affects the design.

Keywords: sequences, synthesis, decimation, parallelism, LFSRs, FCSRs.

1 Introduction

The synthesis of shift registers consists in finding the smallest automaton able to generate a given sequence. This problem has many applications in cryptography, sequences and electronics. The synthesis of a single sequence with the smallest linear feedback shift register is achieved by the Berlekamp-Massey [1] algorithm. There exists also an equivalent of Berlekamp-Massey in the case of multiple sequences [2,3]. In the case of FCSRs, we can use algorithms based on lattice approximation [4] or on Euclid's algorithm [5]. This paper addresses the following issue in the synthesis of shift registers: *given an automaton generating a sequence S, how to find an automaton which generates in parallel the sub-sequences associated to S*. Throughout this paper, we will refer to this problem as *the sub-sequences generator problem*. We aim to find the best solution to transform a 1-bit output pseudo-random generator into a multiple outputs generator. In particular, we investigate this problem when S is generated by a feedback

S.W. Golomb et al. (Eds.): SETA 2008, LNCS 5203, pp. 299–312, 2008.

with carry shit register (FCSR) with a maximal period, *i.e.* S is an ℓ-sequence. This class of pseudo-random generators was introduced by Klapper and Goresky in [6]. FCSRs and LFSRs are very similar in terms of properties [7,8]. However, FCSRs have a non-linear feedback which is a significant property to thwart algebraic attacks [9] in cryptographic applications [10].

The design of sub-sequences generators has been investigated in the case of LFSRs [11,12] and two solutions have been proposed. The first solution [13,14] is based on the classical synthesis of shift registers, *i.e.* the Berlekamp-Massey algorithm, to define each sub-sequence. The second solution [11] is based on a multiple steps design of the LFSR. We have applied those two solutions to FCSRs. The contributions of the paper are as follows:

- We explore the decimation properties of ℓ-sequences for the design of a sub-sequences generator by using an FCSR synthesis algorithm.
- We show how to implement a multiple steps FCSR in Fibonacci and Galois configuration.

The next section presents the motivation of this work and recalls the different representations of LFSRs and FCSRs. In Section 3, the existing results on LFSRs are described and we show multiple steps implementations of the Galois and the Fibonacci configuration. We describe in Section 4 our main results on the synthesis of sub-sequences generators in the case of ℓ-sequences. Then, we give some conclusions in Section 5.

2   Motivation and Preliminaries

The decimation is the main tool to transform a 1-bit output generator into a sub-sequences generator. This allows us to increase the throughput of a pseudo-random sequence generator (PRSG). Let $S = (s_0, s_1, s_2, \cdots)$ be an infinite binary sequence of period T, thus $s_j \in \{0, 1\}$ and $s_{j+T} = s_j$ for all $j \geq 0$. For a given integer d, a d–*decimation* of S is the set of sub-sequences defined by:

$$S_d^i = (s_i, s_{i+d}, s_{i+2d}, \cdots, s_{i+jd}, \cdots)$$

where $i \in [0, d-1]$ and $j = 0, 1, 2, \cdots$. Hence, a sequence S is completely described by the sub-sequences:

$$
\begin{aligned}
S_d^0 &= (s_0, s_d, \cdots) \\
S_d^1 &= (s_1, s_{1+d}, \cdots) \\
&\vdots \\
S_d^{d-2} &= (s_{d-2}, s_{2d-2}, \cdots) \\
S_d^{d-1} &= (s_{d-1}, s_{2d-1}, \cdots) \,.
\end{aligned}
$$

A single automaton is often used to generate the pseudo-random sequence S. In this case, it is difficult to achieve parallelism. The decomposition into sub-sequences overcomes this issue as shown by Lempel and Eastman in [11]. Each sub-sequence is associated to an automaton. Then, the generation of the d sub-sequences of S uses d automata which operate in parallel. Parallelism has two benefits, it can increase the throughput or reduce the power consumption of the automaton generating a sequence.

Throughput — The throughput $\mathcal{T}$ of a PRSG is defined by: $\mathcal{T} = n \times f$, with n is the number of bits produced every cycle and f is the clock frequency of the PRSG. Usually, we have $n = 1$, which is often the case with LFSRs. The decimation achieves a very interesting tradeoff for the throughput: $\mathcal{T}_d = d \times \gamma f$ with $0 < \gamma \leq 1$ the degradation factor of the original automaton frequency. The decimation provides an improvement of the throughput if and only if $\gamma d > 1$. It is then highly critical to find good automata for the generation of the sub-sequences. In an ideal case, we would have $\gamma = 1$ and then a d-decimation would imply a multiplication of the throughput by d.

Power consumption — The power consumption of a CMOS device can be estimated by the following equation: $P = C \times V_{dd}^2 \times f$, with C the capacity of the device and V_{dd} the supply voltage. The sequence decimation can be used to reduce the frequency of the device by interleaving the sub-sequences. The sub-sequences generator will be clocked at frequency $\frac{\gamma f}{d}$ and the outputs will be combined with a d-input multiplexer clocked at frequency γf. The original power consumption can then be reduced by the factor $\frac{\gamma}{d}$, where γ must be close to 1 to guarantee that the final representation of S is generated at frequency f.

The study of the γ parameter is out of the scope of this paper since it is highly related to the physical characteristics of the technology used for the implementation. In the following, we consider m-sequences and ℓ-sequences which are produced respectively by LFSRs and FCSRs.

Throughout the paper, we detail different representations of several automata. We denote by x_i a memory cell and by $(x_i)_t$ the content of the cell x_i at time t. The internal state of an automaton at time t is denoted by X_t.

2.1   LFSRs

An LFSR is an automaton which generates linear recursive sequences. A detailed description of this topic can be found in the monographs of Golomb and McEliece [15,16]. Let $s(x) = \sum_{i=0}^{\infty} s_i x^i$ define the power series of the sequence $S = (s_0, s_1, s_2, \ldots)$ produced by an LFSR. Then, there exists two polynomials, such that

$$s(x) = \frac{p(x)}{q(x)}$$

where $q(x)$ is the *connection polynomial* defined by the feedback positions of the automaton. Let m be the degree of $q(x)$, then the reciprocal polynomial $Q(x) = x^m q(1/x)$ is named the *characteristic polynomial*. An output sequence of an LFSR is called an *m-sequence* if it has the maximal period of $2^m - 1$. This is the case if and only if $q(x)$ is a primitive polynomial. There exists two different representations of an LFSR, the so-called Galois and Fibonacci setup,

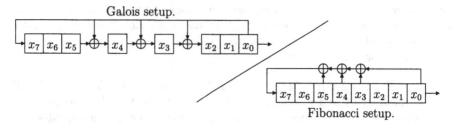

Fig. 1. Galois and Fibonacci LFSR

as we show in Figure 1. Both setups use the addition modulo 2. The Fibonacci setup is characterized by *a multiple inputs and single output feedback function*, while the Galois setup has *multiple feedback functions with a common input*. In both setups, we denote by x_0 the cell corresponding to the output of the LFSR. The same sequence can be produced by an LFSR in Fibonacci or Galois setup with the same characteristic polynomial but with different initializations. The *linear complexity* of a sequence S is defined as the size of the smallest LFSR which is able to produce this sequence [17].

2.2   FCSRs

FCSRs were introduced by Klapper and Goresky in [6]. Instead of addition modulo 2, FCSRs use additions with carry, which means that they need additional memory to store the carry. Their non–linear update function makes them particularly interesting for areas where linearity is an issue, like for instance stream ciphers. The output of a binary FCSR corresponds to the 2–adic expansion of the rational number:

$$\frac{h}{q} \leq 0 \ .$$

As for the LFSRs, there exists a Fibonacci and a Galois setup [7]. Their different structures can be seen in Figure 2, where $\sum$ represents the hamming weight of the incoming bits, $+$ an integer addition, c the additional memory for the carry, and $\boxplus$ an addition with carry of two bits where the carry is stored in $\square$. The

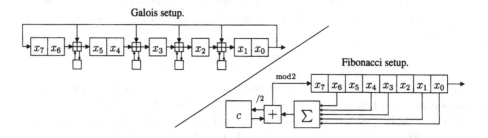

Fig. 2. Galois and Fibonacci FCSR

value q determines the feedback positions of the automata, depending on the setup used. In this article, we consider *ℓ-sequences*, *i.e.* sequences with maximal period $\varphi(q)$, where φ denotes Euler's phi function. This is equivalent to the case that q is a prime power and 2 has multiplicative order $\varphi(q)$ modulo q. For simplicity reasons, we restrict ourselves also to the case where the generated sequence is strictly periodic. This property is equivalent to $-q \le h \le 0$, which is always the case for Galois FCSRs but not necessarily for Fibonacci FCSRs. The *2-adic complexity* [5] of a sequence is defined as the size of the smallest FCSR which produces this sequence. In the periodic case, this is equivalent to the bit length of q.

3   Sub-sequences Generators and m-Sequences

The decimation of LFSR sequences has been used in cryptography in the design of new stream ciphers [18]. There exists two approaches to use decimation theory to define the automata associated to the sub-sequences.

Construction using LFSR synthesis. This first solution associates an LFSR to each sub-sequence. It is based on well-known results on the decimation of LFSR sequences. It can be applied to both Fibonacci and Galois representation without any distinction.

Theorem 1 ([19,13]). *Let S be a sequence produced by an LFSR whose characteristic polynomial $Q(x)$ is irreducible in $\mathbf{F}_2$ of degree m. Let α be a root of $Q(x)$ and let T be the period of $Q(x)$. Let S_d^i be a sub-sequence resulting from the d-decimation of S. Then, S_d^i can be generated by an LFSR with the following properties:*

- *The minimum polynomial of α^d in $\mathbf{F}_{2^m}$ is the characteristic polynomial $Q^\star(x)$ of the resulting LFSR.*
- *The period $T^\star$ of $Q^\star(x)$ is equal to $\frac{T}{\gcd(d,T)}$.*
- *The degree $m^\star$ of $Q^\star(x)$ is equal to the multiplicative order of $Q(x)$ in $\mathbf{Z}_{T^\star}$.*

In practice, the characteristic polynomial $Q^\star(x)$ can be determined using the Berlekamp-Massey algorithm [1]. The sub-sequences are generated using d LFSRs defined by the characteristic polynomial $Q^\star(x)$ but initialized with different values. In the case of LFSRs, the degree $m^\star$ must always be smaller or equal to m.

Construction using a multiple steps LFSR. This method was first proposed by Lempel and Eastman [11]. It consists in clocking the LFSR d times in one clock cycle by changing the connections between the memory cells and by some duplications of the feedback function. We obtain a network of linearly interconnected shift registers. This method differs for Galois and Fibonacci setup. The transformation of an m-bit Fibonacci LFSR into an automaton which generates d bits per cycle is achieved using the following equations:

$$next^d(x_i) = x_{i-d \bmod m} \tag{1}$$

$$(x_i)_{t+d} = \begin{cases} f(X_{t+i-m+d}) & \text{if } m - d \le i < m \\ (x_{i+d})_t & \text{if } i < m - d \end{cases} \tag{2}$$

where $next^d(x_i)$ is the cell connected to the output of x_i and f is the feedback function. The Equation 1 corresponds to the transformation of the connections between the memory cells. All the cells x_i of the original LFSR, such that $i \bmod d = k$, are gathered to form a sub-shift register, where $0 \le k \le d - 1$. This is the basic operation to transform a LFSR into a sub-sequences generator with a multiple steps solution. The content of the last cell of the k-th sub-shift registers corresponds to the k-th sub-sequence S_d^k. The Equation 2 corresponds to the transformation of the feedback function. It must be noticed that the synthesis requires to have only relations between the state of the register at time $t + d$ and t. The Figure 3 shows an example of such a synthesis for a Fibonacci setup defined by the connection polynomial $q(x) = x^8 + x^5 + x^4 + x^3 + 1$ with the decimation factor $d = 3$. The transformation of a Galois setup is described by the Equations 1 and 3:

$$(x_i)_{t+d} = \begin{cases} (x_0)_{t+d-m+i} \oplus \bigoplus_{k=0}^{m-2-i} a_{i+k}\,(x_0)_{t+d-k-1} & \text{if } m - d \le i < m \\ (x_{i+d})_t \oplus \bigoplus_{k=0}^{d-1} a_{i+d-1-k}\,(x_0)_{t+k} & \text{if } i < m - d \end{cases} \tag{3}$$

with $q(x) = 1 + a_0 x + a_1 x^2 + \cdots + a_{m-2} x^{m-1} + x^m$. The Equation 3 does not provide a direct relation between the state of the register at time $t + d$ and t. However, this equation can be easily derived to obtain more practical formulas as shown in Figure 4.

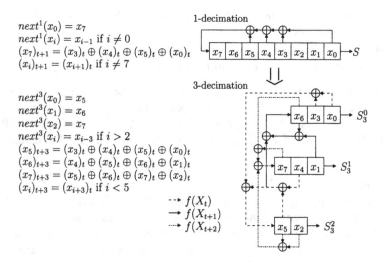

$next^1(x_0) = x_7$
$next^1(x_i) = x_{i-1}$ if $i \ne 0$
$(x_7)_{t+1} = (x_3)_t \oplus (x_4)_t \oplus (x_5)_t \oplus (x_0)_t$
$(x_i)_{t+1} = (x_{i+1})_t$ if $i \ne 7$

$next^3(x_0) = x_5$
$next^3(x_1) = x_6$
$next^3(x_2) = x_7$
$next^3(x_i) = x_{i-3}$ if $i > 2$
$(x_5)_{t+3} = (x_3)_t \oplus (x_4)_t \oplus (x_5)_t \oplus (x_0)_t$
$(x_6)_{t+3} = (x_4)_t \oplus (x_5)_t \oplus (x_6)_t \oplus (x_1)_t$
$(x_7)_{t+3} = (x_5)_t \oplus (x_6)_t \oplus (x_7)_t \oplus (x_2)_t$
$(x_i)_{t+3} = (x_{i+3})_t$ if $i < 5$

$\dashrightarrow f(X_t)$
$\longrightarrow f(X_{t+1})$
$\cdots\rightarrow f(X_{t+2})$

Fig. 3. Multiple steps generator for a Fibonacci LFSR

Table 1. Comparison of the two methods for the synthesis of a sub-sequences generator

Method	Memory cells	Logic Gates
LFSR synthesis	$d \times m^\star$	$d \times wt(Q^\star)$
Multiple steps LFSRs [11]	m	$d \times wt(Q)$

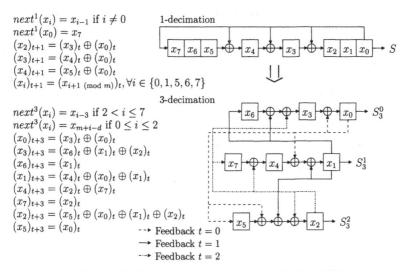

$next^1(x_i) = x_{i-1}$ if $i \neq 0$ 1-decimation
$next^1(x_0) = x_7$
$(x_2)_{t+1} = (x_3)_t \oplus (x_0)_t$
$(x_3)_{t+1} = (x_4)_t \oplus (x_0)_t$
$(x_4)_{t+1} = (x_5)_t \oplus (x_0)_t$
$(x_i)_{t+1} = (x_{i+1 \ (mod \ m)})_t, \forall i \in \{0, 1, 5, 6, 7\}$

3-decimation

$next^3(x_i) = x_{i-3}$ if $2 < i \leq 7$
$next^3(x_i) = x_{m+i-d}$ if $0 \leq i \leq 2$
$(x_0)_{t+3} = (x_3)_t \oplus (x_0)_t$
$(x_3)_{t+3} = (x_6)_t \oplus (x_1)_t \oplus (x_2)_t$
$(x_6)_{t+3} = (x_1)_t$
$(x_1)_{t+3} = (x_4)_t \oplus (x_0)_t \oplus (x_1)_t$
$(x_4)_{t+3} = (x_2)_t \oplus (x_7)_t$
$(x_7)_{t+3} = (x_2)_t$
$(x_2)_{t+3} = (x_5)_t \oplus (x_0)_t \oplus (x_1)_t \oplus (x_2)_t$
$(x_5)_{t+3} = (x_0)_t$

- -▸ Feedback $t = 0$
→ Feedback $t = 1$
- ·▸ Feedback $t = 2$

Fig. 4. Multiple steps generator for a Galois LFSR

Comparison. We have summarized in the Table 1 the two methods used to synthesize the sub-sequences generator. By $wt(Q(x))$, we mean the Hamming weight of Q, *i.e.* the number of non-zero monomials. The method based on LFSR synthesis proves that there exists a solution for the synthesis of the sub-sequences generator. With this solution, both memory cost and gate number depends on the decimation factor d. The method proposed by Lempel and Eastman [11] uses a constant number of memory cells for the synthesis of the sub-sequences generator.

The sub-sequences generators defined with the Berlekamp-Massey algorithm are not suitable to reduce the power consumption of an LFSR. Indeed, d LFSRs will be clocked to produce the sub-sequences, however the power consumption of such a sub-sequence generator is given by:

$$P = d \times \left(C_d \times V_{dd}^2 \times \frac{\gamma f}{d} \right)$$
$$= \lambda C \times V_{dd}^2 \times \gamma f$$

with $C_d = \lambda C$ and C the capacity of the original LFSR. We can achieve a better result with a multiple steps LFSR:

$$P = \lambda' C \times V_{dd}^2 \times \frac{\gamma f}{d}$$

with $C_d = \lambda' C$.

4   Sub-sequences Generators and ℓ-Sequences

This section presents the main contribution of the paper. We apply the two methods used in the previous section on the case of ℓ-sequences.

Construction using FCSR synthesis. There exists algorithms based on Euclid's algorithm [5] or on lattice approximation [4], which can determine the smallest FCSR to produce S_d^i. These algorithms use the first k bits of S_d^i to find h^* and q^* such that h^*/q^* is the 2-adic representation of the sub-sequence, $-q^* < h^* \leq 0$ and $\gcd(q^*, h^*) = 1$. Subsequently, we can find the feedback positions and the initial state of the FCSR in Galois or Fibonacci architecture. The value k is in the range of twice the linear 2-adic complexity of the sequence. For our new sequence S_d^i, let h^* and q^* define the values found by one of the algorithms mentioned above. By T^* and T, we mean the periods of respectively S_d^i and S.

For the period of the decimated sequences, we can make the following statement, which is true for all periodic sequences.

Lemma 1. *Let $S = (s_0, s_1, s_2, \ldots)$ be a periodic sequence with period T. For a given $d > 1$ and $0 \leq i \leq d - 1$, let S_d^i be the decimated sequence with period T^*. Then, it must hold:*

$$T^* \left| \frac{T}{\gcd(T, d)} \right. \quad (4)$$

If $\gcd(T, d) = 1$ then $T^* = T$.

Proof. The first property is given by:

$$s_{d[j+T/\gcd(T,d)]+i} = s_{dj+i+T[d/\gcd(T,d)]} = s_{dj+i} .$$

For the case $\gcd(T, d) = 1$, there exits $x, y \in \mathbb{Z}$ such that $xT + yd = 1$ due to Bezout's lemma. Since S is periodic, we define for any $j < 0$ and $k \geq 0$, $s_j = s_k$ such that $j \pmod{T} = k$. Thus, we can write for any j:

$$s_j \quad = s_{i+(j-i)xT+(j-i)yd} \qquad\qquad = s_{i+(j-i)yd} ,$$

$$s_{j+T^*} = s_{i+(j-i)xT+T^*xT+(j-i)yd+T^*yd} = s_{i+(j-i)yd+T^*yd} .$$

However, since T^* is the period of S_d^i we get:

$$s_{j+T^*} = s_{j+(j-i)yd} = s_j .$$

Therefore, it must hold that $T|T^*$ which together with (4) proves that $T^* = T$ if $\gcd(T, d) = 1$. □

In the case of $\gcd(T, d) > 1$, *the real value of T^* might depend on i*, e.g. for S being the 2-adic representation of $-1/19$ and $d = 3$ we have $T/\gcd(T, d) = 6$, however, for S_3^0 the period $T^* = 2$ and for S_3^1 the period $T^* = 6$.

A critical point in this approach is that the size of the new FCSR can be exponentially bigger than the original one. In general, we only know that for the new q^* it must hold that $q^* | 2^{T^*} - 1$. From the previous paragraph we know that T^* can be as big as $T/\gcd(T, d)$. In the case of an *allowable decimation* [20], *i.e.* a decimation where d and T are coprime, we have more informations:

Corollary 1 ([21]). *Let S be the 2-adic representation of h/q, where $q = p^e$ is a prime power with p prime, $e \geq 1$ and $-q \leq h \leq 0$. Let $d > 0$ be relatively prime to $T = p^e - p^{e-1}$, the period of S. Let S_d^i be a d-decimation of S with $0 \leq i < d$ and let h^*/q^* be its 2-adic representation such that $-q^* \leq h^* \leq 0$ and $\gcd(q^*, h^*) = 1$. Then q^* divides*

$$2^{T/2} + 1.$$

If the following conjecture is true, we have more information on the new value q^*.

Conjecture 1 ([21]). Let S be an ℓ-sequence with connection number $q = p^e$ and period T. Suppose p is prime and $q \notin \{5, 9, 11, 13\}$. Let d_1 and d_2 be relatively prime to T and incongruent modulo T. Then for any i and j, $S_{d_1}^i$ and $S_{d_2}^j$ are cyclically distinct, *i.e.* there exists no $\tau \geq 0$ such that $S_{d_1}^i$ is equivalent to $S_{d_2}^j$ shifted by τ positions to the left.

This conjecture has already been proved for many cases [20,22] but not yet for all. If it holds, this implies that for any $d > 1$, S_d^i is cyclically distinct from our original ℓ-sequence. We chose q such that the period was maximal, thus, any other sequence with the same period which is cyclically distinct must have a value $q^* > q$. This means that the complexity of the FCSR producing the subsequence S_d^i will be larger than the original FCSR, *if d and T are relative prime.*

Remark 1. Let us consider the special case where q is prime and the period $T = q - 1 = 2p$ is twice a prime number $p > 2$, as recommended for the stream cipher proposed in [23]. The only possibilities in this case for $\gcd(d, T) > 1$ is $d = 2$ or $d = T/2$.

For $d = T/2$, we will have $T/2$ FCSRs where each of them outputs either $0101\ldots$ or $1010\ldots$, since for an ℓ-sequence the the second half of the period is the inverse of the first half [21]. Thus, the size of the sub-sequences generator will be in the magnitude of T which is exponentially bigger than the 2-adic complexity of S which is $\log_2(q)$.

In the case of $d = 2$, we get two new sequences with period $T^* = p$. As for any FCSR, it must hold that $T^* | ord_{q^*}(2)$, where $ord_{q^*}(2)$ is the multiplicative order of 2 modulo q^*. Let $\varphi(n)$ denote Euler's function, *i.e.* the number of integers smaller than n which are relative prime to n. It is well known, *e.g.* [16], that $ord_{q^*}(2) | \varphi(q^*)$ and if q^* has the prime factorization $p_1^{e_1} p_2^{e_2} \cdots p_r^{e_r}$ then $\varphi(q^*) = \prod_{i=1}^{r} p_i^{e_i - 1}(p_i - 1)$. From this follows that $p \neq \varphi(q^*)$, because otherwise $(p + 1)$ must be a prime number, which is not possible since $p > 2$ is a prime. We also know that $T^* = p | \varphi(q^*)$, thus $2 \times p = q - 1 \leq \varphi(q^*)$. This implies that $q^* > q$, since from $T^* = T/2$ follows that $q \neq q^*$.

Together with Conjecture 1, we obtain that for such an FCSR any decimation would have a larger complexity than the original one. This is also interesting from the perspective of stream ciphers, since any decimated subsequence of such an FCSR has larger 2-adic complexity than the original one, except for the trivial case with $d = T/2$.

Construction using a multiple steps FCSR. A multiple steps FCSR is a network of interconnected shift registers with a carry path: the computation of the feedback at time t depends directly on the carry generated at time $t-1$. The transformation of an m-bit FCSR into a d sub-sequences generator uses first Equation 1 to modify the mapping of the shift register. For the Fibonacci setup, the transformation uses the following equations:

$$(x_i)_{t+d} = \begin{cases} g\left(X_{t+d-m+i}, c_{t+d-m+i}\right) \bmod 2 & \text{if } m-d \le i < m \\ (x_{i+d})_t & \text{if } i < m-d \end{cases} \tag{5}$$

$$c_{t+d} = g(X_{t+d-1}, c_{t+d-1})/2 \tag{6}$$

with $g(X_t, c_t) = h(X_t) + c_t$ the feedback function of an FCSR in Fibonacci setup and

$$h(X_t) = \sum_{i=0}^{m-1} a_i(x_i)_t . \tag{7}$$

Due to the nature of the function g, we can split the automaton into two parts. The first part handles the computation related to the shift register X_t and the other part is the carry path as shown in Figure 5 for $q = 347$.

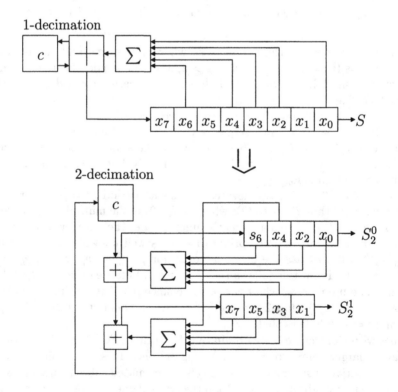

Fig. 5. Multiple steps generator for a Fibonacci FCSR

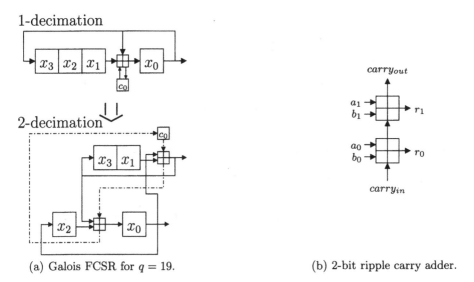

(a) Galois FCSR for $q = 19$.   (b) 2-bit ripple carry adder.

Fig. 6. Example for a Galois FCSR with $q = 19$

The case of Galois FCSRs is more difficult because the circuit can not be split into two parts: each bit of the carry register must be handle separately. The modification of the basic operator of a Galois FCSR, *i.e.* addition with carry, is the key transformation to obtain a sub-sequences generator. Let us consider a Galois FCSR with $q = 19$. This automaton has a single addition with carry as shown in Figure 6(a). The sub-sequences generator for $d = 2$ associated to this FCSR is defined by:

$$t + 1 \begin{cases} (x_0)_{t+1} = (x_0)_t \oplus (x_1)_t \oplus (c_0)_t \\ (c_0)_{t+1} = [(x_0)_t \oplus (x_1)_t] \, [(x_0)_t \oplus (c_0)_t] \oplus (x_0)_t \end{cases} \tag{8}$$

$$t + 2 \begin{cases} (x_0)_{t+2} = (x_0)_{t+1} \oplus (x_2)_t \oplus (c_0)_{t+1} \\ (c_0)_{t+2} = [(x_0)_{t+1} \oplus (x_2)_t] \, [(x_0)_{t+1} \oplus (c_0)_{t+1}] \oplus (x_0)_{t+1} \end{cases} \tag{9}$$

with c_0 the carry bit of the FCSR. The previous equations correspond to the description of the addition with carry at the bit-level (and represented by $\boxplus$ in the figures). This operator is also known as a full adder. The Equations corresponding to time $t + 2$ depend on the carry, $(c_0)_{t+1}$, generated at time $t + 1$. This dependency between full adders is a characteristic of a well-known arithmetic circuit: the n-bit ripple carry adder (Figure 6(b)).

Thus, all the full adders in a d sub-sequences generator are replaced by d-bit ripple carry adders as shown in Figure 7.

We can derive a more general representation of a multiple steps Galois FCSR from the previous formula:

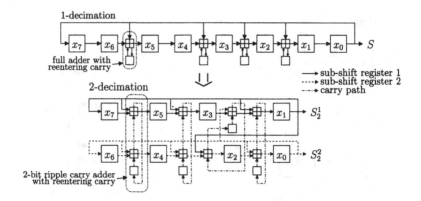

Fig. 7. Multiple steps generator for a Galois FCSR

$$(x_i)_{t+d} = \begin{cases} (x_0)_{t+d-m+i} \oplus \bigoplus_{k=0}^{m-2-i} a_{i+k} \left[(x_0)_{t+d-k-1} \oplus (c_{i+k})_{t+d-k-1} \right] \\ \qquad\qquad\qquad\qquad\qquad\qquad\qquad\qquad\qquad \text{if } m-d \le i < m \\ (x_{i+d})_t \oplus \bigoplus_{k=0}^{d-1} a_{i+k} \left[(x_0)_{t+d-k-1} \oplus (c_{i+k})_{t+d-k-1} \right] \\ \qquad\qquad\qquad\qquad\qquad\qquad\qquad\qquad\qquad \text{if } i < m-d \end{cases} \tag{10}$$

$$(c_i)_{t+d} = \left[(x_0)_{t+d-1} \oplus (x_{i+1})_{t+d-1} \right] \left[(x_0)_{t+d-1} \oplus (c_i)_{t+d-1} \right] \oplus (x_0)_{t+d-1} \tag{11}$$

The Equation 11 shows the dependencies between the carries which corresponds to the propagation of the carry in a ripple carry adder. The Equation 10 corresponds to the path of the content of a memory cell $(x_i)_t$ through the different levels of the ripple carry adders.

Comparison The construction using FCSR synthesis is more complicated than in the case of LFSRs. The size of the minimal generator producing S_d^i can depend on i, and we can only give upper bounds for $q^\star$ namely that $q^\star | 2^{T\star} - 1$ in the general case and that $q^\star | 2^{T/2} + 1$ if $\gcd(d, T) = 1$. Based on Conjecture 1, we saw that $q^\star > q$ if $\gcd(T, d) = 1$. Moreover, in the case $p = 2p + 1$ with p and q prime, the resulting sub-sequences generator is always larger than the original one.

Apart from the carry path and the cost of the addition with carry, the complexity of a multiple steps implementation of a FCSR is very similar to the multiple steps implementation of an LFSR. There is no overhead in memory and the number of logic gates for a Galois FCSR is $5d \times wt(q)$ where $wt(q)$ is the number of ones in the binary representation of q and the number 5 corresponds to the five gates required for a full-adder (four Xors and one And *cf.* Equation 8). In the case of Fibonacci setup, the number of logic gates is given by:

$$d \times (5 \times (wt(q) - \lceil \log_2(wt(q) + 1) \rceil) +$$
$$+2 \times (\lceil \log_2(wt(q) + 1) \rceil - wt(wt(q)) + 5 \times size(c))$$

with $(5 \times (wt(q) - \lceil \log_2(wt(q) + 1) \rceil)) + 2 \times (\lceil \log_2(wt(q) + 1) \rceil - wt(wt(q)))$ the cost of the implementation of a parallel bit counter [24], *i.e.* the h function (*cf.*

Equation 7) and $5 \times size(c)$ is the cost of a ripple carry adder which adds the content of $size(c)$ bits of the carry with the output of h.

5   Conclusion

We have presented in this paper how to transform an FCSR into a sub-sequences generator to increase its throughput or to reduce its power consumption. Our results emphasize the similarity between LFSRs and FCSRs in terms of implementation properties. In both cases, the solution based on the classical synthesis algorithm fails to provide a minimal solution. Even worse, in the case of FC-SRs the memory needed is in practice exponentially bigger than for the original FCSR. Thus, we need to use multiple steps implementations as for LFSRs. The propagation of carries is the main problem in multiple steps implementations of FCSRs: if we consider d sub-sequences, we obtain d-bit ripple carry adders with carry instead of additions with carry in a Galois setup. In the case of a Fibonacci setup, the situation is different since we can split the feedback function into two parts. This new representation can significantly improve the hardware implementation of FCSRs but it may also be possible to improve software implementations.

References

1. Massey, J.L.: Shift-register synthesis and BCH decoding. IEEE Transactions on Information Theory 15, 122–127 (1969)
2. Feng, G.L., Tzeng, K.: A Generalization of the Berlekamp-Massey Algorithm for Multisequence Shift-Register Synthesis with Applications to Decoding Cyclic Codes. IEEE Transactions on Information Theory 37(5), 1274–1287 (1991)
3. Schmidt, G., Sidorenko, V.: Linear Shift-Register Synthesis for Multiple Sequences of Varying Length. In: IEEE International Symposium on Information Theory - ISIT 2006, pp. 1738–1742. IEEE, Los Alamitos (2006)
4. Klapper, A., Goresky, M.: Feedback shift registers, 2-adic span, and combiners with memory. Journal of Cryptology 10, 111–147 (1997)
5. Arnault, F., Berger, T.P., Necer, A.: Feedback with Carry Shift Registers synthesis with the Euclidean Algorithm. IEEE Transactions on Information Theory 50(5) (2004)
6. Klapper, A., Goresky, M.: 2-adic shift registers. In: Anderson, R. (ed.) FSE 1993. LNCS, vol. 809, pp. 174–178. Springer, Heidelberg (1994)
7. Goresky, M., Klapper, A.: Fibonacci and Galois representations of feedback-with-carry shift registers. IEEE Transactions on Information Theory 48(11) (2002)
8. Goresky, M., Klapper, A.: Algebraic Shift Register Sequences (preprint)
9. Courtois, N., Meier, W.: Algebraic Attacks on Stream Ciphers with Linear Feedback. In: Biham, E. (ed.) EUROCRYPT 2003. LNCS, vol. 2656, pp. 345–359. Springer, Heidelberg (2003)
10. Arnault, F., Berger, T.P., Minier, M.: On the security of FCSR-based pseudorandom generators. In: State of the Art of Stream Ciphers - SASC (2007), http://sasc.crypto.rub.de/program.html

11. Lempel, A., Eastman, W.L.: High Speed Generation of Maximal Length Sequences. IEEE Transactions on Computer 2, 227–229 (1971)
12. Smeets, B.J.M., Chambers, W.G.: Windmill Generators: A Generalization and an Observation of How Many There Are. In: Günther, C.G. (ed.) EUROCRYPT 1988. LNCS, vol. 330, pp. 325–330. Springer, Heidelberg (1988)
13. Rueppel, R.A.: Analysis and Design of Stream Ciphers. Springer, Heidelberg (1986)
14. Filiol, E.: Decimation Attack of Stream Ciphers. In: Roy, B., Okamoto, E. (eds.) INDOCRYPT 2000. LNCS, vol. 1977, pp. 31–42. Springer, Heidelberg (2000)
15. Golomb, S.W.: Shift Register Sequences. Aegean Park Press (1981)
16. McEliece, R.J.: Finite field for scientists and engineers. Kluwer Academic Publishers, Dordrecht (1987)
17. Cusick, T.W., Ding, C., Renvall, A.: Stream Ciphers and Number Theory. North-Holland, Amsterdam (1998)
18. Massey, J.L., Rueppel, R.A.: Linear ciphers and random sequence generators with multiple clocks. In: Beth, T., Cot, N., Ingemarsson, I. (eds.) EUROCRYPT 1984. LNCS, vol. 209, pp. 74–87. Springer, Heidelberg (1985)
19. Zierler, N.: Linear recurring Sequences. Journal of the Society for Industrial and Applied Mathematics 2, 31–48 (1959)
20. Goresky, M., Klapper, A., Murty, R., Shparlinski, I.: On Decimations of ℓ-Sequences. SIAM Journal of Discrete Mathematics 18(1), 130–140 (2004)
21. Goresky, M., Klapper, A.: Arithmetic crosscorrelations of feedback with carry shift register sequences. IEEE Transactions on Information Theory 43(4), 1342–1345 (1997)
22. Xu, H., Qi, W.: Further Results on the Distinctness of Decimations of ℓ-Sequences. IEEE Transactions on Information Theory 52(8), 3831–3836 (2006)
23. Arnault, F., Berger, T.P.: F-fcsr: Design of a new class of stream ciphers. In: Gilbert, H., Handschuh, H. (eds.) FSE 2005. LNCS, vol. 3557, pp. 83–97. Springer, Heidelberg (2005)
24. Muller, D.E., Preparata, F.P.: Bounds to complexities of networks for sorting and switching. J. ACM 22, 1531–1540 (1975)
25. Hurd, W.: Efficient Generation of Statistically Good Pseudonoise by Linearly Interconnected Shift Registers. IEEE Transactions on Computer 2, 146–152 (1974)
26. Rueppel, R.A.: When Shift Registers Clock Themselves. In: Price, W.L., Chaum, D. (eds.) EUROCRYPT 1987. LNCS, vol. 304, pp. 53–64. Springer, Heidelberg (1988)
27. Key, E.L.: An Analysis of the Structure and Complexity of Nonlinear Binary Sequence Generators. IEEE Transactions Information Theory 22(4), 732–736 (1976)
28. Berger, T.P., Minier, M.: Two Algebraic Attacks Against the F-FCSRs Using the IV Mode. In: Maitra, S., Veni Madhavan, C.E., Venkatesan, R. (eds.) INDOCRYPT 2005. LNCS, vol. 3797, pp. 143–154. Springer, Heidelberg (2005)
29. Arnault, F., Berger, T.P.: Design and Properties of a New Pseudorandom Generator Based on a Filtered FCSR Automaton. IEEE Transaction on Computers. 54(11), 1374–1383 (2005)

Design of M-Ary Low Correlation Zone Sequence Sets by Interleaving⋆

Jin-Ho Chung and Kyeongcheol Yang

Dept. of Electronics and Electrical Engineering
Pohang University of Science and Technology (POSTECH)
Pohang, Kyungbuk 790-784, Korea
{jinho, kcyang}@postech.ac.kr

Abstract. In this paper we present a new method to construct an M-ary low correlation zone (LCZ) sequence set from an M-ary sequence with good autocorrelation by using the interleaved technique, where M is an even integer. We also show that the constructed LCZ sequence sets are optimal or nearly optimal with respect to the Tang-Fan-Matsufuji bound. Due to the flexibility in the choice of alphabet size, LCZ size, and period, our construction may be applied to various situations in quasi-synchronous code-division multiple access environment.

Keywords: low correlation zone (LCZ) sequences, quasi-synchronous code-division multiple access (QS-CDMA), M-ary sequences, interleaved sequences.

1 Introduction

In quasi-synchronous code-division multiple access (QS-CDMA) systems [?, 10], relative chip time delays among the signals of different users are restricted in a certain time interval. For this reason, the performance of a QS-CDMA system is determined by the correlation of spreading sequences around the origin [4]. Low correlation zone (LCZ) sequences are good candidates for spreading sequences in such systems. For applications in QS-CDMA systems, we need an LCZ sequence set which contains many sequences and has a wide low correlation zone. However, there is some trade-off between the set size and the LCZ size of an LCZ sequence set, which is formulated by the bound established by Tang, Fan, and Matsufuji [12]. Therefore, designing an optimal LCZ sequence set with respect to the Tang-Fan-Matsufuji bound is a very important problem.

There have been many research results for design of LCZ sequence sets [3, 4, 5, 6, 7, 8, 11, 13, 14]. Almost all of the previous constructions were based on period $p^n - 1$, and LCZ size $(p^n - 1)/(p^m - 1)$ for some $m|n$ (see [3] for a unified theory on this type of LCZ sequence sets). Recently, Kim et al. presented a new design of binary LCZ sequence sets of period $2(2^n - 1)$ with flexible LCZ size

---

⋆ This work was supported by the Korea Science and Engineering Foundation(KOSEF) grant funded by the Korea government(MEST) (No. R01-2008-000-10669-0).

S.W. Golomb et al. (Eds.): SETA 2008, LNCS 5203, pp. 313–321, 2008.

and near-optimality [8]. However, no design method for M-ary LCZ sequence sets for general M has been reported.

In this paper we present a new method to construct an M-ary LCZ sequence set from an M-ary sequence with good autocorrelation by using the interleaved technique, where M is an even integer. Our construction method is flexible in the sense of period, LCZ size, and alphabet size. We also show that the constructed LCZ sequence sets are optimal or nearly optimal with respect to the Tang-Fan-Matsufuji bound by giving some construction examples.

The outline of the paper is as follows. Section II gives some preliminaries for our presentation. In Section III, we present a new design of an M-ary LCZ sequence set from an M-ary sequence with good autocorrelation. In Section IV, we discuss the optimality of the parameters obtained from our construction. Finally, some concluding remarks are given in Section V.

2   Preliminaries

Let $\{s(t)\}$ be an M-ary sequence over $\mathbb{Z}_M = \{0, 1, ..., M-1\}$. It is called an N-periodic M-ary sequence or an M-ary sequence of period N if it satisfies

$$s(t + N) = s(t)$$

for all t and some positive integer N. Let $\{s_1(t)\}$ and $\{s_2(t)\}$ be two M-ary sequences of period N. If there is no integer τ such that $s_2(t) = s_1(t+\tau)$ for all t, they are said to be *cyclically distinct*. Otherwise, they are said to be *cyclically equivalent*. The *crosscorrelation* $C_{s_1,s_2}(\tau)$ of $\{s_1(t)\}$ and $\{s_2(t)\}$ is defined as

$$C_{s_1,s_2}(\tau) = \sum_{t=0}^{N-1} \tilde{s}_1(t)\tilde{s}_2^*(t + \tau),$$

where $\tilde{s}(t) = \omega_M^{s(t)}$ and $\omega_M = \exp\left(\frac{2\pi\sqrt{-1}}{M}\right)$. If $s_1(t) = s_2(t)$ for all t, then $C_{s_1,s_1}(\tau)$ is called the *autocorrelation* of $\{s_1(t)\}$, simply denoted by $C_{s_1}(\tau)$.

If an N-periodic binary sequence $\{b(t)\}$ has

$$C_b(\tau) = \begin{cases} N, & \tau \equiv 0 \mod N \\ \delta, & \tau \not\equiv 0 \mod N, \end{cases}$$

$\{b(t)\}$ is called a *binary sequence with two-level autocorrelation*. In the particular case that $N \equiv -1 \mod 4$, if $C_b(\tau) = -1$ for any $\tau \not\equiv 0 \mod N$, $\{b(t)\}$ is called a *binary sequence with ideal autocorrelation*.

Let $\mathcal{S} = \{\{s_i(t)\} \mid 0 \le i \le L-1\}$ be a set of L sequences with period N. For a positive integer $Z \ge 2$, let the crosscorrelation $C_{s_i,s_j}(\tau)$ between two sequences $\{s_i(t)\}$ and $\{s_j(t)\}$ in $\mathcal{S}$ satisfy

$$|C_{s_i,s_j}(\tau)| \le \eta$$

for $0 < |\tau| < Z$ when $i = j$, and $|\tau| < Z$ when $i \ne j$. Then we call $\mathcal{S}$ an (N, L, Z, η) LCZ sequence set, and Z the LCZ size of $\mathcal{S}$.

3   Design of $2T$-Periodic M-Ary LCZ Sequence Sets

The interleaved technique [2] may be applied to construct nT-periodic sequences with low correlation from a T-periodic sequence with good autocorrelation, where both n and T are positive integers. The technique may also be used to construct $2T$-periodic M-ary LCZ sequences from a T-periodic M-ary sequence with good autocorrelation.

Let $\{s(t)\}$ be a T-periodic M-ary sequence satisfying

$$|C_s(\tau)| \leq \epsilon$$

for some nonnegative constant ϵ. Assume that M is even and $1 \leq d \leq \lfloor \frac{T-1}{2} \rfloor$. Let

$$f = \begin{cases} \lfloor \frac{2T-1}{4d} \rfloor & \text{if } d \nmid \frac{T-1}{2} \\ \lfloor \frac{2T-3}{4d} \rfloor & \text{if } d \mid \frac{T-1}{2}. \end{cases} \tag{1}$$

The set $\mathcal{IS}$ of $2T$-periodic sequences is defined as

$$\mathcal{IS} = \{ \{s_{(i,m)}(t)\} \mid 0 \leq i \leq f - 1, m = 0, 1\} \tag{2}$$

where

$$s_{(i,0)}(2t) = s(t - id), \quad s_{(i,0)}(2t + 1) = s(t + 1 + (i+1)d),$$
$$s_{(i,1)}(2t) = s(t - id) + \frac{M}{2}, \quad s_{(i,1)}(2t + 1) = s(t + 1 + (i+1)d)$$

for $d \nmid \frac{T-1}{2}$, and

$$s_{(i,0)}(2t) = s(t - id), \quad s_{(i,0)}(2t + 1) = s(t + 2 + (i+1)d),$$
$$s_{(i,1)}(2t) = s(t - id) + \frac{M}{2}, \quad s_{(i,1)}(2t + 1) = s(t + 2 + (i+1)d)$$

for $d \mid \frac{T-1}{2}$. Note that the set $\mathcal{IS}$ contains $2f$ M-ary sequences of period $2T$. The crosscorrelation between any two sequences in $\mathcal{IS}$ is derived in the following lemma.

Lemma 1. *Let* $\tau = 2\tau_1 + \tau_0$, $\tau_0 \in \{0,1\}$, *and* $0 \leq \tau_1 \leq T - 1$. *For* $0 \leq i, j \leq f - 1$, $m, n \in \{0, 1\}$, *let* $a = j - i$ *and* $b = j + i + 1$. *Then the crosscorrelation* $C_{(i,m),(j,n)}(\tau)$ *between* $\{s_{(i,m)}(t)\}$ *and* $\{s_{(j,n)}(t)\}$ *is given as follows:*

Case i) $d \nmid \frac{T-1}{2}$:

$$C_{(i,0),(j,0)}(\tau) = \begin{cases} C_s(\tau_1 + ad) + C_s(\tau_1 - ad), & \text{if } \tau_0 = 0; \\ C_s(\tau_1 + bd + 1) + C_s(\tau_1 - bd), & \text{if } \tau_0 = 1; \end{cases}$$

$$C_{(i,1),(j,1)}(\tau) = \begin{cases} C_s(\tau_1 + ad) + C_s(\tau_1 - ad), & \text{if } \tau_0 = 0; \\ -C_s(\tau_1 + bd + 1) - C_s(\tau_1 - bd), & \text{if } \tau_0 = 1; \end{cases}$$

$$C_{(i,0),(j,1)}(\tau) = \begin{cases} -C_s(\tau_1 + ad) + C_s(\tau_1 - ad), & \text{if } \tau_0 = 0; \\ C_s(\tau_1 + bd + 1) - C_s(\tau_1 - bd), & \text{if } \tau_0 = 1; \end{cases}$$

$$C_{(i,1),(j,0)}(\tau) = \begin{cases} -C_s(\tau_1 + ad) + C_s(\tau_1 - ad), & \text{if } \tau_0 = 0; \\ -C_s(\tau_1 + bd + 1) + C_s(\tau_1 - bd), & \text{if } \tau_0 = 1; \end{cases}$$

Case ii) $d\,|\,\frac{T-1}{2}$:

$$C_{(i,0),(j,0)}(\tau) = \begin{cases} C_s(\tau_1 + ad) + C_s(\tau_1 - ad), & \text{if } \tau_0 = 0; \\ C_s(\tau_1 + bd + 2) + C_s(\tau_1 - bd - 1), & \text{if } \tau_0 = 1; \end{cases}$$

$$C_{(i,1),(j,1)}(\tau) = \begin{cases} C_s(\tau_1 + ad) + C_s(\tau_1 - ad), & \text{if } \tau_0 = 0; \\ -C_s(\tau_1 + bd + 2) - C_s(\tau_1 - bd - 1), & \text{if } \tau_0 = 1; \end{cases}$$

$$C_{(i,0),(j,1)}(\tau) = \begin{cases} -C_s(\tau_1 + ad) + C_s(\tau_1 - ad), & \text{if } \tau_0 = 0; \\ C_s(\tau_1 + bd + 2) - C_s(\tau_1 - bd - 1), & \text{if } \tau_0 = 1; \end{cases}$$

$$C_{(i,1),(j,0)}(\tau) = \begin{cases} -C_s(\tau_1 + ad) + C_s(\tau_1 - ad), & \text{if } \tau_0 = 0; \\ -C_s(\tau_1 + bd + 2) + C_s(\tau_1 - bd - 1), & \text{if } \tau_0 = 1. \end{cases}$$

Proof. Case i) $d \nmid \frac{T-1}{2}$: When $m = n = 0$, we have

$$C_{(i,0),(j,0)}(\tau) = \sum_{t=0}^{2T-1} \tilde{s}_{(i,0)}(t) \tilde{s}^*_{(j,0)}(t + \tau)$$

$$= \sum_{t=0}^{T-1} \tilde{s}_{(i,0)}(2t) \tilde{s}^*_{(j,0)}(2t + \tau) + \sum_{t=0}^{T-1} \tilde{s}_{(i,0)}(2t + 1) \tilde{s}^*_{(j,0)}(2t + 1 + \tau)$$

$$= \sum_{t=0}^{T-1} \tilde{s}_{(i,0)}(2t) \tilde{s}^*_{(j,0)}(2(t + \tau_1) + \tau_0)$$

$$+ \sum_{t=0}^{T-1} \tilde{s}_{(i,0)}(2t + 1) \tilde{s}^*_{(j,0)}(2(t + \tau_1) + \tau_0 + 1)$$

$$= \begin{cases} \tilde{s}(t - id)\tilde{s}(t + \tau_1 - jd) \\ \quad + \tilde{s}(t + (i+1)d + 1)\tilde{s}(t + \tau_1 + (j+1)d + 1), & \text{if } \tau_0 = 0; \\ \tilde{s}(t - id)\tilde{s}(t + \tau_1 + (j+1)d + 1) \\ \quad + \tilde{s}(t + (i+1)d + 1)\tilde{s}(t + \tau_1 + 1 - jd), & \text{if } \tau_0 = 1 \end{cases}$$

$$= \begin{cases} C_s(\tau_1 + ad) + C_s(\tau_1 - ad), & \text{if } \tau_0 = 0; \\ C_s(\tau_1 + bd + 1) + C_s(\tau_1 - bd), & \text{if } \tau_0 = 1 \end{cases}$$

where $a = j - i$ and $b = j + i + 1$.

When $(m, n) = (1, 1)$, we have

$$C_{(i,1),(j,1)}(\tau) = \begin{cases} (-\tilde{s}(t - id))(-\tilde{s}(t + \tau_1 - jd)) \\ \quad + \tilde{s}(t + (i+1)d + 1)\tilde{s}(t + \tau_1 + (j+1)d + 1), & \text{if } \tau_0 = 0; \\ (-\tilde{s}(t - id))\tilde{s}(t + \tau_1 + (j+1)d + 1) \\ \quad + \tilde{s}(t + (i+1)d + 1)(-\tilde{s}(t + \tau_1 + 1 - jd)), & \text{if } \tau_0 = 1 \end{cases}$$

$$= \begin{cases} C_s(\tau_1 + ad) + C_s(\tau_1 - ad), & \text{if } \tau_0 = 0; \\ -C_s(\tau_1 + bd + 1) - C_s(\tau_1 - bd), & \text{if } \tau_0 = 1. \end{cases}$$

When $(m, n) = (0, 1)$, we have

$$C_{(i,0),(j,1)}(\tau) = \begin{cases} \tilde{s}(t - id))(-\tilde{s}(t + \tau_1 - jd)) \\ \quad + \tilde{s}(t + (i+1)d + 1)\tilde{s}(t + \tau_1 + (j+1)d + 1), \text{ if } \tau_0 = 0; \\ \tilde{s}(t - id)\tilde{s}(t + \tau_1 + (j+1)d + 1) \\ \quad + \tilde{s}(t + (i+1)d + 1)(-\tilde{s}(t + \tau_1 + 1 - jd)), \quad \text{if } \tau_0 = 1 \end{cases}$$

$$= \begin{cases} -C_s(\tau_1 + ad) + C_s(\tau_1 - ad), \text{ if } \tau_0 = 0; \\ C_s(\tau_1 + bd + 1) - C_s(\tau_1 - bd), \text{ if } \tau_0 = 1. \end{cases}$$

When $(m, n) = (1, 0)$, we have

$$C_{(i,1),(j,0)}(\tau) = \begin{cases} (-\tilde{s}(t - id))\tilde{s}(t + \tau_1 - jd) \\ \quad + \tilde{s}(t + (i+1)d + 1)\tilde{s}(t + \tau_1 + (j+1)d + 1), \text{ if } \tau_0 = 0; \\ (-\tilde{s}(t - id))\tilde{s}(t + \tau_1 + (j+1)d + 1) \\ \quad + \tilde{s}(t + (i+1)d + 1)\tilde{s}(t + \tau_1 + 1 - jd), \quad \text{if } \tau_0 = 1 \end{cases}$$

$$= \begin{cases} -C_s(\tau_1 + ad) + C_s(\tau_1 - ad), \quad \text{if } \tau_0 = 0; \\ -C_s(\tau_1 + bd + 1) + C_s(\tau_1 - bd), \text{ if } \tau_0 = 1. \end{cases}$$

Case ii) $d \mid \frac{T-1}{2}$: Similar to the Proof of Case i). □

Remark: In Case i) of Lemma 1, it is easily checked that $ad \neq -ad \mod T$ if $0 \leq i, j \leq f - 1$ and $(i, j) \neq (0, 0)$. The condition $d \nmid \frac{T-1}{2}$ implies $bd + 1 \neq -bd$ mod T for any $0 \leq i, j \leq f - 1$. Hence, $\{s_{(i,m)}(t)\}$ and $\{s_{(j,n)}(t)\}$ are cyclically distinct if $(i, m) \neq (j, n)$. Cyclic distinctness for Case ii) can also be checked similarly. Therefore, the set $\mathcal{IS}$ in (2) contains $2f$ cyclically distinct sequences of period $2T$. □

Theorem 2. Let $1 \leq d \leq \frac{T-1}{2}$. The set $\mathcal{IS}$ in (2) is a $(2T, 2f, 2d, 2\epsilon)$ M-ary LCZ sequence set, where

$$f = \begin{cases} \lfloor \frac{2T-1}{4d} \rfloor \text{ for } d \nmid \frac{T-1}{2} \\ \lfloor \frac{2T-3}{4d} \rfloor \text{ for } d \mid \frac{T-1}{2}. \end{cases}$$

Proof. Consider the case that $d \nmid \frac{T-1}{2}$. Without loss of generality, we can assume that $i \leq j$ for $C_{(i,m),(j,n)}(\tau)$. By Lemma 1, we have

$$|C_{(i,m),(j,n)}(\tau)| \leq 2\epsilon$$

except for

$$(\tau_0, \tau_1) = (0, ad), (0, T - ad), (1, bd), (1, T - bd - 1)$$

which are equivalent to

$$\tau = 2ad, 2T - 2ad, 2bd + 1, 2T - 2bd - 1,$$

respectively. If $0 \leq i \neq j \leq f - 1 = \lfloor \frac{2T-1}{4d} \rfloor - 1$, we have

$$2d \leq 2ad, 2T - 2ad \leq 2T - 2d,$$

$$2d + 1 \leq 2bd + 1, 2T - 2bd - 1 \leq 2T - 2d - 1.$$

Thus

$$|C_{(i,m),(j,n)}(\tau)| \leq 2\epsilon$$

for $0 \leq |\tau| < 2d$ and $i \neq j$. If $i = j$, we have

$$2ad \equiv 2T - 2ad \equiv 0 \mod 2T,$$

$$2d + 1 \leq 2bd + 1, 2T - 2bd - 1 \leq 2T - 2d - 1.$$

Therefore, it is easily checked that

$$|C_{(i,m),(i,n)}(\tau)| \leq 2\epsilon$$

for $0 < |\tau| < 2d$ and $m = n$, and for $0 \leq |\tau| < 2d$ and $m \neq n$.

The case that $d | \frac{T-1}{2}$ may be proved in a similar way.   □

For any even integer M, we showed that an M-ary LCZ sequence set of period $2T$ can be constructed from a T-periodic M-ary sequence with good autocorrelation. In addition, it is possible to select any even integer $2 \leq Z \leq T - 1$ as the LCZ size.

Remark: It is easily checked that the construction in [8] is equivalent to replacing the component sequences in (2) by

$$s_{(u,0)}(2t) = s(2t - u), \ s_{(u,0)}(2t + 1) = s(2t + u),$$
$$s_{(u,1)}(2t) = s(2t - u) + 1, \ s_{(u,1)}(2t + 1) = s(2t + u)$$

for some integers $1 \leq u < 2^{n-1} - 2$ and a $(2^n - 1)$-periodic binary sequence $\{s(t)\}$ with ideal autocorrelation. Thus, our construction is a modified generalization of the construction in [8] in the sense that the alphabet size and period are more flexible.   □

4   Optimality of Constructed LCZ Sequence Sets

Tang, Fan, and Matsufuji derived a bound on LCZ sequence sets as in the following theorem [12].

Theorem 3 (Tang, Fan, and Matsufuji, [12]). *Let S be an (N, L, Z, η) LCZ sequence set. Then we have*

$$LZ - 1 \leq \frac{N - 1}{1 - \eta^2/N}.$$

Table 1. Some possible parameters for our construction and corresponding possible set sizes for the construction in [8] for $M = 2, N = 254$, and $\eta = 2$ (Here L^* denotes the optimal set size)

N	Z	L	Possible set size in [8]	L^*	η
254	4	62	62	64	2
	6	40		43	
	8	30	30	32	
	10	24		25	
	12	20	20	21	
	14	16		18	
	16		14	16	
	18	14		14	
	19		12	13	
	20	12		12	
	22		10	11	
	24	10		10	
	28		8	9	
	30	8		8	
	36		6	7	
	40	6		6	
	50		4	5	
	62	4		4	

By Theorem 3, the optimal set size L^* for an (N, L, Z, η) LCZ sequence set is given by

$$L^* = \left\lfloor \frac{1}{Z} \cdot \frac{N^2 - \eta^2}{N - \eta^2} \right\rfloor .$$

If $N = 2T$, $Z = 2d$, and $\eta = 2\epsilon$, then we have

$$L^* = \left\lfloor \frac{1}{2d} \cdot \frac{(2T)^2 - 4\epsilon^2}{2T - 4\epsilon^2} \right\rfloor$$

which goes to $\lfloor \frac{T}{d} \rfloor$ when ϵ is sufficiently small compared to T. Therefore, the $(2T, 2\lfloor \frac{2T-1}{4d} \rfloor, 2d, 2\epsilon)$ or $(2T, 2\lfloor \frac{2T-3}{4d} \rfloor, 2d, 2\epsilon)$ LCZ sequence set $\mathcal{IS}$ given in Theorem 2 may be optimal or nearly optimal with respect to the bound in Theorem 3.

From the construction given in Section III, it is possible to get a $(2(2^n - 1), 2f, 2d, 2)$ binary LCZ sequence set by selecting a binary sequence with ideal autocorrelation as the component sequence $\{s(t)\}$, where $2 \leq d \leq 2^{n-1} - 1$ and f is given in (1). Table 1 compares some possible parameters in our construction with those by the constructions in [8] when $M=2$, $N = 254$, and $\eta = 2$.

Remark: Table 1 shows that our construction has an LCZ size greater than or equal to that of the constructions in [8] for the same set size. Note that there is no case which has an optimal set size for the constructions in [8]. However, our construction gives an optimal set size for some cases. $\quad\square$

Table 2. Some possible parameters and optimal set sizes L^* when $M = 4$ and $N = 6248$

N	Z	L	L^*	η	N	Z	L	L^*	η
6248	4	1560	1570	$4\sqrt{2}$	6248	24	260	261	$4\sqrt{2}$
	6	1040	1046			26	240	241	
	8	780	785			28	222	224	
	10	624	628			30	208	209	
	12	520	523			50	124	125	
	14	446	449			100	62	62	
	16	390	392			300	20	20	
	18	346	348			500	12	12	
	20	312	314			1000	6	6	
	22	282	285			1500	4	4	

Sidel'nikov sequences are well-known as M-ary sequences with good autocorrelation property [9]. Thus we can also construct M-ary LCZ sequences using an M-ary Sidel'nikov sequence by the construction in Section III. Table 2 shows some possible parameters of quaternary LCZ sequence sets of period 6248 constructed from a quaternary Sidel'nikov sequence of period 3124. Note that there is no construction for the cases such that $M \neq 2$ or $N = 6248$ in [8].

Remark: Table 2 tells us that there is a gap between the optimal set size and the possible set size when the LCZ size Z is very small compared to the period, since the maximum magnitude of the out-of-phase autocorrelation of a Sidel'nikov sequence is greater than 1. However, for larger LCZ sizes, we can get LCZ sequence sets having an optimal set size with respect to the Tang-Fan-Matsufuji bound as in the case that $N = 2(2^n - 1)$. □

5   Conclusion

We proposed a new method to construct M-ary LCZ sequence sets from any M-ary sequence with good autocorrelation for even M. The proposed design gives M-ary LCZ sequence sets which are optimal or nearly optimal with respect to the Tang-Fan-Matsufuji bound. Furthermore, our design is very flexible in the selection of period, LCZ size, and alphabet size. Therefore, our new design can be applied to various situations in the QS-CDMA environment.

References

1. De Gaudenzi, R., Elia, C., Viola, R.: Bandlimited quasi-synchronous CDMA: A novel satellite access technique for mobile and personal communication systems. IEEE J. Sel. Areas Commun. 10, 328–343 (1992)
2. Gong, G.: Theory and applications of q-ary interleaved sequences. IEEE Trans. Inform. Theory 41, 400–411 (1995)

3. Gong, G., Golomb, S.W., Song, H.-Y.: A note on low-correlation zone signal sets. IEEE Trans. Inform. Theory 53, 2575–2581 (2007)
4. Long, B., Zhang, P., Hu, J.: A generalized QS-CDMA system and the design of new spreading codes. IEEE Trans. Veh. Tech. 47, 1268–1275 (1998)
5. Jang, J.-W., No, J.-S., Chung, H.: A new construction of optimal p^2-ary low correlation zone sequences using unified sequences. IEICE Trans. Fundamentals E89-A, 2656–2661 (2006)
6. Jang, J.-W., No, J.-S., Chung, H., Tang, X.H.: New sets of optimal p-ary low-correlation zone sequences. IEEE Trans. Inform. Theory 53, 815–821 (2007)
7. Kim, S.-H., Jang, J.-W., No, J.-S., Chung, H.: New constructions of quaternary low correlation zone sequences. IEEE Trans. Inform. Theory 51, 1469–1477 (2005)
8. Kim, Y.-S., Jang, J.-W., No, J.-S., Chung, H.: New design of low correlation zone sequence sets. IEEE Trans. Inform. Theory 52, 4607–4616 (2006)
9. Sidel'nikov, V.M.: Some k-valued pseudo-random sequences and nearly equidistant codes. Probl. Inf. Transm. 5(1), 12–16 (1969)
10. Suehiro, N.: Approximately synchoronized CDMA system without cochannel using pseudo-periodic sequences. In: Proc. Int. Symp. Personal Communication 1993, Nanjing, China, pp. 179–184 (July 1994)
11. Tang, X.H., Fan, P.Z.: A class of pseudonoise sequences over $GF(p)$ with low correlation zone. IEEE Trans. Inform. Theory 47, 1644–1649 (2001)
12. Tang, X.H., Fan, P.Z., Matsufuji, S.: Lower bounds on the maximum correlation of sequence set with low or zero correlation zone. Electron. Lett. 36, 551–552 (2000)
13. Tang, X.H., Udaya, P.: New construction of low correlation zone sequences from Hadamard matrices. In: Proc. 2005 IEEE Inter. Symp. Inform. Theory (ISIT 2005), Adelaide, Australia, September 4–9, pp. 482–486 (2005)
14. Tang, X.H., Udaya, P.: New recursive construction of low correlation zone sequences. In: Proc. 2nd Inter. Workshop on Sequence Design and its Application in Communication (IWSDA 2005), Shimonoseki, Japan, October 10–14, pp. 86–89 (2005)

The Peak to Sidelobe Level of the Most Significant Bit of Trace Codes over Galois Rings

Patrick Solé* and Dimitrii Zinoviev

CNRS-I3S,

Les Algorithmes, Euclide B, 2000, route des Lucioles, 06 903 Sophia Antipolis, France
sole@unice.fr
http://www.i3s.unice.fr/~sole/

Abstract. Weighted degree trace codes over even characteristic Galois rings give binary sequences by projection on their most significant bit (MSB). Upper bounds on the aperiodic correlation, peak to sidelobe level (PSL), partial period imbalance and partial period pattern imbalance of these sequences are derived. The proof technique involves estimates of incomplete character sums over Galois rings, combining Weil-like bounds with Fourier transform estimates.

Keywords: aperiodic autocorrelation, partial period distribution, Galois rings, MSB, PSL.

1 Introduction

The aperiodic autocorrelation of binary sequences is an important design criterion of binary spreading sequences in a CDMA environment. It is also a fascinating mathematical invariant in relation with the merit factor [3]. In a seminal paper [9] Sarwate used Fourier coefficient estimates to derive an upper bound on what would be called later the PSL (Peak to Sidelobe Level) of binary $M-$sequences [2]. This bound was extended recently, using similar techniques, to the Most Significant Bit of M-sequences over rings [10] in the terminology of Z-D. Dai [1]. Since the proof involves estimates of incomplete character sums it is very natural to find applications in partial period statistics showing that the distribution of symbols (or the $r-$tuples of symbols) in the sequences are close to uniform. In the present work we generalize the results of [10] to weighted degree trace codes in primitive length. The case of [10] corresponds in that setting to a linear polynomial argument of the trace. Our work is also the analogue for the MSB map of the work [4] for the Gray map.

The material is organized in the following way. Section 2 collects some well-known definitions on Galois rings. Section 3 contains our version of Sarwate's DFT bounding technique. Section 4 recalls properties of polynomials over Galois rings and establishes an important technical lemma (Lemma 4.3). Section 5 studies the uniformity of distribution of symbols zero and one in the above

---

* Supported in part by ANR grant NUGET.

mentioned sequences. Section 6 is dedicated to bounds on the aperiodic auto-correlation. Section 7 extends Section 5 to $r-$tuples of symbols.

2   Preliminaries

Let $R = GR(2^l, m)$ denote the Galois ring of characteristic 2^l. It is the unique Galois extension of degree m of $\mathbb{Z}_{2^l}$, with 2^{lm} elements.

$$R = GR(2^l, m) = \mathbb{Z}_{2^l}[X]/(h(X)).$$

where $h(X)$ is a basic irreducible polynomial of degree m. Let ξ be an element in $GR(2^l, m)$ that generates the Teichmüller set $\mathcal{T}$ of $GR(2^l, m)$ which reduces to $\mathbb{F}_{2^m}$ modulo 2. Specifically, let $\mathcal{T} = \{0, 1, \xi, \xi^2, \dots, \xi^{2^m - 2}\}$ and $\mathcal{T}^* = \{1, \xi, \xi^2, \dots, \xi^{2^m - 2}\}$. We use the *convention* that $\xi^\infty = 0$.

The 2-adic expansion of $x \in GR(2^l, m)$ is given by

$$x = x_0 + 2x_1 + \cdots + 2^{l-1}x_{l-1},$$

where $x_0, x_1, \dots, x_{l-1} \in \mathcal{T}$. The Frobenius operator F is defined for such an x as

$$F(x_0 + 2x_1 + \cdots + 2^{l-1}x_{l-1}) = x_0^2 + 2x_1^2 + \cdots + 2^{l-1}x_{l-1}^2,$$

and the trace Tr, from $GR(2^l, m)$ to $\mathbb{Z}_{2^l}$, as

$$\text{Tr} := \sum_{j=0}^{m-1} F^j.$$

We also define another trace tr from $\mathbb{F}_{2^m}$ to $\mathbb{F}_2$ as

$$\text{tr}(x) := \sum_{j=0}^{m-1} x^{2^j}.$$

Throughout this note, we put $n = 2^m$ and $R^* = R \backslash 2R$. Let MSB : $\mathbb{Z}_{2^l}^n \to \mathbb{Z}_2^n$ be the most-significant-bit map, i.e.

$$\text{MSB}(y) = y_{l-1}, \text{where } y = y_0 + 2y_1 + \dots + 2^{l-1}y_{l-1} \in \mathbb{Z}_{2^l},$$

is its 2-adic expansion.

3   DFT and the Local Weil Bound

We assume henceforth in the whole paper that $l \geq 3$. Let l be a positive integer and $\omega = e^{2\pi i/2^l}$ be a primitive 2^l-th root of 1 in $\mathbb{C}$. Let ψ_k be the additive character of $\mathbb{Z}_{2^l}$ such that

$$\psi_k(x) = \omega^{kx}.$$

Let $\mu : \mathbb{Z}_{2^l} \to \{\pm 1\}$ be the mapping $\mu(t) = (-1)^c$, where c is the most significant bit of $t \in \mathbb{Z}_{2^l}$; in other words it maps $0, 1, ..., 2^{l-1} - 1$ to $+1$ and $2^{l-1}, 2^{l-1} + 1, ..., 2^l - 1$ to -1. Our goal is to express this map as a linear combination of characters. Recall the Fourier transformation formula on $\mathbb{Z}_{2^l}$:

$$\mu = \sum_{j=0}^{2^l-1} \mu_j \psi_j, \text{ where } \mu_j = \frac{1}{2^l} \sum_{x=0}^{2^l-1} \mu(x)\psi_j(-x). \tag{1}$$

Combining Lemma 4.1 and Corollary 7.4 of [5], we obtain

Lemma 3.1. *Let* $l = 3$. *For the constants* $\mu_j = (1 + \omega^{-j} + \omega^{-2j} + \omega^{-3j})/4$ $j = 1, 3, 5, 7$ *we have*

$$\mu = \mu_1\psi_1 + \mu_3\psi_3 + \mu_5\psi_5 + \mu_7\psi_7,$$

and $\mu_j = 0$, *for even* j. *Furthermore*

$$(|\mu_1| + |\mu_3| + |\mu_5| + |\mu_7|)^2 = 2 + \sqrt{2}.$$

Let $q = 2^l$ *where* $l \geq 4$. *Then*

$$\sum_{j=0}^{q-1} |\mu_j| < \frac{2}{\pi}\ln(q) + 1.$$

For all $\beta \neq 0$ in the ring $R = GR(2^l, m)$, we denote by Ψ_β the additive character

$$\Psi_\beta : R \to \mathbb{C}^*, x \mapsto \omega^{\text{Tr}(\beta x)}.$$

Note that for the defined above ψ_k and Ψ_β, we have:

$$\psi_k(\text{Tr}(\beta x)) = \Psi_{\beta k}(x).$$

Let $f(X)$ denote a polynomial in $R[X]$ and let

$$f(X) = F_0(X) + 2F_1(X) + \ldots + 2^{l-1}F_{l-1}(X)$$

denote its 2-adic expansion. Let d_i be the degree in x of F_i. Let Ψ be an arbitrary additive character of R, and set D_f to be the *weighted degree* of f, defined as

$$D_f = \max\{d_0 2^{l-1}, d_1 2^{l-2}, \ldots, d_{l-1}\}.$$

With the above notation, and for any integers k, H such that $0 \leq k < k+H-1 \leq 2^m - 2$, we have, under mild technical conditions, the bound

$$\left| \sum_{j=k}^{k+H-1} \Psi(f(\xi^j)) \right| \leq \left(\frac{2}{\pi}\ln\frac{4(2^m - 1)}{\pi} + 1 \right) D_f \sqrt{2^m}. \tag{2}$$

See [10] for details.

4   Polynomials over the Galois Ring $GR(2^l, m)$

Recall that $R = GR(2^l, m)$. A polynomial

$$f(X) = \sum_{j=0}^{d} c_j x^j \in R[X]$$

is called **canonical** if $c_j = 0$ for all even j. Given an integer $D \geq 4$, define

$$S_D = \{f(X) \in R[X] \mid D_f \leq D, f \text{ is canonical}\},$$

where D_f is the weighted degree of f. Observe that S_D is an $GR(2^l, m)$−module. Recall [7, Lemma 4.1]. For a weaker condition on D see [6, Theorem 6.13].

Lemma 4.1. *For any integer $D \geq 4$, we have:*

$$|S_D| = 2^{(D - \lfloor D/2^l \rfloor)m},$$

where $\lfloor x \rfloor$ is the largest integer $\leq x$.

Recall the following property of the weighted degree [8, Lemma 3.1].

Lemma 4.2. *Let $f(X) \in R[X]$ and $\alpha \in R^* = R \backslash 2R$ is a unit of R and let $g(X) = f(\alpha X) \in R[X]$. Then*

$$D_g = D_f,$$

where D_f, D_g are respectively the weighted degrees of the polynomials $f(X)$ and $g(X)$.

We will need the following technical result.

Lemma 4.3. *Let $f(X) \in R[X]$ and assume that $f \in S_D$ with a non-zero linear term. If we fix any r integers $0 \leq s_1 < s_2 < \ldots < s_r = n - 1$ then $f(\xi^{s_1} X), f(\xi^{s_2} X), \ldots, f(\xi^{s_r} X)$, are linearly independent over $\mathbb{Z}_{2^l}$, i.e. for any integers $j_1, j_2, \ldots, j_r$ the equality*

$$j_1 f(\xi^{s_1} X) + j_2 f(\xi^{s_2} X) + \ldots + j_r f(\xi^{s_r} X) = 0$$

implies $j_1 = j_2 = \ldots = j_r = 0$.

Proof. Suppose $f(X) \in R[X]$ is of degree $d \leq D$ and

$$f(X) = \alpha_1 X + \ldots + \alpha_{d-1} X^d,$$

where $\alpha_1 \neq 0$ and $\alpha_k = 0$ for even k. Fix the integers $s_1, s_2, \ldots, s_r$, then for any r integers $j_1, j_2, \ldots, j_r$ let

$$g(X) = g_{j_1 \ldots j_r}(X) = \sum_{i=0}^{r} j_i f(\xi^{s_i} X).$$

From there, we can write

$$g(X) = \beta_0 + \beta_1 X + \ldots + \beta_{d-1} X^d, \text{ where } \beta_k = \alpha_k \left(\sum_{i=1}^{r} j_i \xi^{k s_i} \right).$$

The condition $g(X) = 0$ implies that in particular $\beta_1 = 0$. So

$$\sum_{i=1}^{r} j_i \xi^{s_i} = 0.$$

Since $\xi^{s_1}, \xi^{s_2}, \ldots, \xi^{s_r}$ are linearly independent over $\mathbb{Z}_{2^l}$, the above equality implies $j_1 = j_2 = \ldots = j_r = 0$. $\square$

5 Partial Period Distributions

In this section we will consider periodic binary sequence of period $2^m - 1$. For any integer $D \geq 4$, let $f \in S_D$ be a polynomial with non-zero linear term, and set $c_t = \text{MSB}(\text{Tr}(f(\xi^t)))$, where $t = 0, \ldots, n-2$, $n = 2^m$.

Theorem 5.1. *With notation as above, let H be an integer in the range $0 < k < n-1$ and for any k in the range $0 < k < n-1$, consider the sequence $c_k, \ldots, c_{k+H-1}$ of length H. For any $\delta \in \{0,1\}$, let N_δ be the number of δ in $c_k, \ldots, c_{k+H-1}$. Then we have*

$$\left| N_\delta - \frac{H}{2} \right| \leq \frac{1}{2} \left(\frac{2l}{\pi} \ln(2) + 1 \right) \left(\frac{2}{\pi} \ln \frac{4(2^m - 1)}{\pi} + 1 \right) D \sqrt{2^m}.$$

Proof. Following [10], we have

$$N_\delta = \sum_{j=k}^{k+H-1} \frac{1}{2} (1 + (-1)^{c_j + \delta}).$$

Thus

$$N_\delta - \frac{H}{2} = \frac{(-1)^\delta}{2} \sum_{j=k}^{k+H-1} (-1)^{c_j}.$$

As we have $c_t = \text{MSB}(\text{Tr}(f(\xi^t)))$ where t ranges between 0 and $n-2$ and by (1), we obtain that $(-1)^{c_t}$ is equal to:

$$\mu(\text{Tr}(f(\xi^t))) = \sum_{j=0}^{2^l - 1} \mu_j \psi_j(\text{Tr}(f(\xi^t))) = \sum_{j=0}^{2^l - 1} \mu_j \Psi_j(f(\xi^t)).$$

Changing the order of summation, we obtain that:

$$\sum_{t=k}^{k+H-1} (-1)^{c_t} = \sum_{j=0}^{2^l - 1} \mu_j \sum_{t=k}^{k+H-1} \Psi_j(f(\xi^t)).$$

Applying (2), we have

$$\left| \sum_{t=k}^{k+H-1} (-1)^{c_t} \right| \le \sum_{j=0}^{2^l-1} |\mu_j| \left| \sum_{t=k}^{k+H-1} \Psi_j(f(\xi^t)) \right|$$

$$< \left(\frac{2l}{\pi} \ln(2) + 1 \right) \left(\frac{2}{\pi} \ln \frac{4(2^m - 1)}{\pi} + 1 \right) D\sqrt{2^m}.$$

The estimate of the Theorem follows.   □

6   Aperiodic Autocorrelation

As in the previous section, consider periodic sequences $c_0, c_1, \ldots$ of period $n-1$, where $n = 2^m$.

Theorem 6.1. *With notation as above, and for any τ in the range $0 < \tau < n-1$,*

$$\Theta(\tau) = \sum_{t=0}^{n-2-\tau} (-1)^{c_t}(-1)^{c_{t+\tau}},$$

where $c_t = \mathrm{MSB}(\mathrm{Tr}(f(\xi^t)))$. We then have the following bound ($l \ge 4$) on PSL:

$$|\Theta(\tau)| \le \left(\frac{2l}{\pi} \ln(2) + 1 \right)^2 \left(\frac{2}{\pi} \ln \frac{4(2^m - 1)}{\pi} + 1 \right) D\sqrt{2^m}.$$

In particular $|\Theta(\tau)| = O(\sqrt{n} \log n)$, for large n.

Proof. As we have $c_t = \mathrm{MSB}(\mathrm{Tr}(f(\xi^t)))$ where t ranges between 0 and $n-2$ and by (1), we obtain that $(-1)^{c_t}$ is equal to:

$$\mu(\mathrm{Tr}(f(\xi^t))) = \sum_{j=0}^{2^l-1} \mu_j \psi_j(\mathrm{Tr}(f(\xi^t))) = \sum_{j=0}^{2^l-1} \mu_j \Psi_j(f(\xi^t)).$$

Changing the order of summation, we obtain that:

$$\Theta(\tau) = \sum_{j_1=0}^{2^l-1} \sum_{j_2=0}^{2^l-1} \mu_{j_1} \mu_{j_2} \sum_{t=0}^{n-\tau-2} \Psi_{j_1}(f(\xi^t)) \Psi_{j_2}(f(\xi^{t+\tau})).$$

By definition of Ψ, we have:

$$\Psi_{j_1}(f(\xi^t)) \Psi_{j_2}(f(\xi^{t+\tau})) = \Psi(g(\xi^t)),$$

where $g(X) = j_1 f(X) + j_2 f(X\xi^\tau)$. By Lemma 4.3, since $1, \xi^\tau$ are linearly independent over $\mathbb{Z}_{2^l}$, we obtain that $g(X) \ne 0$. Note that if $f(X) \in S_D$ then $f(X\xi^\tau) \in S_D$ since, by Lemma 4.2 the change of variable $X \to X\xi^\tau$ does not increase the weighted degree. Moreover S_D is an R-module. Thus the polynomial $g(X)$ belongs to S_D. Applying (2), we obtain:

$$\left| \sum_{t=0}^{n-\tau-2} \Psi_{j_1}(f(\xi^t))\Psi_{j_2}(f(\xi^{t+\tau})) \right| = \left| \sum_{t=0}^{n-\tau-2} \Psi(g(\xi^t)) \right|$$

$$\leq \left(\frac{2}{\pi} \ln \frac{4(2^m - 1)}{\pi} + 1 \right) D_f \sqrt{2^m}.$$

Applying Lemma 3.1, we obtain

$$\sum_{j_1=0}^{2^l-1} \sum_{j_2=0}^{2^l-1} |\mu_{j_1}\mu_{j_2}| = \left(\sum_{j=0}^{2^l-1} |\mu_j| \right)^2 \leq \left(\frac{2l\ln(2)}{\pi} + 1 \right)^2.$$

Combining the two estimates the result follows.   □

Remarks: If we compare with the results on binary M-sequences of [9], we see that the PSL of our MSB sequences is also $O(\sqrt{N}\log N)$ with N the sequence period and the implied constant depending on D, l. It would be very interesting to lead an experimental study similar to that of [2] to see how tight this bound is.

7   Partial Period r-Pattern Distributions

Fix $k \geq 0$ and some $0 \leq H \leq 2^m - 1$ so that $k + H \leq 2^m$. For a positive integer r, fix $0 \leq \tau_1 < \ldots < \tau_r \leq 2^m - 1$ and let $v = (v_1, \ldots, v_r) \in \mathbb{Z}_2^r$. Then define $N(v)$ to be the number of integers $t \in [k, k + H - 1]$ such that

$$c_{t+\tau_i} = v_i, \ 1 \leq i \leq r.$$

If $u = (u_1, \ldots, u_r) \in \mathbb{Z}_2^r$ and $v = (v_1, \ldots, v_r) \in \mathbb{Z}_2^r$, let

$$\langle u \cdot v \rangle = \sum_{i=1}^{r} u_i v_i.$$

To study the distribution of r-patterns we need the following result.

Lemma 7.1. *With notation as above, and for any r integers $0 \leq \tau_1 < \tau_2 < \ldots < \tau_r \leq n - 1$, where $n = 2^m$, let*

$$\Theta(\tau_1, \ldots, \tau_r) = \sum_{t=k}^{k+H-1} (-1)^{c_{t+\tau_1} + c_{t+\tau_2} + \ldots + c_{t+\tau_r}},$$

where $c_t = \mathrm{MSB}(\mathrm{Tr}(f(\xi^t)))$, $f \in S_D$ with non-zero linear term. We then we have

$$|\Theta(\tau_1, \ldots, \tau_r)| \leq \left(\frac{2l}{\pi}\ln(2) + 1 \right)^r \left(\frac{2}{\pi}\ln\frac{4(2^m - 1)}{\pi} + 1 \right) D\sqrt{2^m}.$$

Proof. As we have $c_t = \text{MSB}(\text{Tr}(f(\xi^t)))$, and by (1), we obtain that $(-1)^{c_t}$ is equal to:

$$\mu(\text{Tr}(f(\xi^t))) = \sum_{j=0}^{2^l-1} \mu_j \psi_j(\text{Tr}(f(\xi^t))) = \sum_{j=0}^{2^l-1} \mu_j \Psi_j(f(\xi^t))).$$

Changing the order of summation, we obtain that:

$$\Theta(\tau_1,\ldots,\tau_r) = \sum_{j_1=0}^{2^l-1} \cdots \sum_{j_r=0}^{2^l-1} \mu_{j_1}\cdots\mu_{j_r} \sum_{t=k}^{k+H-1} \Psi(g(\xi^t)),$$

where

$$g(X) = j_1 f(X\xi^{\tau_1}) + j_2 f(X\xi^{\tau_2}) + \ldots + j_r f(X\xi^{\tau_r}).$$

By Lemma 4.3 $g(X) \neq 0$, and moreover $g(X) \in S_D$. Applying Lemma 2, we have:

$$\left| \sum_{j=k}^{k+H-1} \Psi(f(\xi^j)) \right| < D2^{m/2} \left(\frac{2}{\pi} \ln \frac{4(2^m - 1)}{\pi} + 1 \right).$$

Applying Corollary 7.4 of [5] (for $l \geq 4$), we have:

$$\sum_{j_1=0}^{2^l-1} \cdots \sum_{j_r=0}^{2^l-1} |\mu_{j_1}\cdots\mu_{j_r}| = \left(\sum_{j=0}^{2^l-1} |\mu_j| \right)^r \leq \left(\frac{2l}{\pi} \ln(2) + 1 \right)^r.$$

The Lemma follows.   □

The main result is the following estimate:

Theorem 7.2. *With notation as above, we have the bound:*

$$\left| N(v) - \frac{H}{2^r} \right| \leq \frac{1}{2^r} \left(\frac{2l}{\pi} \ln(2) + 1 \right)^r \left(\frac{2}{\pi} \ln \frac{4(2^m - 1)}{\pi} + 1 \right) D\sqrt{2^m}.$$

Proof. For any $t \in [k, k + H - 1]$, let $\mathbf{c}_t = (c_{t+\tau_1}, \ldots, c_{t+\tau_r}) \in \mathbb{Z}_2^r$. Let $u = (u_1, \ldots, u_r) \in \mathbb{Z}_2^r$. Then

$$\langle u \cdot \mathbf{c}_t \rangle = \sum_{i=1}^{r} u_i c_{t+\tau_i}$$

and by definition of $N(v)$ we have

$$S(u) = \sum_{t=k}^{k+H-1} (-1)^{\langle u \cdot \mathbf{c}_t \rangle} = \sum_{v \in \mathbb{Z}_2^r} (-1)^{\langle u \cdot v \rangle} N(v).$$

Thus we have

$$N(v) = \frac{1}{2^r} \sum_{u \in \mathbb{Z}_2^r} (-1)^{\langle u \cdot v \rangle} S(u).$$

Note that $S(0) = H$, and we obtain

$$N(v) = \frac{1}{2^r} \left(H + \sum_{0 \neq u \in \mathbb{Z}_2^r} (-1)^{\langle u \cdot v \rangle} S(u) \right).$$

It implies

$$\left| N(v) - \frac{H}{2^r} \right| = \frac{1}{2^r} \left| \sum_{0 \neq u \in \mathbb{Z}_2^r} (-1)^{\langle u \cdot v \rangle} S(u) \right| < \frac{1}{2^r} \times \max_{0 \neq u \in \mathbb{Z}_2^r} \{|S(u)|\}. \qquad (3)$$

For any non-zero vector $u = (u_1, \ldots, u_r) \in \mathbb{Z}_2^r$, we have

$$\langle u \cdot \mathbf{c}_t \rangle = \sum_{i=1}^{r} u_i c_{t+\tau_i} = \sum_{i \in I} c_{t+\tau_i},$$

where $I = \{i \in [1, r] : u_i = 1\}$. Let I be $\{i_1, \ldots, i_s\}$ then

$$S(u) = \Theta(\tau_{i_1}, \ldots, \tau_{i_s}),$$

where $s \leq r$, thus we can apply Lemma 7.1 to obtain the bound

$$\max_{0 \neq u \in \mathbb{Z}_2^r} \{|S(u)|\} \leq \left(\frac{2l}{\pi} \ln(2) + 1 \right)^r \left(\frac{2}{\pi} \ln \frac{4(2^m - 1)}{\pi} + 1 \right) D\sqrt{2^m}.$$

Substituting this estimate into (3) the Theorem follows. $\square$

References

1. Dai, Z.D.: Binary sequences derived from ML-sequences over rings I: period and minimal polynomial. J. Cryptology 5, 193–507 (1992)
2. Dmitriev, D., Jedwab, J.: Bounds on the growth rate of the peak sidelobe level of binary sequences. Advances in Mathematics of Communications 1, 461–475 (2007)
3. Jedwab, J.: A survey of the merit factor problem for binary sequences. In: Helleseth, T., Sarwate, D., Song, H.-Y., Yang, K. (eds.) SETA 2004. LNCS, vol. 3486, pp. 30–55. Springer, Heidelberg (2005)
4. Koponen, S., Lahtonen, J.: On the aperiodic and odd correlations of the binary Shanbhag-Kumar-Helleseth sequences. IEEE Trans. Inform. Theory IT-43, 1593–1596 (1997)
5. Lahtonen, J., Ling, S., Solé, P., Zinoviev, D.: $\mathbb{Z}_8$-Kerdock codes and pseudo-random binary sequences. Journal of Complexity 20(2-3), 318–330 (2004)
6. Schmidt, K.U.: PhD thesis, On spectrally bounded codes for multicarrier communications. Vogt Verlag, Dresden, Germany (2007),
http://www.ifn.et.tu-dresden.de/~schmidtk/#publications
7. Solé, P., Zinoviev, D.: The most significant bit of maximum length sequences over $\mathbb{Z}_{2^l}$: autocorrelation and imbalance. IEEE Transactions on Information Theory IT-50, 1844–1846 (2004)

8. Solé, P., Zinoviev, D.: Low correlation, high nonlinearity sequences for multi-code CDMA. IEEE Transactions on Information Theory IT-52, 5158–5163 (2006)

9. Sarwate, D.V.: Upper bound on the aperiodic autocorrelation function for a maximal-length sequence. IEEE Trans. Inform. Theory IT-30, 685–687 (1984)

10. Hu, H., Feng, D., Wu, W.: Incomplete exponential sums over Galois rings with applications to some binary sequences derived from $\mathbb{Z}_{2^l}$. IEEE Trans. Inform. Theory, IT-52, 2260–2265 (2006)

On Partial Correlations of Various Z_4 Sequence Families

Parampalli Udaya[1] and Serdar Boztaş[2]

[1] Department of Computer Science and Software Engineering
University of Melbourne, Melbourne 3010, Australia
udaya@csse.unimelb.edu.au
[2] School of Mathematical and Geospatial Sciences, RMIT University,
Melbourne 3001, Australia
serdar.boztas@ems.rmit.edu.au

Abstract. Galois ring $m-$sequences were introduced in the late 1980s and early 1990s, and have near-optimal full periodic correlations. They are related to Z_4-linear codes, and are used in CDMA communications. We consider periodic correlation and obtain algebraic expressions of the first two partial period correlation moments of the sequences belonging to families A, B and C. These correlation moments have applications in synchronisation performance of CDMA systems using Galois ring sequences. The use of Association Schemes provides us with a new uniform technique for analyzing the sequence families A, B and C.

Keywords: CDMA, spread spectrum sequence design, Z_4-sequences, autocorrelation, crosscorrelation, partial period correlation, Galois rings, coding theory, association schemes.

1   Introduction and Background

The aim of *Code Division Multiple Access* (CDMA) in wireless networks is to enable wireless transmitters to successfully exchange information in the presence of potential conflicts which lead to interference. There are two main flavours of CDMA, *Frequency Hopping* (FH) and *Direct Sequence* (DS). For details of CDMA networks, we refer the interested reader to the recent survey in the *Spread Spectrum Communications Handbook* by Simon et al. [8].

In this paper, we shall be concerned with the so-called "spreading codes" in DS-CDMA, and specifically their performance in synchronisation, when it is convenient to use partial period correlations to acquire the correct phase of the chip sequence which is used for spreading the transmitted signal. For a detailed survey on pseudorandom sequence design, please see the chapter by Helleseth and Kumar in the *Handbook of Coding Theory* [4]. Very briefly, a CDMA communication system with phase-shift keying (PSK) modulation assigns unique-phase code sequences to each transmitter-receiver pair. The traditional design methods for sequence families relied on Galois field theory. More recently, Galois rings have been used (by Solé, Boztaş, Hammons, Kumar, Udaya and Siddiqi) to design CDMA sequence families, both for DS-CDMA [1,5,6] and FH-CDMA [10].

S.W. Golomb et al. (Eds.): SETA 2008, LNCS 5203, pp. 332–344, 2008.

Here, we restrict our attention to DS-CDMA. It is algebraically convenient to design sequence families with good periodic correlation properties and there are benchmarks to measure how good such a design is, namely the Welch [12] and Sidelnikov bounds [7]. The aperiodic correlation properties also play a significant part in system performance for the case of Galois ring sequences, the aperiodic correlation was investigated in [11]. Another significant contributor to the performance, especially in the current wireless environment where longer and longer sequence periods are necessary to support an increasing number of users (the family size is typically an increasing function of the period), is the partial period correlation, which is the main focus of this paper.

In this paper we have obtained new results on the partial period correlations of families A, B and C. This substantially extends the results we have obtained in [2].

The paper is organised as follows. In Section 2, we provide a brief overview of the structure of Galois rings and properties of the Galois ring trace function, after introducing some definitions and notation for general sequence designs. This is followed by the definition of *Families A, B and C*. This section concludes with a discussion of a related Cayley table and its properties. In Section 3 we obtain the first moment of the partial correlation function of the Galois ring $m-$sequences in families A, B, and C. In Section 4 we obtain so-called local and global second moments of the partial correlation function of *Family A*. Section 5 concludes the paper.

2   Rings, Trace Functions and Sequences

2.1   Galois Ring Preliminaries

We will be quite informal, highlighting the details we need, and ask the reader unfamiliar with the topic to consult [1].

We denote the Galois ring as $R = GR(4, n)$ and note that it is a Galois extension of Z_4, defined by $R = Z_4[\beta]$ where β has multiplicative order $2^n - 1$ and is a root of a primitive basic irreducible polynomial (i.e., a basic irreducible polynomial whose modulo 2 reduction is a primitive polynomial over Z_2). It is always possible to construct such a polynomial. Note that the ring R contains 4^n elements, and $R = \langle 1, \beta, \ldots, \beta^{n-1} \rangle$ as a Z_4-module.

Every element $c \in R$ has a unique 2-adic representation $c = a + 2b$, where a, b belong to the Teichmuller set

$$\mathcal{T} = \{0, 1, \beta, \ldots, \beta^{2^n - 2}\}$$

and the map $\alpha : c \mapsto a$ is given by $\alpha(c) = c^{2^n}$. Given c, after determining a, as above, b can then be solved for. If we denote the modulo 2 reduction function by μ and extend it to polynomials in the obvious way, then $\mu(\mathcal{T}) = \{0, 1, \theta, \ldots, \theta^{2^n - 2}\} = GF(2^n)$. The set of invertible elements of R is denoted $R^* = R \setminus 2R$ where $2R$ is the set of zero divisors and is the unique maximal ideal in R. Every element in R^* has a unique representation of the form $\beta^r(1 + 2z), 0 \leq$

$r \leq 2^n - 2, t \in T$. Also, R^* is a multiplicative group of order $2^n(2^n - 1)$ which is a direct product $G_1 \times \mathcal{E}$ where G_1 is a cyclic group of order $2^n - 1$ (made up of the nonzero elements in the Teichmuller set) generated by β and $\mathcal{E}$ is made up of elements of the form $1 + 2t$ where $t \in T$.

The *Frobenius map* from R to R is the ring automorphism that takes any element $c = a + 2b$ in the 2-adic representation to the element $c^f = a^2 + 2b^2$ and it generates the Galois group of R over Z_4 with f^m the identity map. The *Trace map* from R to Z_4 is defined by

$$T(c) = c + c^f + c^{f^2} + \cdots + c^{f^{m-1}}, \qquad c \in R.$$

The trace is onto and has nice equidistribution properties. It will play a role in the moment calculations we shall use later in the paper. If we let $f_2(c) = c^2$ be the squaring map defined on the finite field $GF(2^m)$ then the finite field trace is given by

$$tr(c) = c + c^2 + c^{2^2} + \cdots + c^{2^{m-1}}, \qquad c \in GF(2^m),$$

and the following commutativity relationships hold:

$$\mu \circ f = f_2 \circ \mu, \qquad \mu \circ T = tr \circ \mu.$$

Let $G_C = T \setminus \{0\}$. Consider the following partition $\mathcal{X}$ of R defined by the equivalence relation $\alpha \cong \beta$ if and only if $\alpha G_C = \beta G_C$. The partition $\mathcal{X}$ consists of the following subsets which partition R:

1. 2^n subsets corresponding to each $a \in G_A$ of $R^* : [a] = a(G_C)$.
2. A subset consisting of proper zero divisors: $[e] = \langle 2 \rangle \setminus \{0\}$.
3. The zero subset: $[\infty] = \{0\}$.

For $a, b, c \in \mathcal{X}$, define $N(a, b; c)$ to be the number of times a fixed element of the class $[c]$ occurs in the Cayley table of $[a] + [b]$. This number is independent of the element of $[c]$ that is chosen, since in $([a] + [b])$, the occurrence of any element of $[c]$ implies the occurrence of all the elements of $[c]$. The commutative property of R implies $N(a, b; c) = N(b, a; c)$. Various structural constants $N(a, b; c)$, a, b, $c \in \mathcal{X}$, are computed in [10], and they are reproduced in the following lemma.

Lemma 1.   *1.* $n(\infty, w; x) = \begin{cases} 0 & \text{if } w \neq x \\ 1 & \text{if } w = x. \end{cases}$

2. $N(e, e; x) = \begin{cases} 0 & \text{if } x \neq e, \infty \\ 2^n - 1 & \text{if } x = \infty \\ 2^n - 2 & \text{if } x = e. \end{cases}$

3. $N(e, a; x) = \begin{cases} 0 & \text{if } x = a, e, \text{ or } \infty \text{ for any } a \in G_A \\ 1 & \text{otherwise.} \end{cases}$

4. $N(a, b; \infty) = \begin{cases} 2^n - 1 & \text{if } b = 3a \text{ for any } a, b \in G_A \\ 0 & \text{otherwise.} \end{cases}$

5. $N(a, b; e) = \begin{cases} 0 & \text{if } b = 3a \text{ for any } a, b \in G_A \\ 1 & \text{otherwise.} \end{cases}$

6. $n(0, 0; 0) = 0$.

7. If $a, b, c, d \in G_A$, then $N(a, b; c) = N(ad, bd; cd)$.

8. Let $a, b \in G_A$. Then $N(a, 3a; b) = \begin{cases} 0 & \text{if } b=a,3a \\ 1 & \text{otherwise.} \end{cases}$

9. Let $a, b \in G_A$, $a \neq b$. Then $N(a, a; b) = \begin{cases} 2 & \text{if } tr(\tilde{b})=tr(\tilde{a}) \\ 0 & \text{otherwise.} \end{cases}$

10. Let $a, b, c \in G_A$, $a \neq b, 3b$. Then

$$N(a, b; c) = \begin{cases} 1 & \text{if } c = a, b \\ 2 & \text{if } c \neq a, b, tr(\tilde{a}\tilde{b} + \tilde{a}\tilde{c} + \tilde{b}\tilde{c}) = tr(\tilde{c}) \\ 0 & \text{otherwise.} \end{cases}$$

We need the following definition in the next section.

Definition 1. Let $\gamma = (1 + 2a) \in \{1 + 2\beta^k, \; k = \infty, 0, \cdots, 2^n - 2\}$. Then the *trace number* of γ is defined as the value of $tr(\mu(a))$. The trace number is always 1 or 0.

2.2 Sequence Families A, B and C

A $q-$ary sequence family made up of M cyclically distinct sequences of length N is defined to be the collection of vectors

$$\{\mathbf{s}_1, \ldots, \mathbf{s}_M\}, \qquad \mathbf{s}_i \in Z_q^N, \; 1 \leq i \leq M$$

with $\mathbf{s}_1 = (s_1(0), \ldots, s_1(N-1))$, where Z_q is the ring of integers modulo q. Here we restrict ourselves to quaternary sequences, i.e., $q = 4$. The (periodic) correlation function between sequences i and j at relative shift τ is defined as

$$C_{i,j}(\tau) = \sum_{t=0}^{N-1} \omega^{s_i(t \oplus \tau) - s_j(t)}$$

where $\omega = exp(2\pi j/4) = \sqrt{-1}$ is a primitive fourth root of unity and where $\oplus$ denotes addition modulo N.

Given a sequence family such as above, the Welch [12] and Sidelnikov [7] lower bounds determine how good such a family can be. For example, if $M \approx N$ then the maximum nontrivial correlation magnitude (sometimes called the maximum sidelobe)

$$C_{max} = \max\{|C_{i,j}(\tau)| : i \neq j \text{ or } \tau \neq 0\}$$

obeys $C_{max} \geq \sqrt{2N}$ for binary sequences and $C_{max} \geq \sqrt{N}$ for nonbinary sequences. *Family A* [1,6] is a large sequence family which delivers the promised improvement for C_{max} for the practically significant (due to the widespread use of quaternary PSK modulation) $q = 4$ value.

Family A comprises a set of $M = 2^n + 1$ cyclically distinct sequences over Z_4 with length $N = 2^n - 1$, which obey a common linear recurrence whose characteristic polynomial is a primitive basic irreducible polynomial of degree n over the ring $Z_4[x]$. Each element $\mathbf{s}_i$ of *Family A* can be expressed as $s_i(t) = T(\gamma\beta^t)$ where β is a generator of the Teichmuller set, and $\gamma \neq 0$. In fact the enumeration of representatives

$$\Gamma_\nu = \{2\} \cup G_A,$$

where $G_A = \{1 + 2\beta^k, \ k = \infty, 0, \cdots, 2^n - 2\}$, can be used to enumerate the cyclically distinct elements in *Family A*, since each member γ_i, $1 \leq i \leq 2^n + 1$ of Γ_ν gives a distinct sequence in the family if we take $s_i(t) = T(\gamma_i \beta^t)$.

Each sequence in *Family A* corresponds to a class in $\mathcal{X}$.

We conclude this section by stating the complete full period correlation distribution for *Family A*, which is obtained by considering the distribution of values taken by sums of the form

$$S(\gamma) = \sum_{x \in G_1} \omega^{T(\gamma x)} = \sum_{t=0}^{2^n - 2} \omega^{T(\gamma \beta^t)},$$

as γ ranges over the ring R, where we count the solutions of $\gamma = \gamma_i \beta^\tau - \gamma_j$.

Families B and C: If sequences generated as trace of powers of $(\gamma \beta)$, $\gamma \in G_A$ and $\gamma \neq 1$, the resulting sequences are of period $2(2^n - 1)$[1,10]. Families of interleaved m-sequences comprises of $2^{n-1} + 1$ sequences which obey a common linear recurrence relation over Z_4 determined by the minimal polynomial corresponding to $(\gamma \beta)$, where $\gamma \in G_A$ and $\gamma \neq 1$. An interleaved sequence is^a can be expressed as $is^a(t) = T(a(\gamma \beta)^t)$ where β is a generator of the Teichmuller set, and $a \neq 0$. We call interleaved family as *Family B* when $\gamma \neq 3$ trace number of γ is 1. *Family C* is obtained when $\gamma = 3$. It can be noticed that each interleaved sequence can be seen as interleaved version of two *Family A* sequences [10]. Using this fact, sequences in *Family B* are enumerated with the following representatives:

$$\{is^a, a \in \ Quotient \ group \ G_A/\{1, \gamma\}\}.$$

And similarly sequences in *Family C* are enumerated with the following representatives:

$$\{is^a, a \in \ Quotient \ group \ G_A/\{1, 3\}\}.$$

In [10], we used an association scheme over R to study these sequence families properties. Here we only use related Cayley table defined on R.

Theorem 2 ([1,10]). *The correlation sum and weight distributions of sequences in Family A of period $2^n - 1$ are given in Table 1. These sequences are grouped under five subsets $\mathcal{P}, \mathcal{Q}, \mathcal{R}, \mathcal{S}$ and $\mathcal{B}$. The trace numbers of sequences within any subset are same. For the first subset $\mathcal{B}$ (binary), $w_2 = 2^{r-1}$, and $w_3 = 0$. For remaining subsets, $w_3 = 2^{r-1} - w_1$ and $w_2 = 2^{r-1} - 1 - w_0$.*

The following theorem describes correlation sum and weight distribution of sequences in *Family B* and *Family C*. It also describes the internal composition of *Family A* sequences.

Theorem 3. *The correlation sum and weight distributions of the families of B and C of period $2(2^r - 1)$ are given in Tables 2 and 3. Like before, the sequences are grouped based on distinct correlation values and named with a ($\tilde{\ }$) to distinguish from sequences in Family A which have half the period. In all these tables the subset $\tilde{\mathcal{B}}$ corresponds to is^2 and for all the items except the last, $w_2 = (2^n - 2) - w_0$, $w_3 = 2^n - w_1$; for the last item (subset $\tilde{\mathcal{B}}$), $w_3 = 0$, $w_2 = 2^n$.*

Table 1. Correlation Sum and weight Distributions of m-sequences of period $2^r - 1$

Subset	$\aleph$	No. of Sequences	Trace Number	w_0	w_1
(a) $n = 2t + 1$ (an odd integer)					
$\mathcal{B}$	-1	1	0	$2^{r-1} - 1$	0
$\mathcal{P}$	$2^t - 1 + \omega\, 2^t$	$2^{t-1}(2^t + 1)$	ξ	$2^{n-2} + 2^{t-1} - 1$	$2^{n-2} + 2^{t-1}$
$\mathcal{Q}$	$-2^t - 1 - \omega\, 2^t$	$2^{t-1}(2^t - 1)$	ξ	$2^{n-2} - 2^{t-1} - 1$	$2^{n-2} - 2^{t-1}$
$\mathcal{R}$	$2^t - 1 - \omega\, 2^t$	$2^{t-1}(2^t + 1)$	$\bar\xi$	$2^{n-2} + 2^{t-1} - 1$	$2^{n-2} - 2^{t-1}$
$\mathcal{S}$	$-2^t - 1 + \omega\, 2^t$	$2^{t-1}(2^t - 1)$	$\bar\xi$	$2^{n-2} - 2^{t-1} - 1$	$2^{n-2} + 2^{t-1}$
(b) $n = 2t$ (an even integer)					
$\mathcal{B}$	-1	1	0	$2^{n-1} - 1$	0
$\mathcal{P}$	$2^t - 1$	$2^{t-1}(2^{t-1} + 1)$	ξ	$2^{n-2} + 2^{t-1} - 1$	2^{n-2}
$\mathcal{Q}$	$-2^t - 1$	$2^{t-1}(2^{t-1} - 1)$	ξ	$2^{n-2} - 2^{t-1} - 1$	2^{n-2}
$\mathcal{R}$	$-1 + \omega\, 2^t$	2^{n-2}	$\bar\xi$	$2^{n-2} - 1$	$2^{n-2} + 2^{t-1}$
$\mathcal{S}$	$-1 - \omega\, 2^t$	2^{n-2}	$\bar\xi$	$2^{n-2} - 1$	$2^{n-2} - 2^{t-1}$

Table 2. Correlation Sum and Weight Distributions of *Family C* $((\gamma = 3))$

Subset	$\aleph$	No. of Sequences	Constituent class	w_0	w_1
(a) $n = 2t + 1$ (an odd integer) ; Period $= 2(2^n - 1)$					
$\bar{\mathcal{P}}$	$2(2^t - 1)$	$2^{t-1}(2^t + 1)$	$\eta \in \mathcal{P}; \eta\gamma \in \mathcal{R}$	$2^{2t} + 2^t - 2$	2^{2t}
$\bar{\mathcal{Q}}$	$-2(2^t + 1)$	$2^{t-1}(2^t - 1)$	$\eta \in \mathcal{Q}; \eta\gamma \in \mathcal{S}$	$2^{2t} - 2^t - 2$	2^{2t}
$\bar{\mathcal{B}}$	-2	1	$\eta \in <2>$	$2^n - 2$	0
(b) $n = 2t$ (an even integer); Period $= 2(2^n - 1)$					
$\bar{\mathcal{P}}$	$2(2^t - 1)$	$2^{t-2}(2^{t-1} + 1)$	$\eta \in \mathcal{P}; \eta\gamma \in \mathcal{P}$	$2^{n-1} + 2^t - 2$	2^{n-1}
$\bar{\mathcal{Q}}$	$-2(2^t + 1)$	$2^{t-2}(2^{t-1} - 1)$	$\eta \in \mathcal{Q}; \eta\gamma \in \mathcal{Q}$	$2^{n-1} - 2^t - 2$	2^{n-1}
$\bar{\mathcal{R}}$	-2	2^{2t-2}	$\eta \in \mathcal{R}; \eta\gamma \in \mathcal{S}$	$2^{n-1} - 2$	2^{n-1}
$\bar{\mathcal{B}}$	-2	1	$\eta \in <2>$	$2^n - 2$	0

3   The Partial Correlation and Its First Moment

Let $\mathbf{s}_i = (s_i(0), \ldots, s_i(2^n - 1))$ be a sequence from *Family A*, thus $s_i(t) = T(\gamma_i \beta^t)$ for $0 \le t \le 2^n - 2$, and where $\gamma_i \in \Gamma_\nu$ for $i = 1, \ldots, 2^n + 1$. Hence $N = 2^n - 1$ and $M = 2^n + 1$ here.

Definition 2. *The periodic partial correlation function of* $\mathbf{s}_i$ *with* $\mathbf{s}_j$ *at shift* τ *and offset* k *with correlation length* $L \le 2^n - 1$ *is given by*

$$P_{i,j}(\tau, k, L) = \sum_{t=k}^{k+L-1} \omega^{s_i(t \oplus \tau) - s_j(t)}, \qquad 1 \le i \le j \le 2^n + 1,$$

where $\oplus$ *denotes addition modulo* $2^n - 1$, *and* $0 \le \tau \le 2^n - 2$.

Table 3. Correlation Sum and Weight Distributions of *Family B* (Trace Number of $\gamma = 1$)

Subset	$\aleph$	No. of Sequences	Constituent class	w_0	w_1
\multicolumn{6}{c}{(a) $n = 2t + 1$ (an odd integer) ; Period $= 2(2^n - 1)$}					
$\bar{P}$	$2(2^t - 1)$	$2^{t-1}(2^{t-1} + 1)$	$\eta \in \mathcal{P}; \eta\gamma \in \mathcal{R}$	$2^{2t} + 2^t - 2$	2^{2t}
$\bar{Q}$	$-2(2^t + 1)$	$2^{t-1}(2^{t-1} - 1)$	$\eta \in \mathcal{Q}; \eta\gamma \in \mathcal{S}$	$2^{2t} - 2^t - 2$	2^{2t}
$\bar{R}$	$-2 + \omega\, 2^{t+1}$	2^{2t-2}	$\eta \in \mathcal{P}; \eta\gamma \in \mathcal{S}$	$2^{2t} - 2$	$2^{2t} + 2^t$
$\bar{S}$	$-2 - \omega\, 2^{t+1}$	2^{2t-2}	$\eta \in \mathcal{Q}; \eta\gamma \in \mathcal{R}$	$2^{2t} - 2$	$2^{2t} - 2^t$
$\bar{B}$	-2	1	$\eta \in< 2 >$	$2^n - 2$	0
\multicolumn{6}{c}{(b) $n = 2t$ (an even integer); Period $= 2(2^n - 1)$}					
$\bar{P}$	$(2^t - 2 + \omega\, 2^t)$	$2^{t-2}(2^{t-1} + 1)$	$\eta \in \mathcal{P}; \eta\gamma \in \mathcal{R}$	$2^{n-1} + 2^{t-1} - 2$	$2^{n-1} + 2^{t-1}$
$\bar{Q}$	$(-2^t - 2 - \omega\, 2^t)$	$2^{t-2}(2^{t-1} - 1)$	$\eta \in \mathcal{Q}; \eta\gamma \in \mathcal{S}$	$2^{n-1} - 2^{t-1} - 2$	$2^{n-1} - 2^{t-1}$
$\bar{R}$	$(2^t - 2 - \omega\, 2^t)$	$2^{t-2}(2^{t-1} + 1)$	$\eta \in \mathcal{P}; \eta\gamma \in \mathcal{S}$	$2^{n-1} + 2^{t-1} - 2$	$2^{n-1} - 2^{t-1}$
$\bar{S}$	$(-2^t - 2 + \omega\, 2^t)$	$2^{t-2}(2^{t-1} - 1)$	$\eta \in \mathcal{Q}; \eta\gamma \in \mathcal{R}$	$2^{n-1} - 2^{t-1} - 2$	$2^{n-1} + 2^{t-1}$
$\bar{B}$	-2	1	$\eta \in< 2 >$	$2^n - 2$	0

Note that the non-trivial (off-peak) values of this function are those for which either $i \neq j$ or $\tau \neq 0$. We recover the usual full period correlation if $L = 2^n - 1$.

Definition 3. *The first moment of the partial correlation function in Definition 1 is given by*

$$\langle P_{i,j}(\tau, k, L)\rangle_k = \frac{1}{2^n - 1} \sum_{k=0}^{2^n - 2} P_{i,j}(\tau, k, L) \triangleq \overline{P_{i,j}(\tau, L)}$$

while its second absolute moment is given by

$$\left\langle |\, P_{i,j}(\tau, k, L)\,|^2 \right\rangle_k = \frac{1}{2^n - 1} \sum_{k=0}^{2^n - 2} |P_{i,j}(\tau, k, L)|^2 .$$

We remark that the correlations are, in general, complex valued. It is quite straightforward to obtain the first moment of the partial periodic correlation (for all possible i, j, τ). In fact this applies to the partial periodic correlation of *any two complex valued sequences provided they have the same length*.

Theorem 4. *The first moment obeys*

$$\langle P_{i,j}(\tau, k, L)\rangle_k = \frac{L}{2^n - 1} C_{i,j}(\tau),$$

and therefore, for Family A, it simply takes on values proportional to the values in Theorem 3 with the same multiplicities.

Proof. We have

$$\overline{P_{i,j}(\tau, L)} = \frac{1}{2^n - 1} \sum_{k=0}^{2^n - 2} P_{i,j}(\tau, k, L)$$

$$= \frac{1}{2^n - 1} \sum_{k=0}^{2^n - 2} \sum_{t=0}^{L-1} \omega^{s_i(k \oplus t \oplus \tau) - s_j(k \oplus t)}$$

$$= \frac{1}{2^n - 1} \sum_{t=0}^{L-1} \sum_{k=0}^{2^n - 2} \omega^{s_i(k \oplus t \oplus \tau) - s_j(k \oplus t)}$$

$$= \frac{1}{2^n - 1} \sum_{t=0}^{L-1} C_{i,j}(\tau) = \frac{L}{2^n - 1} C_{i,j}(\tau),$$

where after interchanging the summations the resulting inner sum is clearly a full period correlation sum which is independent of t. □

It is of interest in applications to consider only the nontrivial periodic auto-correlation function, i.e., $i = j, \tau \neq 0 \pmod{2^n - 1}$, for estimating the false self-synchronisation probability. Before addressing this, we need a definition.

Definition 4. *We define the Standard Normalized Correlation Distribution for a quantity $\theta(\tau)$ as:*

1. *If $n = 2t + 1$, then*

$$\theta(\tau) = \begin{cases} 2^n - 1, & 1 \ time, \\ -1 + 2^t + \omega 2^t, \ 2^{n-2} + 2^{t-1} & times, \\ -1 + 2^t - \omega 2^t, \ 2^{n-2} + 2^{t-1} & times, \\ -1 - 2^t + \omega 2^t, \ 2^{n-2} - 2^{t-1} & times, \\ -1 - 2^t - \omega 2^t, \ 2^{n-2} - 2^{t-1} & times. \end{cases}$$

2. *If $n = 2t$, then*

$$\theta(\tau) = \begin{cases} 2^n - 1, & 1 \ time, \\ -1 + 2^t, & 2^{n-2} + 2^{t-1} & times, \\ -1 - 2^t, & 2^{n-2} - 2^{t-1} & times, \\ -1 + \omega 2^t, \ 2^{n-2} & times, \\ -1 - \omega 2^t, \ 2^{n-2} & times. \end{cases}$$

This definition is used in the proof below. The result follows from arguments along the lines of [1] but the autocorrelation distribution was never computed there; in that paper, the focus was on the global correlation distribution.

Lemma 5. *For Family A the full period autocorrelation function obeys:*

1. *If we consider the zero divisor sequence (binary m-sequence), then*

$$C_{i,i}(\tau) = \begin{cases} 2^n - 1, 1 \ time, \\ -1, \quad 2^n - 2 \ times. \end{cases}$$

2. *Otherwise, the autocorrelation distribution obeys the Standard Normalized Correlation Distribution from Definition 3, except that the value $S(\gamma_i)$ and its complex conjugate occur with frequency one less than that specified in Definition 3.*

Proof. For the binary m-sequence corresponding to a coset leader chosen from the maximal ideal $2R$ the autocorrelation takes on the value -1 for $\tau \neq 0$, and $2^n - 1$ otherwise. For the rest of the proof, we restrict ourselves to the sequences which are not all zero divisors. Let $n = 2t + 1$ and consider correlation between S^a and τ^{th} shift of itself. Because of the linearity, this value is correlation sum of some S^b, where $b \in G_A$. These b's are exactly those in Cayley table of $N(a, 3a; b)$. From Lemma 1, $N(a, 3a; b)$ takes all values in G_A except a and $3a$. The correlation sum of S^a and S^{3a} are conjugate of each other. This proves the result. $\qquad\square$

The following follows immediately.

Theorem 6. *For Family A the first moment of the autocorrelation function obeys*

$$\langle P_{i,i}(\tau, k, L)\rangle_k = \frac{L}{2^n - 1}C_{i,i}(\tau),$$

and therefore it simply takes on values proportional to the values in Lemma 3 with the same multiplicities.

Proof. The proof is similar to that of Theorem 4. $\qquad\square$

Lemma 7. *Let n be an odd number, then for Family C the full period autocorrelation function obeys:*

1. *If we consider the zero divisor sequence (binary m-sequence), then*

$$\theta(\tau) = \begin{cases} 2(2^n - 1), & 2 \text{ times,} \\ -2, & 2^{n+1} - 4 \text{ times.} \end{cases}$$

2. *Otherwise, if $is^a \in \bar{P}$, then*

$$\theta(\tau) = \begin{cases} 2(2^n - 1), & 1 \text{ time,} \\ -2, & 1 \text{ time,} \\ 2(2^t - 1), & 2^{t+1}(2^t + 1) - 4 \text{ times,} \\ -2(2^t + 1), & 2^{t+1}(2^t - 1) \text{ times,} \end{cases}$$

3. *Otherwise, if $is^a \in \bar{Q}$, then*

$$\theta(\tau) = \begin{cases} 2(2^n - 1), & 1 \text{ time,} \\ -2, & 1 \text{ time,} \\ 2(2^t - 1), & 2^{t+1}(2^t + 1) \text{ times,} \\ -2(2^t + 1), & 2^{t+1}(2^t - 1) - 4 \text{ times,} \end{cases}$$

Proof. From Lemma 9 of [10], the autocorrelation of is^a at τ^{th} shift is given by $S(\eta) + S(\gamma\eta)$, where $\eta = (a - a(\gamma\beta)^\tau)$. When $\tau = 2^n - 1$, η is multiple of 2 and hence -2 will occur only once in the distribution. The case of $\tau = 0$ leads to trivial correlation of $2(2^n - 1)$. For the rest of the proof we assume $is^a \in \bar{\mathcal{P}}$. For even values of τ, $(\gamma)^\tau = 1$. This leads to correlations of $S(b) + S(\gamma b)$, where b is the Cayley table of $N(a, 3a, b)$. These have been computed in Lemma 1 and the distribution of $S(\eta) + S(\gamma\eta)$ have been computed in Table 2. Here the correlations take values from G_A except a and $3a$. For odd values of τ, $(\gamma)^\tau = \gamma$. This results in correlation sums of c's in $N(a, 3a\gamma, b)$. In this case, $\gamma = 3$. Then this corresponds to values in $N(a, a, b)$. Combining the two cases gives the result. The case for $is^a \in \bar{\mathcal{Q}}$ can be similarly proved.   □

Lemma 8. *Let n be an odd number, then for Family B the full period autocorrelation function obeys:*

1. If we consider the zero divisor sequence (binary m-sequence), then

$$\theta(\tau) = \begin{cases} 2(2^n - 1), & 2 \ times, \\ -2, & 2^{n+1} - 3 \ times. \end{cases}$$

2. Otherwise, if $is^a \in \bar{\mathcal{P}}$, then

$$\theta(\tau) = \begin{cases} 2(2^n - 1), & 1 \ time, \\ -2, & 1 \ time, \\ 2(2^t - 1), & 2^{t+1}(2^{t-1} + 1) - 4 \ times, \\ -2(2^t + 1), & 2^{t+1}(2^{t-1} - 1) \ times, \\ -2 + \omega \ 2^{t+1}, 2^{2t} \ times, \\ -2 - \omega \ 2^{t+1}, 2^{2t} \ times, \end{cases}$$

3. Otherwise, if $is^a \in \bar{\mathcal{Q}}$, then

$$\theta(\tau) = \begin{cases} 2(2^n - 1), & 1 \ time, \\ -2, & 1 \ time, \\ 2(2^t - 1), & 2^{t+1}(2^{t-1} + 1) \ times, \\ -2(2^t + 1), & 2^{t+1}(2^{t-1} - 1) - 4 \ times, \\ -2 + \omega \ 2^{t+1}, 2^{2t} \ times, \\ -2 - \omega \ 2^{t+1}, 2^{2t} \ times. \end{cases}$$

4. Otherwise, if $is^a \in \bar{\mathcal{R}}$ or $\bar{\mathcal{S}}$, then

$$\theta(\tau) = \begin{cases} 2(2^n - 1), & 1 \ time, \\ -2, & 1 \ time, \\ 2(2^t - 1), & 2^{t+1}(2^{t-1} + 1) \ times, \\ -2(2^t + 1), & 2^{t+1}(2^{t-1} - 1) \ times, \\ -2 + \omega \ 2^{t+1}, 2^{2t} - 2 \ times, \\ -2 - \omega \ 2^{t+1}, 2^{2t} - 2 \ times, \end{cases}$$

Proof. The proof follows as in Lemma 7 except here the results of $N(a, 3a, b)$ and $N(a, 3\gamma a, c)$ of Lemma 1 need to be used.   □

Remark: Result similar to Theorem 6 holds good for *Family B* and *Family C*.

4   The Second Moment of the Partial Correlation Function

In this section we proceed in stages. First we observe the (well known) second moment of the partial autocorrelation function of a binary m-sequence (i.e., the zero divisor sequence in *Family A*) which follows from the "shift-and-add" property of binary m-sequences.

Proposition 9. *For the binary m-sequence s_1, we have*

$$\langle | P_{1,1}(\tau,k,L) |^2 \rangle_k = L \left(1 - \frac{L-1}{2^n - 1} \right)$$

when $\tau \neq 0 \pmod{2^n - 1}$.

We next proceed to the second moment of the partial period autocorrelation function. We first state an intermediate result, which holds for both odd and even n:

Lemma 10. *Consider the sum defined as*

$$S(\gamma) = \sum_{x \in G_1} \omega^{T(\gamma x)}.$$

Then, for all ν in R^* we have

$$\sum_{\gamma_i, \gamma_j \in \Gamma_\nu} \sum_{\tau=0}^{2^n - 2} \mathcal{R}e \left\{ S(\gamma_i \beta^\tau - \gamma_j) \right\} = 1.$$

Proof. We directly apply the distributions from Table 1.   □

We are now ready to prove our "local" second moment for partial period correlations in *Family A*. We perform an additional averaging along τ since the dependence of the second moment on τ is quite complicated and seems to be intractable analytically.

Definition 5. *We define the local second partial correlation moment for Family A **with respect to sequence** s_i as:*

$$\left\langle | P(L)^{(i)} |^2 \right\rangle = \frac{1}{(2^n - 1)^2 (2^n + 1)} \sum_{k,\tau=0}^{2^n - 2} \sum_{\gamma_j \in \Gamma_\nu} | P_{i,j}(\tau,k,L) |^2 .$$

Theorem 11. *Let n be odd. The local second partial correlation moment for Family A is given by*

$$\left\langle | P(L)^{(i)} |^2 \right\rangle = L \pm \frac{L(L-1) 2^{(n-1)/2}}{(2^n - 1)^2 (2^n + 1)}.$$

where if $\gamma_i \in \mathcal{Q} \cup \mathcal{S}$ the second term is positive, and if $\gamma_i \in \mathcal{P} \cup \mathcal{R}$ the second term is negative.

Proof. We evaluate the unnormalized sum in the above definition, where for convenience we denote the coset representatives by $\gamma_0, \ldots, \gamma_{2^n}$, and drop summation limits whenever convenient. We have:

$$\sum_{k,\tau=0}^{2^n-2} \sum_{j=0}^{2^n} P_{i,j}(\tau, k, L)\,[P_{i,j}(\tau, k, L)]^*$$

which can be rewritten as

$$\sum_{t,\tau=0}^{2^n-2} \sum_{j=0}^{2^n} \sum_{k,l=0}^{L-1} \omega^{s_i(k\oplus t\oplus \tau)-s_j(k\oplus t)-s_i(l\oplus t\oplus \tau)+s_j(l\oplus t)} =$$

$$= \sum_{j=0}^{2^n} \sum_{\tau=0}^{2^n-2} \sum_{k,l=0}^{L-1} \sum_{t=0}^{2^n-2} \omega^{T[(\beta^k-\beta^l)(\beta^\tau\gamma_i-\gamma_j)\beta^t]}.$$

We now separate the case $k = l$, and note that we can use complex conjugate symmetry of the $(\beta^k - \beta^l)$ terms to rewrite the sum as

$$L(2^n-1)\left(\sum_{j=0}^{2^n}\sum_{\tau=0}^{2^n-2}1\right) +$$

$$+ \sum_{0\leq k\neq l\leq L-1}\sum_{j,\tau}\mathcal{R}e\left\{S((\beta^k-\beta^l)(\beta^\tau\gamma_i-\gamma_j))\right\}$$

$$= L(2^n-1)^2(2^n+1) + 2\sum_{0\leq l<k\leq L-1}\sum_{j,\tau}\mathcal{R}e\left\{S(\beta^\tau\gamma_i-\gamma_j)\right\}$$

where the argument of the sum $S(\cdot)$ simplifies since it is invariant under multiplication by a nonzero unit. The sum on the right hand side can be evaluated by considering the equation $\gamma = \beta^\tau\gamma_i - \gamma_j$, and asking how many solutions this equation has for a fixed γ_i. By following the argument in the proof of Theorem 6 in [1], it can be seen that a modified distribution will occur where the term corresponding to γ_j is missing from the Standard Normalized Correlation Distribution from Definition 3. Using this and normalizing we get the claimed result.   □

Note that from the local second partial correlation moment, it is straightforward to obtain the global partial correlation moment defined below, and its distribution stated in the theorem below, after a renormalization of sums.

Definition 6. *We define the global second partial correlation moment for Family A as:*

$$\langle\,|\,P(L)\,|^2\,\rangle = \frac{1}{(2^{2n}-1)^2}\sum_{k,\tau=0}^{2^n-2}\sum_{\gamma_i,\gamma_j\in\Gamma_v}|\,P_{i,j}(\tau,k,L)\,|^2$$

Theorem 12. *The global second partial correlation moment for Family A is given by*

$$\langle\,|\,P(L)\,|^2\,\rangle = L + \frac{L(L-1)}{(2^{2n}-1)^2}.$$

5   Conclusions and Discussion

In this paper we have studied the first two moments of the periodic partial correlation functions of Galois ring m-sequences which form *Families A, B, C* and obtained complete results for the first partial correlation moment. For the case of *Family A*, we have also obtained local and global second partial correlation moments. The local second moment characterizes the variability of the interference a given signal experiences with respect to all the other signals in a CDMA system.

It is of interest to extend this work to further sequence families, such as $S(2)$ in[5] which is a superset of Family A. The extension of the second moment results to *Family B* from [1] and *Family C* and *Family D* from [9] is challenging and is left for future work.

References

1. Boztaş, S., Hammons, A.R., Vijay Kumar, P.: 4-phase sequences with near-optimum correlation properties. IEEE Trans. Inform. Theory 38(3), 1101–1113 (1992)
2. Boztaş, S., Udaya, P.: Partial Correlations of Galois Ring Sequences. In: Proceedings of IEEE IWSDA Workshop, Chengdu, China, 23-27 September, pp. 157–161 (2007)
3. Golomb, S.W., Gong, G.: Signal Design for Good Correlation. Cambridge University Press, Cambridge (2005)
4. Helleseth, T., Vijay Kumar, P.: Sequences with low correlation. In: Pless, V., Huffman, W.C. (eds.) Handbook of Coding Theory, vol. II, Elsevier, Amsterdam (1998)
5. Vijay Kumar, P., Helleseth, T., Calderbank, A.R., Hammons Jr., A.R.: Large families of quaternary sequences with low correlation. IEEE Trans. Inform. Theory 42(2), 579–592 (1996)
6. Solé, P.: A quaternary cyclic code, and a family of quadriphase sequences with low correlation properties. In: Cohen, G., Wolfmann, J. (eds.) Proceedings of Coding Theory and Applications. LNCS, vol. 388, pp. 193–201. Springer, Heidelberg (1989)
7. Sidelnikov, V.M.: On mutual correlation of sequences. Probl. Kybern. 24, 537–545 (1978)
8. Simon, M.K., Omura, J.K., Scholtz, R.A., Levitt, B.K.: Spread Spectrum Communications Handbook (revised ed.). McGraw Hill, New York (1994)
9. Tang, X.H., Udaya, P.: A Note on the Optimal Quadriphase Sequence Families. IEEE Trans. Inform. Theory 53(1), 433–436 (2007)
10. Udaya, P., Siddiqi, M.U.: Optimal and suboptimal quadriphase sequences derived from maximal length sequences over $\mathbb{Z}_4$. J. Appl. Algebra Eng., Commun. 9, 161–191 (1998)
11. Udaya, P., Boztaş, S.: On the aperiodic correlation function of Galois ring m-Sequences. In: Bozta, S., Sphparlinski, I. (eds.) AAECC 2001. LNCS, vol. 2227, pp. 229–238. Springer, Heidelberg (2001)
12. Welch, L.R.: Lower bounds on the maximum cross correlation of signals. IEEE Trans. Inform. Theory 20(2), 397–399 (1974)

On the Higher Order Nonlinearities of Boolean Functions and S-Boxes, and Their Generalizations
(Invited Paper)

Claude Carlet

University of Paris 8, Department of Mathematics (MAATICAH)
2 rue de la liberté, 93526 Saint-Denis Cedex, France
claude.carlet@inria.fr

Abstract. The r-th order nonlinearity of a Boolean function $f : F_2^n \to F_2$ is its minimum Hamming distance to all functions of algebraic degrees at most r, where r is a positive integer. The r-th order nonlinearity of an S-box $F : F_2^n \to F_2^m$ is the minimum r-th order nonlinearity of its component functions $v \cdot F$, $v \in F_2^m \setminus \{0\}$. The role of this cryptographic criterion against attacks on stream and block ciphers has been illustrated by several papers. Its study is also interesting for coding theory and is related to the covering radius of Reed-Muller codes (i.e. the maximum multiplicity of errors that have to be corrected when maximum likelihood decoding is used on a binary symmetric channel). We give a survey of what is known on this parameter, including the bounds involving the algebraic immunity of the function, the bounds involving the higher order nonlinearities of its derivatives, and the resulting bounds on the higher order nonlinearities of the multiplicative inverse functions (used in the S-boxes of the AES). We show an improvement, when we consider an S-box instead of a Boolean function, of the bounds on the higher order nonlinearity expressed by means of the algebraic immunity. We study a generalization (for S-boxes) of the notion and we give new results on it.

Keywords: Block cipher, Boolean function, Covering radius, Cryptography, Higher-order nonlinearity, Reed-Muller code, S-box, Stream cipher.

1 Introduction

The invention, in the late 70's, of public key cryptography (which has the important advantage that the sender and the receiver do not need to exchange a private key on a secure channel before safely communicating) has not eliminated conventional[1] cryptography because of the lower efficiency of public key cryptosystems and the much larger size of the encryption and decryption keys (some public key cryptosystems are fast, but they have still larger keys and they also have some additional drawbacks). Both techniques must then be combined to

---

[1] Also called symmetric, or private key.

S.W. Golomb et al. (Eds.): SETA 2008, LNCS 5203, pp. 345–367, 2008.

exchange large size data securely. The present paper is devoted to symmetric cryptography and more precisely to the Boolean functions it often uses for making the systems as nonlinear as possible, allowing them to resist known attacks and hopefully future attacks. These are central objects for the design and the security of symmetric cryptosystems (stream ciphers and block ciphers), see e.g. [5,6]. In cryptography, the most usual representation of these functions is the *algebraic normal form* (ANF):

$$f(x_1, \ldots, x_n) = \sum_{I \subseteq \{1,\ldots,n\}} a_I \prod_{i \in I} x_i,$$

where the a_I's are in F_2. The terms $\prod_{i \in I} x_i$ are called *monomials*. The *algebraic degree* $d°f$ of a Boolean function f equals the global degree of its (unique) ANF, that is, the maximum degree of those monomials whose coefficients are nonzero. *Affine functions* are those Boolean functions of algebraic degrees at most 1. The Boolean functions used in stream or block ciphers must have high degrees to avoid the Berlekamp-Massey attack on stream ciphers and the higher order differential cryptanalysis on block ciphers.

Another possible representation of Boolean functions uses the identification between the vector-space F_2^n and the field F_{2^n}. It represents any Boolean function as a polynomial in one variable $x \in F_{2^n}$ of the form $f(x) = \sum_{i=0}^{2^n-1} f_i x^i$, where the f_i's are elements of the field. This representation exists for every function from F_{2^n} to F_{2^n} (this is easy to prove; note that the polynomial $\sum_{i=0}^{2^n-1} f_i x^i$ can be obtained by using the so-called Mattson-Solomon polynomial [43,5]) and such function f is Boolean if and only if f_0 and f_{2^n-1} belong to F_2 and $f_{2i} = f_i^2$ for every $i \neq 0, 2^n - 1$, where $2i$ is taken mod $2^n - 1$. This allows representing $f(x)$ in the form $\sum_{k \in \Gamma(n)} tr_{n_k}(f_k x^k) + f_{2^n-1} x^{2^n-1}$, where $\Gamma(n)$ is the set obtained by choosing one element in each cyclotomic class of 2 modulo $2^n - 1$ (the most usual choice for k is the smallest element in its cyclotomic class, called the coset leader of the class), where n_k is the size of the cyclotomic class containing k and where tr_{n_k} is the trace function from $F_{2^{n_k}}$ to F_2: $tr_{n_k}(x) = x + x^2 + x^{2^2} + \cdots + x^{2^{n_k-1}}$. This representation is called the *trace representation*. Note that, for every $k \in \Gamma(n)$ and every $x \in F_{2^n}$, we have $f_k \in F_{2^{n_k}}$ (since $f_k^{2^{n_k}} = f_k$) and $x^k \in F_{2^{n_k}}$ as well. A slightly different representation, often called also the trace representation, has the form $f(x) = tr(\sum_{i=0}^{2^n-1} u_i x^i)$, where $tr = tr_n$ and where the $u_i's$ are elements of the field F_{2^n} (it can be easily obtained from $f(x) = \sum_{i=0}^{2^n-1} f_i x^i$ since $f(x) = tr(\lambda f(x))$, when $tr(\lambda) = 1$). The former representation is unique for every Boolean function and the latter is not (see more e.g. in [5]). Recall that the 2-weight $w_2(i)$ of an integer i equals by definition the number of 1's in its binary expansion. The algebraic degree of the function is then equal to the maximum 2-weight of the exponents i with nonzero coefficients f_i in the former representation and is upper bounded by the maximum 2-weight of the exponents i with nonzero coefficients u_i in the latter.

A characteristic of Boolean functions, called their nonlinearity profile, plays an important role with respect to the security of the cryptosystems in which they are involved. For every non-negative integer $r \leq n$, we denote by $nl_r(f)$ the minimum Hamming distance between f and all functions of algebraic degrees at most r (in the case of $r = 1$, we shall simply write $nl(f)$). In other words, $nl_r(f)$ equals the distance from f to the Reed-Muller code $RM(r, n)$ of length 2^n and of order r, that is, the number of bits to change in the truth table of f to get a Boolean function of algebraic degree at most r). This parameter is called the r-th order nonlinearity of f (simply the nonlinearity in the case $r = 1$). The maximum r-th order nonlinearity of all Boolean functions in n variables equals by definition the covering radius of $RM(r, n)$ [19]. The nonlinearity profile of a function f is the sequence of those values $nl_r(f)$ for r ranging from 1 to $n - 1$.

The same notion can be defined for S-boxes in stream or block ciphers as well, that is, for vectorial Boolean functions $F : F_2^n \rightarrow F_2^m$ (also called (n, m)-functions). Such functions are used in stream ciphers in the place of Boolean functions to speed up the enciphering and deciphering processes (since they output m bits instead of one at each clock cycle). They are used as well (and more systematically) in block ciphers to bring confusion [49] into the system. The algebraic degree of an S-box is the maximum algebraic degree of its coordinate functions. We shall denote by $nl_r(F)$ the minimum r-th order nonlinearity of all the component functions $\ell \circ F$, where ℓ ranges over the set of all the nonzero linear forms over F_2^m (hence, the component functions are those linear combinations of coordinate functions whose coefficients are not all-zero). Equivalently, $nl_r(F)$ is the minimum r-th order nonlinearity of all the functions $v \cdot F$, $v \in F_2^m \setminus \{0\}$, where "$\cdot$" denotes the usual inner product in F_2^m (or any other inner product). If F_2^m is endowed with the structure of the field F_{2^m}, then $nl_r(F)$ is the minimum r-th order nonlinearity of all the functions $tr(vF(x))$, $v \in F_{2^m}^*$. The algebraic degree of an S-box is also the maximum degree of its component functions.

The cryptographic relevance of the higher order nonlinearity has been illustrated by several papers [20,33,34,40,44,47,50]. Computing the r-th order nonlinearity of a given Boolean function with algebraic degree strictly greater than r is a hard task for $r > 1$. In the case of the first order, much is known in theory and also algorithmically since the nonlinearity is related to the Walsh transform, which can be computed by the algorithm of the Fast Fourier Transform (FFT). Recall that the Walsh transform of f is the Fourier transform of the "sign" function $(-1)^f$, defined at any vector $a \in F_2^n$ as $W_f(a) = \sum_{x \in F_2^n}(-1)^{f(x)+x \cdot a}$ (where $x \cdot a$ is an inner product in F_2^n - when the vector space F_2^n is identified to the field F_{2^n}, we can take $x \cdot a = tr(xa)$). The relation between the nonlinearity and the Walsh transform is well-known: $nl(f) = 2^{n-1} - \frac{1}{2}\max_{a \in F_2^n}|W_f(a)|$. But for $r > 1$, very little is known. Even the second order nonlinearity is known only for a few peculiar functions and for functions in small numbers of variables. A nice algorithm due to G. Kabatiansky and C. Tavernier and improved and implemented by Fourquet et al. [29,30,31,35,28] works well for $r = 2$ and $n \leq 11$ (in some cases, $n \leq 13$). But for $r \geq 3$, it is inefficient, even for $n = 8$ (the number of variables in the sub-S-boxes of the AES). No better algorithm is known.

Proving lower bounds on the r-th order nonlinearity of functions (and therefore proving their good behavior with respect to this criterion) is also a quite difficult task, even for the second order. Until recently, there had been only one attempt, by Iwata-Kurosawa [34], to construct functions with lower bounded r-th order nonlinearity. But the obtained value, $2^{n-r-3}(r+5)$, of the lower bound was small.

In the present paper, we give a survey of the state of the art in the domain: some simple observations, the lower bounds on the higher order nonlinearity of a Boolean function with given algebraic immunity, a recent recursive lower bound (a lower bound on the r-th order nonlinearity of a given function f, knowing a lower bound on the $(r-1)$-th order nonlinearity of the derivatives of f), and the deduced lower bounds on the whole nonlinearity profile of the multiplicative inverse function. We also give new results. We show that, when considering an S-box instead of a Boolean function, we can prove better bounds involving the algebraic immunity, on the higher order nonlinearity. We begin the study of a more general notion of higher order nonlinearity of S-boxes, which poses many interesting problems.

2   Higher Order Nonlinearity of Boolean Functions and S-Boxes

2.1   Some Simple Facts

• Adding to a function f a function of algebraic degree at most r clearly does not change the r-th order nonlinearity of f.

• Since $RM(r,n)$ is invariant under any affine automorphism, composing a Boolean function by an affine automorphism does not change its r-th order nonlinearity (i.e. the characteristic nl_r is affine invariant). And composing an S-box by affine automorphisms on the left and on the right does not change its r-th order nonlinearity.

• The minimum distance of $RM(r,n)$ being equal to 2^{n-r} for every $r \leq n$, we have $nl_r(f) \geq 2^{n-r-1}$ for every function f of algebraic degree exactly $r+1 \leq n$. Moreover, any minimum weight function f of algebraic degree $r+1$ (that is, the indicator - i.e. the characteristic function - of any $(n-r-1)$-dimensional flat, see [43]), has r-th order nonlinearity equal to 2^{n-r-1} since a closest function of algebraic degree at most r to f is clearly the null function.

• As observed by Iwata and Kurosawa [34] (for instance), if f_0 is the restriction of f to the linear hyperplane H of equation $x_n = 0$ and f_1 the restriction of f to the affine hyperplane H' of equation $x_n = 1$ (these two functions will be viewed as $(n-1)$-variable functions), then we have $nl_r(f) \geq nl_r(f_0) + nl_r(f_1)$ since, for every function g of algebraic degree at most r, the restrictions of g to H and H' having both algebraic degree at most r, we have $d_H(f, g) \geq nl_r(f_0) + nl_r(f_1)$ where d_H denotes the Hamming distance (obviously, this inequality is more generally valid if f_0 is the restriction of f to any linear hyperplane H and f_1 its restriction to the complement of H).

- Moreover, if $f_0 = f_1$, then there is equality since if g is the best approximation of algebraic degree at most r of $f_0 = f_1$, then g now viewed as an n-variable function lies at distance $2nl_r(f_0)$ from f.

- Since nl_r is affine invariant, this implies that, if there exists a nonzero vector $a \in F_2^n$ such that $f(x+a) = f(x)$, then the best approximation of f by a function of algebraic degree r is achieved by a function g such that $g(x + a) = g(x)$ and $nl_r(f)$ equals twice the r-th order nonlinearity of the restriction of f to any linear hyperplane H excluding a.

- Note that the equality $nl_r(f) = 2nl_r(f_0)$ is also true if f_0 and f_1 differ by a function of algebraic degree at most $r-1$ since the function $x_n(f_0 + f_1)$ has then algebraic degree at most r.

• Conversely, the r-th order nonlinearity of the restriction of a function f to a hyperplane is lower bounded by means of the r-th order nonlinearity of f:

Proposition 1. *[9] Let f be any n-variable Boolean function, r a positive integer smaller than n and H an affine hyperplane of F_2^n. Then the r-th order nonlinearity of the restriction f_0 of f to H (viewed as an $(n-1)$-variable function) satisfies:*

$$nl_r(f_0) \geq nl_r(f) - 2^{n-2}.$$

2.2 On the Highest Possible r-th Order Nonlinearity (the Covering Radius of the Reed-Muller Code of Order r)

The best known general upper bound on the first order nonlinearity of any Boolean function is:

$$nl(f) \leq 2^{n-1} - 2^{\frac{n}{2}-1}.$$

It can be directly deduced from the Parseval relation $\sum_{a \in F_2^n} W_f^2(a) = 2^{2n}$. It is tight for n even and untight for n odd (some lower bounds on the covering radius exist then, see e.g. [5]). This bound is obviously valid for (n, m)-functions as well and it is then the best known bound when $m < n$. But, as proved by K. Nyberg, it is tight (achieved with equality by the so-called bent functions) only when n is even and $m \leq n/2$ (see e.g. [6]). When $m = n$, we have the better bound [16]:

$$nl(F) \leq 2^{n-1} - 2^{\frac{n-1}{2}},$$

which is achieved with equality (by the so-called almost bent functions) for every odd n. For $m > n$ further (better) bounds are given in [12] but no such bound is tight. Improving upon all these bounds when they are not tight is a series of difficult open problems.

The best known upper bound on the r-th order nonlinearity of any Boolean function for $r \geq 2$ is given in [14] and has asymptotic version:

$$nl_r(f) = 2^{n-1} - \frac{\sqrt{15}}{2} \cdot (1 + \sqrt{2})^{r-2} \cdot 2^{n/2} + O(n^{r-2}).$$

This bound is obviously also valid for vectorial functions. It would be nice to improve it, for S-boxes, as the bound has been improved in the case $r = 1$. We leave this as an open problem.

It can be proved [19,7] that, for every positive real c such that $c^2 \log_2(e) > 1$, where e is the base of the natural logarithm, (e.g. for $c = 1$), almost all Boolean functions satisfy

$$nl_r(f) \geq 2^{n-1} - c\sqrt{\sum_{i=0}^{r}\binom{n}{i}} \ 2^{\frac{n-1}{2}} \approx 2^{n-1} - \frac{c\,n^{r/2}\,2^{n/2}}{\pi^{1/4}\,r^{(2r+1)/4}\,2^{3/4}}, \qquad (1)$$

and that the probability that $nl_r(f)$ is smaller than this expression is asymptotically at most $O(2^{(1-\log_2 e)}\sum_{i=0}^{r}\binom{n}{i})$ when n tends to ∞.

This proves that the best possible r-th order nonlinearity of n-variable Boolean functions is asymptotically equivalent to 2^{n-1}, and that its difference with 2^{n-1} is polynomially (in n, for every fixed r) proportional to $2^{n/2}$. The proof of this fact is obtained by counting the number of functions having upper bounded r-th order nonlinearity (or more precisely by upper bounding this number) and it does not help obtaining explicit functions with non-weak r-th order nonlinearity.

2.3   Lower Bounds on the Higher Order Nonlinearity of a Boolean/Vectorial Boolean Function with Given Algebraic Immunity

The algebraic immunity [45] of a Boolean function f quantifies the resistance to the standard algebraic attack of the pseudo-random generators using it as a nonlinear function.

Definition 1. *Let $f : F_2^n \to F_2$ be an n-variable Boolean function. We call an annihilator of f any n-variable function g whose product with f is null (i.e. whose support is included in the support of $f + 1$, or in other words any function which is a multiple of $f + 1$). The algebraic immunity of f is the minimum algebraic degree of all the nonzero annihilators of f or of $f + 1$.*

We shall denote the algebraic immunity of f by $AI(f)$. Clearly, since f is an annihilator of $f + 1$ (and $f + 1$ is an annihilator of f) we have $AI(f) \leq d^\circ f$.

As shown in [22], we always have $AI(f) \leq \lceil \frac{n}{2} \rceil$. This bound is tight. Also, we know that almost all Boolean functions have algebraic immunities close to this optimum; more precisely, for all $a < 1$, $AI(f)$ is almost surely greater than $\frac{n}{2} - \sqrt{\frac{n}{2}\ln\left(\frac{n}{a\ln 2}\right)}$ when n tends to infinity, see [27].

Remark. Few functions are known (up to affine equivalence) with provably optimum algebraic immunities: the functions whose construction is introduced in [26] (see in [11] their further properties) and some functions which are symmetric (that is, whose outputs depend only on the Hamming weights of their inputs) [25,4] or almost symmetric (see [10]). These functions have some drawbacks: all of them have insufficient nonlinearities and many are non-balanced (i.e. their output is not uniformly distributed over F_2). Moreover, the functions studied in [25,4], and to a slightly smaller extent the functions introduced in [26], have not a good behavior against fast algebraic attacks, see [1]. But the research in this domain is active and better examples of functions will be found in the future. A first one is available in [13].

The weight of a function f with given algebraic immunity satisfies the relation: $\sum_{i=0}^{AI(f)-1} \binom{n}{i} \leq wt(f) \leq \sum_{i=0}^{n-AI(f)} \binom{n}{i}$. This shows that if n is odd and f has optimum algebraic immunity, then f is balanced. This has also led to a bound on the first order nonlinearity by Dalai et al., which has been later improved by Lobanov [42], who has obtained the tight lower bound: $nl(f) \geq 2\sum_{i=0}^{AI(f)-2} \binom{n-1}{i}$. In [11], a generalization of the Dalai et al. bound to the r-th order nonlinearity has been derived: $nl_r(f) \geq \sum_{i=0}^{AI(f)-r-1} \binom{n}{i}$. In [8], another bound has been obtained, which improves upon the bound of [11] for all values of $AI(f)$ when the number of variables is smaller than or equal to 12, and for most values of $AI(f)$ when the number of variables is smaller than or equal to 22 (which covers the practical situation of stream ciphers), and which also improves asymptotically upon it: $nl_r(f) \geq \max_{r' \leq n} (\min(\lambda_{r'}, \mu_{r'}))$, where:

$$\lambda_{r'} = 2 \max \left(\sum_{i=0}^{r'-1} \binom{n}{i}, \sum_{i=0}^{AI(f)-r-1} \binom{n-r}{i} \right) \quad \text{if } r' \leq AI(f) - r - 1,$$

$$= 2 \sum_{i=0}^{AI(f)-r-1} \binom{n}{i} \quad \text{if } r' > AI(f) - r - 1,$$

$$\mu_{r'} = \sum_{i=0}^{AI(f)-r-1} \binom{n-r}{i} + \sum_{i=0}^{AI(f)-r'} \binom{n-r'+1}{i}.$$

Finally, S. Mesnager [46] has obtained a simpler bound, which improves upon the bounds of [11] and [8] for low values of r (which play the most important role for attacks):

$$nl_r(f) \geq \sum_{i=0}^{AI(f)-r-1} \binom{n}{i} + \sum_{i=AI(f)-2r}^{AI(f)-r-1} \binom{n-r}{i}.$$

These results have the interest of showing that, *automatically*, a function with good algebraic immunity has not a very bad nonlinearity profile. However, all the bounds above are rather weak with respect to the asymptotic bound (1).

A generalization to S-boxes. The algebraic immunity of an (n, m)-function F can be defined as follows (see e.g. [2]). We call an annihilator of a subset A of F_2^n any n-variable Boolean function g whose restriction to A is null. We denote by $An_k(A)$ the vectorspace of annihilators of A of degrees at most k. The algebraic immunity $AI(A)$ of A equals the minimum algebraic degree of all the nonzero annihilators of A and the algebraic immunity $AI(F)$ of F is the minimum algebraic immunity of all the sets $F^{-1}(z) = \{x \in F_2^n / F(x) = z\}$, where z ranges over F_2^m (note that $AI(F)$ equals 0 if and only if F is not onto F_2^m

and that, if F is surjective and if there exists z such that $F^{-1}(z)$ is a singleton - if F is a permutation, for instance - then $AI(F)$ equals 1; other generalizations of the notion of algebraic immunity to S-boxes can be given, which correspond to diverse variants of algebraic attacks, see e.g. [2]). Note that, for every $z \in F_2^m$, we have $|F^{-1}(z)| \geq \sum_{i=0}^{AI(F)-1} \binom{n}{i}$ since otherwise there would exist a nonzero annihilator of degree at most $AI(F) - 1$ of $F^{-1}(z)$, a contradiction (indeed, a function is a nonzero annihilator of degree at most $AI(F) - 1$ of $F^{-1}(z)$ if and only if the coefficients in its ANF constitute a non-trivial solution of a system of $|F^{-1}(z)|$ equations in $\sum_{i=0}^{AI(F)-1} \binom{n}{i}$ variables and if $|F^{-1}(z)| < \sum_{i=0}^{AI(F)-1} \binom{n}{i}$ then this system must have a non-trivial solution). This property will be used in the proof of Theorem 1 below.

The results of [8] and [46] nicely generalize to S-boxes. We give in Proposition 2 below a lower bound on the size of the intersection between a set of a given algebraic immunity and the support of a function of given algebraic degree, which is a straightforward adaptation of several lemmas of these two papers. We deduce in Theorem 1 the generalization to S-boxes of the bounds of [8] and [46].

Proposition 2. *Let A be a subset of F_2^n and let r be a positive integer and r' a non-negative integer such that $r' \leq r$ and $AI(A) - r - 1 \geq 0$. For every n-variable Boolean function h of degree r and of algebraic immunity r', we have:*

$$|A \cap supp(h)| \geq \max \left(\sum_{i=0}^{r'-1} \binom{n}{i}, \sum_{i=0}^{AI(A)-r-1} \binom{n-r}{i} \right) \text{ if } r' \leq AI(A) - r - 1,$$

$$\geq \sum_{i=0}^{AI(A)-r-1} \binom{n}{i} \text{ if } r' > AI(A) - r - 1,$$

where $supp(h)$ denotes the support of h, and:

$$|A \cap supp(h)| \geq dim \left(An_{AI(A)-1}(supp(h+1)) \right) \geq dim \left(Mul_{AI(A)-r-1}(h) \right),$$

where $Mul_{AI(A)-r-1}(h)$ is the set of all products of h with functions of degrees at most $AI(A) - r - 1$. In all cases, we have:

$$|A \cap supp(h)| \geq \sum_{i=0}^{AI(A)-r-1} \binom{n-r}{i}.$$

Proof. Let k be any non-negative integer. A Boolean function of degree at most k belongs to $An_k(A \cap supp(h))$ if and only if the coefficients in its ANF satisfy a system of $|A \cap supp(h)|$ equations in $\sum_{i=0}^{k} \binom{n}{i}$ variables (the coefficients of the monomials of degrees at most k in the ANF of the function). Hence we have: $dim(An_k(A \cap supp(h))) \geq \sum_{i=0}^{k} \binom{n}{i} - |A \cap supp(h)|$.

According to Lemma 1 of [8], we have:

$$dim(An_k(supp(h))) \leq \min \left(\sum_{i=r'}^{k} \binom{n}{i}, \sum_{i=0}^{k} \binom{n}{i} - \sum_{i=0}^{k} \binom{n-r}{i} \right).$$

If $\dim(An_k(A \cap supp(h))) > \dim(An_k(supp(h)))$, then there exists an annihilator g of degree at most k of $A \cap supp(h)$ which is not an annihilator of h. Then, gh is a nonzero annihilator of A and has degree at most $k+r$. Thus, if $k = AI(A) - r - 1 \geq 0$, we arrive to a contradiction. We deduce that $\dim(An_{AI(A)-r-1}(A \cap supp(h))) \leq \dim(An_{AI(A)-r-1}(supp(h)))$. This implies:

$$\sum_{i=0}^{AI(A)-r-1} \binom{n}{i} - |A \cap supp(h)| \leq$$

$$\min\left(\sum_{i=r'}^{AI(A)-r-1} \binom{n}{i}, \sum_{i=0}^{AI(A)-r-1} \binom{n}{i} - \sum_{i=0}^{AI(A)-r-1} \binom{n-r}{i} \right),$$

that is:

$$|A \cap supp(h)| \geq \max\left(\sum_{i=0}^{r'-1} \binom{n}{i}, \sum_{i=0}^{AI(A)-r-1} \binom{n-r}{i} \right) \quad \text{if } r' \leq AI(A) - r - 1,$$

and

$$|A \cap supp(h)| \geq \sum_{i=0}^{AI(A)-r-1} \binom{n}{i} \quad \text{if } r' > AI(A) - r - 1.$$

The inequality $|A \cap supp(h)| \geq \dim\left(An_{AI(A)-1}(supp(h+1)) \right)$ comes from the facts that the system of equations expressing that a given function of degree at most $AI(A) - 1$ (given by its ANF) is an annihilator of A (we shall denote this system by S_A) has full rank by definition of the algebraic immunity, and that the system $S_{supp(h+1)}$ has $|A \cap supp(h+1)|$ equations in common with S_A, which implies that the rank of $S_{supp(h+1)}$ is at least equal to the rank of S_A minus $|A \cap supp(h)|$ (see a more detailed proof in Lemma 9 of [46]). The inequality $\dim\left(An_{AI(A)-1}(supp(h+1)) \right) \geq \dim\left(Mul_{AI(A)-r-1}(h) \right)$ is straightforward.$\square$

This implies:

Theorem 1. *Let F be any (n, m)-function and let r be any positive integer such that $AI(F) - r - 1 \geq 0$. Then*

$$nl_r(F) \geq$$

$$\max\left(2^m \sum_{i=0}^{AI(F)-r-1} \binom{n-r}{i}, 2^{m-1}\left(\sum_{i=0}^{AI(F)-r-1} \binom{n}{i} + \sum_{i=AI(F)-2r}^{AI(F)-r-1} \binom{n-r}{i} \right) \right).$$

Moreover, we have $nl_r(F) \geq \max\limits_{r' \leq n} (\min(\lambda_{r'}, \mu_{r'}))$, *where:*

$$\lambda_{r'} = 2^m \max\left(\sum_{i=0}^{r'-1} \binom{n}{i}, \sum_{i=0}^{AI(F)-r-1} \binom{n-r}{i}\right) \quad \text{if } r' \leq AI(F) - r - 1,$$

$$= 2^m \sum_{i=0}^{AI(F)-r-1} \binom{n}{i} \quad \text{if } r' > AI(F) - r - 1,$$

$$\mu_{r'} = 2^{m-1} \sum_{i=0}^{AI(F)-r-1} \binom{n-r}{i} + 2^{m-1} \sum_{i=0}^{AI(F)-r'} \binom{n-r'+1}{i}.$$

Proof. Let h be any n-variable function of degree at most r and let ℓ be any m-variable nonzero linear function. We have $d_H(\ell \circ F, h) = \sum_{z \in supp(\ell)} |F^{-1}(z) \cap supp(h+1)| + \sum_{z \in supp(\ell+1)} |F^{-1}(z) \cap supp(h)|$.

If h is constant, then $d_H(\ell \circ F, h)$ equals $w_H(\ell \circ F) = \sum_{z \in supp(\ell)} |F^{-1}(z)|$ or $w_H(\ell \circ F + 1) = \sum_{z \in supp(\ell+1)} |F^{-1}(z)|$ which, according to the property recalled before Proposition 2, are lower bounded by $2^{m-1} \sum_{i=0}^{AI(F)-1} \binom{n}{i}$. It is a simple matter to check that the bound of Theorem 1 is then satisfied.

If h is not constant, then Proposition 2 (last line of) implies $d_H(\ell \circ F, h) \geq 2^m \sum_{i=0}^{AI(F)-r-1} \binom{n-r}{i}$.

The inequality

$$nl_r(F) \geq 2^{m-1} \left(\sum_{i=0}^{AI(F)-r-1} \binom{n}{i} + \sum_{i=AI(F)-2r}^{AI(F)-r-1} \binom{n-r}{i} \right)$$

is a direct consequence of Proposition 2 and of the inequality

$$\dim \left(Mul_{AI(F)-r-1}(h)\right) + \dim \left(Mul_{AI(F)-r-1}(h+1)\right) \geq$$

$$\sum_{i=0}^{AI(F)-r-1} \binom{n}{i} + \sum_{i=AI(F)-2r}^{AI(F)-r-1} \binom{n-r}{i}$$

which is a direct consequence of Lemma 2 and Corollary 6 in [46].

Let r' be any nonnegative integer. If $AI(h) \geq r'$, then Proposition 2 shows that, for every $z \in F_2^m$, $|F^{-1}(z) \cap supp(h)|$ and $|F^{-1}(z) \cap supp(h+1)|$ are lower bounded by

$$\max\left(\sum_{i=0}^{r'-1} \binom{n}{i}, \sum_{i=0}^{AI(F)-r-1} \binom{n-r}{i}\right) \quad \text{if } r' \leq AI(F) - r - 1,$$

$$\sum_{i=0}^{AI(F)-r-1} \binom{n}{i} \quad \text{if } r' > AI(F) - r - 1.$$

If $AI(h) < r'$, then there exists $g \neq 0$ such that either $g \in An_{r'-1}(h+1)$, and therefore $supp(g) \subseteq supp(h)$, or $g \in An_{r'-1}(h)$, and therefore $supp(g) \subseteq supp(h+1)$. If $supp(g) \subseteq supp(h)$, then we apply Proposition 2 (last sentence of) to the set $F^{-1}(z)$ (where z ranges over F_2^m) and to the function g (resp. $h+1$) and we get $|F^{-1}(z) \cap supp(h)| \geq |F^{-1}(z) \cap supp(g)| \geq \sum_{i=0}^{AI(F)-r'} \binom{n-r'+1}{i}$ and $|F^{-1}(z) \cap supp(h+1)| \geq \sum_{i=0}^{AI(F)-r-1} \binom{n-r}{i}$. The case where $supp(g) \subseteq supp(h+1)$ is similar. □

Remark. In [8], the possibility that h could be constant was not taken into account. The statement of Proposition 5 and the proof of Theorem 1 in this reference are therefore incomplete. However, the result of [8, Theorem 1] is valid (it is implied by Theorem 1 of the present paper, reduced to the case $m = 1$).

2.4   Recent Recursive Lower Bounds

Lower bounds on the nonlinearity profile of a Boolean function by means of the nonlinearity profiles of its derivatives. We denote by $D_a f$ the so-called derivative of any Boolean function f in the direction of $a \in F_2^n$:

$$D_a f(x) = f(x) + f(x+a).$$

The addition is performed mod 2 (i.e. $D_a f$ is a Boolean function too). Applying such discrete derivation several times to a function f leads to the so-called higher order derivatives $D_{a_1} \cdots D_{a_k} f(x) = \sum_{u \in F_2^k} f(x + \sum_{i=1}^k u_i a_i)$.

A simple tight lower bound on the r-th order nonlinearity of any function f, knowing a lower bound on the $(r-1)$-th order nonlinearity of at least one of its derivatives (in a nonzero direction) is (see [9]):

$$nl_r(f) \geq \frac{1}{2} \max_{a \in F_2^n} nl_{r-1}(D_a f).$$

When applied repeatedly, the resulting bound

$$nl_r(f) \geq \frac{1}{2^i} \max_{a_1, \ldots, a_i \in F_2^n} nl_{r-i}(D_{a_1} \cdots D_{a_i} f)$$

is also tight, but can not lead to bounds equivalent to 2^{n-1} and so is weak with respect to (1). A potentially stronger lower bound, is valid when a lower bound on the $(r-1)$-th order nonlinearity is known for all the derivatives (in nonzero directions) of the function.

Proposition 3. *[9] Let f be any n-variable function and r a positive integer smaller than n. We have:*

$$nl_r(f) \geq 2^{n-1} - \frac{1}{2} \sqrt{2^{2n} - 2 \sum_{a \in F_2^n} nl_{r-1}(D_a f)}.$$

This bound also is tight.

Corollary 1. *[9] Let f be any n-variable function and r a positive integer smaller than n. Assume that, for some non-negative integers K and k, we have $nl_{r-1}(D_a f) \geq 2^{n-1} - K 2^k$ for every nonzero $a \in F_2^n$, then*

$$nl_r(f) \geq 2^{n-1} - \frac{1}{2}\sqrt{(2^n - 1)K 2^{k+1} + 2^n}$$

$$\approx 2^{n-1} - \sqrt{K} 2^{(n+k-1)/2}.$$

As we shall see below, Corollary 1 can allow proving that a given infinite class of functions has r-th order nonlinearity asymptotically equivalent to 2^{n-1} for every r.

Applying two times Proposition 3, we obtain the bound

$$nl_r(f) \geq$$
$$2^{n-1} - \frac{1}{2}\sqrt{\sum_{a \in F_2^n} \sqrt{2^{2n} - 2 \sum_{b \in F_2^n} nl_{r-2}(D_a D_b f)}}. \tag{2}$$

These bounds are applied in [9] to some quadratic functions (whose low algebraic degree makes them inappropriate for use in most cryptosystems) with good first order nonlinearities to check that the approximation given by the bounds can be good. Of most interest is what these bounds give in the case of the function used as the S-box in the AES:

The multiplicative inverse function is defined as $F_{inv}(x) = x^{2^n-2}$, where n is any positive integer and x ranges over the field F_{2^n}. We denote $f_\lambda(x) = tr(\lambda x^{2^n-2})$, where λ is any element of $F_{2^n}^*$. All the Boolean functions f_λ, $\lambda \neq 0$, are affinely equivalent to each others. We shall write f_{inv} for f_1. But we shall need however the notation f_λ in the calculations below. We have $f_\lambda(x) = tr\left(\frac{\lambda}{x}\right)$, with the convention that $\frac{\lambda}{0} = 0$ (we shall always assume this kind of convention in the sequel). Recall that the component functions of the Substitution boxes (S-boxes) of the Advanced Encryption Standard (AES) - the current non-military standard for block encryption [24] - are all of the form f_λ (with $n = 8$).

The first order nonlinearity of f_{inv} is quite good, according to the extension of the Weil bound first obtained by Carlitz and Uchiyama [15] and extended by Shanbhag, Kumar and Helleseth [48] (see also the additional information on the Walsh spectrum of the function by Lachaud and Wolfmann [41]). However, the extension of the Weil bound is efficient only for $r = 1$. Indeed, already for $r = 2$, the degree of a quadratic function in trace representation form can be upper bounded by $2^{\lfloor n/2 \rfloor} + 1$ only and this gives a bound in 2^n on the maximum magnitude of the Walsh transform and therefore no information on the nonlinearity. However, an effective bound can be obtained by using the recursive bound of Proposition 3.

For every nonzero $a \in F_{2^n}$, we have $(D_a f_\lambda)(ax) = tr\left(\frac{\lambda}{ax} + \frac{\lambda}{ax+a}\right) = tr\left(\frac{\lambda/a}{x^2+x}\right)$ $= f_{\lambda/a}(x^2 + x)$ if $x \notin F_2$ and $(D_a f_\lambda)(ax) = tr(\lambda/a)$ if $x \in F_2$. We deduce, denoting $g_{\lambda/a}(x) = f_{\lambda/a}(x^2 + x)$ that, for every r, we have $nl_r(D_a f_\lambda) = nl_r(g_{\lambda/a})$ if

$tr(\lambda/a) = 0$ and $nl_r(D_a f_\lambda) \geq nl_r(g_{\lambda/a}) - 2$ otherwise. Note that $g_{\lambda/a}$ is such that $g_{\lambda/a}(x+1) = g_{\lambda/a}(x)$. We have seen in Section 2.1 that this implies that $nl_r(g_{\lambda/a})$ equals twice the r-th order nonlinearity of the restriction of $g_{\lambda/a}$ to any linear hyperplane H excluding 1. Since the function $x \rightarrow x^2 + x$ is a linear isomorphism from H to the hyperplane $\{x \in F_{2^n} / tr(x) = 0\}$, we see that $nl_r(g_{\lambda/a})$ equals twice the r-th order nonlinearity of the restriction of $f_{\lambda/a}$ to this hyperplane. Applying then Proposition 1, we deduce that

$$nl_r(D_a f_\lambda) \geq 2\, nl_r(f_{\lambda/a}) - 2^{n-1} - 2\, tr(\lambda/a) \tag{3}$$

(where $tr(\lambda/a)$ is viewed here as an integer equal to 0 or 1). The first order nonlinearity of the inverse function is lower bounded by $2^{n-1} - 2^{n/2}$ (it equals this value if n is even). It has been more precisely proven in [41] that the character sums $\sum_{x \in F_{2^n}} (-1)^{f_\lambda(x) + tr(ax)}$, called Kloosterman sums, can take any value divisible by 4 in the range $[-2^{n/2+1} + 1, 2^{n/2+1} + 1]$. This leads to:

Proposition 4. *[9] Let $F_{inv}(x) = x^{2^n-2}$, $x \in F_{2^n}$. Then we have:*

$$nl_2(F_{inv}) \geq 2^{n-1} - \frac{1}{2}\sqrt{(2^n - 1)2^{n/2+2} + 3 \cdot 2^n}$$
$$\approx 2^{n-1} - 2^{3n/4}.$$

In Table 1, for n ranging from 4 to 12 (for smaller values of n, the bound gives negative numbers), we indicate the values given by this bound, compared with the actual values computed by Fourquet et al. [29,35,28]. Note that Proposition 4 gives an approximation of the actual value which is proportionally better and better when n increases. The difference between 2^{n-1} and our bound is in average 1.5 times the difference between 2^{n-1} and the actual value (for these values of n). In Table 2 we give, for $n = 13, 14$ and 15, the values given by our bound, compared with upper bounds obtained by Fourquet et al. [28,29,35].

Table 1. The values of the lower bound on $nl_2(F_{inv})$ given by Proposition 4, the actual values and the ratio

n	4	5	6	7	8	9	10	11	12
bound	0	2	9	25	63	147	329	718	1534
values	2	6	14	36	82	182	392	842	1760
%	0	33	52	69	76	80	84	85	87

Table 2. The values of the lower bound on $nl_2(F_{inv})$ given by Proposition 4, an overestimation of the actual values and the ratio

n	13	14	15
the lower bound	3232	6740	13944
overestimation of the values	3696	7580	15506
%	87	89	90

Table 3. The values of the lower bound on $nl_2(F_{inv})$ given by Proposition 3 and the FFT algorithm

n	4	5	6	7	8	9	10	11	12
Prop. 3	2	5	12	30	69	156	340	731	1551

Table 4. The values of the lower bounds on $nl_3(F_{inv})$ given by Proposition 5

n	7	8	9	10	11	12	13
bound of Proposition 5	5	5	20	58	149	358	827	1859

Proposition 3 gives nicer results than those deduced from Proposition 4 and listed in Table 1, when using the fast FFT algorithm to compute (exactly, instead of having a lower bound) the nonlinearities of the derivatives of the inverse function: see Table 3.

A bound for the whole nonlinearity profile of the inverse function. Thanks to Relations (2) and (3), we have:

Proposition 5. *[9] Let $F_{inv}(x) = x^{2^n-2}$. Then we have:* $nl_3(F_{inv}) \geq 2^{n-1} - \frac{1}{2}\sqrt{(2^n - 1)\sqrt{2^{3n/2+3} + 3 \cdot 2^{n+1} - 2^{n/2+3} + 16}} \approx 2^{n-1} - 2^{7n/8-1/4}$.

In Table 4, for n ranging from 7 to 13, we indicate the values given by this bound (for $n < 7$ it gives nothing better than 0).

The process leading to Proposition 5 can be iteratively applied, giving a lower bound on the r-th order nonlinearity of the inverse function for $r \geq 4$. The expression of this lower bound is:

$$nl_r(F_{inv}) \geq 2^{n-1} - l_r,$$

where, according to Relation (3) and to Corollary 1, the sequence l_r is defined by $l_1 = 2^{n/2}$ and $l_r = \sqrt{(2^n - 1)(l_{r-1} + 1)} + 2^{n-2}$. The expression of l_r is more and more complex when r increases. Its value is approximately equal to 2^{k_r}, where $k_1 = n/2$ and $k_r = \frac{n+k_{r-1}}{2}$, and therefore $k_r = (1 - 2^{-r}) n$. Hence, $nl_r(F_{inv})$ is approximately lower bounded by $2^{n-1} - 2^{(1-2^{-r})n}$ and asymptotically equivalent to 2^{n-1}, whatever is r.

3   Generalizations of the Higher Order Nonlinearity for S-Boxes

There are two natural extensions of the notion of higher order nonlinearity of S-boxes:

Definition 2. *For every S-box* $F : F_2^n \rightarrow F_2^m$, *for every positive integers* $s \leq m$ *and* $t \leq n + m$, *and every non-negative integer* $r \leq n$, *we define:*

$$nl_{s,r}(F) = \min\{nl_r(f \circ F); f \in \mathcal{B}_m, d^\circ f \leq s, f \neq cst\}$$
$$= \min\{d_H(g, f \circ F); f \in \mathcal{B}_m, d^\circ f \leq s, f \neq cst, g \in \mathcal{B}_n, d^\circ g \leq r\}$$

and

$$NL_t(F) = \min\{w_H(h(x, F(x))); h \in \mathcal{B}_{n+m}, d^\circ h \leq t, h \neq cst\}$$

where w_H *denotes the Hamming weight,* d_H *denotes the Hamming distance and* $\mathcal{B}_n$ *denotes the set of n-variable Boolean functions.*

Note that when $s \cdot d^\circ(F), r$ and $t \cdot d^\circ(F)$ are strictly smaller than n, then $f \circ F$, g and $h(x, F)$ have even weights and $nl_{s,r}(F)$, $NL_t(F)$ are therefore even. We have also that $nl_{s,r}(F)$ is even when F is a non-bijective balanced function and r is strictly smaller than n, and when F is a permutation and s, r are strictly smaller than n, since $f \circ F$, g and $f \circ F + g$ have then even weights too.

Definition 2 excludes obviously $f = cst$ and $h = cst$ because the knowledge of the distance $d_H(g, f \circ F)$ or of the weight $w_H(h(x, F(x)))$ when f or h is constant gives no information specific to F and usable in an attack against a stream or block cryptosystem using F as an S-box. Note that once the value of $nl_{s,r}(F)$ (resp. $NL_t(F)$) has been determined, the number of linearly independent pairs (f, g) such that $d^\circ f \leq s$, $f \neq cst$, $d^\circ g \leq r$ and $nl_{s,r}(F) = d_H(g, f \circ F)$ (resp. of linearly independent functions h such that $d^\circ h \leq t$, $h \neq cst$ and $NL_t(F) = w_H(h(x, F(x)))$) is also important (see [36] for linear attacks and [18,21] for algebraic attacks).

T. Shimoyama and T. Kaneko have exhibited in [50] several quadratic functions h and pairs (f, g) of quadratic functions showing that the nonlinearities NL_2 and $nl_{2,2}$ of some sub-S-boxes of the DES are null (and therefore that the global S-box of the DES has the same property). They deduced a "higher-order non-linear" attack (an attack using the principle of the linear attack by Matsui but with non-linear approximations, as introduced by Knudsen and Robshaw in [40]) which needs 26% less data than Matsui's attack. This improvement is not very significant, practically, but some recent studies [32], not yet published, seem to show that the notions of NL_t and $nl_{s,r}$ can be related to potentially more powerful attacks. Note that we have $NL_{\max(s,r)}(F) \leq nl_{s,r}(F)$ by taking $h(x, y) = g(x) + f(y)$ (since $f \neq cst$ implies then $h \neq cst$) and the inequality can be strict if $s > 1$ or $r > 1$ since a function $h(x, y)$ of low degree and such that $w_H(h(x, F(x)))$ is small can exist while no such function exists with separated variables x and y, that is, of the form $g(x) + f(y)$. This is the case, for instance, of the S-box of the AES for $s = 1$ and $r = 2$ (see below). It is not yet quite clear whether better attacks can be achieved when h has separated variables than when it does not. We need therefore to study both notions for the eventuality that such better attack could be found in the future.

Note that Theorem 1 generalizes to $nl_{s,r}(F)$, but the case where h is constant reduces the bound to at most $2^{m-s} \sum_{i=0}^{AI(F)-1} \binom{n}{i}$ and when s grows, it is this term which reduces the lower bound. We show now some straightforward upper bounds:

Proposition 6. *For every positive integers m, n, $s \leq m$ and $r \leq n$ and every (n, m)-function F, we have: $NL_s(F) \leq 2^{n-s}$ and $nl_{s,r}(F) \leq 2^{n-s}$. These inequalities are strict if F is not balanced (that is, if its output is not uniformly distributed over F_2^m).*

Proof. There necessarily exists an $(m - s)$-dimensional affine subspace A of F_2^m such that the size of $F^{-1}(A)$ is at most 2^{n-s} (more precisely, for every $(m - s)$-dimensional vector subspace E of F_2^m, there exists $a \in F_2^m$ such that the size of $F^{-1}(a + E)$ is at most 2^{n-s}, since the sets $F^{-1}(a + E)$ constitute a partition of F_2^n when a ranges over a subspace of F_2^m supplementary to E, and the number of the elements of this partition equals 2^s). Taking for f the indicator of A, for g the null function and defining $h(x, y) = f(y)$, we have $w_H(h(x, F(x)) = w_H(f \circ F) \leq 2^{n-s}$, and the degree of f equals s. This proves that $NL_s(F) \leq 2^{n-s}$ and $nl_{s,r}(F) \leq 2^{n-s}$. Moreover, according to the observations above, equality is possible only if, for every $(m - s)$-dimensional affine subspace A of F_2^m, the size of $F^{-1}(A)$ equals 2^{n-s}. This implies that for every affine hyperplane H, the size of $F^{-1}(H)$ equals 2^{n-1}. And this is equivalent to saying that for every nonzero $v \in F_2^m$, the Boolean function $v \cdot F$ is balanced, which is equivalent to F balanced (see e.g. [6]).   □

We shall see in Proposition 8 that the bound $nl_{s,r}(F) \leq 2^{n-s}$ is asymptotically approximately tight for permutations when $r \leq s \leq .227\, n$.

Remark. The inequality $nl_{s,r}(F) \leq 2^{n-s}$ probably implies that we have $nl_{s,r}(F) \neq nl_{r,s}(F)$, in general and for many cases such that $r \neq s$ - at least for $s = 1$ and $r > 1$ and $m = o(n^2)$. We do not have a rigorous proof of this, except for a few functions and cases (the inverse function for $s = 1$ and $r > 1$, the Welch function for $s = 1$ and $r = 2$, ...). We give a very informal argument: applying (1) to each component function $v \cdot F$; $v \in F_2^{m*}$ (assuming that the component functions of a random S-box behave as random functions, which is not true), we have that a function $F : F_2^n \to F_2^m$ has $nl_{1,r}$ smaller than $2^{n-1} - \sqrt{\sum_{i=0}^r \binom{n}{i}}\ 2^{\frac{n-1}{2}}$ with a probability in $O((2^m - 1)2^{(1-\log_2 e)\sum_{i=0}^r \binom{n}{i}})$ which tends to 0 also, when $r > 1$, if $m = o(n^2)$. Hence, it seems that almost all functions F have an $nl_{1,r}$ greater than $2^{n-1} - \sqrt{\sum_{i=0}^r \binom{n}{i}}\ 2^{\frac{n-1}{2}}$. This kind of "argument" still works when s is greater than 1 (and is reasonably small with regard to r) and m is small enough with regard to n. It would be nice to have a formal proof of all this.

3.1   The Inverse S-Box

For $F_{inv}(x) = x^{2^n - 2}$ and $f_{inv}(x) = tr(F_{inv}(x))$, $NL_1(F_{inv}) = nl(f_{inv})$ equals $2^{n-1} - 2^{n/2}$ when n is even and is close to this number when n is odd. We have $NL_t(F_{inv}) = 0$ for all $t \geq 2$, since we have $w_H(h(x, F_{inv}(x))) = 0$ for the bilinear function $h(x, y) = tr(axy)$ where a is any nonzero element of null trace and xy denotes the product of x and y in F_{2^n}. Indeed we have $x\,F_{inv}(x) = 1$ for every nonzero x. As observed in [23], we have also $w_H(h(x, F_{inv}(x))) = 0$ for the bilinear functions $h(x, y) = tr(a(x + x^2 y))$ and $h(x, y) = tr(a(y + y^2 x))$ where a is any nonzero element, and for the quadratic functions $h(x, y) = tr(a(x^3 + x^4 y))$ and $h(x, y) = tr(a(y^3 + y^4 x))$. These properties are the core properties used in the tentative algebraic attack on the AES by Courtois and J. Pieprzyk [23].

Let us study now $nl_{s,r}(F_{inv})$:

Proposition 7. *For every ordered pair* (s, r) *of strictly positive integers, we have:*

- $nl_{s,r}(F_{inv}) = 0$ *if* $r + s \geq n$;
- $nl_{s,r}(F_{inv}) > 0$ *if* $r + s < n$.

In particular, for every ordered pair (s, r) *of positive integers such that* $r + s = n - 1$, *we have* $nl_{s,r}(F_{inv}) = 2$.

Proof. If $r + s \geq n$ then there exists an integer $d \in \{1, \cdots, 2^n - 2\}$ whose 2-weight $w_2(d)$ is at most r and such that $w_2(2^n - 1 - d) = n - w_2(d) \leq s$ (taking for instance $w_2(d) = r$ which implies $w_2(2^n - 1 - d) = n - r \leq s$). Then taking $f(x) = tr(ax^{2^n - 1 - d})$, $a \in F_{2^n}^*$ and $g(x) = tr(ax^d)$, we have, by construction $f \circ F_{inv} = g$ and since f has degree at most s and g has degree at most r, we deduce that $nl_{s,r}(F_{inv}) = 0$ (indeed, there exists at least one $a \in F_{2^n}^*$ such that f is not constant, since otherwise we would have $tr(ax^{2^n - 1 - d}) = 0$ for every $a \in F_{2^n}$ and every $x \in F_{2^n}$ while we know that $tr(ax^{2^n - 1 - d}) = 0$ for every $a \in F_{2^n}$ is impossible when $x \neq 0$).

If $r + s < n$, then consider any Boolean functions f of degree at most s and g of degree at most r. Let us consider their representation as polynomials in one variable:

$$f(x) = \sum_{i=0}^{2^n - 1} f_i x^i; \quad g(x) = \sum_{j=0}^{2^n - 1} g_j x^j; \quad f_i, g_j \in F_{2^n}$$

where $f_i = 0$ for every i such that $w_2(i) > s$ and $g_j = 0$ for every j such that $w_2(j) > r$. Assume that $g = f \circ F_{inv}$. We have then $f_0 = g_0$. Without loss of generality, we can assume that $f_0 = g_0 = 0$. Since the composition of the function x^i with F_{inv} (on the right) equals the function $x^{2^n - 1 - i}$ for every $i \neq 0$, $2^n - 1$, and since $s < 2^n - 1$, the equality $f \circ F_{inv} = g$ is then equivalent to the fact that the function $\sum_{i=0}^{2^n - 1} f_i x^{2^n - 1 - i} + \sum_{j=0}^{2^n - 1} g_j x^j$ is null. For every $i \neq 0$ such that $w_2(i) \leq s$ and every $j \neq 0$ such that $w_2(j) \leq r$, the inequalities $w_2(2^n - 1 - i) \geq n - s > r$ and $w_2(j) \leq r$ imply that $2^n - 1 - i \neq j$. According to the uniqueness of the representation as a polynomial in one variable, this implies that $f = g = 0$. Hence we have $nl_{s,r}(F_{inv}) > 0$.

In particular, if $r + s = n - 1$, then let:

$$f(x) = \sum_{0 < i < 2^n - 1 \,/\, w_2(i) \leq s} x^i \text{ and } g(x) = \sum_{0 \leq i < 2^n - 1 \,/\, w_2(i) \leq r} x^i.$$

Note that f and g are both Boolean, since $f_{2i} = f_i^2$ and $g_{2i} = g_i^2$ for every $i \neq 0$, $2^n - 1$ and $f_0, f_{2^n - 1}, g_0, g_{2^n - 1} \in F_2$. Then we have $(f \circ F_{inv})(x) +

$$g(x) = \sum_{0 \leq i < 2^n - 1} x^i \text{ (since } n - s = r + 1\text{). Hence } (f \circ F_{inv})(0) + g(0) = 1,$$

$(f \circ F_{inv})(1) + g(1) = 1$ and, if $x \notin F_2$, $(f \circ F_{inv})(x) + g(x) = \dfrac{1 + x^{2^n - 1}}{1 + x} = 0$.
Hence we have $nl_{s,r}(F_{inv}) \leq 2$ and since we know that $nl_{s,r}(F_{inv})$ is strictly positive and that it is even, we deduce that $nl_{s,r}(F_{inv}) = 2$. $\square$

3.2   Existence of Permutations with Lower Bounded Higher Order Nonlinearities

We investigate now, with a simple counting argument, the existence of permutations F such that $nl_{s,r}(F) > D$ for some values of n, s, r and D.

Lemma 1. *Let n and s be positive integers and let r be a non-negative integer. If $2^{\sum_{i=0}^{s} \binom{n}{i} + \sum_{i=0}^{r} \binom{n}{i}} \leq \binom{2^n}{2^{n-s}}$ then there exist permutations F from F_2^n to itself whose higher order nonlinearity $nl_{s,r}(F)$ is strictly greater than D for every D such that $\sum_{t=0}^{D} \binom{2^n}{t} \leq \dfrac{\binom{2^n}{2^{n-s}}}{2^{\sum_{i=0}^{s} \binom{n}{i} + \sum_{i=0}^{r} \binom{n}{i}}}$.*

Proof. For every integers $i \in [0, 2^n]$ and r, let us denote by $A_{r,i}$ the number of codewords of Hamming weight i in the Reed-Muller code of order r. Given a number D, a permutation F and two Boolean functions f and g, if we have $d_H(f \circ F, g) \leq D$ then F^{-1} maps the support $supp(f)$ of f to the symmetric difference $supp(g) \Delta E$ between $supp(g)$ and a set E of size at most D (equal to the symmetric difference between $F^{-1}(supp(f))$ and $supp(g)$). And F^{-1} maps $F_2^n \setminus supp(f)$ to the symmetric difference between $F_2^n \setminus supp(g)$ and E. Given f, g and E and denoting by i the size of $supp(f)$ (with $0 < i < 2^n$, since $f \neq cst$), the number of permutations, whose restriction to $supp(g) \Delta E$ is a one-to-one function onto $supp(f)$ and whose restriction to $(F_2^n \setminus supp(g)) \Delta E$ is a one-to-one function onto $F_2^n \setminus supp(f)$, equals $i! (2^n - i)!$. Denoting by j the size of $supp(g)$, then $d_H(g, f \circ F) \leq D$ implies $|i - j| \leq D$. We deduce that the number of permutations F such that $nl_{s,r}(F) \leq D$ is upper bounded by

$$\sum_{t=0}^{D} \binom{2^n}{t} \sum_{0 < i < 2^n} \sum_{j \,/\, |i-j| \leq D} A_{s,i} A_{r,j} \, i! \, (2^n - i)!$$

Since the nonconstant codewords of the Reed-Muller code of order s have weights between 2^{n-s} and $2^n - 2^{n-s}$, we deduce that the probability $P_{s,r,D}$ that a permutation F chosen at random (with uniform probability) satisfies $nl_{s,r}(F) \leq D$ is upper bounded by

$$\sum_{t=0}^{D} \binom{2^n}{t} \sum_{j=0}^{2^n} A_{r,j} \sum_{2^{n-s} \leq i \leq 2^n - 2^{n-s}} A_{s,i} \frac{i! \, (2^n - i)!}{2^n!} =$$

$$\sum_{t=0}^{D} \binom{2^n}{t} \sum_{j=0}^{2^n} A_{r,j} \sum_{2^{n-s} \leq i \leq 2^n - 2^{n-s}} \frac{A_{s,i}}{\binom{2^n}{i}}$$

$$< \frac{\left(\sum_{t=0}^{D} \binom{2^n}{t}\right) 2^{\sum_{i=0}^{s} \binom{n}{i} + \sum_{i=0}^{r} \binom{n}{i}}}{\binom{2^n}{2^{n-s}}}. \tag{4}$$

If this upper bound is at most 1, then we deduce that $P_{s,r,D} < 1$ and this proves that there exist permutations F from F_2^n to itself whose higher order nonlinearity $nl_{s,r}(F)$ is strictly greater than D. This completes the proof.   □

We give in Table 5 below the values obtained from Lemma 1, such that, for each of them, there surely exists a permutation F whose higher nonlinearity $nl_{s,r}(F)$ is not smaller. Note that since the weight of a Boolean function of algebraic degree strictly smaller than n is even, we could increase any odd number in this table by 1 (but we prefer leaving the values as they are given by Lemma 1). Note also that these lower bounds may be improved by using the results by Kasami et al. [37,38,39] on the values of $A_{s,i}$ when $i \leq 2.5 \cdot 2^{n-s}$.

We are also interested in what happens when n tends to ∞. Let $H_2(x) = -x \log_2(x) - (1-x) \log_2(1-x)$ be the entropy function.

Proposition 8. *Let s_n be a sequence of integers tending to ∞ and such that, asymptotically, $s_n \leq .227\, n$. Then for every sequence r_n and D_n of positive*

Table 5. Lower bounds, deduced from Lemma 1, on the highest possible values of $nl_{s,r}(F)$, when F is a permutation and for $n \leq 12$

n	s	$r=1$	2	3	4	5	6	7	8	9
4	1	2								
5	1	6	3							
	2	2								
6	1	16	9	4						
	2	6	2							
7	1	39	26	13	4					
	2	16	10	3						
8	1	89	66	39	17	5				
	2	42	31	15	2					
	3	7	2							
9	1	196	158	105	54	21	5			
	2	99	83	52	19					
	3	25	17	3						
10	1	422	359	261	156	73	25	6		
	2	219	196	145	77	20				
	3	72	60	33						
11	1	890	788	614	408	223	95	30	7	
	2	466	436	356	229	100	13			
	3	178	163	119	48					
	4	15	8							
12	1	1849	1686	1389	1004	615	308	121	35	7
	2	969	931	814	597	332	112			
	3	409	388	324	197	44				
	4	82	71	36						

integers such that $2^{-n(1-H_2(r_n/n))} + H_2\left(\frac{D_n}{2^n}\right) = o(2^{-s_n})$, *almost all permutations* F *of* F_2^n *satisfy* $nl_{s_n,r_n}(F) > D_n$.

If $\frac{s_n}{n}$ *tends to a limit* $\rho \le .227$ *when* n *tends to* ∞ *and if* $r_n \le \mu n$ *for every* n, *where* $1 - H_2(\mu) > \rho$ *(e.g. if* r_n/s_n *tends to a limit strictly smaller than 1), then for every* $\rho' > \rho$, *almost all permutations* F *of* F_2^n *satisfy* $nl_{s_n,r_n}(F) \ge 2^{(1-\rho')n}$.

Proof. We know (see e.g. [43], page 310) that, for every integer n and every $\lambda \in [0, 1/2]$, we have $\sum_{i \le \lambda n} \binom{n}{i} \le 2^{nH_2(\lambda)}$. According to the Stirling formula, we have also, when i and j tend to ∞: $i! \sim i^i e^{-i}\sqrt{2\pi i}$ and $\binom{i+j}{i} \sim \frac{(\frac{i+j}{i})^i(\frac{i+j}{j})^j}{\sqrt{2\pi}}\sqrt{\frac{i+j}{ij}}$. For $i + j = 2^n$ and $i = 2^{n-s_n}$, this gives

$$\binom{2^n}{2^{n-s_n}} \sim \frac{(2^{s_n})^{2^{n-s_n}}}{\sqrt{2\pi}(1 - 2^{-s_n})^{2^n - 2^{n-s_n}}}\sqrt{\frac{2^{s_n}}{2^n - 2^{n-s_n}}}$$

$$= \frac{2^{s_n 2^{n-s_n}}}{\sqrt{2\pi}\, 2^{(2^n - 2^{n-s_n})\ln(1-2^{-s_n})\log_2 e}}\sqrt{\frac{2^{s_n}}{2^n - 2^{n-s_n}}}.$$

We deduce then from inequality (4):

$$\log_2 P_{s_n,r_n,D_n} = O\left(2^n\left[H_2\left(\frac{D_n}{2^n}\right) + 2^{-n(1-H_2(s_n/n))} + 2^{-n(1-H_2(r_n/n))}\right.\right.$$

$$\left.\left. -2^{-s_n+\log_2(s_n)} - 2^{-s_n}(1 - 2^{-s_n})\log_2 e\right]\right)$$

(we omit $-\frac{s_n}{2^{n+1}} + \frac{n}{2^{n+1}}\log_2(1 - 2^{-s_n})$ inside the brackets above since it is negligible).

For $\rho \le .227$, we have $1 - H_2(\rho) > \rho$ and therefore, if asymptotically we have $s_n \le .227\, n$, then $2^{-n(1-H_2(s_n/n))}$ is negligible with respect to 2^{-s_n} (and therefore to $2^{-s_n+\log_2(s_n)}$ and to $2^{-s_n}(1-2^{-s_n})\log_2 e$). This completes the proof that if $2^{-n(1-H_2(r_n/n))} + H_2\left(\frac{D_n}{2^n}\right) = o(2^{-s_n})$ and $s_n \le .227\, n$, then almost all permutations F of F_2^n satisfy $nl_{s_n,r_n}(F) > D_n$.

If $\lim \frac{s_n}{n} = \rho \le .227$ then there exists $\rho' > \rho$ such that $1 - H_2(\rho') > \rho'$ and such that asymptotically we have $s_n \le \rho'\, n$; hence $2^{-n(1-H_2(s_n/n))}$ is negligible with respect to 2^{-s_n}. And if $r_n \le \mu n$ where $1 - H_2(\mu) > \rho$, then we have $2^{-n(1-H_2(r_n/n))} = o(2^{-s_n})$ and for $D_n = 2^{(1-\rho')n}$ where ρ' is any number strictly greater than ρ, we have $H_2\left(\frac{D_n}{2^n}\right) = H_2\left(2^{-\rho'n}\right) = \rho'n\,2^{-\rho'n} - (1 - 2^{-\rho'n})\log_2(1 - 2^{-\rho'n}) = o(2^{-\rho n}) = o(2^{-s_n})$. We obtain that, asymptotically, $nl_{s_n,r_n}(F) > 2^{(1-\rho')n}$ for every $\rho' > \rho$. □

References

1. Armknecht, F., Carlet, C., Gaborit, P., Künzli, S., Meier, W., Ruatta, O.: Efficient computation of algebraic immunity for algebraic and fast algebraic attacks. In: Vaudenay, S. (ed.) EUROCRYPT 2006. LNCS, vol. 4004, pp. 147–164. Springer, Heidelberg (2006)

2. Armknecht, F., Krause, M.: Constructing single- and multi-output boolean functions with maximal immunity. In: Bugliesi, M., Preneel, B., Sassone, V., Wegener, I. (eds.) ICALP 2006. LNCS, vol. 4052, pp. 180–191. Springer, Heidelberg (2006)
3. Baignères, T., Junod, P., Vaudenay, S.: How far can we go beyond linear cryptanalysis? In: Lee, P.J. (ed.) ASIACRYPT 2004. LNCS, vol. 3329, pp. 432–450. Springer, Heidelberg (2004)
4. Braeken, A., Preneel, B.: On the Algebraic Immunity of Symmetric Boolean Functions. In: Maitra, S., Veni Madhavan, C.E., Venkatesan, R. (eds.) INDOCRYPT 2005. LNCS, vol. 3797, pp. 35–48. Springer, Heidelberg (2005), http://homes.esat.kuleuven.be/~abraeken/thesisAn.pdf
5. Carlet, C.: The monograph Boolean Methods and Models. In: Crama, Y., Hammer, P. (eds.) Boolean Functions for Cryptography and Error Correcting Codes. Cambridge University Press, Cambridge (to appear), http://www-rocq.inria.fr/codes/Claude.Carlet/pubs.html
6. Carlet, C.: The monography Boolean Methods and Models. In: Crama, Y., Hammer, P. (eds.) Vectorial (multi-output) Boolean Functions for Cryptography. Cambridge University Press, Cambridge (to appear), http://www-rocq.inria.fr/codes/Claude.Carlet/pubs.html
7. Carlet, C.: The complexity of Boolean functions from cryptographic viewpoint. Dagstuhl Seminar. Complexity of Boolean Functions (2006), http://drops.dagstuhl.de/portals/06111/
8. Carlet, C.: On the higher order nonlinearities of algebraic immune functions. In: Dwork, C. (ed.) CRYPTO 2006. LNCS, vol. 4117, pp. 584–601. Springer, Heidelberg (2006)
9. Carlet, C.: Recursive Lower Bounds on the Nonlinearity Profile of Boolean Functions and Their Applications. IEEE Trans. Inform. Theory 54(3), 1262–1272 (2008)
10. Carlet, C.: A method of construction of balanced functions with optimum algebraic immunity. In: The Proceedings of the International Workshop on Coding and Cryptography, The Wuyi Mountain, Fujiang, China, June 11-15, 2007. Series of Coding and Cryptology, vol. 4, World Scientific Publishing Co., Singapore (2008)
11. Carlet, C., Dalai, D., Gupta, K., Maitra, S.: Algebraic Immunity for Cryptographically Significant Boolean Functions: Analysis and Construction. IEEE Trans. Inform. Theory 52(7), 3105–3121 (2006)
12. Carlet, C., Ding, C.: Nonlinearities of S-boxes. Finite Fields and its Applications 13(1), 121–135 (2007)
13. Carlet, C., Feng, K.: New balanced Boolean functions satisfying all the main cryptographic criteria. IACR cryptology e-print archive 2008/244
14. Carlet, C., Mesnager, S.: Improving the upper bounds on the covering radii of binary Reed-Muller codes. IEEE Trans. on Inform. Theory 53, 162–173 (2007)
15. Carlitz, L., Uchiyama, S.: Bounds for exponential sums. Duke Math. Journal 1, 37–41 (1957)
16. Chabaud, F., Vaudenay, S.: Links between Differential and Linear Cryptanalysis. In: De Santis, A. (ed.) EUROCRYPT 1994. LNCS, vol. 950, pp. 356–365. Springer, Heidelberg (1995)
17. Charpin, P., Helleseth, T., Zinoviev, V.: Propagation characteristics of $x \to x^{-1}$ and Kloosterman sums. Finite Fields and their Applications 13(2), 366–381 (2007)
18. Cheon, J.H., Lee, D.H.: Resistance of S-Boxes against Algebraic Attacks. In: Roy, B., Meier, W. (eds.) FSE 2004. LNCS, vol. 3017, pp. 83–94. Springer, Heidelberg (2004)
19. Cohen, G., Honkala, I., Litsyn, S., Lobstein, A.: Covering codes. North-Holland, Amsterdam (1997)

20. Courtois, N.: Higher order correlation attacks, XL algorithm and cryptanalysis of Toyocrypt. In: Lee, P.J., Lim, C.H. (eds.) ICISC 2002. LNCS, vol. 2587, pp. 182–199. Springer, Heidelberg (2003)
21. Courtois, N., Debraize, B., Garrido, E.: On exact algebraic [non-]immunity of S-boxes based on power functions. IACR e-print archive 2005/203
22. Courtois, N., Meier, W.: Algebraic attacks on stream ciphers with linear feedback. In: Biham, E. (ed.) EUROCRYPT 2003. LNCS, vol. 2656, pp. 345–359. Springer, Heidelberg (2003)
23. Courtois, N., Pieprzyk, J.: Cryptanalysis of block ciphers with overdefined systems of equations. In: Zheng, Y. (ed.) ASIACRYPT 2002. LNCS, vol. 2501, pp. 267–287. Springer, Heidelberg (2002)
24. Daemen, J., Rijmen, V.: AES proposal: Rijndael (1999), http://csrc.nist.gov/encryption/aes/rijndael/Rijndael.pdf
25. Dalai, D.K., Maitra, S., Sarkar, S.: Basic Theory in Construction of Boolean Functions with Maximum Possible Annihilator Immunity. Designs, Codes Cryptogr. 40(1), 41–58 (2006), http://eprint.iacr.org/
26. Dalai, D.K., Gupta, K.C., Maitra, S.: Cryptographically Significant Boolean functions: Construction and Analysis in terms of Algebraic Immunity. In: Gilbert, H., Handschuh, H. (eds.) FSE 2005. LNCS, vol. 3557, pp. 98–111. Springer, Heidelberg (2005)
27. Didier, F.: A new upper bound on the block error probability after decoding over the erasure channel. IEEE Trans. Inform. Theory 52, 4496–4503 (2006)
28. Dumer, I., Kabatiansky, G., Tavernier, C.: List decoding of Reed-Muller codes up to the Johnson bound with almost linear complexity. In: Proceedings of ISIT 2006, Seattle, USA (2006)
29. Fourquet, R.: Une FFT adaptée au décodage par liste dans les codes de Reed-Muller d'ordres 1 et 2. Master-thesis of the University of Paris VIII, Thales communication, Bois Colombes (2006)
30. Fourquet, R.: Private communication (2007)
31. Fourquet, R., Tavernier, C.: An improved list decoding algorithm for the second order Reed-Muller codes and its applications. Des. Codes Cryptogr (to appear, 2008)
32. Fourquet, R., Tavernier, C.: Private communication (2008)
33. Golic, J.: Fast low order approximation of cryptographic functions. In: Maurer, U.M. (ed.) EUROCRYPT 1996. LNCS, vol. 1070, pp. 268–282. Springer, Heidelberg (1996)
34. Iwata, T., Kurosawa, K.: Probabilistic higher order differential attack and higher order bent functions. In: Lam, K.-Y., Okamoto, E., Xing, C. (eds.) ASIACRYPT 1999. LNCS, vol. 1716, pp. 62–74. Springer, Heidelberg (1999)
35. Kabatiansky, G., Tavernier, C.: List decoding of second order Reed-Muller codes. In: Proc. 8th Int. Symp. Comm. Theory and Applications, Ambleside, UK (July 2005)
36. Kaliski, B., Robshaw, M.: Linear cryptanalysis using multiple approximations. In: Desmedt, Y.G. (ed.) CRYPTO 1994. LNCS, vol. 839, pp. 26–38. Springer, Heidelberg (1994)
37. Kasami, T., Tokura, N.: On the weight structure of Reed-Muller codes. IEEE Trans. Inform. Theory IT-16 (6), 752–825 (1970)
38. Kasami, T., Tokura, N., Azumi, S.: On the weight enumeration of weights less than 2.5d of Reed-Muller codes. Information and Control 30, 380–395 (1976)

39. Kasami, T., Tokura, N., Azumi, S.: On the weight enumeration of weights less than 2.5d of Reed-Muller codes. Report of faculty of Eng. Sci, Osaka Univ., Japan
40. Knudsen, L.R., Robshaw, M.J.B.: Non-linear approximations in linear cryptanalysis. In: Maurer, U.M. (ed.) EUROCRYPT 1996. LNCS, vol. 1070, pp. 224–236. Springer, Heidelberg (1996)
41. Lachaud, G., Wolfmann, J.: The Weights of the Orthogonals of the Extended Quadratic Binary Goppa Codes. IEEE Trans. Inform. Theory 36, 686–692 (1990)
42. Lobanov, M.: Tight bound between nonlinearity and algebraic immunity. Paper 2005/441 http://eprint.iacr.org/
43. MacWilliams, F.J., Sloane, N.J.A.: The theory of error-correcting codes. North-Holland, Amsterdam (1977)
44. Maurer, U.M.: New approaches to the design of self-synchronizing stream ciphers. In: Davies, D.W. (ed.) EUROCRYPT 1991. LNCS, vol. 547, pp. 458–471. Springer, Heidelberg (1991)
45. Meier, W., Pasalic, E., Carlet, C.: Algebraic attacks and decomposition of Boolean functions. In: Cachin, C., Camenisch, J.L. (eds.) EUROCRYPT 2004. LNCS, vol. 3027, pp. 474–491. Springer, Heidelberg (2004)
46. Mesnager, S.: Improving the lower bound on the higher order nonlinearity of Boolean functions with prescribed algebraic immunity. IEEE Trans. Inform. Theory 54(8) (August 2008); Preliminary version available at Cryptology ePrint Archive, no. 2007/117
47. Millan, W.: Low order approximation of cipher functions. In: Dawson, E.P., Golić, J.D. (eds.) Cryptography: Policy and Algorithms 1995. LNCS, vol. 1029, pp. 144–155. Springer, Heidelberg (1996)
48. Shanbhag, A., Kumar, V., Helleseth, T.: An upper bound for the extended Kloosterman sums over Galois rings. Finite Fields and their Applications 4, 218–238 (1998)
49. Shannon, C.E.: Communication theory of secrecy systems. Bell system technical journal 28, 656–715 (1949)
50. Shimoyama, T., Kaneko, T.: Quadratic Relation of S-box and Its Application to the Linear Attack of Full Round DES. In: Krawczyk, H. (ed.) CRYPTO 1998. LNCS, vol. 1462, pp. 200–211. Springer, Heidelberg (1998)

On a Class of Permutation Polynomials over $\mathbb{F}_{2^n}$

Pascale Charpin[1] and Gohar M. Kyureghyan[2]

[1] INRIA, SECRET research Team, B.P. 105, 78153 Le Chesnay Cedex, France
pascale.charpin@inria.fr
[2] Department of Mathematics, Otto-von-Guericke-University Magdeburg,
Universitätsplatz 2, 39106 Magdeburg, Germany
gohar.kyureghyan@ovgu.de

Abstract. We study permutation polynomials of the shape $F(X) = G(X) + \gamma \, Tr(H(X))$ over $\mathbb{F}_{2^n}$. We prove that if the polynomial $G(X)$ is a permutation polynomial or a linearized polynomial, then the considered problem can be reduced to finding Boolean functions with linear structures. Using this observation we describe six classes of such permutation polynomials.

Keywords: Permutation polynomial, linear structure, linearized polynomial, trace, Boolean function.

1 Introduction

Let $\mathbb{F}_{2^n}$ be the finite field with 2^n elements. A polynomial $F(X) \in \mathbb{F}_{2^n}[X]$ is called a permutation polynomial (PP) of $\mathbb{F}_{2^n}$ if the associated polynomial mapping

$$F : \mathbb{F}_{2^n} \to \mathbb{F}_{2^n},$$
$$x \mapsto F(x)$$

is a permutation of $\mathbb{F}_{2^n}$. There are several criteria ensuring that a given polynomial is a PP, but those conditions are, however, rather complicated, cf. [7]. PP are involved in many applications of finite fields, especiallly in cryptography, coding theory and combinatorial design theory. Finding PP of a special type is of great interest for the both theoretical and applied aspects.

In this paper we study PP of the following shape

$$F(X) = G(X) + \gamma \, Tr(H(X)), \tag{1}$$

where $G(X), H(X) \in \mathbb{F}_{2^n}[X]$, $\gamma \in \mathbb{F}_{2^n}$ and $Tr(X) = \sum_{i=0}^{n-1} X^{2^i}$ is the polynomial defining the absolute trace function of $\mathbb{F}_{2^n}$. Examples of such polynomials are obtained in [3],[6] and [9]. We show that in the case the polynomial $G(X)$ is a PP or a linearized polynomial the considered problem can be reduced to finding Boolean functions with linear structures. We use this observation to describe six classes of PP of type (1).

S.W. Golomb et al. (Eds.): SETA 2008, LNCS 5203, pp. 368–376, 2008.
© Springer-Verlag Berlin Heidelberg 2008

2   A Linear Structure of a Boolean Function

A Boolean function from $\mathbb{F}_{2^n}$ to $\mathbb{F}_2$ can be represented as $Tr(R(x))$ for some (not unique) mapping $R : \mathbb{F}_{2^n} \to \mathbb{F}_{2^n}$. A Boolean function $Tr(R(x))$ is said to have a linear structure $\alpha \in \mathbb{F}_{2^n}^*$ if

$$Tr(R(x)) + Tr(R(x + \alpha)) = Tr(R(x) + R(x + \alpha))$$

is a constant function. We call a linear structure c-linear structure if

$$Tr(R(x) + R(x + \alpha)) \equiv c,$$

where $c \in \mathbb{F}_2$. Given $\gamma \in \mathbb{F}_{2^n}^*$ and $c \in \mathbb{F}_2$, let $H_\gamma(c)$ denote the affine hyperplane defined by the equation $Tr(\gamma x) = c$, i.e.,

$$H_\gamma(c) = \{x \in \mathbb{F}_{2^n} \mid Tr(\gamma x) = c\}.$$

Then $\alpha \in \mathbb{F}_{2^n}^*$ is a c-linear structure for $Tr(R(x))$ if and only if the image set of the mapping $R(x) + R(x + \alpha)$ is contained in the affine hyperplane $H_1(c)$.

The Walsh transform of a Boolean function $Tr(R(x))$ is defined as follows

$$\mathcal{W} : \mathbb{F}_{2^n} \to \mathbb{Z}, \lambda \mapsto \sum_{x \in \mathbb{F}_{2^n}} (-1)^{Tr(R(x) + \lambda x)}.$$

Whether a Boolean function $Tr(R(x))$ has a linear structure can be recognized from its Walsh transform.

Proposition 1 ([2,8]). *Let $c \in \mathbb{F}_2$ and $R : \mathbb{F}_{2^n} \to \mathbb{F}_{2^n}$. An element $\alpha \in \mathbb{F}_{2^n}^*$ is a $(c + 1)$-linear structure for $Tr(R(x))$ if and only if*

$$\mathcal{W}(\lambda) = \sum_{x \in \mathbb{F}_{2^n}} (-1)^{Tr(R(x) + \lambda x)} = 0$$

for all $\lambda \in H_\alpha(c)$.

In [5] all Boolean functions assuming a linear structure are characterized as follows.

Theorem 1 ([5]). *Let $R : \mathbb{F}_{2^n} \to \mathbb{F}_{2^n}$. Then the Boolean function $Tr(R(x))$ has a linear structure if and only if there is a non-bijective linear mapping $L : \mathbb{F}_{2^n} \to \mathbb{F}_{2^n}$ such that*

$$Tr(R(x)) = Tr(H \circ L(x) + \beta x) + c,$$

where $H : \mathbb{F}_{2^n} \to \mathbb{F}_{2^n}$, $\beta \in \mathbb{F}_{2^n}$ and $c \in \mathbb{F}_2$.

Clearly, any element from the kernel of L is a linear structure of $Tr(R(x))$ considered in Theorem 1. Moreover, those are the only ones if the mapping $Tr(H(x))$ has no linear structure belonging to the image of L. We record this observation in the following lemma to refer it later.

Lemma 1. *Let $H : \mathbb{F}_{2^n} \to \mathbb{F}_{2^n}$ be an arbitrary mapping. Then $\gamma \in \mathbb{F}_{2^n}^*$ is a linear structure of*

$$Tr\big(H(x^2 + \gamma x) + \beta x\big)$$

for any $\beta \in \mathbb{F}_{2^n}$.

Next lemma describes another family of Boolean functions having a linear structure. Its proof is straightforward.

Lemma 2. *Let $F : \mathbb{F}_{2^n} \to \mathbb{F}_{2^n}$ and $\alpha \in \mathbb{F}_{2^n}^*$. Then α is a linear structure of $Tr(F(x) + F(x + \alpha) + \beta x)$ for any $\beta \in \mathbb{F}_{2^n}$.*

In general, for a given Boolean function it is difficult to recognize whether it admits a linear structure. Slightly extending results from [4], we characterize all monomial Boolean functions assuming a linear structure. More precisely, for a given nonzero $a \in \mathbb{F}_{2^n}$, we describe all exponents s and nonzero $\delta \in \mathbb{F}_{2^n}$ such that a is a linear structure for the Boolean function $Tr(\delta x^s)$.

Let $0 \leq s \leq 2^n - 2$. We denote by C_s the cyclotomic coset modulo $2^n - 1$ containing s:

$$C_s = \{s, 2s, \ldots, 2^{n-1}s\} \pmod{2^n - 1}.$$

Note that if $|C_s| = l$, then $\{x^s \mid x \in \mathbb{F}_{2^n}\} \subseteq \mathbb{F}_{2^l}$ and $\mathbb{F}_{2^l}$ is the smallest such subfield.

The next lemma is an extension of Lemma 2 from [4].

Lemma 3. *Let $0 \leq s \leq 2^n - 2$, $\delta \in \mathbb{F}_{2^n}^*$ be such that the Boolean function $Tr(\delta x^s)$ is a nonzero function. Then $a \in \mathbb{F}_{2^n}^*$ is a linear structure of the Boolean function $Tr(\delta x^s)$ if and only if*

(a) $s = 2^i$ *and a is arbitrary*
(b) $s = 2^i + 2^j$ $(i \neq j)$ *and* $(\delta a^{2^i + 2^j})^{2^{n-i}} + (\delta a^{2^i + 2^j})^{2^{n-j}} = 0$.

Proof. Let $a \in \mathbb{F}_{2^n}^*$ be a linear structure for $Tr(\delta x^s)$. Then

$$Tr(\delta(x^s + (x + a)^s)) \equiv c \tag{2}$$

holds for all $x \in \mathbb{F}_{2^n}$ and a fixed $c \in \mathbb{F}_2$. In [4] it is shown that in the case $|C_s| = n$ the identity (2) can be satisfied only if the binary weight of s does not exceed 2. On the other side it is easy to see that for an s of binary weight 1 the corresponding Boolean function $Tr(\delta x^s)$ is linear and thus any nonzero element is a linear structure. If $s = 2^i + 2^j$, then

$$Tr(\delta(x^{2^i+2^j} + (x+a)^{2^i+2^j})) = Tr\left(\delta a^{2^i+2^j}\left(\left(\frac{x}{a}\right)^{2^i} + \left(\frac{x}{a}\right)^{2^j}\right)\right) + Tr\left(\delta a^{2^i+2^j}\right)$$

$$= Tr\left(\left((\delta a^{2^i+2^j})^{2^{n-i}} + (\delta a^{2^i+2^j})^{2^{n-j}}\right)\frac{x}{a}\right)$$

$$+ Tr\left(\delta a^{2^i+2^j}\right),$$

implying (b). To complete the proof, we need to consider the case $|C_s| = l < n$. Let $n = lm$. Then

$$Tr(\delta(x^s + (x+a)^s)) = Tr(\beta(y^s + (y+1)^s)),$$

where $y = x/a$ and $\beta = \delta a^s$. We write $i \prec s$ if $i \neq s$ and the binary representation of i is covered by the one of s. Then

$$Tr(\beta(y^s + (y+1)^s)) = \sum_{i \prec s} Tr(\beta y^i) = \sum_{k \prec s,\ k \text{ is a coset repr.}} Tr(\beta_k y^k).$$

Note that the exponents in the monomial summands $Tr(\beta_k y^k)$ are from different cyclotomy cosets. Hence to have

$$\sum_{k \prec s,\ k \text{ is a coset repr.}} Tr(\beta_k y^k) \equiv c$$

it is necessary that $c = Tr(\beta)$ and $Tr(\beta_k y^k) \equiv 0$ for all $k \neq 0$. Consider $k_0 \prec s$ such that $k_0 = s - 2^i$. Lemma 3 of [1] implies that $|C_{k_0}| = n$, and therefore $Tr(\beta_{k_0} y^{k_0}) \equiv 0$ holds only if $\beta_{k_0} = 0$. Further $\beta_{k_0} = \beta + \beta^{2^l} + \dots \beta^{2^{l(m-1)}} = Tr_l^n(\beta)$, where Tr_v^u denotes the trace function from $\mathbb{F}_{2^u}$ onto its subfield $\mathbb{F}_{2^v}$. Hence necessarily $Tr_l^n(\beta) = Tr_l^n(\delta a^s) = a^s Tr_l^n(\delta) = 0$, and thus the Boolean function

$$Tr(\delta x^s) = Tr_1^l(x^s Tr_l^n(\delta))$$

is the zero function. □

Observe that $\delta = a^{-(2^i + 2^j)}$ satisfies condition (b) of Lemma 3.

3   Permutation Polynomials

In this section we study permutation polynomials of the shape

$$F(X) = G(X) + \gamma Tr(H(X)),$$

where $G(X), H(X) \in \mathbb{F}_{2^n}[X]$, $\gamma \in \mathbb{F}_{2^n}$. Firstly we observe the following necessary property of $G(X)$.

Claim. Let $G(X), H(X) \in \mathbb{F}_{2^n}[X]$ and $\gamma \in \mathbb{F}_{2^n}$. If

$$F(X) = G(X) + \gamma Tr(H(X))$$

is a PP of $\mathbb{F}_{2^n}$, then for any $\beta \in \mathbb{F}_{2^n}$ there are at most 2 elements $x_1, x_2 \in \mathbb{F}_{2^n}$ such that $G(x_1) = G(x_2) = \beta$.

Proof. Suppose there are different x_1, x_2, x_3 with $G(x_1) = G(x_2) = G(x_3) = \beta$. Then F cannot be a PP, since $F(x_i) \in \{\beta, \beta + \gamma\}$ for $i = 1, 2, 3$. □

Proposition 2. *Let* $G(X), H(X) \in \mathbb{F}_{2^n}[X]$ *and* $\gamma \in \mathbb{F}_{2^n}$. *Then*

$$F(X) = G(X) + \gamma \, Tr(H(X))$$

is a PP of $\mathbb{F}_{2^n}$ *if and only if for any* $\lambda \in \mathbb{F}_{2^n}^*$ *it holds*

$$\sum_{x \in \mathbb{F}_{2^n}} (-1)^{Tr(\lambda G(x))} = 0 \quad \text{if } Tr(\gamma \lambda) = 0 \tag{3}$$

$$\sum_{x \in \mathbb{F}_{2^n}} (-1)^{Tr(\lambda G(x) + H(x))} = 0 \quad \text{if } Tr(\gamma \lambda) = 1. \tag{4}$$

Proof. Recall that $F(X)$ is a PP if and only if

$$\sum_{x \in \mathbb{F}_{2^n}} (-1)^{Tr(\lambda F(x))} = 0$$

for all $\lambda \in \mathbb{F}_{2^n}^*$, cf. [7]. Since

$$Tr(\lambda F(x)) = Tr(\lambda G(x)) + Tr(H(x))Tr(\gamma \lambda) = Tr(\lambda G(x) + H(x)Tr(\gamma \lambda)),$$

it must hold

$$\sum_{x \in \mathbb{F}_{2^n}} (-1)^{Tr(\lambda F(x))} = \begin{cases} \sum_{x \in \mathbb{F}_{2^n}} (-1)^{Tr(\lambda G(x))} = 0 & \text{if } Tr(\gamma \lambda) = 0 \\ \sum_{x \in \mathbb{F}_{2^n}} (-1)^{Tr(\lambda G(x) + H(x))} = 0 & \text{if } Tr(\gamma \lambda) = 1. \end{cases}$$

□

Next we consider polynomials $F(X) = G(X) + \gamma \, Tr(H(X))$, where $G(X)$ is a PP or a linearized polynomial.

3.1 $G(X)$ Is a Permutation Polynomial

Firstly we establish a connection of the considered problem with the Boolean functions assuming a linear structure.

Theorem 2. *Let* $G(X), H(X) \in \mathbb{F}_{2^n}[X]$, $\gamma \in \mathbb{F}_{2^n}$ *and* $G(X)$ *be a PP. Then*

$$F(X) = G(X) + \gamma \, Tr(H(X)) \tag{5}$$

is a PP of $\mathbb{F}_{2^n}$ *if and only if* $H(X) = R(G(X))$, *where* $R(X) \in \mathbb{F}_{2^n}[X]$ *and* γ *is a 0-linear structure of the Boolean function* $Tr(R(x))$.

Proof. Since $G(X)$ is a PP, condition (3) is satisfied. Let G^{-1} be the inverse mapping of the associated mapping of G. Then condition (4) is equivalent to

$$\sum_{x \in \mathbb{F}_{2^n}} (-1)^{Tr(\lambda G(x) + H(x))} = \sum_{y \in \mathbb{F}_{2^n}} (-1)^{Tr(\lambda y + H(G^{-1}(y)))} = 0$$

for all $\lambda \in \mathbb{F}_{2^n}$ with $Tr(\gamma \lambda) = 1$. Proposition 1 completes the proof.   □

From Theorem 2 it follows that any PP of type (5) is obtained by substituting $G(X)$ into a PP of shape $X + \gamma Tr(R(X))$. The next theorem describes two classes of such polynomials.

Theorem 3. *Let* $\gamma, \beta \in \mathbb{F}_{2^n}$ *and* $H(X) \in \mathbb{F}_{2^n}[X]$.

(a) *Then the polynomial*

$$X + \gamma Tr \left(H(X^2 + \gamma X) + \beta X \right)$$

is PP if and only if $Tr(\beta\gamma) = 0$.
(b) *Then the polynomial*

$$X + \gamma Tr \left(H(X) + H(X + \gamma) + \beta X \right)$$

is PP if and only if $Tr(\beta\gamma) = 0$.

Proof. (a) By Theorem 2 the considered polynomial is a PP if and only if γ is a 0-linear structure of $Tr \left(H(x^2 + \gamma x) + \beta x \right)$. To complete the proof note that

$$Tr \left(H((x + \gamma)^2 + \gamma(x + \gamma)) + \beta(x + \gamma) \right) + Tr \left(H(x^2 + \gamma x) + \beta x \right) = Tr(\beta\gamma).$$

(b) The proof follows from Lemma 2 and Theorem 2 similarly to the previous case.   □

Our next goal is to characterize all permutation polynomials of shape $X + \gamma Tr(\delta X^s + \beta X)$. Firstly, observe that if $s = 2^i$, then Theorem 2 yields that $X + \gamma Tr(\delta X^{2^i} + \beta X)$ is a PP if and only if $Tr(\delta\gamma^{2^i} + \beta\gamma) = 0$. The remaining cases are covered in the following theorem.

Theorem 4. *Let* $\gamma, \beta \in \mathbb{F}_{2^n}$ *and* $3 \le s \le 2^n - 2$ *be of binary weight* ≥ 2. *Let* $\delta \in \mathbb{F}_{2^n}$ *be such that the Boolean function* $x \mapsto Tr(\delta x^s)$, $x \in \mathbb{F}_{2^n}$, *is not the zero function. Then the polynomial*

$$X + \gamma Tr(\delta X^s + \beta X)$$

is PP if and only if $s = 2^i + 2^j$, $(\delta\gamma^{2^j})^{2^{n-i}} + (\delta\gamma^{2^i})^{2^{n-j}} = 0$ *and* $Tr(\delta\gamma^{2^i + 2^j} + \beta\gamma) = 0$.

Proof. By Theorem 2 the polynomial $X + \gamma Tr(\delta X^s + \beta X)$ defines a permutation if and only if γ is a 0-linear structure of $Tr(\delta x^s + \beta x)$. Then Lemma 3 implies that the binary weight of s must be 2. Note that for $s = 2^i + 2^j$ it holds

$$Tr(\delta(x + \gamma)^{2^i + 2^j} + \beta(x + \gamma)) + Tr(\delta x^{2^i + 2^j} + \beta x)$$
$$= \quad Tr(\delta x^{2^i}\gamma^{2^j} + \delta x^{2^j}\gamma^{2^i} + \delta\gamma^{2^i + 2^j} + \beta\gamma)$$
$$= Tr \left(((\delta\gamma^{2^j})^{2^{n-i}} + (\delta\gamma^{2^i})^{2^{n-j}})x \right) + Tr(\delta\gamma^{2^i + 2^j} + \beta\gamma).$$

Thus γ is a 0-linear structure of $Tr(\delta x^s + \beta x)$ if and only if $(\delta\gamma^{2^j})^{2^{n-i}} + (\delta\gamma^{2^i})^{2^{n-j}} = 0$ and $Tr(\delta\gamma^{2^i + 2^j} + \beta\gamma) = 0$.   □

As an application of Theorem 4 we get the complete characterization of PP of type $X^d + Tr(X^t)$.

Corollary 1. *Let $1 \leq d, t \leq 2^n - 2$. Then*

$$X^d + Tr(X^t)$$

is PP over $\mathbb{F}_{2^n}$ if and only if the following conditions are satisfied:

- n *is even*
- $\gcd(d, 2^n - 1) = 1$
- $t = d \cdot s \pmod{2^n - 1}$ *for some s such that $1 \leq s \leq 2^n - 2$ and has binary weight 1 or 2.*

Proof. By Claim 3 the considered polynomial defines a permutation on $\mathbb{F}_{2^n}$ only if X^d does it, which forces $\gcd(d, 2^n - 1) = 1$. Let d^{-1} be the multiplicative inverse of d modulo $2^n - 1$. Then $X^d + Tr(X^t)$ is PP if and only if $X + Tr(X^{d^{-1} \cdot t})$ is PP. Theorems 2 and 4 with $\gamma = \delta = 1$ and $\beta = 0$ imply that the later polynomial is PP if and only if $d^{-1} \cdot t = 2^i + 2^j \pmod{2^n - 1}$ with $i \geq j$ and $Tr(1) = 0$. Finally note that $Tr(1) = 0$ if and only if n is even.   □

3.2 $G(X)$ Is a Linearized Polynomial

Let $G(X) = L(X)$ be a linearized polynomial over $\mathbb{F}_{2^n}$. In this subsection we characterize elements $\gamma \in \mathbb{F}_{2^n}$ and polynomials $H(X) \in \mathbb{F}_{2^n}[X]$ for which $L(X) + \gamma Tr(H(X))$ is PP. By Claim 3 the mapping defined by L must necessarily be bijective or 2-to-1. Since the case of bijective L is covered in the previous subsection, we consider here 2-to-1 linear mappings.

Lemma 4. *Let $L : \mathbb{F}_{2^n} \to \mathbb{F}_{2^n}$ be a linear 2-to-1 mapping with kernel $\{0, \alpha\}$ and $H : \mathbb{F}_{2^n} \to \mathbb{F}_{2^n}$. If for some $\gamma \in \mathbb{F}_{2^n}$ the mapping*

$$N(x) = L(x) + \gamma Tr(H(x))$$

is a permutation of $\mathbb{F}_{2^n}$, then γ does not belong to the image set of L. Moreover, for such an element γ the mapping $N(x)$ is a permutation if and only if α is a 1-linear structure for $Tr(H(x))$.

Proof. Note that if γ belongs to the image set of L, then the image set of N is contained in that of L. In particular, N is not a permutation. We suppose now γ does not belong to the image set of L. It holds

$$N(x) = \begin{cases} L(x) & \text{if } Tr(H(x)) = 0 \\ L(x) + \gamma & \text{if } Tr(H(x)) = 1, \end{cases}$$

and for all $x \in \mathbb{F}_{2^n}$ we have

$$N(x) + N(x + \alpha) = \gamma Tr(H(x) + H(x + \alpha)).$$

Thus, if N is a permutation, then $Tr(H(x) + H(x + \alpha)) = 1$ for all x, *i.e.*, α is a 1-linear structure for $Tr(H(x))$. Conversely, assume that

$$Tr(H(x) + H(x + \alpha)) = 1 \quad \text{for all} \quad x \in \mathbb{F}_{2^n}. \tag{6}$$

Let $y, z \in \mathbb{F}_{2^n}$ be such that $N(y) = N(z)$. If $Tr(H(y) + H(z)) = 0$ then

$$N(y) + N(z) = L(y + z) = 0,$$

and hence $y + z \in \{0, \alpha\}$. Further, (6) forces $y = z$. To complete the proof, observe that $Tr(H(y) + H(z)) = 1$ is impossible, since it implies

$$N(y) + N(z) = L(y + z) + \gamma = 0,$$

which contradicts the assumption that γ is not in the image set of L.   □

Lemmas 1, 2 in combination with Lemma 4 imply the following classes of PP.

Theorem 5. *Let $L \in \mathbb{F}_{2^n}[X]$ be a linearized polynomial, defining a 2- to -1 mapping with kernel $\{0, \alpha\}$. Further let $H \in \mathbb{F}_{2^n}[X]$, $\beta \in \mathbb{F}_{2^n}$ and $\gamma \in \mathbb{F}_{2^n}$ be not in the image set of L.*

(a) *The polynomial*

$$L(X) + \gamma\, Tr\left(H(X^2 + \alpha X) + \beta X\right)$$

is PP if and only if $Tr(\beta\alpha) = 1$.

(b) *The polynomial*

$$L(X) + \gamma\, Tr\left(H(X) + H(X + \alpha) + \beta X\right)$$

is PP if and only if $Tr(\beta\alpha) = 1$.

Remark 1. To apply Theorem 5 we need to have a linearized 2- to -1 polynomial with known kernel and image set. An example of such a polynomial is $X^{2^k} + \alpha^{2^k-1}X$ where $1 \le k \le n - 1$ with $\gcd(k, n) = 1$ and $\alpha \in \mathbb{F}_{2^n}^*$. The kernel of its associated mapping is $\{0, \alpha\}$ and the image set is $H_{\alpha^{-2^k}}(0)$. Moreover, any linear 2- to -1 mapping with kernel $\{0, \alpha\}$ (or image set $H_{\alpha^{-2^k}}(0)$) can be obtained as a left (or right) composition of this mapping with an appropriate bijective linear mapping.

The next result is a direct consequence of Lemmas 3 and 4.

Theorem 6. *Let $L \in \mathbb{F}_{2^n}[X]$ be a linearized polynomial defining a 2- to -1 mapping with kernel $\{0, \alpha\}$. Let $\beta, \gamma \in \mathbb{F}_{2^n}$ and γ do not belong to the image set of L. If $3 \le s \le 2^n - 2$ is of binary weight ≥ 2, then the polynomial*

$$L(X) + \gamma\, Tr(\delta X^s + \beta X)$$

is PP if and only if $s = 2^i + 2^j$, $(\delta\alpha^{2^j})^{2^{n-i}} + (\delta\alpha^{2^i})^{2^{n-j}} = 0$ and $Tr(\delta\alpha^{2^i+2^j} + \beta\alpha) = 1$.

Theorem 6 yields the complete characterization of PP of type $X^{2^k} + X + Tr(X^s)$.

Corollary 2. *Let $1 \leq k \leq n-1$ and $1 \leq s \leq 2^n - 2$. Then*

$$X^{2^k} + X + Tr(X^s)$$

is PP over $\mathbb{F}_{2^n}$ if and only if the following conditions are satisfied:

- *n is odd*
- *$gcd(k, n) = 1$*
- *s has binary weight 1 or 2.*

Proof. Firstly observe that the polynomial $X^{2^k} + X$ has at least two zeros, 0 and 1. Hence from Claim 3 it follows that if $X^{2^k} + X + Tr(X^s)$ is PP then necessarily the mapping $L(x) = x^{2^k} + x$ is 2-to-1. This holds if and only if $gcd(k, n) = 1$. Further note that the image set of such an L is the hyperplane $H_1(0)$. Hence $\gamma = 1$ does not belong to the image set of L if and only if $Tr(1) = 1$, equivalently if n is odd. The rest of the proof follows from Lemma 4 and Theorem 6 with $\alpha = \delta = 1$ and $\beta = 0$. □

Remark 2. Some results of this paper are valid also in the finite fields of odd characteristic. In a forthcoming paper we will report more accurately on that.

References

1. Bierbrauer, J., Kyureghyan, G.: Crooked binomials. Des. Codes Cryptogr. 46, 269–301 (2008)
2. Dubuc, S.: Characterization of linear structures. Des. Codes Cryptogr. 22, 33–45 (2001)
3. Hollmann, H.D.L., Xing, Q.: A class of permutation polynomials of $\mathbb{F}_{2^n}$ related to Dickson polynomials. Finite Fields Appl. 11(1), 111–122 (2005)
4. Kyureghyan, G.: Crooked maps in $\mathbb{F}_{2^n}$. Finite Fields Appl. 13(3), 713–726 (2007)
5. Lai, X.: Additive and linear structures of cryptographic functions. In: Preneel, B. (ed.) FSE 1994. LNCS, vol. 1008, pp. 75–85. Springer, Heidelberg (1995)
6. Laigle-Chapuy, Y.: A note on a class of quadratic permutations over F_{2^n}. In: Boztaş, S., Lu, H.-F(F.) (eds.) AAECC 2007. LNCS, vol. 4851, pp. 130–137. Springer, Heidelberg (2007)
7. Lidl, R., Niederreiter, H.: Finite Fields. Encyclopedia of Mathematics and its Applications 20
8. Yashchenko, V.V.: On the propagation criterion for Boolean functions and bent functions. Problems of Information Transmission 33(1), 62–71 (1997)
9. Yuan, J., Ding, C., Wang, H., Pieprzyk, J.: Permutation polynomials of the form $(x^p - x + \delta)^s + L(x)$. Finite Fields Appl. 14(2), 482–493 (2008)

On 3-to-1 and Power APN S-Boxes

Deepak Kumar Dalai*

Applied Statistics Unit, Indian Statistical Institute,
203, B T Road, Calcutta 700 108, India
deepak_r@isical.ac.in

Abstract. Almost Perfect Nonlinear (APN) S-boxes are used in block ciphers to prevent differential attacks. The non-evidence of permutation APN S-box on even number of variables and the efficiency of power functions bring the importance of power APN S-boxes to use in block ciphers. We present a special class of 3-to-1 S-box (named as S3-to-1 S-box) on even number of variables. The power APN S-boxes on even number of variables fall in this class. Further, another important class of APN functions $X^3 + tr(X^9)$ too falls in this class. We study some results of S3-to-1 S-boxes. In another section we present a necessary condition for power functions to be APN. Using this necessary condition we can filter out some non-APN power functions. Specifically, if the number of variables is multiple of small primes, then one can filter out many non-APN functions.

Keywords: S-box, Power Function, APN Function, Differential Cryptanalysis.

1   Introduction

We denote by V_m, the field $GF(2^m)$ of all m-dimensional binary vectors. The multi-output Boolean functions of the form $F : V_m \mapsto V_m$ are used by many block ciphers (e.g., AES, DES, RC6) for the confusion part of the round function, which are called as the substitution box (in short, S-box). Therefore, most of the cryptanalytic techniques on block ciphers are based on the analysis of cryptographic strengths of underlying S-boxes. Differential cryptanalysis is one of the important techniques to verify the strength of S-box against differential attack [2]. Differential attack can be applied successfully if the number of solutions of $F(x + a) + F(x) = b$ for $a \neq 0, b \in V_m$ are non uniform. Hence, to prevent differential attack the output derivative $F(x+a)+F(x), a \neq 0$ should be uniformly distributed. In the binary case the best that can be expected is that half the values occur twice; the S-boxes satisfy this property are called Almost Perfect Nonlinear (APN) S-boxes [7]. Apart from the application in cryptography, the APN functions have great interests in the study of coding theory and some other areas of telecommunications.

* A substantial amount of this work was done when the author was in Project CODES, INRIA, Rocquencourt, France as a postdoctoral researcher. The author is very much thankful to INRIA for providing fund to work there.

S.W. Golomb et al. (Eds.): SETA 2008, LNCS 5203, pp. 377–389, 2008.

The randomness criteria demands to use permutation S-boxes in the design of block ciphers. At the same time, the S-boxes on even number of variables are being preferred in the design of some block ciphers for the reason of easy implementation and hardware friendliness of even variable S-box. Unfortunately, there is no evidence of existence of permutation APN S-box on even number of variables. Further, the power functions are being preferred in the design of block ciphers for the reason of fast implementation of S-boxes. For an example, the most popular block cipher AES uses inverse function $(X^{-1} = X^{2^m-2})$ as underlying S-box. However, the power APN functions on even number of variables are of the form X^{3d} [1]. The APN S-boxes on even number of variables of the form X^{3d} are 3-to-1 functions (i.e, each nonzero element has either 3 or 0 pre-images and zero maps to zero). For an instance, the function X^3 is 3-to-1 function and APN when m is even. Therefore, 3-to-1 functions on even number of variables have an important role in the study of APN S-boxes. In Section 3 we have studied on a special type of 3-to-1 functions which is named as S3-to-1 functions. The Power APN functions and the function $X^3 + tr(X^9)$ falls in this category. Then we have studied for some more results on the APN property of this class of functions.

Since the power functions are being used as underlying S-boxes in many popular block ciphers, the identification of APN power functions is an important topic in the study of design of block ciphers. The complete identification of APN power functions is an exciting open problem. In Section 4, we present a necessary condition for a power function to be APN. Using the necessary condition we can filter out some non-APN power functions. The necessary condition shows that if m is multiple of small primes, one can filter out many power non-APN functions. In the following section we present some preliminary information which is required for our results.

2   Preliminary

In this paper, we always consider the S-boxes are of the form $F : V_m \mapsto V_m$. The derivative of F with respect to $a \in V_m$ is defined as follows.

Definition 1. *Let $F : V_m \mapsto V_m$ be a S-box. The derivative of F with respect to $a \in V_m$ is the function $D_a F : V_m \mapsto V_m$ is defined as*

$$D_a F(x) = F(x) + F(x + a), \forall x \in V_m.$$

δ is an integer valued function from $V_m \times V_m$ is defined as

$$\delta(a, b) = |\{x \in V_m, D_a F(x) = b\}| \text{ for } a, b \in V_m.$$

Abusing the notation δ, we define

$$\delta(F) = \max_{a \neq 0, b \in V_m} \delta(a, b).$$

$\delta(F)$ needs to be as low as possible to resist differential attacks on block ciphers [7]. Since $D_a F(x) = D_a F(x + a)$ for $a \neq 0 \in V_m$, we have $\delta(F) \geq 2$ and even. The S-boxes for which the equality holds are the best choices against the differential attack.

Definition 2. *An S-box* $F : V_m \mapsto V_m$ *is called* Almost Perfect Nonlinear *(in short, APN) if* $\delta(F) = 2$.

Lemma 1. $F : V_m \mapsto V_m$ *is APN iff there do not exist different* $x, y, z \in V_m$ *such that* $F(x) + F(y) + F(z) + F(x + y + z) = 0$.

Now we define a class of functions as following.

Definition 3. *Consider* m *is even. An S-box* $F : V_m \mapsto V_m$ *is defined as 3-to-1 S-box if* $F^{-1}(0) = \{0\}$ *and* $|F^{-1}(a)| = 3$ *or, 0 for* $a \in V_m^* = V_m \setminus \{0\}$.

Note that, the 3-to-1 S-box is defined for even variable S-boxes because $2^m - 1$ is not divisible by 3 when m is odd.

The APN property is preserved by Extended Affine (EA) transformation [8] and CCZ transformation [4]. Two S-boxes F and F' are EA-equivalent if there exist two affine permutations A_1, A_2 and an affine function A such that $F' = A_1 \circ F \circ A_2 + A$. CCZ-equivalence corresponds to the affine equivalence of the graphs of two S-boxes [4]. EA-equivalence is a particular case of CCZ-equivalence. In [3] one can find a list of classes of APN functions which are EA-inequivalent and CCZ-inequivalent to power functions. In this paper we will show that an important class of even variable S-boxes are 3-to-1 S-boxes.

We denote $e \subseteq d$ for two non-negative integers e and d if $e \wedge d = e$ where $\wedge$ is the bitwise logical *AND* operation i.e., $e_i \leq d_i, 0 \leq i < n$ where e_i and d_i's are ith bit of the n-bit binary representation of e and d respectively. From Lucas' theorem [6, page 79], we have $\binom{d}{e} = 1 \bmod 2$ iff $e \subseteq d$ for two non-negative integers d and e.

3   S3-to-1 APN S-Box

In this section, we present a special class of 3-to-1 APN S-box and show that some of the known constructions of S-boxes on even variables are of this type. Note that, in this section we always consider m as even positive integer and $k = \frac{2^m - 1}{3}$.

Construction 1 (S3-to-1 function). *Let* V_m *be partitioned into disjoint parts,* $P_0 = \{0\}, P_1, P_2, \ldots, P_k$, *such that each set* P_i, $1 \leq i \leq k$ *contains 3 different elements* a, b, c *where* $a + b = c$ *(i.e.,* $P_i \cup P_0, 1 \leq i \leq k$ *are 2-dimensional flats). Further, let* $U \subset V_m$ *be an ordered set and* $|U| = k+1$. *Then the S-box* $F : V_m \mapsto V_m$ *is constructed as* $F(x) = u_i$ *where* $x \in P_i$ *and* u_i *is the* ith *element in* U.

We name this special class of 3-to-1 functions as S3-to-1 functions.

Notation 1. *Referring to the partitions* P_i *in Construction 1, we denote (1)* $F(P_i) = y$ *where* $F(x) = y$ *for* $x \in P_i$} *and (2) for* $x \in V_m$, $P(x) = P_i$ *where* $x \in P_i$.

Proposition 1. *Let F be a S3-to-1 function. Referring to the Construction 1, if there are no four distinct elements $w, x, y, z \in U$ such that $w + x + y + z = 0$, then F is APN.*

Proof. To prove it, we have to show that for any distinct $a, b, c, d \in V_m$, $a + b + c + d = 0$ implies $F(a) + F(b) + F(c) + F(d) \neq 0$. If a, b, c, d are from four distinct partitions then from the supposition $F(a) + F(b) + F(c) + F(d) \neq 0$. If at least two elements, say a and b, are from the same partition, then $F(a) + F(b) + F(c) + F(d) = F(c) + f(d)$. To be $F(c) + F(d) = 0$, c and d have to be in one partition. c and d can not be in the same partition where a and b belong because each partition contains 3 elements or 1 element. If c and d are in another partition then $a + b + c + d = (a + b) + (c + d)$ can not be 0 because $(a + b) \in P(a)$ and $(c + d) \in P(c)$ are two different elements.   □

If one can choose $k + 1$ elements from V_m of size $2^m = 3k + 1$ such that no four elements from them can add to 0, then it is possible to construct an APN function. The question is whether such type of set exists for some even m ? For $m = 4$ the set $U = \{0, a_1, a_2, a_3, a_4, a_1 + a_2 + a_3 + a_4\}$ where $\{a_1, a_2, a_3, a_4\}$ is a basis of vector space V_4, satisfies the condition. Here we present an example of APN function when $m = 4$.

Example 1. Represent the m-dimensional vectors $(e_0, e_1, \ldots, e_{m-1})$ by the integers $\sum_{i=0}^{m-1} e_i 2^i$. The partition of V_4 is as $P_0 = \{0\}, P_1 = \{1, 2, 3\}, P_2 = \{4, 8, 12\}, P_3 = \{5, 10, 15\}, P_4 = \{6, 11, 13\}, P_5 = \{7, 9, 14\}$ and the set $U = \{0, 1, 2, 4, 8, 15\}$. Now, define F as $F(P_0) = 0; F(P_1) = 1; F(P_2) = 2; F(P_3) = 4; F(P_4) = 8; F(P_5) = 15$. Then F is APN.

Unfortunately, the following theorem tells about the non-existence of such set when $m \geq 6$.

Theorem 1. *For $m \geq 6$ (m even), there does not exist a set $U \subset V_m$ of size $k + 1$ such that $w + x + y + z \neq 0$ for all distinct $w, x, y, z \in U$.*

Proof. We use induction to prove it. For the base case (i.e., $m = 6$), we will construct a largest set $W \subset V_6$ such that there is no distinct $w, x, y, z \in W$ such that $w + x + y + z = 0$. Since W is largest, it must contain a basis of V_6. Without loss of generality, we consider that W contains the unit basis $e_1 = (0, 0, 0, 0, 0, 1), e_2 = (0, 0, 0, 0, 1, 0), \ldots, e_6 = (1, 0, 0, 0, 0, 0)$ and all zero element 0 (otherwise, one can use an affine transformation to get the unit basis and 0). Now we will start with $W = \{0, e_1, \ldots, e_6\}$ and try to add more to W. Now, we can not add any vector of weight 2 and 3 with W. Further, it can be checked that the addition of the vector of weight 6 will not allow to add any vector of weight 4 or 5. Then one can not add two or more vectors of weight 5, because addition of two vectors of weight 5 results in a vector of weight 2. Finally, one can not add more than 2 vectors of weight 4, since there always exist 2 vectors out of 3 vectors of weight 4 such that their sum results in a vector of weight 2. So, one can add at most three more (two of weight 4 and one of weight 5). The size of W will be at most 10, but we need of a set of size $\frac{2^6-1}{3} = 21$. Therefore, the theorem is true for $m = 6$.

Now we suppose it is true for $m = t$ and we will prove for $m = t+2$. We prove it by contradiction. Consider that such a set U of size $\frac{2^{t+2}-1}{3}$ exists for $n = t+2$. Now divide U into 4 parts U_{00}, U_{01}, U_{10} and U_{11} such that $U_{ij} \subset U$ contains the vectors having $(t+1)$th and $(t+2)$th co-ordinates i and j respectively. Hence from pigeonhole principle, there must be a U_{ij} such that $|U_{ij}| \geq \frac{1}{4} \times \frac{2^{t+2}-1}{3} = \frac{2^t - \frac{1}{4}}{3} \geq \frac{2^t-1}{3}$. Now one can use this U_{ij} (excluding $t+1$ and $t+2$ th coordinates) for $m = t$, which contradicts our supposition. $\qquad\square$

Using Lemma 1, we have the following result for when a S3-to-1 function will be APN.

Theorem 2. *Let F be a S3-to-1 function. F is APN iff for all $x, y \in V_m$ such that $P(x) \neq P(y)$, $F(x) + F(y) + F(z) + F(x+y+z) \neq 0$ for all $z \in V_m$, $z \notin P(x) \cup P(y) \cup P(x+y)$.*

Proof. For the proof, we use Lemma 1. For this class of S-boxes the search domains of y and z are decreased by putting some extra conditions. Now we will show that the discarded y's and z's will not satisfy the condition $F(x) + F(y) + F(z) + F(x+y+z) = 0$.

Let $y \in P(x)$. Then from the construction of P_i's, we have $x + y \in P(x)$. $F(x)+F(y)+F(z)+F(x+y+z) = 0$ i.e., $F(z)+F(x+y+z) = 0$ implies $x+y+z \in P(z)$ i.e, $x+y \in P(z)$. This implies $x, y, z, x+y+z \in P(x)$, a contradiction that x, y, z and $x+y+z$ are all distinct. Hence, $F(x)+F(y)+F(z)+F(x+y+z) \neq 0$. The exclusion of z's can be proved in similar way. $\qquad\square$

Note 1. In Construction 1, each set $P_i \cup \{0\}$, $0 < i \leq k$, forms a subspace of dimension 2. If we wanted to use subspaces with dimension greater than 2 then for the following reason F can not be APN. Consider P_i such that $P_i \cup \{0\}$ is of dimension greater than 2. Then P_i will contain a subset $\{x, y, x+y, z, x+z, y+z, x+y+z\}$. Here the last 4 elements add to zero. This implies that we can not consider the partitions $P_i, 0 < i \leq k$, such that $P_i \cup \{0\}$ is 3 dimensional.

Definition 4. *A S3-to-1 function $F : V_m \mapsto V_m$ is called identity S3-to-1 (in short, IDS3-to-1) function if $F(P_i) \in P_i$ for $0 \leq i \leq k$.*

IDS3-to-1 functions are similar to identity function. There are 3^k IDS3-to-1 functions. Like identity function, in the following theorem we show that IDS3-to-1 functions are not APN.

Theorem 3. *IDS3-to-1 functions are not APN when $m \geq 6$.*

Proof. Let $P_i = \{u_i, u_i', u_i''\}$ where $F(P_i) = u_i$ for $1 \leq i \leq k$. Let $s_{ij} = u_i + u_j$ for $1 \leq i < j \leq k$. If all s_{ij}'s are not distinct then there is $s_{i_1 i_2} = s_{i_3 i_4}$ i.e., $u_{i_1} + u_{i_2} + u_{i_3} + u_{i_4} = 0$, which implies that F is not APN. Hence, consider all s_{ij}s are distinct. There are $\binom{k}{2} = \frac{k(k-1)}{2}$ many s_{ij}s and $2k$ many u_i's and u_i''s. For $m \geq 6$, we have $\frac{k(k-1)}{2} > 2k$. Hence there is a $s_{i_1 i_2} = u_{i_3}$ i.e., $u_{i_1} + u_{i_2} + u_{i_3} = 0$. Hence F is not APN. $\qquad\square$

Since $\frac{k(k-1)}{2} = 2k = 10$ for $m = 4$, there exist some IDS3-to-1 APN functions and experimentally we found some of them. In the remaining part of this section we will show that some known important APN functions are S3-to-1 functions.

1. Power APN Functions

The power APN functions on even number of variables are of the form $F(X) = X^{3d}$ where $\gcd(d, k) = 1$ and $k = \frac{2^m - 1}{3}$. Let α be a primitive element of V_m. Since m is even, V_2 is a subfield of V_m with $\beta = \alpha^k$ a generator of $V_2^* = \{1, \beta, \beta^2\}$. Now consider $P_{i+1} = \alpha^i V_2^* = \{\alpha^i, \alpha^i\beta, \alpha^i\beta^2\}$ for $0 \leq i < k$ and $P_0 = \{0\}$. $\{P_i, 0 \leq i \leq k\}$ makes a disjoint partition over V_m and $\alpha^i + \alpha^i\beta + \alpha^i\beta^2 = \alpha^i(1 + \beta + \beta^2) = \alpha^i.0 = 0$ for $0 \leq i < k$. Now for the S-box $F(X) = X^{3d}$, $F(P_{i+1}) = \{\alpha^{3di}, \alpha^{3di}\beta^{3d}, \alpha^{3di}\beta^{6d}\} = \alpha^{3di}$ for $0 \leq i < k$. Since $\gcd(d, k) = 1$, $\alpha^{3di} \neq \alpha^{3dj}$ for $0 \leq i < j < k$ i.e., $|F^{-1}(\alpha^{3di})| = 3$. Here $U = \{0\} \bigcup \{\alpha^{3i}, 0 \leq i \leq k-1\}$. Therefore, the APN power functions i.e., $X^{3d}, \gcd(d, k) = 1$ satisfies Construction 1. Hence, the power APN functions follow the restriction imposed on a S-box to be APN in Theorem 2.

2. $F(X) = X^3 + tr(X^9)$

The function $F(X) = X^3 + tr(X^9)$ (where $tr(X) = \sum_{i=0}^{m-1} X^{2^i}$ is the trace function from V_m to V_1) is APN function and when $m \geq 7$ and $m > 2p$ where p is the smallest positive integer such that $m \neq 1, m \neq 3$ and $\gcd(m, p) = 1$, $X^3 + tr(X^9)$ is CCZ-inequivalent to all power functions on V_m [3]. Similar to the power function case (i.e., Item 1), one can easily prove that $F(\alpha^i) = F(\alpha^{i+k}) = F(\alpha^{i+2k})$ for $0 \leq i < k = \frac{2^m - 1}{3}$ and $F(0) = 0$ where α is a primitive element in V_m. Let $x = \alpha^i$ and $y = \alpha^j$ where $0 \leq i < j < k$. Now we will show that $F(x) \neq F(y)$ i.e., $x^3 + tr(x^9) \neq y^3 + tr(y^9)$ i.e., $tr(x^9 + y^9) \neq x^3 + y^3$. If $tr(x^9+y^9) = 0$ then we are done because $x^3 \neq y^3$. Now consider $tr(x^9+y^9) = 1$. If $x^3+y^3 = 1$ i.e., $y^3 = 1+x^3$ then $tr(x^9+y^9) = tr(x^9+(1+x^3)^3) = tr(1+x^3+x^6) = \sum_{i=0}^{m-1}(1 + x^3 + x^6)^{2^i} = \sum_{i=0}^{m-1}(1 + x^{3.2^i} + x^{6.2^i}) = \sum_{i=0}^{m-1}(1 + x^{3.2^i} + x^{3.2^{i+1}}) = x^3 + x^{3.2^m} = x^3 + x^3 = 0$ which is a contradiction. Therefore $F(x) \neq F(y)$ implies $X^3 + tr(X^9)$ is S3-to-1 function.

The general study of APN property of S3-to-1 functions can give clearer picture to generalize the APN power functions on even number of variables and constructions of new class of APN functions. Overall the S3-to-1 functions covers many interesting parts of the studies of APN functions. The study on finding the exact relation of the ordered set U and the partitions P_i (or, flats $P_i \cup P_0$) which makes F APN will be very interesting.

4   Power Function

In this section we present a necessary condition for a power function, $F : X \mapsto X^d, X \in V_m$, to be APN. Unlike the previous section, in this section we study for general m unless it is specified as even or odd. The complete characterization of all APN power functions is not known. Some results on the necessary conditions for a power function to be APN are available in recent literature. If F is APN then $\gcd(d, 2^m - 1) = 1$ for odd m and $\gcd(d, 2^m - 1) = 3$ for even m [1]. For another instance, if there is h which divides m and $d = l(2^h - 1) + 2^r$ for some l and r then F is not APN [5,4]. Therefore, to be an APN S-box, a power function must satisfy the necessary conditions. For more details one can refer [1]. In this section we have presented another necessary condition for F to be an APN functions. Therefore, while searching for APN function one can simple discard those functions. If F is a power function then the Lemma 1 can be simplified as in the following lemma.

Lemma 2. *A power S-box $F : V_m \mapsto V_m$ is APN iff there do not exist different $x \neq 1$ and $y \neq 1$ in V_m such that $1 + S(x) + S(y) + S(1 + x + y) = 0$.*

The proof is simple, since for different $x \neq 0, y, z \in V_m$, $x^d + y^d + z^d + (x+y+z)^d = x^{-d}[1 + (\frac{y}{x})^d + (\frac{z}{x})^d + (1 + \frac{y}{x} + \frac{z}{x})^d]$. Now Lemma 2 can be written in terms of primitive elements as following.

Proposition 2. *For a primitive element $\alpha \in V_m$, a power S-box $F : V_m \mapsto V_m$ is APN iff there do not exist*

1. *$0 < i < 2^m - 1$ such that $1 + F(\alpha^i) + F(1 + \alpha^i) \neq 0$ (the case when one of $x, y, 1 + x + y$ is zero), and*
2. *$0 < i < j < 2^m - 1$ such that $1 + F(\alpha^i) + F(\alpha^j) + F(1 + \alpha^i + \alpha^j) \neq 0$ (the case when none of $x, y, 1 + x + y$ is zero).*

Therefore, the APN property of a power function (say, X^d) can be checked by solving Conditions 1 and 2 in the Proposition 2 for given m. It will be simpler if Condition 2 can be reduced to Condition 1. Now consider the question: does Condition 1 imply Condition 2 for some d and m? In the following theorem we find an instance where it is possible.

Lemma 3. *Suppose $wt(d) = 2$. Then $a^d + b^d + c^d + (a + b + c)^d = (a + b)^d + (a + c)^d + (b + c)^d$ for $a, b, c \in V_m$ and any $m > 0$.*

Since d is of the form $2^p + 2^q$, the proof simply follows from Lucas' theorem.

Theorem 4. *A quadratic S-box $F : V_m \mapsto V_m$ such that $F(X) = X^d$ (where $wt(d) = 2$) is APN iff there does not exist $0 < i < 2^m - 1$ such that $1 + \alpha^{id} + (1 + \alpha^i)^d = 0$ where α is a primitive element in F_{2^m}.*

Proof. Here we reduce Item 2 to Item 1 in Definition 2. For $0 < i < j < 2^m - 1$,
$1 + \alpha^{id} + \alpha^{jd} + (1 + \alpha^i + \alpha^j)^d \neq 0$
i.e., $(1 + \alpha^i)^d + (1 + \alpha^j)^d + (\alpha^i + \alpha^j)^d \neq 0$ (according to Lemma 3)
i.e., $1 + [(1 + \alpha^i)^{-1}(1 + \alpha^j)]^d + (1 + (1 + \alpha^i)^{-1}(1 + \alpha^j))^d \neq 0$
i.e., $1 + \alpha^{ld} + (1 + \alpha^l)^d \neq 0$ for some $0 < l < 2^m - 1$.   □

Hence a quadratic function $F(X) = X^d$ is APN iff the equation $1 + x^d + (1 + x)^d = 0$ has no solution in $V_m \setminus \{0, 1\}$. Using Lucas' theorem we have $1 + x^d + (1 + x)^d = x^{2^p} + x^{2^q} = 0$, iff $gcd(2^p - 2^q, 2^m - 1) \neq 1$ where $d = 2^p + 2^q$. Hence X^d is APN iff $gcd(2^p - 2^q, 2^m - 1) = 1$ i.e., $gcd(p - q, m) = 1$. Certainly, this is not a new result. It has been done by Nyberg in [8]. But the motivation is to find some other situations where Condition 2 can be reduced to Condition 1 which could be solved easily (because Condition 1 is dependent on one variable).

Since the Condition 1 is easier to solve, now we shall study the S-boxes X^d using Condition 1 to find some situations when they are not APN. Since X^d is APN iff X^{2d} is APN, we consider d is an odd positive number. We can write every odd positive integer d of the form

$$(2^{a_0} - 1) + 2^{a_0+b_0}(2^{a_1} - 1) + \cdots + 2^{\sum_{i=0}^{q-2}(a_i+b_i)}(2^{a_{q-1}} - 1) + 2^{\sum_{i=0}^{q-1}(a_i+b_i)}(2^{a_q} - 1)$$

where $a_i, 0 \leq i \leq q$ and $b_i, 0 \leq i < q$ are the length of ith contiguous 1's and 0's in the binary representation of d. For example, $(77)_{10} = (1001101)_2$ where $a_0 = 1, b_0 = 1, a_1 = 2, b_1 = 2, a_2 = 1$ and $q = 2$.

Theorem 5. *Let d be of the form $d = (2^{a_0} - 1) + 2^{a_0+b_0}(2^{a_1} - 1) + \cdots + 2^{\sum_{i=0}^{q-2}(a_i+b_i)}(2^{a_{q-1}} - 1) + 2^{\sum_{i=0}^{q-1}(a_i+b_i)}(2^{a_q} - 1)$. Denote $l_0 = d$,*
$l_1 = 2^{a_0+b_0}(2^{a_1} - 1) + 2^{a_0+b_0+a_1+b_1}(2^{a_2} - 1) + \cdots + 2^{\sum_{i=0}^{q-1}(a_i+b_i)}(2^{a_q} - 1)$,
$l_j = 2^{a_{j-1}+b_{j-1}}(2^{a_j}-1) + 2^{a_{j-1}+b_{j-1}+a_j+b_j}(2^{a_{j+1}}-1) + \cdots + 2^{\sum_{i=j-1}^{q-1}(a_i+b_i)}(2^{a_q}-1)$
(that is, $l_j = l_{j-1} >> (a_{j-2} + b_{j-2}) - (2^{a_{j-1}} - 1)$, where "$t >> n$" is bit wise right shift of integer t by n places) for $1 < j \leq q$. If

$- \gcd(l_{i+1} - 1, 2^m - 1) \neq 1$ *or,* $\gcd(a_i, m) \neq 1$ *for* $0 \leq i < q$,
$-$ *and* $\gcd(a_q - 1, m) \neq 1$.

then $F(X) = X^d$ is not APN.

Proof. Denote $s_j = \sum_{i=0}^{j}(a_i + b_i), 0 \leq j < q$. Similar to l_j, denote k_j as $k_0 = 2^{a_0} - 1$ and $k_j = k_{j-1} + 2^{s_{j-1}}(2^{a_j} - 1), 0 \leq j < q$. Now we open $(1 + x)^d$ using Lucas' theorem.

$$(1 + x)^d = \sum_{i:i \subseteq d} x^i$$

$$= \sum_{i:i \subseteq k_0} x^i + \sum_{i:i \subseteq k_0} x^i \sum_{0 \neq i \subseteq 2^{a_1}-1} x^{i2^{s_0}} + \cdots + \sum_{i:i \subseteq k_{q-1}} x^i \sum_{0 \neq i \subseteq 2^{a_q}-1} x^{i2^{s_{q-1}}}$$

$$= \sum_{i=0}^{2^{a_0}-1} x^i + \sum_{i:i \subseteq k_0} x^i \sum_{i=1}^{2^{a_1}-1} x^{i2^{s_0}} + \cdots + \sum_{i:i \subseteq k_{q-1}=i} x^i \sum_{i=1}^{2^{a_q}-1} x^{i2^{s_{q-1}}}$$

$$= 1 + \sum_{i=1}^{2^{a_0}-1} x^i + (1 + x)^{k_0} \sum_{i=1}^{2^{a_1}-1} (x^{2^{s_0}})^i + \cdots + (1 + x)^{k_{q-1}} \sum_{i=1}^{2^{a_q}-1} (x^{2^{s_{q-1}}})^i$$

$$= 1 + \sum_{i=1}^{2^{a_0}-1} x^i + (1 + x)^{k_0} \sum_{i=1}^{2^{a_1}-1} (x^{2^{s_0}})^i + \cdots + (1 + x)^{k_{q-2}} \sum_{i=1}^{2^{a_{q-1}}-1} (x^{2^{s_{q-2}}})^i$$

$$+ (1 + x)^{k_{q-1}} \sum_{i=1}^{2^{a_q}-2} (x^{2^{s_{q-1}}})^i + (1 + x)^{k_{q-1}} (x^{2^{s_{q-1}}})^{2^{a_q}-1}$$

$$= 1 + x\frac{1+x^{2^{a_0}-1}}{1+x} + (1+x)^{k_0}x^{2^{s_0}}\frac{1+(x^{2^{s_0}})^{2^{a_1}-1}}{1+x^{2^{s_0}}} + \cdots$$

$$+ (1+x)^{k_q-2}x^{2^{s_q-2}}\frac{1+(x^{2^{s_q-2}})^{2^{a_q-1}-1}}{1+x^{2^{s_q-2}}} + (1+x)^{k_q-1}x^{2^{s_q-1}}\frac{1+(x^{2^{s_q-1}})^{2^{a_q}-2}}{1+x^{2^{s_q-1}}}$$

$$+ (1+x)^{k_q-1}x^{2^{s_q-1}(2^{a_q}-1)}. \tag{1}$$

Opening $(1+x)^{k_q-1}$ in the last term of Equation 1 to separate out x^d we have

$$(1+x)^{k_q-1}x^{2^{s_q-1}(2^{a_q}-1)} = x^d + x^{2^{s_q-1}(2^{a_q}-1)}(1+x)^{k_q-2}\sum_{i=0}^{2^{a_q-1}-2}x^{i2^{s_q-2}}$$

$$+ x^{2^{s_q-1}(2^{a_q}-1)+2^{s_q-2}(2^{a_q-1}-1)}(1+x)^{k_q-3}\sum_{i=0}^{2^{a_q-2}-2}x^{i2^{s_q-3}} + \cdots$$

$$+ x^{2^{s_q-1}(2^{a_q}-1)+\cdots+2^{s_0}(2^{a_1}-1)}\sum_{i=0}^{2^{a_0}-2}x^{i}$$

$$= x^d + x^{l_q 2^{s_q-2}}(1+x)^{k_q-2}\frac{1+(x^{2^{s_q-2}})^{2^{a_q-1}-1}}{1+x^{2^{s_q-2}}}$$

$$+ x^{l_q-1 2^{s_q-3}}(1+x)^{k_q-3}\frac{1+(x^{2^{s_q-3}})^{2^{a_q-2}-1}}{1+x^{2^{s_q-3}}} + \cdots + x^{l_1}\frac{1+x^{2^{a_0}-1}}{1+x}. \tag{2}$$

Now from Equations 1 and 2 we have

$$1 + x^d + (1+x)^d = x\frac{1+x^{2^{a_0}-1}}{1+x} + (1+x)^{k_0}x^{2^{s_0}}\frac{1+(x^{2^{s_0}})^{2^{a_1}-1}}{1+x^{2^{s_0}}} + \cdots$$

$$+ (1+x)^{k_q-2}x^{2^{s_q-2}}\frac{1+(x^{2^{s_q-2}})^{2^{a_q-1}-1}}{1+x^{2^{s_q-2}}} + (1+x)^{k_q-1}x^{2^{s_q-1}}\frac{1+(x^{2^{s_q-1}})^{2^{a_q}-2}}{1+x^{2^{s_q-1}}}$$

$$+ x^{l_q 2^{s_q-2}}(1+x)^{k_q-2}\frac{1+(x^{2^{s_q-2}})^{2^{a_q-1}-1}}{1+x^{2^{s_q-2}}}$$

$$+ x^{l_q-1 2^{s_q-3}}(1+x)^{k_q-3}\frac{1+(x^{2^{s_q-3}})^{2^{a_q-2}-1}}{1+x^{2^{s_q-3}}} + \cdots + x^{l_1}\frac{1+x^{2^{a_0}-1}}{1+x}$$

$$= (x+x^{l_1})\frac{1+x^{2^{a_0}-1}}{1+x} + (1+x)^{k_0}(x^{2^{s_0}}+x^{l_2 2^{s_0}})\frac{1+(x^{2^{s_0}})^{2^{a_1}-1}}{1+x^{2^{s_0}}} + \cdots$$

$$+ (1+x)^{k_q-2}(x^{2^{s_q-2}}+x^{l_q 2^{s_q-2}})\frac{1+(x^{2^{s_q-2}})^{2^{a_q-1}-1}}{1+x^{2^{s_q-2}}}$$

$$+ (1+x)^{k_q-1}x^{2^{s_q-1}}\frac{1+(x^{2^{s_q-1}})^{2^{a_q}-2}}{1+x^{2^{s_q-1}}}. \tag{3}$$

If each term in Equation 3 is 0 then $1 + x^d + (1+x)^d = 0$. That is, x^d is not APN if there is a $x \in V_m \setminus \{0,1\}$ such that

$$(x+x^{l_1})\frac{1+x^{2^{a_0}-1}}{1+x} = (1+x)^{k_0}(x^{2^{s_0}}+x^{l_2 2^{s_0}})\frac{1+(x^{2^{s_0}})^{2^{a_1}-1}}{1+x^{2^{s_0}}} = \cdots$$

$$= (1+x)^{k_q-2}(x^{2^{s_q-2}}+x^{l_q 2^{s_q-2}})\frac{1+(x^{2^{s_q-2}})^{2^{a_q-1}-1}}{1+x^{2^{s_q-2}}}$$

$$= (1+x)^{k_q-1} x^{2^s q-1} \frac{1 + (x^{2^s q-1})^{2^{a_q}-2}}{1 + x^{2^s q-1}} = 0.$$

Hence, x^d is not APN if

$(gcd(l_1 - 1, 2^m - 1) \neq 1$ or, $gcd(2^{a_0} - 1, 2^m - 1) \neq 1)$ and

$(gcd(l_2 - 1, 2^m - 1) \neq 1$ or, $gcd(2^{a_1} - 1, 2^m - 1) \neq 1)$ and

$\cdots$ and

$(gcd(l_q - 1, 2^m - 1) \neq 1$ or, $gcd(2^{a_{q-1}} - 1, 2^m - 1) \neq 1)$ and

$gcd(2^{a_q} - 2, 2^m - 1) \neq 1$.

That is, x^d is not APN if

$(gcd(l_1 - 1, 2^m - 1) \neq 1$ or, $gcd(a_0, m) \neq 1)$ and

$(gcd(l_2 - 1, 2^m - 1) \neq 1$ or, $gcd(a_1, m) \neq 1)$ and

$\cdots$ and

$(gcd(l_q - 1, 2^m - 1) \neq 1$ or, $gcd(a_{q-1}, m) \neq 1)$ and

$gcd(a_q - 1, m) \neq 1$.   □

Table 1. $F(X) = X^d, d = 2^p - 1, p \geq 0$

p	d	Result: m such that X^d is not APN
1	1	X can not be APN for any m
2	3	no information for X^3
3	7	X^7 can not be APN if m is even
4	15	X^{15} can not be APN if $3 \mid m$
5	31	X^{31} can not be APN if m is even
6	63	X^{63} can not be APN if $5 \mid m$
7	127	X^{127} can not be APN if $2 \mid m$ or $3 \mid m$
8	255	X^{255} can not be APN if $7 \mid m$
9	511	X^{511} can not be APN if m is even
10	1023	X^{1023} can not be APN if $3 \mid m$

Table 2. $F(X) = X^d, d \neq 2^p - 1$ and odd

d	Result: m such that X^d is not APN
5	X^5 can not be APN if $3 \mid (2^m - 1)$ i.e., m is even
9	X^9 can not be APN if $7 \mid (2^m - 1)$ i.e., $3 \mid m$
11	X^{11} can not be APN if $2 \mid m$ or $3 \mid m$
13	no information for X^{13}
17	X^{17} can not be APN if $gcd(15, 2^m - 1) \neq 1$ i.e., m is even
19	X^{19} can not be APN if $2 \mid m$
21	X^{21} can not be APN if $19 \mid (2^m - 1)$ and $2 \mid m$
23	X^{23} can not be APN if $2 \mid m$ or $3 \mid m$
25	no information for X^{25}
27	no information for X^{27}
29	X^{29} can not be APN if $2 \mid m$ and $gcd(27, 2^m - 1) \neq 1$ (i.e., $2 \mid m$ is enough)

Following Theorem 5, we can specify some m such that X^d is not APN for given $d > 0$. At first we present a simple case when $d = 2^p - 1$. X^{2^p-1} is not APN if

Table 3. Some odd d: X^d is not APN in V_m

m	Result: d such that $X^{2^l d}, l \geq 0$, is not APN
2	1
3	1
4	$1, 5, 7, 11$
5	1
6	$1, 5, 7, 9, 11, 15, 17, 19, 23, 29, 31, 35, 37, 39, 41, 43, 47, 59$
7	1
8	$1, 5, 7, 11, 17, 19, 23, 29, 31, 35, 41, 43, 47, 57, 59, 65, 67, 71, 77, 79, 83, 89, 91, 95, 113,$ $115, 117, 119, 125, 127, 131, 137, 139, 143, 155, 161, 163, 167, 173, 175, 177, 179, 183,$ $185, 187, 189, 191, 227, 233, 235, 237, 239, 251,$
9	$1, 9, 11, 15, 23, 37, 39, 65, 67, 71, 79, 93, 95, 121, 123, 127, 135, 149, 151, 177, 179,$ $183, 191, 247, 261, 263, 289, 291, 295, 303, 317, 319, 359, 373, 375, 485, 487$

Table 4. Comparison of the number of filtered X^d's and X^e's with the number of all elements

m even	(#filtered $X^d, 2^m - 2$), (#filtered X^e, t) where $\gcd(e, 2^m - 1) = 3$, $t = |\{d : \gcd(d, 2^m - 1) = 3\}|$ $1 \leq d \leq 2^m - 2$ and $1 \leq e \leq 2^m - 2$	m odd	(#filtered $X^d, 2^m - 2$), (#filtered X^e, t) where $\gcd(e, 2^m - 1) = 1$, $t = |\{d : \gcd(d, 2^m - 1) = 1\}$ $1 \leq d \leq 2^m - 2$ and $1 \leq e \leq 2^m - 2$
2	$(2, 2), (0, 0)$	3	$(3, 6), (3, 6)$
4	$(10, 14), (0, 4)$	5	$(5, 30), (5, 30)$
6	$(51, 62), (6, 12)$	7	$(7, 126), (7, 126)$
8	$(200, 254), (32, 64)$	9	$(189, 510), (189, 432)$
10	$(820, 1022), (150, 300)$	11	$(11, 2046), (11, 1936)$
12	$(3842, 4094), (468, 576)$	13	$(13, 8190), (13, 8190)$
14	$(10885, 16382), (1036, 5292)$	15	$(10785, 32766), (9615, 27000)$
16	$(50424, 65534), (10384, 16384)$	17	$(17, 131070), (17, 131070)$
18	$(228573, 262142), (36540, 46656)$	19	$(19, 524286), (19, 524286)$
20	$(827884, 1048574), (178660, 240000)$	21	$(297612, 2097150), (290997, 1778112)$
22	$(2101099, 4194302), (199386, 1320352)$	23	$(23, 8388606), (23, 8210080)$
24	$(15070558, 16777214),$ $(1961688, 2211840)$	25	$(223750, 33554430),$ $(223100, 32400000)$
26	$(24603358, 67108862),$ $(53326, 22358700)$	27	$(8321670, 134217726),$ $(8160102, 113467392)$

$\gcd(p-1, m) \neq 1$. In Table 1, we present the results for $p \leq 10$. Further, in Table 2 we present some initial examples for general case i.e., for all odd d (because the APN property is being preserved for $2^l d, l \geq 0$) which are not of the form $2^p - 1$. In Table 3, we present some odd d such that $X^{2^l d}, l \geq 0$ are not APN for a given m. There are some d, for which Theorem 5 can not say anything. Now we discuss some of such cases and some other observations in the following notes.

Note 2. 1. If $a_q = 2$ then $\gcd(a_q - 1, m) = 1$ for each m. Hence, we do not learn any information. For example, we can not say anything about $X^3, X^{13}, X^{25}, X^{27}$ etc.

2. If $a_q > 1$ and m is prime then $\gcd(a_q - 1, m) = 1$. Hence, no information in this case. For example, we can not say anything about X^{11} in V_7, V_{11} etc.

3. If m is Mersenne prime (i.e., $2^m - 1$ is also prime) then all gcd functions in the conditions return 1 except when $d = 1$. Thus, we can say something only for linear functions (i.e., X^{2^i}).

4. If m is a multiple of small primes then we can filter out many X^d's as non-APN because the chance of co-primeness with m and $2^m - 1$ decreases.

5. There are some X^d's, which can not be filtered using the necessary condition but $X^e, e = 2^l d \mod (2^m - 1)$ for some $l > 0$ may satisfy the conditions and can be shown as non-APN. In this case, X^d can too be filtered out. For example, X^{13} can not be shown as non-APN directly in V_4 but X^7 can be used to show X^{13} as non-APN because $13 = (2^2 * 7) \mod 15$ and X^7 is non-APN according to Theorem 5.

We have an interesting fact that we can filter out some power S-boxes X^d, $\gcd(d, 2^m - 1) = 3$ when m is even and $\gcd(d, 2^m - 1) = 1$ when m is odd. In Table 4, we present two types of comparision on the number of filtered S-boxes for both the even and odd number of variable cases. In first tuple, we compare the number X^ds that can be filtered out from the total $2^m - 2$ non-constant power functions. In second tuple, we compare the number of filtered power functions from the total number of power functions satisfying $\gcd(d, 2^m - 1) = 3$ when m is even and $\gcd(d, 2^m - 1) = 1$ when m is odd. In some cases, specifically when m is a multiple of small primes, we can filter out a big percentage of non-APN functions using our necessary conditions. However, using the necessary condition in Theorem 5 and the observation in Note 5 we can filter out some X^d's which are not APN. Therefore, while searching for a good APN S-box, one can simply discard those functions and save some searching time.

Acknowledgment. The author is very much thankful to Prof. Pascale Charpin for her excellent guidance and suggestions that provided me to gain the ideas and to improve the quality of this paper.

References

1. Berger, T.P., Canteaut, A., Charpin, P., Laigle-Chapuy, Y.: Almost Perfect Nonlinear functions. IEEE Trans. Inform. Theory 52(9), 4160–4170 (2006)
2. Biham, E., Shamir, A.: Differential Cryptanalysis of DES-like Cryptosystem. Journal of Cryptology 4(1), 3–72 (1991)
3. Budaghyan, L., Carlet, C., Leander, G.: Constructing new APN functions from known ones. Cryptology ePrint Archive: report 2007/063
4. Carlet, C., Charpin, P., Zinoviev, V.: Codes, Bent Functions and Permutations Suitable For DES-like Cryptosystems. Des. Codes Cryptogr. 15(2), 125–156 (1998)
5. Charpin, P., Tietävämen, A., Zinoviev, V.: On binary cyclic codes with minimum distance $d = 3$. Problems Inform. Transmission 33(4), 287–296 (1997)

6. Comtet, L.: Advanced combinatorics. Reidel Publication (1974)
7. Nyberg, K., Knudsen, L.R.: Provable security against differential cryptanalysis. In: Brickell, E.F. (ed.) CRYPTO 1992. LNCS, vol. 740, pp. 566–574. Springer, Heidelberg (1993)
8. Nyberg, K.: Differentially uniform mappings for cryptography. In: Helleseth, T. (ed.) EUROCRYPT 1993. LNCS, vol. 765, pp. 55–64. Springer, Heidelberg (1994)

Negabent Functions
in the Maiorana–McFarland Class

Kai-Uwe Schmidt[1], Matthew G. Parker[2], and Alexander Pott[3]

[1] Department of Mathematics, Simon Fraser University
Burnaby, BC V5A 1S6, Canada
kuschmidt@sfu.ca
[2] The Selmer Center, Department of Informatics, University of Bergen
N-5020 Bergen, Norway
matthew.parker@ii.uib.no
[3] Institute for Algebra and Geometry, Faculty of Mathematics,
Otto-von-Guericke-University Magdeburg,
D-39016 Magdeburg, Germany
alexander.pott@ovgu.de

Abstract. Boolean functions which are simultaneously bent and ne-gabent are studied. Transformations that leave the bent-negabent property invariant are presented. A construction for infinitely many bent-negabent Boolean functions in $2mn$ variables ($m > 1$) and of algebraic degree at most n is described, this being a subclass of the Maiorana–McFarland class of bent functions. Finally it is shown that a bent-negabent function in $2n$ variables from the Maiorana–McFarland class has algebraic degree at most $n - 1$.

1 Introduction

Bent Boolean functions are the class of Boolean functions whose spectral values have equal magnitude with respect to the *Hadamard transform* [1]. The construction and classification of bent functions is of significant and active interest to designers of cryptographic primitives [2], as such functions have maximum distance to the set of affine functions and, therefore, are not well-approximated by affine functions. It is natural also to consider spectral values with respect to the *nega-Hadamard transform*. If these spectral values of a Boolean function are all equal in magnitude, then we call the function *negabent*.

In this paper we consider how to construct Boolean functions that are simultaneously bent and negabent. Such a problem has been previously considered in [3], providing constructions for quadratic functions. Here we present a new and infinite construction for functions of more general algebraic degree, thereby answering and generalizing a conjecture made in [3]. More precisely, we construct a subclass of the Maiorana–McFarland class of bent functions in which all functions are also negabent. This construction generalizes the construction of quadratic bent-negabent functions described in [4,5]. These functions are in $2mn$ variables and have algebraic degree at most n, where $m > 1$.

S.W. Golomb et al. (Eds.): SETA 2008, LNCS 5203, pp. 390–402, 2008.

We also enlarge the class of symmetry operations over which the bent-negabent property of a Boolean function is preserved. In particular, we show that the bent-negabent property is an invariant with respect to the action of the orthogonal group on the input vector space. Finally we provide an upper bound on the algebraic degree of any bent-negabent Boolean function from the Maiorana-McFarland class.

2   Notation

Let V_n be an n-dimensional vector space over $\mathbb{F}_2$. Let $f : V_n \to \mathbb{F}_2$ be a Boolean function. The *Hadamard transform* of f is defined to be

$$\mathcal{H}(f)(u) := (-1)^{-\frac{n}{2}} \sum_{x \in V_n} (-1)^{f(x)+u \cdot x}, \quad u \in V_n.$$

The *nega-Hadamard transform* of f is defined to be

$$\mathcal{N}(f)(u) := (-1)^{-\frac{n}{2}} \sum_{x \in V_n} (-1)^{f(x)+u \cdot x} i^{\mathrm{wt}(x)}, \quad u \in V_n,$$

where $i := \sqrt{-1}$ and $\mathrm{wt}(.)$ denotes the Hamming weight. The function f is called *bent* if

$$|\mathcal{H}(f)(u)| = 1 \quad \text{for all} \quad u \in V_n.$$

Similarly, f is called *negabent* if

$$|\mathcal{N}(f)(u)| = 1 \quad \text{for all} \quad u \in V_n.$$

If f is both bent and negabent, we say that f is *bent-negabent*.
  Now let $f : V_n \oplus V_n \to \mathbb{F}_2$ be a Boolean function of the form

$$f(x, y) = \sigma(x) \cdot y + g(x),$$

where $\sigma : V_n \to V_n$ and $g : V_n \to \mathbb{F}_2$. It is well known that this function is bent if and only if σ is a permutation. The whole set of such bent functions forms the *Maiorana–McFarland class*.
  In the remainder of this section we will introduce some further notation and a useful lemma. Write $V_n = U \oplus W$, where $\dim W = k$ and $k \le n$, so that $\dim U = n - k$. Let $f : V_n \to \mathbb{F}_2$ be a Boolean function. For each fixed $x \in U$ we may view $f(x, \cdot)$ as a Boolean function on W. We define the *partial Hadamard transform* of f with respect to W as

$$\mathcal{H}_W(f)(x, v) := 2^{-\frac{k}{2}} \sum_{y \in W} (-1)^{f(x,y)+v \cdot y}, \quad v \in W.$$

We say that f is *bent with respect to W* if

$$|\mathcal{H}_W(f)(x, v)| = 1 \quad \text{for each} \quad x \in U, v \in W.$$

If f is bent with respect to W, the *partial dual* $\tilde{f}_W$ of f with respect to W is defined by the relation

$$\mathcal{H}_W(f)(x, v) = (-1)^{\tilde{f}_W(x,v)}.$$

Note that in the special case where $n = k$, $\tilde{f}_W$ is the usual dual of f, which we will denote by $\tilde{f}$.

In the remainder of this paper we shall make frequent use of the following lemma.

Lemma 1. *For any $u \in V_n$ we have*

$$\sum_{x \in V_n} (-1)^{u \cdot x} i^{\text{wt}(x)} = 2^{\frac{n}{2}} \omega^n i^{-\text{wt}(u)},$$

where $\omega = (1 + i)/\sqrt{2}$ is a primitive 8th root of unity.

Proof. Write $u = (u_1, u_2, \ldots, u_n)$. By successively factoring out terms, we obtain

$$\sum_{x \in V_n} (-1)^{u \cdot x} i^{\text{wt}(x)} = \prod_{k=1}^{n} (1 + i(-1)^{u_k})$$

$$= 2^{\frac{n}{2}} \prod_{k=1}^{n} \omega^{(-1)^{u_k}}$$

$$= 2^{\frac{n}{2}} \omega^{n - 2\,\text{wt}(u)}$$

$$= 2^{\frac{n}{2}} \omega^n i^{-\text{wt}(u)}. \qquad \square$$

Note that the preceding lemma shows that all affine functions $f : V_n \to \mathbb{F}_2$ are negabent (see also [3, Prop. 1]).

3   Transformations Preserving Bent-Negabentness

Several transformations that preserve the bent-negabent property have been presented in [3]. Here we provide two new transformations.

It is known that, if $f : V_n \to \mathbb{F}_2$ is a bent function, then the function given by

$$f(Ax + b) + c \cdot x + d, \quad \text{where} \quad A \in GL(2, n), \ b, c \in V_n, \ d \in V_1,$$

is also bent. Here, $GL(2, n)$ is the general linear group of $n \times n$ matrices over $\mathbb{F}_2$. These operations define a group whose action on f leaves the bent property of f invariant. Counterexamples show that these operations generally do not preserve the negabent property of a Boolean function. It is therefore interesting to find a subgroup of the bent-preserving operations that preserves also the negabent property. The following theorem shows that, if we replace $GL(2, n)$ by $O(2, n)$, the orthogonal group of $n \times n$ matrices over $\mathbb{F}_2$, we obtain such a subgroup.

Theorem 2. *Let $f, g : V_n \to \mathbb{F}_2$ be two Boolean functions. Suppose that f and g are related by*

$$g(x) = f(Ax + b) + c \cdot x + d, \quad \text{where} \quad A \in O(2, n),\ b, c \in V_n,\ d \in V_1.$$

Then, if f is bent-negabent, g is also bent-negabent.

Proof. As discussed above, g is bent if f is bent. It remains to show that g is negabent. From [3, Lem. 2] we know that, if $f(Ax)$ is negabent, so is $f(Ax + b) + c \cdot x + d$. It is therefore sufficient to assume that b and c are all-zero vectors and $d = 0$. Observe that

$$\mathrm{wt}(x) = x^T I x,$$

where I is the $n \times n$ identity matrix and the matrix operations are over $\mathbb{Z}$. We therefore have

$$\mathcal{N}(g)(u) = 2^{-\frac{n}{2}} \sum_{x \in V_n} (-1)^{f(Ax) + u \cdot x} i^{x^T I x}.$$

Now, since A is invertible by assumption, there exists B such that $AB = I$. Moreover, when x ranges over V_n, so does Bx. Thus,

$$\mathcal{N}(g)(u) = 2^{-\frac{n}{2}} \sum_{x \in V_n} (-1)^{f(x) + u \cdot Bx} i^{(Bx)^T I(Bx)}.$$

Since $A \in O(2, n)$, we have $B \in O(2, n)$ and so $B^T I B = I$. Hence,

$$(Bx)^T I(Bx) = x^T (B^T I B) x = x^T I x.$$

We conclude

$$\begin{aligned}
\mathcal{N}(g)(u) &= 2^{-\frac{n}{2}} \sum_{x \in V_n} (-1)^{f(x) + u \cdot Bx} i^{x^T I x} \\
&= 2^{-\frac{n}{2}} \sum_{x \in V_n} (-1)^{f(x) + B^T u \cdot x} i^{\mathrm{wt}(x)} \\
&= \mathcal{N}(f)(B^T u),
\end{aligned}$$

which proves the theorem. $\qquad\qquad\qquad\qquad\qquad\qquad\qquad\qquad\qquad\qquad \square$

It is known that the dual of a bent function is again a bent function, and it was proved in [3, Thm. 11] that the dual of a bent-negabent function is also bent-negabent. The following theorem generalizes this concept by showing that, if a bent-negabent function is bent with respect to certain subspaces, then the corresponding partial duals are also bent-negabent.

Theorem 3. *Write $V_n = U \oplus W$, where $\dim W = k$ and $k \le n$, so that $\dim U = n - k$. Let $f : V_n \to \mathbb{F}_2$ be a bent-negabent function that is bent with respect to U and bent with respect to W. Then $\tilde{f}_W$ is also bent-negabent.*

Proof. We first prove that $\tilde{f}_W$ is bent. By direct calculation,

$$\mathcal{H}(\tilde{f}_W)(u,w) = 2^{-\frac{n}{2}} \sum_{x \in U} \sum_{v \in W} (-1)^{\tilde{f}_W(x,v)+u \cdot x + v \cdot w}$$

$$= 2^{-\frac{n+k}{2}} \sum_{x \in U} \sum_{v \in W} \sum_{y \in W} (-1)^{f(x,y)+v \cdot y + u \cdot x + v \cdot w}$$

$$= 2^{-\frac{n+k}{2}} \sum_{x \in U} \sum_{y \in W} (-1)^{f(x,y)+u \cdot x} \sum_{v \in W} (-1)^{v \cdot (y+w)}.$$

The inner sum is zero unless $y = w$, in which case it is 2^k. Hence,

$$\mathcal{H}(\tilde{f}_W)(u,w) = 2^{-\frac{n-k}{2}} \sum_{x \in U} (-1)^{f(x,w)+u \cdot x}$$

$$= \mathcal{H}_U(f)(u,w).$$

By assumption, $|\mathcal{H}_U(f)(u,w)| = 1$ for each $u \in U$ and each $w \in W$. Therefore, $\tilde{f}_W$ is bent.

Next we prove that $\tilde{f}_W$ is negabent. We have

$$\mathcal{N}(\tilde{f}_W)(u,w) = 2^{-\frac{n}{2}} \sum_{x \in U} \sum_{v \in W} (-1)^{\tilde{f}_W(x,v)+u \cdot x + v \cdot w} i^{\mathrm{wt}(v)+\mathrm{wt}(x)}$$

$$= 2^{-\frac{n+k}{2}} \sum_{x \in U} \sum_{v \in W} \sum_{y \in W} (-1)^{f(x,y)+v \cdot y + u \cdot x + v \cdot w} i^{\mathrm{wt}(v)+\mathrm{wt}(x)}$$

$$= 2^{-\frac{n+k}{2}} \sum_{x \in U} \sum_{y \in W} (-1)^{f(x,y)+u \cdot x} i^{\mathrm{wt}(x)} \sum_{v \in W} (-1)^{v \cdot (y+w)} i^{\mathrm{wt}(v)}.$$

The inner sum can be computed with Lemma 1. We therefore obtain

$$\mathcal{N}(\tilde{f}_W)(u,w) = 2^{-\frac{n}{2}} \omega^k \sum_{x \in U} \sum_{y \in W} (-1)^{f(x,y)+u \cdot x} i^{\mathrm{wt}(x)} i^{-\mathrm{wt}(y+w)}$$

$$= 2^{-\frac{n}{2}} \omega^k i^{-\mathrm{wt}(w)} \sum_{x \in U} \sum_{y \in W} (-1)^{f(x,y)+u \cdot x + y \cdot w} i^{\mathrm{wt}(x)-\mathrm{wt}(y)}$$

$$= \omega^k i^{-\mathrm{wt}(w)} \mathcal{N}(f)(u,\bar{w}),$$

where $\omega = (1+i)/\sqrt{2}$ and $\bar{w}$ is the complement of w. Since f is negabent, this shows that $\tilde{f}_W$ is also negabent.   □

4   Constructions

Throughout this section we use the following notation. Define V to be an *mn*-dimensional vector space over $\mathbb{F}_2$, so that

$$V = \underbrace{V_n \oplus V_n \oplus \cdots \oplus V_n}_{m \text{ times}}.$$

Let the Boolean function $f : V \oplus V \to \mathbb{F}_2$ be given by

$$f(x_1,\dots,x_m,y_1,\dots,y_m) = \sigma(x_1,\dots,x_m) \cdot (y_1,\dots,y_m) + g(x_1,\dots,x_m), \quad (1)$$

where $\sigma : V \to V$ is of the form

$$\sigma(x_1,\dots,x_m) = (\psi_1(x_1), \phi_1(x_1) + \psi_2(x_2),\dots,\phi_{m-1}(x_{m-1}) + \psi_m(x_m))$$

and $g : V \to \mathbb{F}_2$ is defined by

$$g(x_1,\dots,x_m) = h_1(x_1) + h_2(x_2) + \cdots + h_m(x_m).$$

Here, $\psi_1,\dots,\psi_m,\phi_1,\dots,\phi_{m-1}$ are permutations on V_n and $h_1,\dots,h_m : V_n \to \mathbb{F}_2$ are arbitrary Boolean functions. Explicitly, f reads

$$f(x_1,\dots,x_m,y_1,\dots,y_m)$$
$$= \psi_1(x_1) \cdot y_1 + h_1(x_1) + \sum_{j=2}^{m}(y_j \cdot [\phi_{j-1}(x_{j-1}) + \psi_j(x_j)] + h_j(x_j)).$$

Since σ is a permutation, f belongs to the Maiorana–McFarland class, and is therefore bent. In the next theorem, we will identify configurations of σ and g so that f is also negabent.

Theorem 4. *Let m be a positive integer satisfying $m \not\equiv 1 \pmod 3$, and let k be an integer satisfying $0 < k < m$ and $k \equiv 0 \pmod 3$ or $(m-k) \equiv 1 \pmod 3$. Let f be as in (1), where*

$$\sigma(x_1,\dots,x_m) = (x_1, x_1 + x_2,\dots,x_{k-1} + \psi(x_k), \phi(x_k) + x_{k+1},\dots,x_{m-1} + x_m)$$
$$g(x_1,\dots,x_m) = h(x_k),$$

ψ, ϕ are permutations on V_n, and $h : V_n \to \mathbb{F}_2$ is an arbitrary Boolean function. (In other words, $\psi_1,\dots,\psi_m,\phi_1,\dots,\phi_{m-1}$ are identity maps except for $\psi := \psi_k$ and $\phi := \phi_k$, and $h_1,\dots,h_m$ are zero except for $h := h_k$.) Then f is bent-negabent.

A lemma is required to prove the theorem.

Lemma 5. *Let s be a nonnegative integer. For any $u_1,\dots,u_s,z_{s+1} \in V_n$ define*

$$E_s(z_{s+1}) := \prod_{j=1}^{s} \sum_{z_j \in V_n} (-1)^{(z_{j+1}+u_j) \cdot z_j} i^{\mathrm{wt}(z_j)},$$

where an empty product is defined to be equal to 1. Then we have

$$E_s(z_{s+1}) = \begin{cases} 2^{sn/2}\omega^c(-1)^{a \cdot z_{s+1}} & \text{if } s \equiv 0 \pmod 3 \\ 2^{sn/2}\omega^c(-1)^{a \cdot z_{s+1}} i^{-\mathrm{wt}(z_{s+1})} & \text{if } s \equiv 1 \pmod 3 \\ 2^{(s+1)n/2}\omega^c \delta_{z_{s+1}+a} & \text{if } s \equiv 2 \pmod 3 \end{cases}$$

for some $c \in \mathbb{Z}_8$ and $a \in V_n$. Here δ_a denotes the Kronecker delta function, i.e., δ_a equals 1 if $a = 0$ and is zero otherwise.

Proof. The lemma is certainly true for $s = 0$. We proceed by induction on s, where we use the lemma as a hypothesis. Observe that for $s > 0$ we have

$$E_s(z_{s+1}) = \sum_{z_s \in V_n} (-1)^{(z_{s+1}+u_s)\cdot z_s} i^{\mathrm{wt}(z_s)} E_{s-1}(z_s).$$

Now assume that the lemma is true for $s \equiv 0 \pmod 3$. Using Lemma 1, we have for $s \equiv 1 \pmod 3$

$$\begin{aligned}
E_s(z_{s+1}) &= 2^{(s-1)n/2}\omega^c \sum_{z_s \in V_n} (-1)^{(z_{s+1}+u_s+a)\cdot z_s} i^{\mathrm{wt}(z_s)} \\
&= 2^{sn/2}\omega^{c+n} i^{-\mathrm{wt}(z_{s+1}+u_s+a)} \\
&= 2^{sn/2}\omega^{c'} (-1)^{a'\cdot z_{s+1}} i^{-\mathrm{wt}(z_{s+1})},
\end{aligned}$$

where $c' = c + n - 2\,\mathrm{wt}(u_s + a)$ and $a' = a + u_s$. This proves the lemma for $s \equiv 1 \pmod 3$ provided that it holds for $s \equiv 0 \pmod 3$. Now assume that the lemma is true for $s \equiv 1 \pmod 3$. Then for $s \equiv 2 \pmod 3$ we obtain

$$\begin{aligned}
E_s(z_{s+1}) &= 2^{(s-1)n/2}\omega^c \sum_{z_s \in V_n} (-1)^{(z_{s+1}+u_s+a)\cdot z_s} \\
&= 2^{(s+1)n/2}\omega^{c'} \delta_{z_{s+1}+a'}.
\end{aligned}$$

where $c' = c$ and $a' = a + u_s$. Assuming that the lemma is true for $s \equiv 2 \pmod 3$, we have for $s \equiv 0 \pmod 3$

$$\begin{aligned}
E_s(z_{s+1}) &= 2^{sn/2}\omega^c \sum_{z_s \in V_n} (-1)^{(z_{s+1}+u_s)\cdot z_s} i^{\mathrm{wt}(z_s)} \delta_{z_s+a} \\
&= 2^{sn/2}\omega^{c'} (-1)^{a'\cdot z_{s+1}},
\end{aligned}$$

where $c' = c + 2\,\mathrm{wt}(a) + 4a\cdot u_s$ and $a' = a$. This completes the induction. $\square$

Proof (of Theorem 4). We define the relabeling $z_{2j} := x_j$ and $z_{2j-1} := y_j$ for $j = 1, 2, \ldots, m$, so that we have

$$f(x_1, \ldots, x_m, y_1, \ldots, y_m) =$$

$$h(z_{2k}) + \sum_{j=1}^{2k-2} z_j \cdot z_{j+1} + z_{2k-1} \cdot \psi(z_{2k}) + \sum_{j=2k+2}^{2m} z_{j-1} \cdot z_j + z_{2k+1} \cdot \phi(z_{2k}).$$

Write

$$\mathcal{N}(f)(u_1, \ldots, u_{2m}) = 2^{-mn} \sum_{z_{2k} \in V_n} (-1)^{h(z_{2k})+z_{2k}\cdot u_{2k}} i^{\mathrm{wt}(z_{2k})} P(z_{2k}) Q(z_{2k}), \quad (2)$$

where

$$P(z_{2k}) = \prod_{j=1}^{2k-2} \sum_{z_j \in V_n} (-1)^{(z_{j+1}+u_j)\cdot z_j} i^{\mathrm{wt}(z_j)} \sum_{z_{2k-1} \in V_n} (-1)^{(\psi(z_{2k})+u_{2k-1})\cdot z_{2k-1}} i^{\mathrm{wt}(z_{2k-1})}$$

$$Q(z_{2k}) = \prod_{j=2k+2}^{2m} \sum_{z_j \in V_n} (-1)^{(z_{j-1}+u_j)\cdot z_j} i^{\mathrm{wt}(z_j)} \sum_{z_{2k+1} \in V_n} (-1)^{(\phi(z_{2k})+u_{2k+1})\cdot z_{2k+1}} i^{\mathrm{wt}(z_{2k+1})}.$$

In what follows, we treat the case $k \equiv 0 \pmod 3$, the case $(m - k) \equiv 1 \pmod 3$ can be proved similarly (essentially, the roles of $P(z_{2k})$ and $Q(z_{2k})$ are exchanged). If $k \equiv 0 \pmod 3$, we have $2k - 1 \equiv 2 \pmod 3$ and from Lemma 5

$$P(z_{2k}) = 2^{kn}\omega^c \delta_{\psi(z_{2k})+a} \tag{3}$$

for some $c \in \mathbb{Z}_8$ and $a \in V_n$. Now $k \equiv 0 \pmod 3$ implies $m - k \equiv m \pmod 3$, so $2(m - k) \equiv 0 \pmod 3$ or $1 \pmod 3$. Hence by Lemma 5

$$Q(z_{2k}) = \begin{cases} 2^{(m-k)n}\omega^d(-1)^{b \cdot \phi(z_{2k})} & \text{if } m \equiv 0 \pmod 3 \\ 2^{(m-k)n}\omega^d(-1)^{b \cdot \phi(z_{2k})} i^{-\text{wt}(\psi(z_{2k}))} & \text{if } m \equiv 2 \pmod 3 \end{cases} \tag{4}$$

for some $d \in \mathbb{Z}_8$ and $b \in V_n$. Combining (2), (3), and (4), we arrive at

$$\mathcal{N}(f)(u_1, \ldots, u_{2m})$$
$$= \begin{cases} \omega^{c+d} \displaystyle\sum_{z_{2k} \in V_n} (-1)^{h(z_{2k})+z_{2k} \cdot u_{2k}+b \cdot \phi(z_{2k})} i^{\text{wt}(z_{2k})} \delta_{\psi(z_{2k})+a} & \text{if } m \equiv 0 \pmod 3 \\ \omega^{c+d} \displaystyle\sum_{z_{2k} \in V_n} (-1)^{h(z_{2k})+z_{2k} \cdot u_{2k}+b \cdot \phi(z_{2k})} \delta_{\psi(z_{2k})+a} & \text{if } m \equiv 2 \pmod 3. \end{cases}$$

In either case the term inside the sum is zero unless $z_{2k} = \psi^{-1}(a)$. Therefore, $|\mathcal{N}(f)(u_1, \ldots, u_{2m})| = 1$, as was claimed. $\qquad\square$

Example 6. Take $m = 2$ and $k = 1$ in Theorem 4. Then f reads

$$f(x_1, x_2, y_1, y_2) = y_1 \cdot \psi(x_1) + \phi(x_1) \cdot y_2 + y_2 \cdot x_2 + h(x_1).$$

In this way we can construct bent-negabent functions in $4n$ variables of degree ranging from 2 to n.

In general, whenever $m \not\equiv 1 \pmod 3$, we can use Theorem 4 to construct bent-negabent functions in $2mn$ variables of degree ranging from 2 to n. This yields bent-negabent functions in $2t$ variables for every $t \geq 2$ and $t \not\equiv 1 \pmod 6$; if $t \not\equiv 1 \pmod 3$, we can take $n = 1$ and $m = t$, and if $t \equiv 1 \pmod 3$ and $t \not\equiv 1 \pmod 6$, we can take $n = 2$ and $m = t/2$.

In the remainder of this section we apply Theorem 3 to construct further bent-negabent functions by taking a partial dual of f given in (1). We therefore have to prove that the partial dual of f exists with respect to certain subspaces of V and to find an explicit expression for this function.

Write $V = U \oplus W$, where $\dim W = k$ and $k \leq mn$. Suppose that we have a function $\tau : V \to V$. We can separate τ on U and W by defining $|W|$ functions $\tau_z : U \to U$ and $|U|$ functions $\tau_x : W \to W$ such that

$$\tau(x, z) = (\tau_z(x), \tau_x(z)), \quad x \in U, z \in W.$$

Lemma 7. *With the notation as above, define* $a : V \oplus V \to \mathbb{F}_2$ *by*

$$a(x, z, y, w) = \tau(x, z) \cdot (y, w) + c(x, z), \quad x, y \in U, \ z, w \in W,$$

where $c : V \to \mathbb{F}_2$. *Then* a *is bent with respect to* $W \oplus W$ *if for every* $x \in U$ *the map* τ_x *is a permutation on* W. *Moreover, in this case, the partial dual of* a *with respect to* $W \oplus W$ *is given by*

$$\tilde{a}_{W \oplus W}(x, u, y, v) = \tau_z(x) \cdot y + u \cdot z + c(x, z), \quad \text{where} \quad z = \tau_x^{-1}(v).$$

Proof. We have

$$\mathcal{H}_{W \oplus W}(a)(x, u, y, v) = 2^{-k} \sum_{z, w \in W} (-1)^{\tau(x,z) \cdot (y,w) + c(x,z) + u \cdot z + v \cdot w}$$

$$= 2^{-k} \sum_{z, w \in W} (-1)^{\tau_z(x) \cdot y + \tau_x(z) \cdot w + c(x,z) + u \cdot z + v \cdot w}$$

$$= 2^{-k} \sum_{z \in W} (-1)^{\tau_z(x) \cdot y + c(x,z) + u \cdot z} \sum_{w \in W} (-1)^{(\tau_x(z) + v) \cdot w}.$$

The inner sum is zero unless $z = \tau_x^{-1}(v)$, in which case the sum is equal to 2^k. Therefore

$$\mathcal{H}_{W \oplus W}(a)(x, u, y, v) = (-1)^{\tilde{a}_{W \oplus W}(x, u, y, v)},$$

where $\tilde{a}_{W \oplus W}$ is given in the lemma. □

Now partition the set $\{1, 2, \ldots, m\}$ into the two subsets

$$S = \{s_1, \ldots, s_k\} \quad \text{and} \quad T = \{t_1, \ldots, t_{m-k}\}.$$

Given $x \in V$, we shall write $x_S = (x_{s_1}, \ldots, x_{s_k})$ and $x_T = (x_{t_1}, \ldots, x_{t_{m-k}})$. As before, let U and W be vector spaces over $\mathbb{F}_2$ such that $V = U \oplus W$ and, if $(x_1, \ldots, x_m) \in V$, we have $x_S \in W$ and $x_T \in U$.

Theorem 8. *With the notation as above,* f, *given in (1), is bent with respect to* $U \oplus U$ *and bent with respect to* $W \oplus W$. *Moreover, the partial dual of* f *with respect to* $W \oplus W$ *is given by*

$$\tilde{f}_{W \oplus W}(x_T, x_S, y_T, y_S) = w_T \cdot y_T + x_S \cdot z_S + g(x_T, z_S),$$

where

$$z_j = \begin{cases} x_j & \text{if } j \notin S \\ \psi_j^{-1}(y_j + \phi_{j-1}(z_{j-1})) & \text{if } j \in S, \end{cases}$$

and

$$w_j = \phi_{j-1}(z_{j-1}) + \psi_j(x_j) \quad \text{for } j \in T.$$

By convention, x_0 *is the all-zero vector and* ϕ_0 *is the identity map.*

Proof. Observe that for every $x_S \in W$ the function $\sigma_{x_S}(x_T)$ is a permutation on U. Similarly, for every $x_T \in U$ the function $\sigma_{x_T}(x_S)$ is a permutation on W. Hence, by Lemma 7, f is bent with respect to $U \oplus U$ and bent with respect to $W \oplus W$. Using Lemma 7, the partial dual of f with respect to $W \oplus W$ can be written as

$$\tilde{f}_{W \oplus W}(x_T, x_S, y_T, y_S) = \sigma_{z_S}(x_T) \cdot y_T + x_S \cdot z_S + g(x_T, z_S), \quad z_S = \sigma_{x_T}^{-1}(y_S).$$

Now we first find z_S by solving the system of k equations implied by

$$\sigma_{x_T}(z_S) = y_S.$$

Then z_S can be used to find $\sigma_{z_S}(x_T)$. The solution is given in the theorem.   □

Starting from Theorem 4, the preceding theorem together with Theorem 3 can be used to construct further bent-negabent functions of the form (1). If $m = 2$, it is easy to check that we do not obtain any new bent-negabent functions. But for larger m the function f and a partial dual of f generally have a different structure. However, explicit expressions for the partial dual of f can look rather cumbersome, so we illustrate the application of Theorem 8 by an example.

Example 9. Take $m = 3$ and $k = 2$ in Theorem 4. Then f reads

$$f(x_1, x_2, x_3, y_1, y_2, y_3) = \sigma(x_1, x_2, x_3) \cdot (y_1, y_2, y_3) + g(x_1, x_2, x_3),$$

where

$$\sigma(x_1, x_2, x_3) = (x_1, x_1 + \psi(x_2), \phi(x_2) + x_3)$$
$$g(x_1, x_2, x_3) = h(x_2).$$

Now set $S = \{0, 1, 2\}$, so that $W = V$, and apply Theorem 8. Then $\tilde{f}_{W \oplus W}$ is the usual dual of f and given by

$$\tilde{f}(x_1, x_2, x_3, y_1, y_2, y_3) = \sigma'(y_1, y_2, y_3) \cdot (x_1, x_2, x_3) + g'(y_1, y_2, y_3),$$

where

$$\sigma'(y_1, y_2, y_3) = (y_1, \psi^{-1}(y_1 + y_2), y_3 + \phi(\psi^{-1}(y_1 + y_2)))$$
$$g'(y_1, y_2, y_3) = h(\psi^{-1}(y_1 + y_2)).$$

The function $\tilde{f}$ is by Theorem 3 negabent.

5   A Bound on the Degree

It is well known that, if $n > 1$, bent Boolean functions in $2n$ variables have a maximum algebraic degree of n [1]. If $n \in \{2, 3\}$, the maximum degree of a bent-negabent function in $2n$ variables is also equal to n. For example, the cubic function $f : V_6 \to \mathbb{F}_2$

$$f(x_1, x_2, x_3, y_1, y_2, y_3) =$$
$$y_1(x_1 x_2 + x_2 x_3 + x_1 + x_2) + y_2(x_1 x_2 + x_2 x_3 + x_3) + y_3(x_1 + x_3)$$

is bent-negabent. Note that f belongs to the Maiorana–McFarland class. In this section we prove that the degree of a Maiorana–McFarland-type bent-negabent function in $2n$ variables is at most $n - 1$ for $n > 3$.

Theorem 10. *Let σ be a permutation on V_n and let $g : V_n \to \mathbb{F}_2$ be an arbitrary Boolean function. Suppose that the function $f : V_n \oplus V_n \to \mathbb{F}_2$ given by*

$$f(x, y) = \sigma(x) \cdot y + g(x)$$

is negabent. Then, if $n > 3$, the degree of f is at most $n - 1$.

The proof of the theorem requires a lemma.

Lemma 11. *The nega-Hadamard transform of a negabent function on V_n contains only values of the form $\omega^n i^k$, where $\omega = (1 + i)/\sqrt{2}$ and $k \in \mathbb{Z}_4$.*

Proof. Let $2^{-\frac{n}{2}} S$ denote an arbitrary value of the nega-Hadamard transform of a negabent function on V_n. Then $\Re(S)$ or $\Im(S)$ must be integers and $|S|^2 = 2^n$ must be a sum of two squares (one of them may be zero). From Jacobi's two-square theorem we know that 2^n has a unique representation as a sum of two squares, namely $2^n = (2^{n/2})^2 + 0^2$ if n is even, and $2^n = (2^{(n-1)/2})^2 + (2^{(n-1)/2})^2$ if n is odd. Hence, if n is even, either $\Re(S)$ or $\Im(S)$ must be zero. If n is odd, we must have $|\Re(S)| = |\Im(S)|$, which proves the lemma. □

Proof (of Theorem 10). Using Lemma 1, we obtain

$$\mathcal{N}(f)(u, v) = 2^{-n} \sum_{x, y \in V_n} (-1)^{\sigma(x) \cdot y + g(x) + u \cdot x + v \cdot y} i^{\text{wt}(x) + \text{wt}(y)}$$

$$= 2^{-n} \sum_{x \in V_n} (-1)^{g(x) + u \cdot x} i^{\text{wt}(x)} \sum_{y \in V_n} (-1)^{(\sigma(x) + v) \cdot y} i^{\text{wt}(y)}$$

$$= 2^{-\frac{n}{2}} \omega^n \sum_{x \in V_n} (-1)^{g(x) + u \cdot x} i^{\text{wt}(x) - \text{wt}(\sigma(x) + v)}$$

$$= 2^{-\frac{n}{2}} \omega^n i^{-\text{wt}(v)} \sum_{x \in V_n} (-1)^{g(x) + u \cdot x + v \cdot \sigma(x)} i^{\text{wt}(x) - \text{wt}(\sigma(x))}$$

$$= 2^{-\frac{n}{2}} \omega^n i^{-\text{wt}(v)} \sum_{x \in V_n} (-1)^{g(x) + u \cdot x + v \cdot \sigma(x)} i^{w(x)},$$

where

$$w(x) = \sum_{j=1}^{n} (x_j + 3\sigma_j(x)) \pmod{4}$$

and $\sigma_j(x)$ is the jth component of $\sigma(x)$, so that $\sigma(x) = (\sigma_1(x), \ldots, \sigma_n(x))$. Now write $w(x)$ in 2-adic expansion, viz $w(x) = l(x) + 2q(x)$ with

$$l(x) = \sum_{j=1}^{n}(x_j + \sigma_j(x)) \pmod 2$$

$$q(x) = \sum_{j=1}^{n}\sigma_j(x) + \sum_{1 \le j < k \le n}[(x_j x_k + \sigma_j(x)\sigma_k(x)] + \sum_{1 \le j,k \le n}x_j\sigma_k(x) \pmod 2.$$

Then we have

$$\Re(\omega^{-n}i^{\mathrm{wt}(v)}\mathcal{N}(f)(u,v)) = 2^{-\frac{n}{2}-1}\sum_{x \in V_n}(-1)^{g(x)+q(x)+v\cdot\sigma(x)+u\cdot x}[1 + (-1)^{l(x)}]$$

$$= \frac{1}{2}[\mathcal{H}(h_v)(u) + \mathcal{H}(h_{\bar{v}})(\bar{u})], \tag{5}$$

where $h_v(x) = g(x) + q(x) + v \cdot \sigma(x)$ and $\bar{u}$ is the complement of u. Similarly we obtain

$$\Im(\omega^{-n}i^{\mathrm{wt}(v)}\mathcal{N}(f)(u,v)) = \frac{1}{2}[\mathcal{H}(h_v)(u) - \mathcal{H}(h_{\bar{v}})(\bar{u})]. \tag{6}$$

By assumption, $|\mathcal{N}(f)(u,v)| = 1$, and by Lemma 11, either the real part or the imaginary part of $\mathcal{N}(f)(u,v)$ must be zero. First suppose that n is even. Then ω^{-n} is a 4th root of unity and either (5) or (6) must be zero. Hence h_v must be bent for every $v \in V_n$, which implies that for $n > 2$ the degree of h_v can be at most $n/2$ [1]. Now let n be odd. Then ω^{-n} is an 8th root of unity and the absolute values of (5) and (6) must be equal. This can only happen if the Hadamard spectrum of h_v contains only the values 0 and $\pm\sqrt{2}$ (such functions are called almost-bent functions [2]). It is known [2, Thm. 1] that the degree of such a function is at most $(n+1)/2$.

For either $n \ge 3$ we conclude that the degree of $h_v(x)$ is at most $\lceil n/2 \rceil$ for every $v \in V_n$. This implies that the degree of $v \cdot \sigma(x)$ is bounded by $\lceil n/2 \rceil$ for every $v \in V_n$ and $n \ge 3$. Note that, since σ is a permutation, $\sigma_j(x)\sigma_k(x) = 1$ has exactly 2^{n-2} solutions in V_n, so for $n \ge 3$ each of the terms $\sigma_j(x)\sigma_k(x)$ cannot have degree equal to n (see, e.g., [6, Ch. 13, Thm. 1]). Therefore, the degree of q is at most $\max\{n-1, \lceil n/2 \rceil + 1\}$. It follows that, if $n > 3$, the degree of q and, therefore, the degree of g is bounded by $n - 1$, which proves the theorem. $\square$

References

1. Rothaus, O.S.: On 'bent' functions. J. Comb. Theory (A) 20, 300–305 (1976)
2. Carlet, C., Charpin, P., Zinoviev, V.: Codes, bent functions and permutations suitable for DES-like cryptosystems. Designs, Codes and Cryptography 15(2), 125–156 (1998)
3. Parker, M.G., Pott, A.: On Boolean functions which are bent and negabent. In: Golomb, S.W., Gong, G., Helleseth, T., Song, H.-Y. (eds.) SSC 2007. LNCS, vol. 4893, pp. 9–23. Springer, Heidelberg (2007)

4. Parker, M.G.: The constabent properties of Golay-Davis-Jedwab sequences. In: Int. Symp. Information Theory, p. 302. IEEE, Sorrento (2000)
5. Riera, C., Parker, M.G.: One and two-variable interlace polynomials: A spectral interpretation. In: Ytrehus, Ø. (ed.) WCC 2005. LNCS, vol. 3969, pp. 397–411. Springer, Heidelberg (2006)
6. MacWilliams, F.J., Sloane, N.J.A.: The Theory of Error-Correcting Codes. North-Holland, The Netherlands (1977)

New Perfect Nonlinear Multinomials
over $\mathbf{F}_{p^{2k}}$ for Any Odd Prime $p^\star$

Lilya Budaghyan and Tor Helleseth

Department of Informatics
University of Bergen
PB 7803, 5020 Bergen, Norway
{Lilya.Budaghyan,Tor.Helleseth}@ii.uib.no

Abstract. We introduce two infinite families of perfect nonlinear Dembowski-Ostrom multinomials over $\mathbf{F}_{p^{2k}}$ where p is any odd prime. We prove that in general these functions are CCZ-inequivalent to previously known PN mappings. One of these families has been constructed by extension of a known family of APN functions over $\mathbf{F}_{2^{2k}}$. This shows that known classes of APN functions over fields of even characteristic can serve as a source for further constructions of PN mappings over fields of odd characteristics.

Besides, we supply results indicating that these PN functions define new commutative semifields. After the works of Dickson (1906) and Albert (1952), these are the firstly found infinite families of commutative semifields which are defined for all odd primes p.

Keywords: Commutative semifield, Equivalence of functions, Perfect nonlinear, Planar function.

1 Introduction

For any positive integer n and any prime p a function F from the field $\mathbf{F}_{p^n}$ to itself is called *differentially δ-uniform* if for every $a \neq 0$ and every b in $\mathbf{F}_{p^n}$, the equation $F(x + a) - F(x) = b$ admits at most δ solutions. Functions with low differential uniformity are of special interest in cryptography (see [3,21]). Differentially 1-uniform functions are called *perfect nonlinear* (PN) or *planar*. PN functions exist only for p odd. For p even differentially 2-uniform functions, called *almost perfect nonlinear* (APN), are those which have the lowest possible differential uniformity.

There are several equivalence relations of functions for which differential uniformity is invariant. First recall that a function F over $\mathbf{F}_{p^n}$ is called *linear* if

$$F(x) = \sum_{0 \leq i < n} a_i x^{p^i}, \qquad a_i \in \mathbf{F}_{p^n}.$$

A sum of a linear function and a constant is called an *affine function*. We say that two functions F and F' are *affine equivalent* (or *linear equivalent*) if

---

$\star$ This work was supported by the Norwegian Research Council.

$F' = A_1 \circ F \circ A_2$, where the mappings A_1, A_2 are affine (resp. linear) permutations. Functions F and F' are called *extended affine equivalent* (EA-equivalent) if $F' = A_1 \circ F \circ A_2 + A$, where the mappings A, A_1, A_2 are affine, and where A_1, A_2 are permutations.

Two mappings F and F' from $\mathbf{F}_{p^n}$ to itself are called *Carlet-Charpin-Zinoviev equivalent* (CCZ-equivalent) if for some affine permutation $\mathcal{L}$ of $\mathbf{F}_{p^n}^2$ the image of the graph of F is the graph of F', that is, $\mathcal{L}(G_F) = G_{F'}$ where $G_F = \{(x, F(x)) \mid x \in \mathbf{F}_{p^n}\}$ and $G_{F'} = \{(x, F'(x)) \mid x \in \mathbf{F}_{p^n}\}$. Differential uniformity is invariant under CCZ-equivalence. EA-equivalence is a particular case of CCZ-equivalence and any permutation is CCZ-equivalent to its inverse. In [5], it is proven that CCZ-equivalence is even more general. In the present paper we prove that for PN functions CCZ-equivalence coincides with EA-equivalence.

Almost all known planar functions are DO polynomials. Recall that a function F is called *Dembowski-Ostrom polynomial* (DO polynomial) if

$$F(x) = \sum_{0 \le k, j < n} a_{kj} x^{p^k + p^j}.$$

When p is odd the notion of planar DO polynomial is closely connected to the notion of *commutative semifield*. A ring with left and right distributivity and with no zero divisors is called a *presemifield*. A presemifield with a multiplicative identity is called a *semifield*. Any finite presemifield can be represented by $\mathbf{S} = (\mathbf{F}_{p^n}, +, \star)$, where $(\mathbf{F}_{p^n}, +)$ is the additive group of $\mathbf{F}_{p^n}$ and $x \star y = \phi(x, y)$ with ϕ a function from $\mathbf{F}_{p^n}^2$ onto $\mathbf{F}_{p^n}$, see [9].

Let $\mathbf{S}_1 = (\mathbf{F}_{p^n}, +, *)$ and $\mathbf{S}_2 = (\mathbf{F}_{p^n}, +, \star)$ be two presemifields. They are called *isotopic* if there exist three linear permutations L, M, N over $\mathbf{F}_{p^n}$ such that

$$L(x * y) = M(x) \star N(y),$$

for any $x, y \in \mathbf{F}_{p^n}$. The triple (M, N, L) is called the *isotopism* between $\mathbf{S}_1$ and $\mathbf{S}_2$. If $M = N$ then $\mathbf{S}_1$ and $\mathbf{S}_2$ are called *strongly isotopic*.

Let $\mathbf{S}$ be a finite semifield. The subsets

$$N_l(\mathbf{S}) = \{\alpha \in \mathbf{S} : (\alpha \star x) \star y = \alpha \star (x \star y) \text{ for all } x, y \in \mathbf{S}\},$$
$$N_m(\mathbf{S}) = \{\alpha \in \mathbf{S} : (x \star \alpha) \star y = x \star (\alpha \star y) \text{ for all } x, y \in \mathbf{S}\},$$
$$N_r(\mathbf{S}) = \{\alpha \in \mathbf{S} : (x \star y) \star \alpha = x \star (y \star \alpha) \text{ for all } x, y \in \mathbf{S}\},$$

are called the *left*, *middle* and *right nucleus* of $\mathbf{S}$, respectively, and the set $N(\mathbf{S}) = N_l(\mathbf{S}) \cap N_m(\mathbf{S}) \cap N_r(\mathbf{S})$ is called the *nucleus*. These sets are finite fields and, if $\mathbf{S}$ is commutative then $N_l(\mathbf{S}) = N_r(\mathbf{S})$. The nuclei measure how far $\mathbf{S}$ is from being associative. *The orders of the respective nuclei are invariant under isotopism* [9].

Let $\mathbf{S} = (\mathbf{F}_{p^n}, +, \star)$ be a commutative presemifield which does not contain an identity. To create a semifield from $\mathbf{S}$ choose any $a \in \mathbf{F}_{p^n}^*$ and define a new multiplication $*$ by

$$(x \star a) * (a \star y) = x \star y$$

for all $x, y \in \mathbf{F}_{p^n}$. Then $\mathbf{S}' = (\mathbf{F}_{p^n}, +, *)$ is a commutative semifield isotopic to $\mathbf{S}$ with identity $a * a$. We say $\mathbf{S}'$ is a commutative semifield *corresponding* to the commutative presemifield $\mathbf{S}$. An isotopism between $\mathbf{S}$ and $\mathbf{S}'$ is a strong isotopism $\big(L_a(x), L_a(x), x\big)$ with a linear permutation $L_a(x) = a \star x$, see [9].

Let F be a planar DO polynomial over $\mathbf{F}_{p^n}$. Then $\mathbf{S} = (\mathbf{F}_{p^n}, +, \star)$, with

$$x \star y = F(x + y) - F(x) - F(y)$$

for any $x, y \in \mathbf{F}_{p^n}$, is a commutative presemifield. We denote by $\mathbf{S}_F = (\mathbf{F}_{p^n}, +, *)$ the commutative semifield corresponding to the commutative presemifield $\mathbf{S}$ with isotopism $\big(L_1(x), L_1(x), x\big)$ and we call $\mathbf{S}_F = (\mathbf{F}_{p^n}, +, *)$ the *commutative semifield defined by the planar DO polynomial F*. Conversely, given a commutative presemifield $\mathbf{S} = (\mathbf{F}_{p^n}, +, \star)$ of odd order, the function given by

$$F(x) = \frac{1}{2}(x \star x)$$

is a planar DO polynomial [9]. We prove in Section 4 that for planar DO polynomials CCZ-equivalence coincides with linear equivalence. This implies that two planar DO polynomials F and F' are CCZ-equivalent if and only if the corresponding commutative semifields $\mathbf{S}_F$ and $\mathbf{S}_{F'}$ are strongly isotopic. It is proven in [9] that for the n odd case two commutative presemifields are isotopic if and only if they are strongly isotopic. There are also some sufficient conditions for the n even case when isotopy of presemifields implies their strong isotopy [9]. Thus, in the case n even it is potentially possible that isotopic commutative presemifields define CCZ-inequivalent planar DO polynomials. However, in practice no such cases are known.

Although commutative semifields have been intensively studied for more than a hundred years, there are only eight distinct cases of known commutative semifields of odd order (see [9]), and only three of them are defined for any odd prime p. The eight distinct cases of known planar DO polynomials and corresponding commutative semifields are the following:

(i)
$$x^2$$

over $\mathbf{F}_{p^n}$ which corresponds to the finite field $\mathbf{F}_{p^n}$;

(ii)
$$x^{p^t+1}$$

over $\mathbf{F}_{p^n}$, with $n/\gcd(t, n)$ odd, which correspond to Albert's commutative twisted fields [1,11,16];

(iii) the functions over $\mathbf{F}_{p^{2k}}$, which correspond to the Dickson semifields [12];

(iv)
$$x^{10} \pm x^6 - x^2$$

over $\mathbf{F}_{3^n}$, with n odd, corresponding to the Coulter-Matthews and Ding-Yuan semifields [8,14];

(v) the function over $\mathbf{F}_{3^{2k}}$, with k odd, corresponding to the Ganley semifield [15];

(vi) the function over $\mathbf{F}_{3^{2k}}$ corresponding to the Cohen-Ganley semifield [7];

(vii) the function over $\mathbf{F}_{3^{10}}$ corresponding to the Penttila-Williams semifield [22];

(viii) the function over $\mathbf{F}_{3^8}$ corresponding to the Coulter-Henderson-Kosick semifield [10].

The representations of the PN functions corresponding to the cases (iii), (v)-(vii), can be found in [18,19]. The only known PN functions which are not DO polynomials are the power functions

$$x^{\frac{3^t+1}{2}}$$

over $\mathbf{F}_{3^n}$, where t is odd and $\gcd(t, n) = 1$ [8,17].

Let p be an odd prime, s and k positive integers, and $n = 2k$. In the present paper we introduce the following new infinite classes of perfect nonlinear DO polynomials over $\mathbf{F}_{p^n}$:

(i*) $(bx)^{p^s+1} - \left((bx)^{p^s+1}\right)^{p^k} + \sum_{i=0}^{k-1} c_i x^{p^i(p^k+1)},$

where $\sum_{i=0}^{k-1} c_i x^{p^i}$ is a permutation over $\mathbf{F}_{p^n}$ with coefficients in $\mathbf{F}_{p^k}$, $b \in \mathbf{F}_{p^n}^*$, and $\gcd(k+s, 2k) = \gcd(k+s, k)$, $\gcd(p^s+1, p^k+1) \neq \gcd\left(p^s+1, (p^k+1)/2\right)$.

(ii*) $bx^{p^s+1} + (bx^{p^s+1})^{p^k} + cx^{p^k+1} + \sum_{i=1}^{k-1} r_i x^{p^{k+i}+p^i},$

where $b \in \mathbf{F}_{p^n}^*$ is not a square, $c \in \mathbf{F}_{p^n} \setminus \mathbf{F}_{p^k}$, and $r_i \in \mathbf{F}_{p^k}$, $0 \leq i < k$, and $\gcd(k+s, n) = \gcd(k+s, k)$.

We show that in general these functions are CCZ-inequivalent to previously known PN functions and define new commutative semifields.

2   A New Family of PN Multinomials

In [20] Ness gives a list of planar DO trinomials over $\mathbf{F}_{p^n}$ for $p \leq 7$, $n \leq 8$ which were found with a computer. Investigation of these functions has led us to the following family of planar DO polynomials.

Theorem 1. *Let p be an odd prime, s and k positive integers such that $\gcd(p^s+1, p^k+1) \neq \gcd\left(p^s+1, (p^k+1)/2\right)$ and $\gcd(k+s, 2k) = \gcd(k+s, k)$. Let also $n = 2k$, $b \in \mathbf{F}_{p^n}^*$, and $\sum_{i=0}^{k-1} c_i x^{p^i}$ be a permutation over $\mathbf{F}_{p^n}$ with coefficients in $\mathbf{F}_{p^k}$. Then the function*

$$F(x) = (bx)^{p^s+1} - \left((bx)^{p^s+1}\right)^{p^k} + \sum_{i=0}^{k-1} c_i x^{p^i(p^k+1)}$$

is PN over $\mathbf{F}_{p^n}$.

Proof. Since F is DO polynomial then it is PN if for any $a \in \mathbf{F}_{p^n}^*$ the equation $F(x+a) - F(x) - F(a) = 0$ has only 0 as a solution. We have

$$\Delta(x) = F(x+a) - F(x) - F(a)$$
$$= b^{p^s+1}(ax^{p^s} + a^{p^s}x) - b^{p^k(p^s+1)}(a^{p^k}x^{p^{k+s}} + a^{p^{k+s}}x^{p^k})$$
$$+ \sum_{i=0}^{k-1} c_i(a^{p^i}x^{p^{k+i}} + a^{p^{k+i}}x^{p^i}).$$

Any solution of the equation $\Delta(x) = 0$ is also a solution of $\Delta(x) + \Delta(x)^{p^k} = 0$ and $\Delta(x) - \Delta(x)^{p^k} = 0$, that is, a solution of

$$\sum_{i=0}^{k-1} c_i(a^{p^i}x^{p^{k+i}} + a^{p^{k+i}}x^{p^i}) = 0, \tag{1}$$

$$b^{p^s+1}(ax^{p^s} + a^{p^s}x) = b^{p^k(p^s+1)}(a^{p^k}x^{p^{k+s}} + a^{p^{k+s}}x^{p^k}). \tag{2}$$

Since $\sum_{i=0}^{k-1} c_i x^{p^i}$ is a permutation then (1) implies

$$ax^{p^k} = -a^{p^k}x. \tag{3}$$

Now we can substitute ax^{p^k} in (2) by $-a^{p^k}x$ and then obtain

$$b^{p^s+1}(ax^{p^s} + a^{p^s}x) = -b^{p^k(p^s+1)}(a^{p^{k+s}+p^k-p^s}x^{p^s} + a^{p^{k+s}+p^k-1}x),$$

that is,

$$(b^{p^s+1}a + b^{p^k(p^s+1)}a^{p^{k+s}+p^k-p^s})x^{p^s} = -(b^{p^s+1}a^{p^s} + b^{p^k(p^s+1)}a^{p^{k+s}+p^k-1})x,$$

and since $a, b \neq 0$ then for $x \neq 0$

$$x^{p^s-1} = -\frac{b^{p^s+1}a^{p^s} + b^{p^k(p^s+1)}a^{p^{k+s}+p^k-1}}{b^{p^s+1}a + b^{p^k(p^s+1)}a^{p^{k+s}+p^k-p^s}} = -a^{p^s-1}, \tag{4}$$

when

$$b^{(p^k-1)(p^s+1)}a^{p^{k+s}+p^k-p^s-1} \neq -1. \tag{5}$$

Now assume that for some nonzero a inequality (5) is wrong, that is,

$$(ba)^{(p^k-1)(p^s+1)} = -1.$$

Then -1 is a power of $(p^k - 1)(p^s + 1)$ which is in contradiction with $\gcd(p^s + 1, p^k + 1) \neq \gcd\left(p^s + 1, (p^k + 1)/2\right)$ since -1 is a power of $(p^n - 1)/2$. From (3) and (4) we get

$$y^{p^k-1} = y^{p^s-1} = -1, \tag{6}$$

where $y = x/a$. Since $n = 2k$ then the first equality in (6) implies $y^{p^{k+s}} = y$, that is, $y \in \mathbf{F}_{p^{k+s}}$. Thus, if $\gcd(k+s, 2k) = \gcd(k+s, k)$ then $y \in \mathbf{F}_{p^{\gcd(k+s,k)}}$ which contradicts the second equality in (6), that is, $y^{p^k-1} = 1 \neq -1$, for any $y \neq 0$. Therefore, the only solution of $\Delta(x) = 0$ is $x = 0$. $\square$

3   Another Family of PN Multinomials

In this section we show that one of the ways to construct PN mappings is to extend a known family of APN functions over $\mathbf{F}_{2^n}$ to a family of PN functions over $\mathbf{F}_{p^n}$ for odd primes p. Below we construct a class of PN functions by following the pattern of APN multinomials over $\mathbf{F}_{2^{2k}}$ presented in [4].

Theorem 2. *Let p be an odd prime, s and k positive integers, $n = 2k$, and $\gcd(k + s, n) = \gcd(k + s, k)$. If $b \in \mathbf{F}_{p^n}^*$ is not a square, $c \in \mathbf{F}_{p^n} \setminus \mathbf{F}_{p^k}$, and $r_i \in \mathbf{F}_{p^k}$, $0 \leq i < k$, then the function*

$$F(x) = \mathrm{Tr}_k^{2k}(bx^{p^s+1}) + cx^{p^k+1} + \sum_{i=1}^{k-1} r_i x^{p^{k+i}+p^i}$$

is PN over $\mathbf{F}_{p^n}$.

Proof. We have to show that for any $a \in \mathbf{F}_{p^n}^*$ the equation $\Delta(x) = 0$ has only 0 as a solution when

$$\Delta(x) = F(x + a) - F(x) - F(a)$$

$$= \mathrm{Tr}_k^{2k}\left(b(x^{p^s}a + xa^{p^s})\right) + c(x^{p^k}a + xa^{p^k}) + \sum_{i=1}^{k-1} r_i(x^{p^{k+i}}a^{p^i} + x^{p^i}a^{p^{k+i}}).$$

After replacing x by ax we get

$$\Delta_1(x) = \Delta(ax) = \mathrm{Tr}_k^{2k}\left(ba^{p^s+1}(x^{p^s} + x)\right) + ca^{p^k+1}(x^{p^k} + x)$$

$$+ \sum_{i=1}^{k-1} r_i a^{p^{k+i}+p^i}(x^{p^{k+i}} + x^{p^i}).$$

Since $\Delta_1(x) = 0$ then $\Delta_1(x) - \Delta_1(x)^{p^k} = 0$, that is,

$$(ca^{p^k+1} - c^{p^k}a^{p^k+1})(x^{p^k} + x) = 0.$$

Thus,

$$(c - c^{p^k})a^{p^k+1}(x^{p^k} + x) = 0$$

and, therefore,

$$x^{p^k} = -x$$

since $c \in \mathbf{F}_{p^{2k}} \setminus \mathbf{F}_{p^k}$.

Substituting $x^{p^k} = -x$ in $\Delta_1(x) = 0$ we obtain

$$\Delta_1(x) = ba^{p^s+1}(x^{p^s} + x) + b^{p^k}a^{p^{s+k}+p^k}(x^{p^{s+k}} + x^{p^k})$$

$$= (ba^{p^s+1} - b^{p^k}a^{p^{s+k}+p^k})(x^{p^s} + x).$$

Hence, if

$$ba^{p^s+1} \neq b^{p^k} a^{p^{s+k}+p^k} \tag{7}$$

then

$$x^{p^s} = -x.$$

Assume that $ba^{p^s+1} = b^{p^k} a^{p^{s+k}+p^k}$ for some nonzero a. Then we get equalities

$$b^{p^k-1} = a^{p^s+1-p^{s+k}-p^k} = a^{-(p^s+1)(p^k-1)} = a^{(p^{k+s}-1)(p^k-1)}$$

which imply that b is a power of $\gcd(p^s+1, p^k+1)$ and of $\gcd(p^{s+k}-1, p^k+1)$. Thus, inequality (7) holds for any $a \neq 0$ if b is not a power of $\gcd(p^s+1, p^k+1)$ or a power of $\gcd(p^{s+k}-1, p^k+1)$. Since $\gcd(p^s+1, p^k+1)$ and $\gcd(p^{s+k}-1, p^k+1)$ are even then we cannot have inequality (7) for any nonzero b but we have this inequality, in particular, when b is not a square in $\mathbf{F}_{p^n}^*$.

Since $x^{p^k} = -x$ and $x^{p^s} = -x$ then $x^{p^k} = x^{p^s}$ and then by taking the p^k-th power we get $x^{p^{k+s}} = x$. Hence, if $\gcd(k+s, 2k) = \gcd(k+s, k)$ then $x \in \mathbf{F}_{p^{\gcd(k+s,k)}}$ and $x^{p^{\gcd(k+s,k)}} = x$. But $x^{p^k} = -x$, which implies $x = 0$.   □

4   On the Equivalence of PN Functions

We prove below that for PN functions CCZ-equivalence coincides with EA-equivalence. In particular it means that PN functions are never permutations.

Proposition 1. *Let F be a PN function and F' be CCZ-equivalent to F. Then F and F' are EA-equivalent.*

Proof. If functions F and F' are CCZ-equivalent then there exists an affine permutation $\mathcal{L}$ over $\mathbf{F}_{p^n}^2$ such that $\mathcal{L}(G_F) = G_{F'}$ where $G_F = \{(x, F(x)) \mid x \in \mathbf{F}_{p^n}\}$ and $G_{F'} = \{(x, F'(x)) \mid x \in \mathbf{F}_{p^n}\}$. The function $\mathcal{L}$ in this case can be introduced as

$$\mathcal{L}(x, y) = (L_1(x, y), L_2(x, y))$$

where $L_1, L_2 : \mathbf{F}_{p^n}^2 \to \mathbf{F}_{p^n}$ are affine and $L_1(x, F(x))$ is a permutation (see [5]). Let us see whether there exists such a function L_1 when F is PN. For some linear functions $L, L' : \mathbf{F}_{p^n} \to \mathbf{F}_{p^n}$ and some $b \in \mathbf{F}_{p^n}^*$ we have

$$L_1(x, y) = L(x) + L'(y) + b.$$

If $L_1(x, F(x))$ is a permutation then for any nonzero a

$$L(x) + L'(F(x)) + b \neq L(x+a) + L'(F(x+a)) + b,$$

that is,

$$L'(F(x+a) - F(x)) \neq -L(a).$$

Since F is PN then $F(x+a) - F(x)$ is a permutation. Thus, the inequality above implies $L'(c) \neq L(a)$ for any c and any nonzero a. First of all we see that L is a permutation since otherwise $L(a') = 0 = L'(0)$ for some nonzero a' and so we get the inequality $L' \circ L^{-1}(c) \neq a$ for any c and any nonzero a which in its turn means $L' \circ L^{-1} = 0$, that is, $L' = 0$.

By the definition of CCZ-equivalence and by the data obtained above we get for CCZ-equivalent functions F and F' that

$$F' = F_2 \circ F_1^{-1},$$

where

$$F_1(x) = L_1(x, F(x)) = L(x) + b,$$
$$F_2(x) = L_2(x, F(x)) = L''(x) + L'''(F(x)) + b'$$

with $b, b' \in \mathbf{F}_{p^n}$, L, L'', L''' linear and L a permutation. Note that

$$F'(x) = L''\big(F_1^{-1}(x)\big) + L'''\Big(F\big(F_1^{-1}(x)\big)\Big) + b' = A(x) + A_1 \circ F \circ A_2(x)$$

where $A_2(x) = F_1^{-1}(x)$ is an affine permutation, $A = L'' \circ F_1^{-1}$ is affine, and we show below that the linear function $A_1 = L'''$ is a permutation. Indeed, the affine function $\mathcal{L}(x, y) = \big(L(x) + b, L''(x) + L'''(y) + b'\big)$ is a permutation, that is, the system of two equations $L(x) = 0$ and $L''(x) + L'''(y) = 0$ only has the solution $(0, 0)$, and then $L'''(y) = 0$ should only have the solution 0 which implies that L''' is a permutation. Thus, F and F' are EA-equivalent. $\square$

From the result above we get the following obvious corollaries.

Corollary 1. *If a PN function F is CCZ-equivalent to a DO polynomial F' then F is also DO polynomial.*

Corollary 2. *Perfect nonlinear DO polynomials F and F' are CCZ-equivalent if and only if they are linear equivalent.*

Now it is obvious that CCZ-equivalence of two DO planar functions implies strong isotopism of the corresponding commutative semifields.

It is also obvious that DO polynomials cannot be CCZ-equivalent to the PN functions $x^{(3^t+1)/2}$ over $\mathbf{F}_{3^n}$ with $\gcd(n, t) = 1$, t odd. Indeed, $x^{(3^t+1)/2}$ is not DO polynomial because $\frac{3^t+1}{2} = 2 + \sum_{i=1}^{t-1} 3^{t-i}$.

5   On the Inequivalence of the Introduced PN Functions with Known PN Mappings

Note that the functions of Theorems 1 and 2 are defined over $\mathbf{F}_{p^{2k}}$ for any odd prime p. Obviously, we can say the same only about PN functions of the cases (i), (ii) and (iii), while the cases (v)-(viii) are defined only for $p = 3$ and cannot cover all the functions of Theorems 1 and 2. So when proving CCZ-inequivalence to the known PN functions we mainly concentrate our attention on the functions (i), (ii), and (iii).

In the proposition below we show that any function which is CCZ-equivalent to x^2 should have some monomial of the form x^{2p^t} for some t, $0 \le t < n$, in its polynomial representation.

Proposition 2. *Let p be an odd prime and n be a positive integer. Any function F of the form*

$$F(x) = \sum_{0 \le k < j < n} a_{kj} x^{p^k + p^j}$$

over $\mathbf{F}_{p^n}$ is CCZ-inequivalent to x^2.

Proof. Since x^2 is a planar DO polynomial then, by Corollary 2, CCZ-equivalence of F to x^2 implies the linear equivalence, that is, the existence of linear permutations L_1 and L_2 such that

$$\big(L_1(x)\big)^2 + L_2\big(F(x)\big) = 0. \tag{8}$$

Let

$$L_1(x) = \sum_{i=0}^{n-1} u_i x^{p^i}, \tag{9}$$

$$L_2(x) = \sum_{i=0}^{n-1} v_i x^{p^i}. \tag{10}$$

Then equality (8) implies

$$0 = \left(\sum_{i=0}^{n-1} u_i x^{p^i} \right)^2 + \sum_{i=0}^{n-1} v_i \left(\sum_{0 \le k < j < n} a_{kj} x^{p^k + p^j} \right)^{p^i}$$

$$= \sum_{i=0}^{n-1} u_i^2 x^{2p^i} + 2 \sum_{0 \le i < j < n} u_i u_j x^{p^i + p^j} + \sum_{0 \le k < j < n, 0 \le i < n} v_i a_{kj}^{p^i} x^{p^i(p^k + p^j)}.$$

Since the identity above takes place for any $x \in \mathbf{F}_{p^n}$ then obviously $u_i^2 = 0$ for all $0 \le i < n$, that is, $L_1(x) = 0$. This contradicts the condition that L_1 is a permutation. Hence F is CCZ-inequivalent to x^2. $\qquad \square$

Corollary 3. *The functions (i*) and (ii*) are CCZ-inequivalent to x^2.*

We give below a sufficient condition on DO polynomials to be CCZ-inequivalent to the PN functions of the case (ii).

Proposition 3. *Let p be an odd prime number, n, n' and t positive integers such that $n' < n$ and $n/\gcd(n,t)$ is odd. Let a function $F : \mathbf{F}_{p^n} \to \mathbf{F}_{p^n}$ be such that*

$$F(x) = \sum_{i=0}^{n'} A_i(x^{p^{s_i}+1}),$$

where $0 < s_i < n$ and $s_i \ne s_j$, for all $i \ne j$, $0 \le i, j \le n'$, and the functions A_i, $0 \le i \le n'$, are linear. If $t \ne s_i$ and $t \ne n - s_i$ for all $0 \le i \le n'$ then the PN function $G(x) = x^{p^t+1}$ is CCZ-inequivalent to F.

Proof. Assume that F and G are CCZ-equivalent. Since G is a planar DO polynomial then, by Corollary 2, CCZ-equivalence implies the existence of linear permutations L_1 and L_2, defined by (9)-(10), such that

$$G(L_1(x)) + L_2(F(x)) = 0.$$

We get

$$0 = \left(\sum_{i=0}^{n-1} u_i x^{p^i}\right)^{p^t+1} + \sum_{i=0}^{n-1} v_i \left(\sum_{i=0}^{n'} A_i(x^{p^{s_i}+1})\right)^{p^i}$$

$$= \sum_{i,j=0}^{n-1} u_i u_j^{p^t} x^{p^i+p^{j+t}} + \sum_{i=0}^{n'} A_i'(x^{p^{s_i}+1}),$$

where A_i', $0 \le i \le n'$, are some linear functions. Since the latter expression is equal to 0 then the terms of the type x^{2p^i}, $0 \le i < n$, should vanish and we get

$$u_i u_{i-t}^{p^t} = 0, \qquad 0 \le i < n. \tag{11}$$

Since $t \ne s_i$ and $t \ne n - s_i$ for all $0 \le i \le n'$ then canceling all terms of the type $x^{p^i(p^t+1)}$, $0 \le i < n$, we get

$$u_i u_i^{p^t} = -u_{i+t} u_{i-t}^{p^t}, \qquad 0 \le i < n. \tag{12}$$

Equalities (11) and (12) imply $L_1 = 0$. Indeed, if $u_i \ne 0$ for some i then from (11) we get $u_{i-t} = 0$ while from (12) we get $u_{i-t} \ne 0$. But L_1 is a permutation and cannot be constantly 0. This contradiction shows that the functions F and x^{p^t+1} are CCZ-inequivalent. □

Corollary 4. *The functions* (i*) *and* (ii*) *are CCZ-inequivalent to* x^{p^t+1} *when* $2k/\gcd(2k, s)$ *is even.*

To prove that functions (i*) and (ii*) are in general CCZ-inequivalent to the functions corresponding to the Dickson semifields we need the following fact which was checked with a computer.

Proposition 4. *The commutative semifields defined by the following planar DO polynomials have the middle nuclei of order* p^2 *and the left nuclei of order* p:

(1) *the functions* (ii*) *with* $p = 3$ *and* $n = 6$;
(2) *the functions* (i*) *with* $p = 3$ *and* $n = 8$;
(3) *the functions* (i*) *with* $p = 5$ *and* $n = 6$.

The work is in progress now to prove that all functions (i*) and (ii*) define commutative semifields with the middle nuclei of order p^2.

Note that the PN functions of case (ii) are not defined for $n = 2^m$ while the PN functions (i*) and (ii*) are. This already shows that in general the semifields defined by the functions of Theorems 1 and 2 cannot be isotopic to Albert's commutative twisted fields. Besides, it is proven in [9] that, for any planar DO function F, isotopism between the commutative semifield defined by F and a commutative twisted field implies strong isotopism. Thus, by Corollary 4, when $2k/\gcd(2k, s)$ is even, the PN functions (i*) and (ii*) define commutative semi-fields nonisotopic to Albert's commutative twisted fields. Proposition 4 shows that the condition on $2k/\gcd(2k, s)$ is not necessary for these commutative semi-fields to be nonisotopic. Indeed, cases (1) and (3) in Proposition 4 are defined also for $2k/\gcd(2k, s)$ odd, and it is proven in [2] that the Albert's commutative twisted fields of order p^n and parameter t (where $n/\gcd(n, t)$ is odd) have the middle and left nuclei of order $p^{\gcd(n,t)}$.

Let $\{1, \beta\}$ be a basis of $\mathbf{F}_{p^{2k}}$ over $\mathbf{F}_{p^k}$. The planar functions over $\mathbf{F}_{p^{2k}}$ derived from the Dickson semifields are

$$x^2 + j\left(\sigma\left(\frac{x^q - x}{\beta^q - \beta}\right)\right)^2 - \beta^2\left(\frac{x^q - x}{\beta^q - \beta}\right)^2,$$

where j is a nonsquare in $\mathbf{F}_{p^k}$, and $1 \neq \sigma \in \mathrm{Aut}(\mathbf{F}_{p^k})$. Different choices of β and σ may give CCZ-inequivalent planar functions. However, as proven in [13], all Dickson commutative semifields of order p^{2k} have the middle nuclei of order p^k. Since the orders of the respective nuclei of semifields are invariant under isotopism then the commutative semifields defined by functions (1)-(3) of Proposition 4 are nonisotopic to all Dickson semifields.

Corollary 5. *Functions (1)-(3) of Proposition 4 define commutative semifields which are nonisotopic to all Dickson semifields.*

Corollary 6. *The functions (i*) with $p = 5$ and $n = 6$ are CCZ-inequivalent to any known PN functions and define commutative semifields nonisotopic to the known ones.*

Note that Proposition 4 also implies that in general the family of Cohen-Ganley semifields is distinct from the families of semifields defined by (i*) and (ii*) with $p = 3$. Indeed, the Cohen-Ganley semifields of order 3^6 and 3^8 have the middle nuclei of order 3^3 and 3^4, respectively. Therefore, they are nonisotopic to the commutative semifields defined by functions (1)-(2) of Proposition 4.

References

1. Albert, A.A.: On nonassociative division algebras. Trans. Amer. Math. Soc. 72, 296–309 (1952)
2. Albert, A.A.: Generalized twisted fields. Pacific J. Math. 11, 1–8 (1961)
3. Biham, E., Shamir, A.: Differential Cryptanalysis of DES-like Cryptosystems. Journal of Cryptology 4(1), 3–72 (1991)
4. Bracken, C., Byrne, E., Markin, N., McGuire, G.: New families of quadratic almost perfect nonlinear trinomials and multinomials. Finite Fields and Applications (to appear, 2008)

5. Budaghyan, L., Carlet, C., Pott, A.: New Classes of Almost Bent and Almost Perfect Nonlinear Functions. IEEE Trans. Inform. Theory 52(3), 1141–1152 (2006)
6. Carlet, C., Charpin, P., Zinoviev, V.: Codes, bent functions and permutations suitable for DES-like cryptosystems. Des., Codes Cryptogr. 15(2), 125–156 (1998)
7. Cohen, S.D., Ganley, M.J.: Commutative semifields, two-dimensional over their middle nuclei. J. Algebra 75, 373–385 (1982)
8. Coulter, R.S., Matthews, R.W.: Planar functions and planes of Lenz-Barlotti class II. Des., Codes Cryptogr. 10, 167–184 (1997)
9. Coulter, R.S., Henderson, M.: Commutative presemifields and semifields. Advances in Math. 217, 282–304 (2008)
10. Coulter, R.S., Henderson, M., Kosick, P.: Planar polynomials for commutative semifields with specified nuclei. Des. Codes Cryptogr. 44, 275–286 (2007)
11. Dembowski, P., Ostrom, T.: Planes of order n with collineation groups of order n^2. Math. Z. 103, 239–258 (1968)
12. Dickson, L.E.: On commutative linear algebras in which division is always uniquely possible. Trans. Amer. Math. Soc. 7, 514–522 (1906)
13. Dickson, L.E.: Linear algebras with associativity not assumed. Duke Math. J. 1, 113–125 (1935)
14. Ding, C., Yuan, J.: A new family of skew Paley-Hadamard difference sets. J. Comb. Theory Ser. A. 133, 1526–1535 (2006)
15. Ganley, M.J.: Central weak nucleus semifields. European J. Combin. 2, 339–347 (1981)
16. Helleseth, T., Rong, C., Sandberg, D.: New families of almost perfect nonlinear power mappings. IEEE Trans. Inf. Theory 45, 475–485 (1999)
17. Helleseth, T., Sandberg, D.: Some power mappings with low differential uniformity. Applic. Alg. Eng., Commun. Comput. 8, 363–370 (1997)
18. Minami, K., Nakagawa, N.: On planar functions of elementary abelian p-group type (submitted)
19. Nakagawa, N.: On functions of finite fields, http://www.math.is.tohoku.ac.jp/~taya/sendaiNC/2006/report/nakagawa.pdf
20. Ness, G.J.: Correlation of sequences of different lengths and related topics. PhD dissertation. University of Bergen, Norway (2007)
21. Nyberg, K.: Differentially uniform mappings for cryptography. In: Helleseth, T. (ed.) EUROCRYPT 1993. LNCS, vol. 765, pp. 55–64. Springer, Heidelberg (1994)
22. Penttila, T., Williams, B.: Ovoids of parabolic spaces. Geom. Dedicata 82, 1–19 (2000)

A New Tool for Assurance of Perfect Nonlinearity

Nuray At* and Stephen D. Cohen

[1] Department of Informatics, University of Bergen, PB 7803, 5020 Bergen, Norway
Nuray.At@ii.uib.no
[2] Department of Mathematics, University of Glasgow, Glasgow G12 8QW, Scotland
sdc@maths.gla.ac.uk

Abstract. Let $f(x)$ be a mapping $f : \mathrm{GF}(p^n) \to \mathrm{GF}(p^n)$, where p is prime and $\mathrm{GF}(p^n)$ is the finite field with p^n elements. A mapping f is called differentially k-uniform if k is the maximum number of solutions $x \in \mathrm{GF}(p^n)$ of $f(x+a) - f(x) = b$, where $a, b \in \mathrm{GF}(p^n)$ and $a \neq 0$. A 1-uniform mapping is called perfect nonlinear (PN). In this paper, we propose an approach for assurance of perfect nonlinearity which involves simply checking a trace condition.

Keywords: perfect nonlinear, equivalence of functions.

1 Introduction

Highly nonlinear mappings have important applications in cryptography, sequences, and coding theory. Let $f(x)$ be a mapping $f : \mathrm{GF}(p^n) \to \mathrm{GF}(p^n)$, where p is prime and $\mathrm{GF}(p^n)$ is the finite field with p^n elements. There are several ways of measuring the nonlinearity of functions, see for example [1]. One robust measure related to differential cryptanalysis uses derivatives. Let $N_f(a, b)$ denote the number of solutions $x \in \mathrm{GF}(p^n)$ of $f(x+a) - f(x) = b$, where $a, b \in \mathrm{GF}(p^n)$. Define

$$\Delta_f = \max\{N_f(a, b) | a, b \in \mathrm{GF}(p^n), a \neq 0\}.$$

The value Δ_f is called the *differential uniformity* of the mapping f. A mapping is said to be differentially k-uniform if $\Delta_f = k$.

For applications in cryptography and coding theory, one would like to find functions for which Δ_f is small. In the binary case, $p = 2$, the solutions of $f(x+a) - f(x) = b$ come in pairs; therefore, $\Delta_f = 2$ is the smallest possible value. If $\Delta_f = 2$, the function f is called *almost perfect nonlinear* (APN). On the other hand, for an odd prime p, there exist mappings with $\Delta_f = 1$. Such mappings are called *perfect nonlinear* (PN), also known as *planar*.

Both APN and PN functions have been investigated for many years. Recently, there is an elicited interest on them. Few functions with small Δ_f, up to equivalence, are known. The monomial power mappings, $f(x) = x^d$, are the most studied

* Research partly supported by the Research Council of Norway; on leave from the Dept of Electrical and Electronics Eng, Anadolu University, 26555, Eskisehir, Turkey.

S.W. Golomb et al. (Eds.): SETA 2008, LNCS 5203, pp. 415–419, 2008.
© Springer-Verlag Berlin Heidelberg 2008

such mappings. A class of PN binomials of the form $f(x) = ux^{p^k+1} + x^2$ in $GF(p^{2k})$ is studied in [8]. These PN binomials are composed with inequivalent monomials and are shown to be equivalent to the monomial x^2. In [5], a new family of PN functions has been constructed that yielded new skew-symmetric Hadamard difference sets, existence of which was an open problem for many years.

A conventional (and not very elegant) way of determining whether a given function is PN over $GF(p^n)$ involves an exhaustive search since the number of solutions $x \in GF(p^n)$ of $f(x+a) - f(x) = b$, where $a, b \in GF(p^n)$ is needed. To this effect, we propose an approach for assurance of perfect nonlinearity. This approach not only reduces the dimension of the search space but also involves simply checking a trace condition. Specifically, we consider binomials of the form

$$f(x) = x^2 + x^{p^k+p^{2k}} \tag{1}$$

over $GF(p^{2k+1})$, where k is a positive integer and p is an odd prime.

2   Preliminaries

In this section, we present some of the basic concepts in finite fields and polynomials over finite fields. We also review results on perfect nonlinear functions and introduce the equivalence notion. The material presented here is fairly standard and can be found, for example, in [9] or [8].

Consider functions $f : GF(p^n) \rightarrow GF(p^n)$ where p is an odd prime. Note that any such function can be described by a polynomial degree at most $p^n - 1$. Conventionally, we must distinguish between polynomials and the associated polynomial functions. However, such a distincion is not needed if all polynomials are reduced to modulo $x^{p^n} - x$.

Note that $f(x+a) - f(x)$ can be thought as derivatives, yet they can also be recognized as difference functions. Then, a function f is called *perfect nonlinear* or *planar* if the difference functions $f(x+a) - f(x)$ are permutations for all $a \in GF(p^n), a \neq 0$. In Table 1, all monomial planar functions which are known so far are listed.

Table 1. Known PN power functions in $GF(p^n)$

function	conditions	proved in
x^2	none	trivial
$x^{\frac{p^k+1}{2}}$	$p = 3$, $\gcd(n,k) = 1$, k is odd	[2], [6]
x^{p^k+1}	$n/\gcd(n,k)$ is odd	[4], [2], [7]

Two more cases of planar functions which are not power mappings are given in Table 2.

The second example in Table 1, i.e., $x^{\frac{p^k+1}{2}}$, is referred to as *Coulter-Matthews* function which was independently found in [6]. Note that all currently known PN functions, except for the Coulter-Matthews monomial, have the form

Table 2. Known nonpower PN functions in $\mathrm{GF}(p^n)$

function	conditions	proved in
$x^{10} - x^6 - x^2$	$p = 3, n$ odd	[5]
$x^{10} + x^6 - x^2$	$p = 3, n$ odd	[2]

$$\sum_{i,j=0}^{n-1} a_{i,j} x^{p^i + p^j}, \quad a_{i,j} \in \mathrm{GF}(p^n).$$

Functions of this type are referred to as *Dembowski-Ostrom* polynomials.

The addition of any PN function and any affine function is clearly still a PN function. This yields the following question: which functions should be considered to be equivalent? In order to distinguish functions, we present the concept of affine equivalence. Two functions $f, g : \mathrm{GF}(p^n) \to \mathrm{GF}(p^n)$ are *affine equivalent* if there are two linearized permutation polynomials L and M and an affine polynomial G such that

$$g = L \circ f \circ M + G,$$

which defines an equivalence relation. Note that all the Coulter-Matthews monomials are affine inequivalent, and they are never affine equivalent to a Dembowski-Ostrom polynomial for $k \neq 1$.

In [3], it is shown that the only mappings $f : \mathrm{GF}(p^n) \to \mathrm{GF}(p^n)$ with linear $f(x + a) - f(x) - f(a)$ are given by a sum of a planar Dembowski-Ostrom polynomial and a linearized polynomial.

We now recall the trace of an element ξ in $\mathrm{GF}(p^n)$ over $\mathrm{GF}(p)$

$$\mathrm{Tr}(\xi) = \sum_{i=0}^{n-1} \xi^{p^i}, \tag{2}$$

which is automatically a member of $\mathrm{GF}(p)$.

3   Main Results

Theorem 1. *Let k be a positive integer, p be an odd prime, and Tr denote the trace mapping defined by (2). Consider a Dembowski-Ostrom binomial of the form (1):*

$$f(x) = x^2 + x^{p^k + p^{2k}}$$

over $\mathrm{GF}(p^{2k+1})$. If $\mathrm{Tr}(a^{2-p^k-p^{2k}}) \neq (-1)^{k+1}(1 + 2^{2k})$ for all $a \in \mathrm{GF}(p^{2k+1})$, then $f(x)$ is PN.

Proof. Since $f(x)$ is Dembowski-Ostrom polynomial then it is PN if the equation $f(x + a) - f(x) - f(a) = 0$ has $x = 0$ as its only solution for any nonzero $a \in \mathrm{GF}(p^{2k+1})$. We have

$$\Delta(x) = f(x + a) - f(x) - f(a) = 2ax + a^{p^{2k}} x^{p^k} + a^{p^k} x^{p^{2k}}.$$

Clearly the number of solutions in x to the equation $\Delta(x) = 0$ is the same as the number of solutions in x to the equation $\Delta(ax) = 0$, so we replace x with ax and get

$$\Delta(ax) = 2a^2x + a^{p^k+p^{2k}}(x^{p^k} + x^{p^{2k}}).$$

Let

$$\Delta_1(x) = x^{p^k} + x^{p^{2k}} + 2cx = 0,$$

where $c = a^{2-p^k-p^{2k}}$. Taking the p-th power of the above equation consecutively $2k$ times, we obtain a set of equations which can be put into a matrix form as follows

$$\begin{pmatrix} 1 & & & 1 & & 2c \\ & & & & \cdot & 1 \\ & \cdot & & \cdot & \cdot & \\ 1 & & \cdot & \cdot & & \\ & \cdot & \cdot & & 1 & \\ & \cdot & \cdot & & & 1 \\ 2c^{p^{2k}} & 1 & & 1 & & \end{pmatrix} \begin{pmatrix} x^{p^{2k}} \\ \vdots \\ \vdots \\ \vdots \\ \vdots \\ x \end{pmatrix} = 0$$

In order this system to have a unique (one) solution, the determinant of the above matrix should be nonzero and this determinant is given by

$$\det = 2\mathrm{Tr}(c) + 2(-1)^k(1 + 2^{2k}).$$

Since $c = a^{2-p^k-p^{2k}}$, the conclusion follows.   □

4   Case Studies

In this section, we present two examples to illustrate our main result.

▷ $f_1(x) = x^2 + x^{56}$ over GF(7^3)
First, $56 = 7^1 + 7^2$, so $k = 1$. Due to the Theorem 1, we need to check for all $a \in \mathrm{GF}(7^3)$ whether $\mathrm{Tr}(a^{2-p^k-p^{2k}}) = \mathrm{Tr}(a^{-54})$ is equal to $(-1)^{k+1}(1+2^{2k}) = 5$. Next, $\gcd(-54, 7^3 - 1) = 18$. That is, for any nonzero a, a^{-54} is a 19th root of unity over GF(7). Moreover, $x^{19} - 1$ factorizes in mod 7 as

$$x^{19} - 1 = (x + 6)(x^3 + 2x + 6)(x^3 + 3x^2 + 3x + 6)(x^3 + 4x^2 + x + 6)$$
$$\cdot (x^3 + 4x^2 + 4x + 6)(x^3 + 5x^2 + 6)(x^3 + 6x^2 + 3x + 6).$$

This means that either $a^{-54} = 1$ (and so has trace 3) or is a root of one of the cubics and so has trace equal to the negative of one of the coefficients of x^2, i.e., 0, 1, 2, 3, or 4. It cannot be 5. Hence, $f_1(x)$ is PN.   □

Note that the function $x^2 + x^{56}$ is PN and it can be shown to be equivalent to the monomial x^2.

▷ $f_2(x) = x^2 + x^{90}$ over GF(3^5)

First, $90 = 3^2 + 3^4$, so $k = 2$. Due to the Theorem 1, we need to check for all $a \in$ GF(3^5) whether $\text{Tr}(a^{2-p^k-p^{2k}}) = \text{Tr}(a^{-88})$ is equal to $(-1)^{k+1}(1+2^{2k}) = -17 \equiv 1 \bmod 3$. Next, $\gcd(-88, 3^5 - 1) = 22$. That is, for any nonzero a, a^{-88} is a 11th root of unity over GF(3). Moreover, $x^{11} - 1$ factorizes in mod 3 as

$$x^{11} - 1 = (x + 2)(x^5 + 2x^3 + x^2 + 2x + 2)(x^5 + x^4 + 2x^3 + x^2 + 2)$$

This means that either $a^{-88} = 1$ (and so has trace 2) or is a root of one of the quintics and so has trace equal to the negative of one of the coefficients of x^4, i.e., 0 or 2. It cannot be 1. Hence, $f_2(x)$ is PN.   □

Note that the function $x^2 + x^{90}$ is PN and it is inequivalent to the known PN functions over GF(3^5), see [8] for inequivalency proof.

5   Conclusions

In this paper, we proposed an approach for assurance of perfect nonlinearity which involves simply checking a trace condition. This approach offers different perspective for viewing nonlinear mappings.

References

1. Carlet, C., Ding, C.: Highly nonlinear mappings. J. of Complexity 20, 205–244 (2004)
2. Coulter, R.S., Matthews, R.W.: Planar functions and planes of Lenz-Barlotti class II. Des., Codes, Cryptogr. 10, 167–184 (1997)
3. Coulter, R.S., Henderson, M.: Commutative presemifields and semifields. Advances in Math. 217, 282–304 (2008)
4. Dembowski, P., Ostrom, T.: Planes of order n with collineation groups of order n^2. Math. Z. 103, 239–258 (1968)
5. Ding, C., Yuan, J.: A family of skew Paley-Hadamard difference sets. J. Comb. Theory Ser. A 113, 1526–1535 (2006)
6. Helleseth, T., Sandberg, D.: Some power mappings with low differential uniformity. Applicable Algebra in Engineering, Communications and Computing 8, 363–370 (1997)
7. Helleseth, T., Rong, C., Sandberg, D.: New families of almost perfect nonlinear power mappings. IEEE Trans. Inform. Theory 52, 475–485 (1999)
8. Helleseth, T., Kyureghyan, G., Ness, G.J., Pott, A.: On a family of perfect nonlinear binomials (submitted)
9. Lidl, R., Niederreiter, H.: Finite Fields, 2nd edn. Encyclopedia of Mathematics and its Applications, vol. 20. Cambridge University Press, Cambridge (1997)

Author Index

Lecture Notes in Computer Science

Sublibrary 1: Theoretical Computer Science and General Issues

For information about Vols. 1– 4934
please contact your bookseller or Springer

Vol. 5072: O. Gervasi, B. Murgante, A. Laganà, D. Taniar, Y. Mun, M.L. Gavrilova (Eds.), Computational Science and Its Applications – ICCSA 2008, Part I. XXIX, 1266 pages. 2008.

Vol. 5065: P. Degano, R. De Nicola, J. Meseguer (Eds.), Concurrency, Graphs and Models. XV, 810 pages. 2008.

Vol. 5062: K.M. van Hee, R. Valk (Eds.), Applications and Theory of Petri Nets. XIII, 429 pages. 2008.

Vol. 5059: F.P. Preparata, X. Wu, J. Yin (Eds.), Frontiers in Algorithmics. XI, 350 pages. 2008.

Vol. 5058: A.A. Shvartsman, P. Felber (Eds.), Structural Information and Communication Complexity. X, 307 pages. 2008.

Vol. 5050: J.M. Zurada, G.G. Yen, J. Wang (Eds.), Computational Intelligence: Research Frontiers. XVI, 389 pages. 2008.

Vol. 5045: P. Hertling, C.M. Hoffmann, W. Luther, N. Revol (Eds.), Reliable Implementation of Real Number Algorithms: Theory and Practice. XI, 239 pages. 2008.

Vol. 5038: C.C. McGeoch (Ed.), Experimental Algorithms. X, 363 pages. 2008.

Vol. 5036: S. Wu, L.T. Yang, T.L. Xu (Eds.), Advances in Grid and Pervasive Computing. XV, 518 pages. 2008.

Vol. 5035: A. Lodi, A. Panconesi, G. Rinaldi (Eds.), Integer Programming and Combinatorial Optimization. XI, 477 pages. 2008.

Vol. 5029: P. Ferragina, G.M. Landau (Eds.), Combinatorial Pattern Matching. XIII, 317 pages. 2008.

Vol. 5028: A. Beckmann, C. Dimitracopoulos, B. Löwe (Eds.), Logic and Theory of Algorithms. XIX, 596 pages. 2008.

Vol. 5022: A.G. Bourgeois, S.Q. Zheng (Eds.), Algorithms and Architectures for Parallel Processing. XIII, 336 pages. 2008.

Vol. 5018: M. Grohe, R. Niedermeier (Eds.), Parameterized and Exact Computation. X, 227 pages. 2008.

Vol. 5015: L. Perron, M.A. Trick (Eds.), Integration of AI and OR Techniques in Constraint Programming for Combinatorial Optimization Problems. XII, 394 pages. 2008.

Vol. 5011: A.J. van der Poorten, A. Stein (Eds.), Algorithmic Number Theory. IX, 455 pages. 2008.

Vol. 5010: E.A. Hirsch, A.A. Razborov, A. Semenov, A. Slissenko (Eds.), Computer Science – Theory and Applications. XIII, 411 pages. 2008.

Vol. 5008: A. Gasteratos, M. Vincze, J.K. Tsotsos (Eds.), Computer Vision Systems. XV, 560 pages. 2008.

Vol. 5004: R. Eigenmann, B.R. de Supinski (Eds.), OpenMP in a New Era of Parallelism. X, 191 pages. 2008.

Vol. 5000: O. Grumberg, H. Veith (Eds.), 25 Years of Model Checking. VII, 231 pages. 2008.

Vol. 4996: H. Kleine Büning, X. Zhao (Eds.), Theory and Applications of Satisfiability Testing – SAT 2008. X, 305 pages. 2008.

Vol. 4988: R. Berghammer, B. Möller, G. Struth (Eds.), Relations and Kleene Algebra in Computer Science. X, 397 pages. 2008.

Vol. 4985: M. Ishikawa, K. Doya, H. Miyamoto, T. Yamakawa (Eds.), Neural Information Processing, Part II. XXX, 1091 pages. 2008.

Vol. 4984: M. Ishikawa, K. Doya, H. Miyamoto, T. Yamakawa (Eds.), Neural Information Processing, Part I. XXX, 1147 pages. 2008.

Vol. 4981: M. Egerstedt, B. Mishra (Eds.), Hybrid Systems: Computation and Control. XV, 680 pages. 2008.

Vol. 4978: M. Agrawal, D.-Z. Du, Z. Duan, A. Li (Eds.), Theory and Applications of Models of Computation. XII, 598 pages. 2008.

Vol. 4975: F. Chen, B. Jüttler (Eds.), Advances in Geometric Modeling and Processing. XV, 606 pages. 2008.

Vol. 4974: M. Giacobini, A. Brabazon, S. Cagnoni, G.A. Di Caro, R. Drechsler, A. Ekárt, A.I. Esparcia-Alcázar, M. Farooq, A. Fink, J. McCormack, M. O'Neill, J. Romero, F. Rothlauf, G. Squillero, A.Ş. Uyar, S. Yang (Eds.), Applications of Evolutionary Computing. XXV, 701 pages. 2008.

Vol. 4973: E. Marchiori, J.H. Moore (Eds.), Evolutionary Computation, Machine Learning and Data Mining in Bioinformatics. X, 213 pages. 2008.

Vol. 4972: J. van Hemert, C. Cotta (Eds.), Evolutionary Computation in Combinatorial Optimization. XII, 289 pages. 2008.

Vol. 4971: M. O'Neill, L. Vanneschi, S. Gustafson, A.I. Esparcia Alcázar, I. De Falco, A. Della Cioppa, E. Tarantino (Eds.), Genetic Programming. XI, 375 pages. 2008.

Vol. 4967: R. Wyrzykowski, J. Dongarra, K. Karczewski, J. Wasniewski (Eds.), Parallel Processing and Applied Mathematics. XXIII, 1414 pages. 2008.

Vol. 4963: C.R. Ramakrishnan, J. Rehof (Eds.), Tools and Algorithms for the Construction and Analysis of Systems. XVI, 518 pages. 2008.

Vol. 4962: R. Amadio (Ed.), Foundations of Software Science and Computational Structures. XV, 505 pages. 2008.

Vol. 4961: J.L. Fiadeiro, P. Inverardi (Eds.), Fundamental Approaches to Software Engineering. XIII, 430 pages. 2008.

Vol. 4960: S. Drossopoulou (Ed.), Programming Languages and Systems. XIII, 399 pages. 2008.

Vol. 4959: L. Hendren (Ed.), Compiler Construction. XII, 307 pages. 2008.

Vol. 4957: E.S. Laber, C. Bornstein, L.T. Nogueira, L. Faria (Eds.), LATIN 2008: Theoretical Informatics. XVII, 794 pages. 2008.

Vol. 4943: R. Woods, K. Compton, C. Bouganis, P.C. Diniz (Eds.), Reconfigurable Computing: Architectures, Tools and Applications. XIV, 344 pages. 2008.

Vol. 4942: E. Frachtenberg, U. Schwiegelshohn (Eds.), Job Scheduling Strategies for Parallel Processing. VII, 189 pages. 2008.

Vol. 4941: M. Miculan, I. Scagnetto, F. Honsell (Eds.), Types for Proofs and Programs. VII, 203 pages. 2008.

Vol. 4935: B. Chapman, W. Zheng, G.R. Gao, M. Sato, E. Ayguadé, D. Wang (Eds.), A Practical Programming Model for the Multi-Core Era. VI, 208 pages. 2008.